Cleaner Petroleum Production and Refining Technologies

Cleaner Petroleum Production and Refining Technologies

Edited by M. R. Riazi and H. W. Yarranton

WILEY

Published by John Wiley & Sons, Inc., Hoboken, New Jersey.

Library of Congress Cataloging-in-Publication Data:

Names: Riazi, M. R. editor | Yarranton, Harvey W. (Harvey William), 1963- editor | John Wiley & Sons publisher
Title: Cleaner petroleum production and refining technologies / edited by M. R. Riazi, Harvey W Yarranton.
Description: Hoboken, New Jersey : John Wiley and Sons, [2026] | Includes index.
Identifiers: LCCN 2025039831 | ISBN 9781394209231 hardback | ISBN 9781394209248 | ISBN 9781394209255 | ISBN 9781394209262
Subjects: LCSH: Petroleum mining—Environmental aspects | Petroleum mining—Technological innovations | Petroleum—Refining—Environmental aspects | Petroleum—Refining—Technological innovations
Classification: LCC TN871 .C55 2026
LC record available at https://lccn.loc.gov/2025039831

Cover Design: Wiley
Cover Image: © dszc/Getty Images

Contents

List of Contributors

Francisco O. Aguirre
Department of Environmental Engineering, Texas A&M University-Kingsville, TX, USA

Clarissa Bergman-Fonte
Centre for Energy and Environmental Economics (Cenergia), Energy Planning Program (PPE), COPPE, Universidade Federal do Rio de Janeiro, Brazil

Lucy Mar Camacho
Department of Environmental Engineering, Texas A&M University-Kingsville, TX, USA

Bruce S. Carey
Peters & Co. Limited, Research, Calgary, Alberta, Canada

Aline T. Carvalho
Centre for Energy and Environmental Economics (Cenergia), Energy Planning Program (PPE), COPPE, Universidade Federal do Rio de Janeiro, Brazil

Gabriela Nascimento da Silva
Centre for Energy and Environmental Economics (Cenergia), Energy Planning Program (PPE), COPPE, Universidade Federal do Rio de Janeiro, Brazil

Marcio Wagner da Silva
Process Engineering, Petrobras, Sao Paulo, Brazil

B. Frank Gupton
Department of Chemical & Life Science, Virginia Commonwealth University, Richmond, VA, USA

George Hoekstra
Hoekstra Trading LLC, Bolingbrook, IL, USA

Elco D. Hollander
Coanda Research & Development Corporation, Tetra Tech Company, Calgary, Alberta, Canada

Amol A. Inamdar
Department of Environmental Engineering, Texas A&M University-Kingsville, TX, USA

Frédéric Lantz
Economic, Energy and Environment, IFP Energies Nouvelles, IFP School, Rueil-Malmaison, France

Eduardo Manrique
Citation Oil & Gas Corp., Lead Reservoir Engineer, Reservoir Department, Houston, Texas, USA

Thomas Murphy
The Whitman School of Management and the College of Law, Syracuse University, Syracuse, NY, USA
Valuation Risk and Strategy LLC, Skaneateles, NY, USA

Olubayo O. Oginni
Department of Environmental Engineering, Texas A&M University-Kingsville, TX, USA

Chike George Okoye-Chine
Department of Chemical & Life Science Engineering, College of Engineering, Virginia Commonwealth University, Richmond, VA, USA
Global R&D, Avery Dennison, Mentor, OH, USA

M.R. Riazi
Department of Chemical and Natural Gas Engineering, Texas A&M University – Kingsville, Kingsville, TX, USA

Stuart Riddick
Department of Science, Engineering and Aviation, University of the Highlands and Islands, Perth, Perthshire, UK

Pedro R.R. Rochedo
Research and Innovation Center on CO_2 and Hydrogen (RICH Center) and Management Science and Engineering Department, Khalifa University, Abu Dhabi, United Arab Emirates
Centre for Energy and Environmental Economics (Cenergia), Energy Planning Program (PPE), COPPE, Universidade Federal do Rio de Janeiro, Brazil

Dicho Stratiev
Institute of Biophysics and Biomedical Engineering-Bulgarian Academy of Sciences, Sofia, Bulgaria

Alexandre Szklo
Centre for Energy and Environmental Economics (Cenergia), Energy Planning Program (PPE), COPPE, Universidade Federal do Rio de Janeiro, Brazil

Danuta M. Sztukowski
Coanda Research & Development Corporation, a Tetra Tech Company, Calgary, Alberta, Canada

Heidi A. Taboada
Department of Environmental Engineering, Texas A&M University-Kingsville, TX, USA

Harvey W. Yarranton
Department of Chemical and Petroleum Engineering, University of Calgary, Calgary, Alberta, Canada

Preface

This book originated from an invitation from Michael Leventhal, editor of the STM division at John Wiley & Sons publishing group in the United States in 2022. In August 2024 and at the time of writing this book, the U.S. crude oil production reached to an all-time high of 13.4 mb/d, up 2% from the 2023 level. This oil production level is the highest a single country has produced in the history of the petroleum industry, despite all efforts to curb the use of fossil-type energy. It is expected that the Oil and Gas global industry will play a major role in the foreseeable future, not only for energy production but mainly as a source of the ever-growing petrochemicals production. However, the industry is facing major challenges to produce cleaner fuels and to control its carbon and GHG emissions during oil production, refining, and the use of petroleum products. The industry must deal with stricter environmental regulations and economic constraints to survive and remain competitive.

This book will discuss the entire industry, considering different methods of production and refining and different types of crude oils to produce quality fuels with minimum environmental impacts. The most significant challenges for minimizing greenhouse gas emissions in oil and gas production are reviewed, including fugitive methane emissions, heavy oil recovery, and bitumen extraction. Methods of carbon capture and storage are presented because this is an important technology to combat carbon emissions in the petroleum industry. Further, the book discusses the configuration of refinery processes to maximize light products (as fuel or feed to petrochemical plants) without producing carbon residue and low-quality fuels. Integration of refineries with petrochemical plants is another aspect of the book, which provides an alternative use of petroleum extraction and refining when demand for petrofuels is in decline. Also discussed is the use of gas-to-liquid processes (e.g., the Fischer–Tropsch process) to produce liquid fuels from syngas or CO_2 to maximize production of middle distillate fuels when the demand is on the rise. Biorefining and biofuels production of alternative fuels with environmental considerations are reviewed.

In general, the book will present a framework for understanding technological needs and opportunities for the future of crude oil production and refining, as well as how the industry would survive under different scenarios. Today's technology choices and the upgrading of petrochemical facilities need to be driven by a completely different logic than the existing industry. Current operations were built on the basis that the economics of carbon dioxide are only due to fuel cost, and that the environmental aspects of fuel use are dominated by sulfur emissions, nitrous oxides, and particulates. The future of regulations and market acceptance looks very different in many countries, and these changes are driving huge financial decisions in the production and refining industries.

Changing regulations may drive different choices of recovery methods in the production of crude oil and greater emphasis on the detection and reduction of methane emissions. Major investments in carbon capture and storage may be required. In addition, instability in crude oil supplies and shifting demand for refined products require greater flexibility at existing refineries. Growth opportunities in chemical intermediates and recycling of post-consumer plastics are needed. Increased environmental regulations and reduced emissions of sulfur, carbon dioxide, methane, and other greenhouse gases are adding financial economic uncertainty at the bottom line and in the availability and sources for investment capital.

We would like to thank the publisher John Wiley and Sons, especially the editor Michael Leventhal, who initially invited the first editor to propose this book for publication. We also thank the publisher, managing editor Elizabeth Rose, Tanisha Sirohi and other staff, as well as all the authors and reviewers who contributed to this book.

M. R. Riazi
Professor & Chair of Chemical and Natural Gas Engineering
Texas A&M University – Kingsville, TX, USA
and
Harvey W. Yarranton
Professor of Chemical and Petroleum Engineering
University of Calgary, Alberta, Canada
April 1, 2025

1

Introduction

M. R. Riazi[1] *and Harvey W. Yarranton*[2]

[1] *Department of Chemical and Natural Gas Engineering, Texas A&M University-Kingsville, Kingsville, TX, USA*
[2] *Department of Chemical and Petroleum Engineering, University of Calgary, Calgary, Alberta, Canada*

Fossil fuels are naturally occurring mixtures of hydrocarbons, some of which contain sulfur, oxygen, nitrogen, and metallic functional groups. Fossil fuels include natural gas, crude oil, and coal. Petroleum refers to crude oils and their refined products. The definition of petroleum is often extended to include natural gas but not coal. In this sense, petroleum refers to flowable fossil fuels. This broader definition includes natural gases, natural gas liquids, condensates, crude oils, kerosenes, diesels, gasolines, and fuel oils. Crude oils themselves are subdivided into the following categories based on their specific gravity (SG): conventional oils (SG < 0.90), heavy oils ($0.90 <$ SG < 1), and bitumen (SG > 1). All petroleums are fluids, but some, such as heavy oils and bitumens, are highly viscous and require heating to flow at practical rates.

An important feature of petroleum is its high energy density. Energy density is defined as the energy content of a material per volume. For fossil fuels and similar materials, the energy density is based on the energy released by combustion. For nuclear materials, it is the energy released by fission. Table 1.1 compares the energy density of naturally occurring fuel sources. Crude oil has the highest energy density of the combustible fuels and is the only abundant naturally occurring fuel source that is a liquid in its natural state. The flowability and high energy density of petroleum are what make it so valuable. It is the most easily transported, efficiently stored energy in existence.

Cleaner Petroleum Production and Refining Technologies, First Edition, Edited by M. R. Riazi and Harvey W. Yarranton.

Table 1.1 Energy density of selected naturally occurring fuel sources.

Material	Energy density MJ/L	Specific energy MJ/kg	Natural state
Uranium	1,540,000,000	80,600,000	Solid
Coal – anthracite	34–43	26–33	Solid
Coal – bituminous	26–49	24–35	Solid
Coal – lignite	—	10–20	Solid
Wood	—	18	Solid
Peat	—	13	Solid
Crude oil	37	42	Liquid
Natural gas	0.04	54	Gas
Hydrogen	4.5–5.3	142	Gas

Source: Adapted from Wikipedia (2024).

1.1 Historical Perspective on Petroleum

Petroleum is typically found in the pore space of rocks deep underground. However, in several places in the world, petroleum seeps to the surface, usually in the form of bitumen. For this reason, humans have been aware of petroleum for tens of thousands of years (Beloe, 2019). Prehistoric people used bitumen to attach spear points to their shafts. Ancient Mesopotamians also used bitumen as an adhesive, in building materials, and to caulk ships. Ancient Egyptians used crude oil for medicinal purposes. Persians used crude oil to make fire arrows, and petroleum was a key ingredient of Greek fire. However, until the 19th century, petroleum only played a small role in human endeavors.

In the 19th century, the industrial revolution arrived, and society was transformed in a few decades. Imagine a home before the 19th century. It was likely small and smoky, heated by the combustion of coal, peat, or wood. Light was provided by burning stinking animal fat or burning clean but expensive whale oil. Sperm whales were hunted almost to extinction to supply this oil. Transportation was slow, either by foot, horse, cart, oars, or sail. Imported items were rare and expensive. Cooking, washing, sewing, repairs, and indeed almost all labor were done by hand. Such had been life for centuries. Today, in the developed world, one can turn on a light and plug in a food processor, a vacuum cleaner, or a washing machine. The electricity, more often than not, is supplied by the combustion of petroleum or coal. Homes and workplaces are heated with natural gas or fuel oil. One can eat a fruit grown on the other side of the planet or drive one's car to the airport and fly across the world to eat it where it was grown.

Since the advent of the petroleum-based economy, the global population has risen from 1 billion in 1800 to over 8 billion today (Roser and Ritchie, 2023), and the poverty rate has decreased from over 80% to approximately 10% (Hasell and Roser, 2019). The world is connected through air, land, and sea as never before thanks to kerosene, gasoline, and fuel oils. Petroleum enables high-intensity agriculture and widespread electrification, lighting, heating and air conditioning, and personal and commercial transportation. It also provides a host of byproducts that touch almost all aspects of our lives, including plastics, lubricants, waxes, textiles, medicines, and fertilizers. The modern population and lifestyle would not be supportable without the cheap, transportable energy provided by petroleum.

The global consumption of crude oil in 2022 was 15.9 million cubic meters per day (Statista, 2024a), adding up to 4.2 billion cubic meters for the year. Our consumption of petroleum is supported by a vast infrastructure consisting of three main segments: upstream (production), midstream (transportation and storage), and downstream (refining). On land, the upstream segment includes: the wells connecting the hydrocarbon reservoir to the surface; pipelines connecting the wellheads to the surface facilities; the surface facilities used to separate oil, water, and gas; and the trucking or small pipelines used to move the products to sales or disposal points. Offshore, the upstream segment is the same in principle, even though the facilities are on an oil rig and the lines from the well head to the oil rig are underwater. The midstream segment includes pipelines or railcars to transport the oil to refineries or ports by land, storage vessels, and oil tankers to transport by sea. Downstream includes the refineries used to convert crude oil into petroleum fractions and other products as well as the infrastructure (tanks, warehouses, trucks, rail, and ships) to store and distribute those products. Table 1.2 provides some statistics on the scale of the petroleum infrastructure in the United States and globally.

Not surprisingly, the large scale of the petroleum infrastructure and end use of petroleum products brings some negative environmental impacts, including oil spills, smog, and greenhouse gas (GHG) emissions. This book focuses on GHG emissions. GHGs are gases in the atmosphere that reflect heat radiated from the earth back to the surface and so raise the temperature of the earth's surface. The major GHGs are water, carbon dioxide, ozone, nitrous oxide, and methane. Water contributes from 40 to 70% of the greenhouse effect, depending on cloud cover (Kiehl and Trenberth, 1997; Schmidt et al., 2010). Carbon dioxide contributes from 18 to 26%, ozone approximately 8%, and nitrous oxide and methane 6%.

The direct combustion of petroleum produces water, carbon dioxide, and to a lesser extent nitrous oxide and sulfur dioxide. In many refineries, the nitrogen is converted to ammonia, and the sulfur dioxide is converted to elemental sulfur.

Table 1.2 Selected parts of the petroleum infrastructure.

Infrastructure	Size
United States	
Oil wells in 2023[a]	>900,000
Midstream pipelines in 2023 – oil[b]	83,000 km
– oil + products[c]	>110,000 km
Oil refineries in 2023[d]	124
Worldwide	
Offshore oil rigs in 2024[e]	240
Midstream pipelines in 2023 – oil[c]	184,000 km
Oil tankers in 2020[f] – VLCC (capacity ~320 · 10^3 m^3)	810
– Aframax (capacity 80–130 · 10^3 m^3)	668
– Suezmax (capacity 80–160 · 10^3 m^3)	571
Oil refineries in 2023[g]	825

a. EIA (2023).
b. API (2024).
c. GEM (2024).
d. Statista (2024b).
e. Statista (2024c).
f. Statista (2024d).
g. Statista (2024e).

Water emissions are not a concern because the water content of the atmosphere is mainly dictated by the oceans and the global temperature and is little affected by manmade emissions. Hence, carbon dioxide is the main GHG emission of concern from the direct combustion of petroleum. The petroleum industry also contributes indirectly to GHG emissions, mainly from methane emissions from leaks. Direct and indirect GHG emissions from oil and gas operations (including all fossil fuels) are estimated to be 33.9 billion tonnes of carbon dioxide equivalent (CO_2-eq) per year (BP Statistical Review of World Energy, 2022). Approximately 60% of these emissions are attributed to petroleum as opposed to coal (C2ES, 2008). The production, transport, and processing of oil and gas resulted in 5.1 billion tons (Gt) CO_2-eq in 2022. These emissions from oil and gas activities are responsible for about 15% of total energy-related GHG emissions (IEA, 2023).

There is now a strong societal drive to reduce and eventually eliminate GHG emissions. This drive was formalized in the December 2015 "Paris Agreement" to limit the increase of global temperatures to 1.5 °C by the end of this century

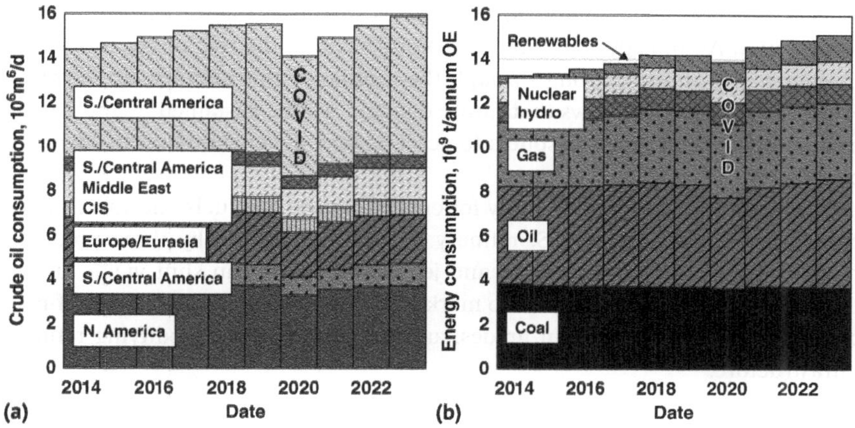

Figure 1.1 Global crude oil consumption (a) and energy consumption by source (b) from 2014 to 2023. *Source*: Adapted from Energy Institute (2024).

by reducing GHG emissions (U.N., 2015). The main path to this objective is to transition away from petroleum to alternative energy sources that do not emit GHGs directly or through their source materials and manufacture. However, despite considerable efforts and financial incentives to develop alternative energy sources, oil consumption continues to increase.

Figure 1.1a shows that crude oil consumption has increased steadily over the past decade with the exception of the COVID shutdown. Despite predictions that oil production would peak in 2000 (BP Statistical World Review, 2022), production has not peaked even by 2024. As noted previously, oil consumption in 2023 was 15.9 million cubic meters per day, exceeding all past highs. Consumption is expected to increase by 1.2 mb/d (190,000 m^3/d) in 2025 (IEA, 2007). In parallel, estimates of proven global oil reserves increase almost each year. According to OGJ annual assessments, proven oil reserves in 2022 were 1,757 billion barrels ($279 \cdot 10^6$ m^3), up from 1,735 billion barrels a year earlier and much higher than a decade ago (Xu and Bell-Hammer, 2022). Figure 1.1b shows that renewables still make up less than 7% of the global energy supply.

1.2 Future Demand for Petroleum and Its Products

Hydrocarbon-type fuels will remain important contributors to the world energy demand at least for the next several decades. Various organizations project different scenarios, but all show petroleum as a major contributor. For example, DNV-GL (DNV, 2024) projects that oil and gas will provide a minimum of about

40–50% of world energy demand by 2050, as shown in Figure 1.2. The Energy Information Administration (EIA, 2023) projects that the role of oil as a source of energy will remain strong through the middle of this century, as shown in Figure 1.3. Figure 1.4 shows that future demand will be sustained by the growth of developing economies, despite a forecast decline in consumption in advanced economies.

Petroleum is in demand not only for energy but also for fuels and value-added chemicals. About 85% of U.S. refinery output is made up by three products: motor gasoline, distillate fuel oil, and jet fuel. Refiners can shift yields among those three products in response to market conditions but are limited by refinery configuration, crude oil input grades, and the high costs of modifying refinery infrastructure.

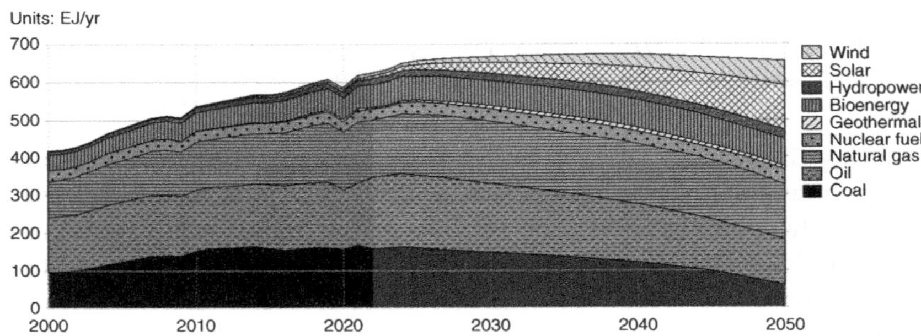

Figure 1.2 World primary energy supply by source. 1 EJ = 23.88 million tons of oil equivalents. Historical data source: IEA WEB (2024). *Source*: Reproduced with permission of DNV(2024)/Det Norske Veritas group.

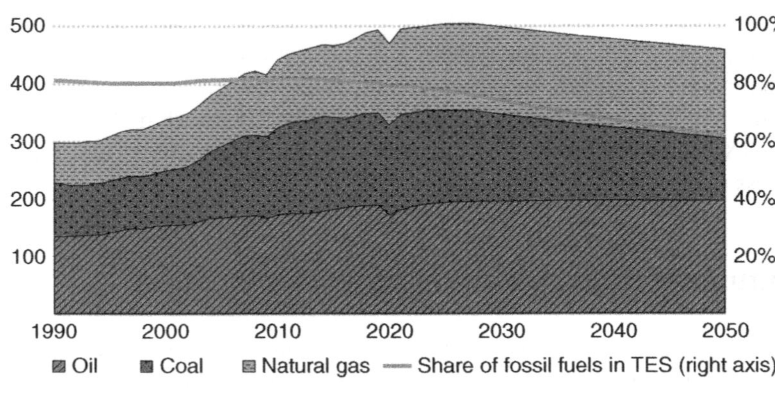

Figure 1.3 Contribution of oil, gas, and coal to world energy supply for coming decades. *Source*: Adapted from EIA (2023).

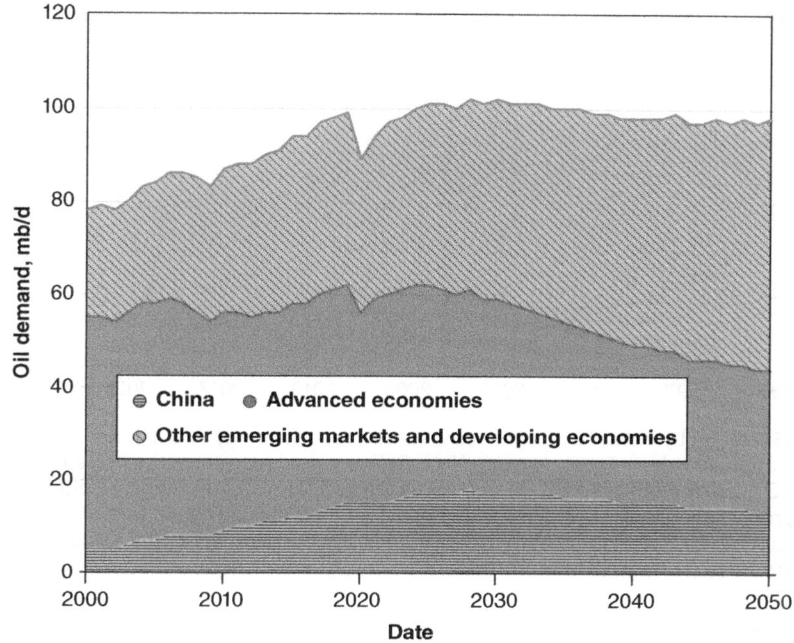

Figure 1.4 Oil demand (in million b/d) by region, 2000–2050. *Source*: Adapted from IEA (2023).

Changes in demand are an important factor driving changes in refinery yields. Increased air travel and passenger volume have increased jet fuel consumption every year after Covid. U.S. refineries produced a record-high share of jet fuel in 2024, reflecting increased demand relative to other transportation fuels, as shown in Figure 1.5. As reported by the U.S. Department of Energy, it is expected jet fuel consumption will reach a record high in 2026 (EIA, 2025a).

The demand for LPG, diesel, and gasoline also continues to rise significantly, as shown in Figure 1.6. According to Brazilian Petroleum Agency, the Brazilian demand for crude oil derivatives will rise by 20% until 2026 (da Silva, 2022). U.S. ethane production, consumption, and exports reached record highs in 2024, as shown in Figure 1.7 (EIA, 2025b). U.S. ethane production increased by 7% in 2024. The Permian Basin and Appalachian Basin in Pennsylvania are two main sources of ethane production in the United States. Crude oil is also used to produce a large number of value-added chemicals, as shown in Figure 1.8.

Based on this analysis, petroleum will play a very important role not only as a source of energy but also to produce many valuable chemicals that cannot be replaced with other alternative sources. Hence, the oil and gas industry will

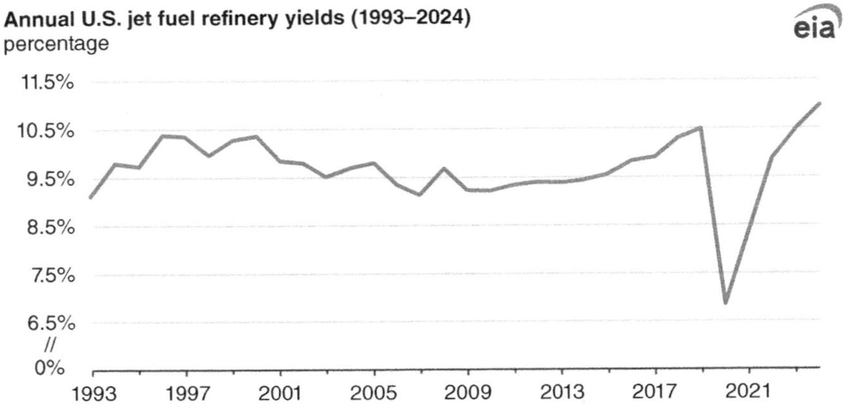

Figure 1.5 Jet fuel yields in U.S. refineries. Data source: U.S. Energy Information Administration, *Petroleum Supply Monthly*. Note: Refinery yield represents the percentage of finished product produced (output) from gross inputs. EIA calculates refinery yield as the net production of a finished petroleum product (output) divided by the sum of the input of crude oil, hydrogen, and other hydrocarbons and the net input of unfinished oils. *Source*: EIA (2025a)/EIA/Public Domain.

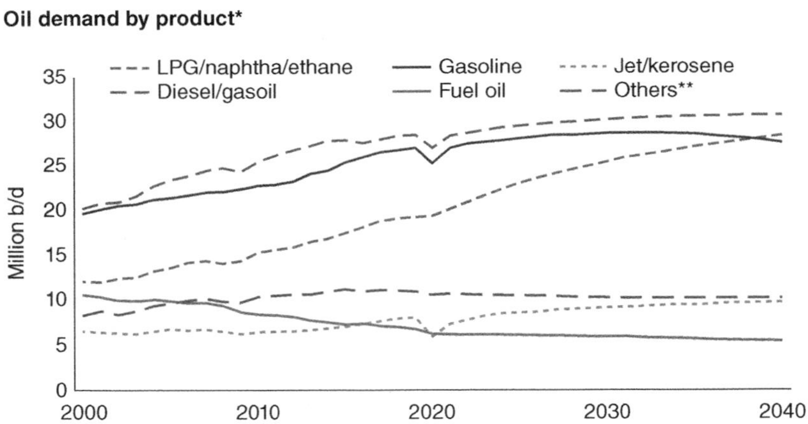

Figure 1.6 Global oil demand of oil derivatives. *Product-level demand is reported on a gross base including backflow. **Includes multiple products such as refinery gas, petroleum coke, bitumen, crude oil, non-specified other products, and backflow (negative figure). *Source*: Adapted from Wood Mackenzie (2020).

remain an integrated part of energy and chemical industry at least for the most part of this century. Since oil consumption will continue, there is a strong incentive to reduce the GHG intensity of the industry. This book presents the current

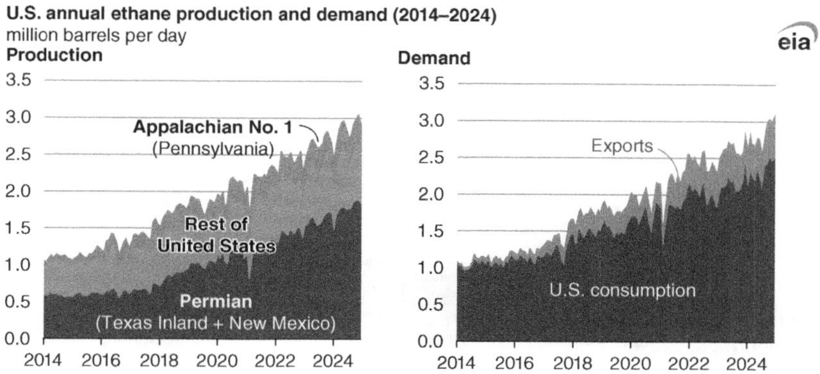

Figure 1.7 U.S. ethane production and exports. Data source: U.S. Energy Information Administration, *Petroleum Supply Monthly.* Note: Demand = consumption plus exports. *Source*: EIA (2025b)/EIA/Public Domain.

and prospective methods of producing and refining petroleum to produce quality fuels with minimum environmental impacts. The focus is on the technologies and combinations of technologies to reduce GHG emissions with an assessment of their pros and cons. In the remainder of the chapter, we give an overview of the major components of the upstream, midstream, and downstream oil industry contributing to GHG emissions and introduce the chapters that discuss the efforts to minimize GHG emissions for each of these components. Carbon capture and storage are also discussed.

1.3 Petroleum Production

As discussed previously, petroleum production includes the flow of petroleum through wells, surface facilities, and pipelines. GHG emissions during production have a number of sources, including natural gas leaks, flares, and the burning of natural gas to provide heat and power for surface processes such as pumps, compressors, and separators. There are two pathways to reducing the impact of these GHG emissions: (1) capture and store (sequester) GHG; (2) reduce GHG emissions from oil and gas processes.

Carbon dioxide is the main GHG emitted by the oil and gas industry, and therefore carbon capture and CO_2 storage are a major focus for the industry. Carbon capture refers to the separation of relatively pure CO_2 from emission sources or directly from the atmosphere. The CO_2 is then transported to an injected into an underground reservoir where it is to be stored. Another option is to inject the CO_2

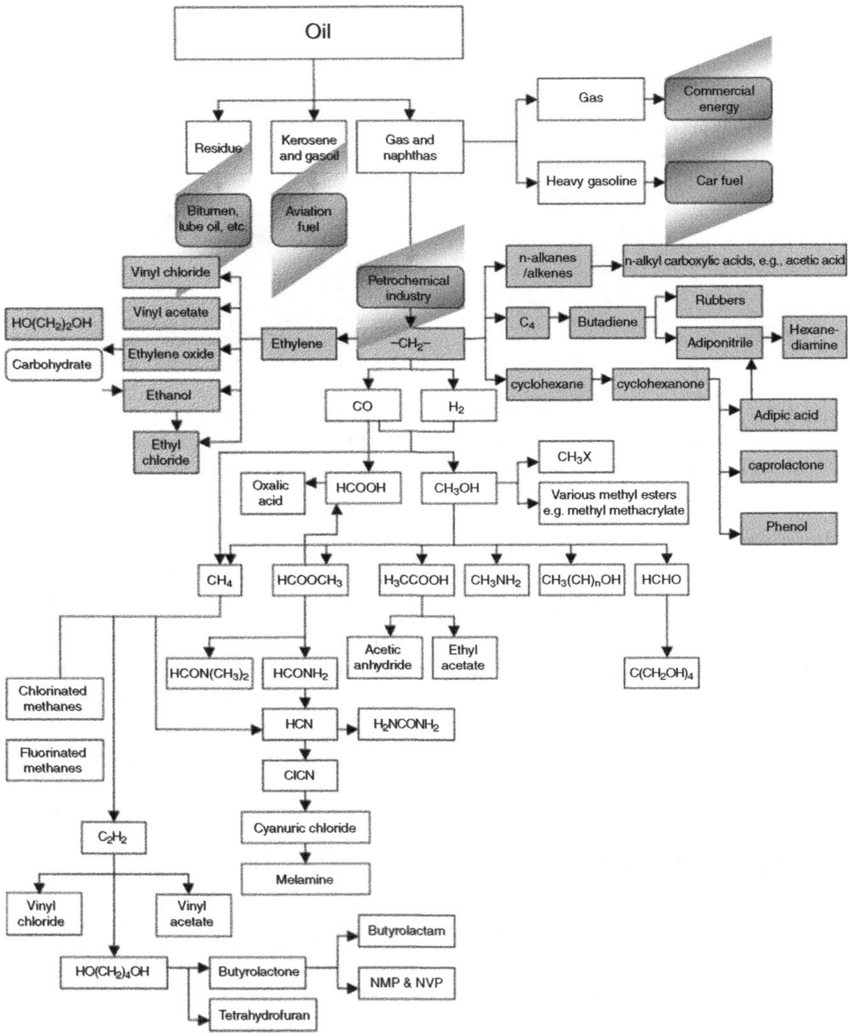

Figure 1.8 General schematic of fuels and chemicals derived from crude oil (Elbersen, 2005). Refinery products indicated by orange signs. *Source*: Elbersen (2005)/John Wiley & Sons.

into an active oil reservoir to improve the oil recovery (enhanced oil recovery). Carbon capture and storage (CCS) can potentially compensate not only for the CO_2 generated in oil and gas processes but also for the CO_2 generated from burning petroleum products. CCS and the use of CO_2 for enhanced oil recovery are discussed in Chapter 2.

Methane is also a significant GHG emitted by the oil industry, mainly through leaks, venting, and flares. Leaks are accidental releases of the natural gases (mainly methane) associated with oil and gas production, typically arising from seal failures and pipe ruptures. The gases released from leaks are termed fugitive emissions. Venting is the permitted release of gases for safety reasons and for equipment blowdowns during maintenance. Some jurisdictions also allow venting of the gas associated with oil production, particularly if the volumes are small. Flaring is the burning of these vented gases and results in the release of water vapor, CO_2, NO_x (nitric oxide and nitrogen dioxide), and in some cases SO_2 (sulfur dioxide). Many jurisdictions require vented gases to be flared. Approximately 140 billion cubic meters of natural gas are vented and flared annually, equivalent to 500 Mt of GHG emissions (IEA, 2024).

Methane has a low concentration and lifetime in the atmosphere relative to CO_2 and capture from the atmosphere is not practical. Therefore, the industry is focused on reducing emissions. The venting and flaring of the natural gases from oil production can be reduced by building the compressors and pipelines necessary to transport the gases to a gas plant. Several jurisdictions now require or are introducing legislation to reduce flaring. Fugitive emissions are more challenging to deal with because they are unintended, can occur anywhere in the production, transport, and processing pathway, and are difficult to detect. Fugitive emissions are discussed in Chapter 3.

Production and surface facilities such as separators and compressors require energy. That energy comes from electricity or from the burning of natural gas. One way to reduce emissions is to convert facilities from natural gas to electricity, as long as the electricity is generated from non-fossil fuel sources. In some cases, electricity from renewable resources is not available. Therefore, the industry is seeking approaches to reduce the energy requirements for petroleum production and surface processing. This drive is particularly acute for in situ heavy oil production because heavy oil is highly viscous. Most in situ heavy oil is recovered with thermal methods, which require steam. The steam is generated by burning natural gas, which emits CO_2. As a result, heavy oil recovery is up to 3 times more GHG intensive than conventional oil production (Century, 2008).

Heavy oil and bitumen reserves make up over half of the current global oil reserves (U.S. Geological Survey, 2003). Most current in situ heavy oil production is in Canada, with total production in 2023 of 0.27 million cubic meters per day (CAP, 2024a). This volume makes up approximately 1.7% of global production, a small but nonetheless significant fraction. Steam generation for heavy oil projects in Alberta, Canada, produced 41 million tonnes of CO_2-eq in 2021 (Government of Canada, 2023). Hence, there is a strong incentive to reduce the GHG intensity of heavy oil production. Heavy oil recovery processes are discussed in Chapter 4.

Some bitumen deposits are near the surface and are too shallow to be recovered with in situ methods. Instead, they are mined, and the bitumen is extracted from the mined ore in a hot-water process. The extraction process produces a froth consisting of bitumen, water, and some fine sand. The bitumen is recovered from the froth in a warm solvent-based treatment process. It requires heat to generate hot water and warm solvent. As with in situ production, this heat is provided by combusting natural gas with the consequent GHG emissions. In addition, the water and fines from the froth treatment process are sent to a tailings pond. Tailings ponds can emit methane and must also be managed to prevent leaks into the groundwater. Most current mined bitumen production is in Canada, with total production in 2023 of 0.25 million cubic meters per day (CAP, 2024a). Current emissions from oil sands mining are estimated to be approximately 15 Mt CO_2-eq/yr, not including upgrading (CAP, 2024b). Oil sands extraction is discussed in Chapter 5. Tailings pond management is discussed in Chapter 6.

1.4 Petroleum Refining

1.4.1 Modification of Refineries to Reduce Emissions

Oil refineries are industrial complexes that upgrade crude oil to fuels and high-value chemical products. A barrel of 42 gallons of crude oil will yield about 45 gallons of petroleum products because of gains during refining processes. In 2021, U.S. refineries produced 20 gallons of gasoline, 12.5 gallons of distillate products, and 3.5 gallons of jet fuel from 42 gallons of crude oil. As of January 2022, the world's largest refinery is now located in Gujarat, India, with the capacity of 1.24 million bbl/d. There are five top refineries in the United States with capacity of 500,000 bbl/d or more, of which three are located in Texas and the other two in Louisiana (EIA, 2024).

In modern refineries, the residue from the atmospheric and vacuum distillation columns, as well as heavy and extra-heavy oils are further processed through thermal and catalytic cracking followed by hydroprocessing to maximize gasoline and middle distillates. The very heavy part of the oil, such as asphaltenes, is gasified, and the produced syngas is converted to light hydrocarbon fuels through a Fischer–Tropsch process, as shown in Figure 1.9. Depending on the market demand when the need for certain fuels is low, a refinery can be re-configurated to produce non-fuel products such as light olefins as a feed for the petrochemical industry, as shown by da Silva (2022). The refining and processing of heavy and

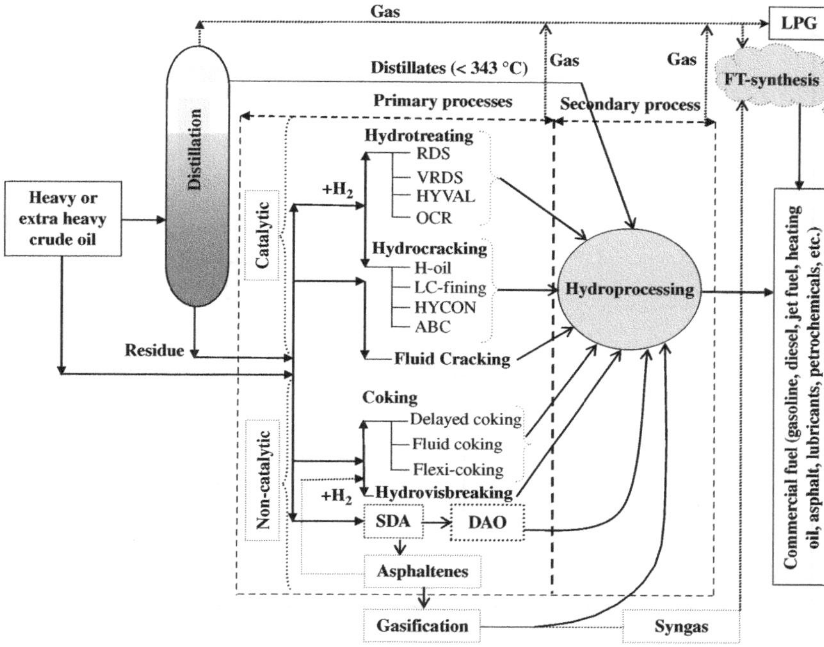

Figure 1.9 A comprehensive schematic flow diagram of typical oil refinery technologies for upgrading of heavy and extra-heavy crude oil and residue. *Source*: ASTM MNL58 (2013)/ASTM International.

extra-heavy oils are discussed in Chapter 7. Upgrading of refinery processes for deep conversion and deep desulfurization is discussed in Chapter 8. Integration of petrochemical industry with the petroleum refining industry is discussed in Chapter 9. The impact of desulfurization on the reduction of pollutant gases is shown in Figure 1.10.

Refineries and related activities are major contributors to CO_2 emission and are subject to environmental regulations. (Madugula et al., 2021). Refineries alone are the second largest contributor of CO_2 emission after electricity generation and heat production plants. Approximately 92% of the refinery CO_2 emissions come from the combustion of oil products, and refinery processes contribute another 5%. A potential solution is to use electricity from non-fossil fuel sources to provide this heat. For example, cracking units for converting heavy oil and residues to lighter products require large quantity of heat generated by burning fossil fuels and are major contributors to refinery CO_2 emissions. On Thursday, 1 September 2022, BASF together with its partner Linde celebrated construction start of the world's first demonstration plant for large-scale electrically heated steam cracker

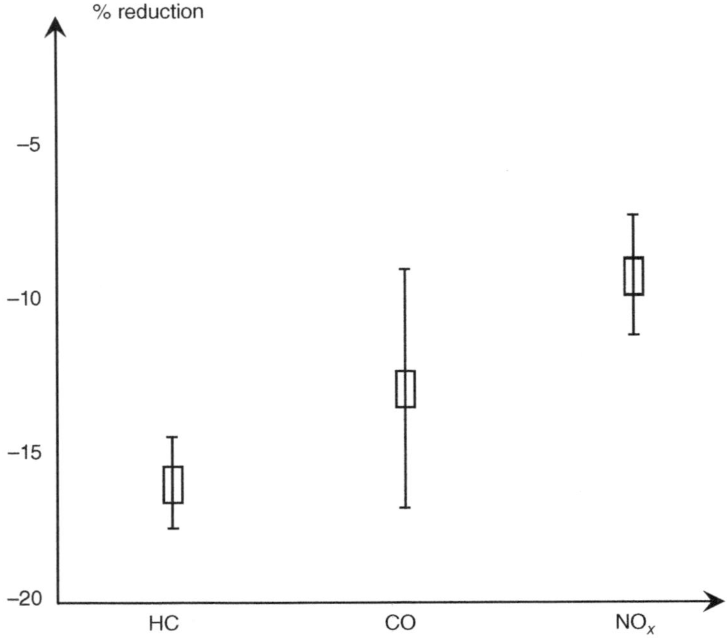

Figure 1.10 Impact of sulfur content in gasoline (from 500 to 50 ppm) on reduction of pollutant gases. *Source*: ASTM MNL50/ASTM International.

furnaces. By using electricity from renewable sources, this fundamentally new technology has the potential to significantly reduce CO_2 emissions of one of the most energy-intensive production processes in the chemical industry. This demonstration plant for large-scale electrically heated steam-cracker furnaces can potentially save at least 90% of the CO_2 emissions emitted by common technologies, according to BASF (BASF, 2022).

Another active area of research to combat CO_2 release is through the conversion of CO_2 and syngas into liquid hydrocarbon fuels. Syngas, a synthesis gas, is a useful and important feedstock for producing chemicals, including ammonia for fertilizer, urea, and methanol as a solvent and chemical intermediary. Chapter 12 will discuss the various methods of syngas production and the further transformation of the produced syngas into long-chain hydrocarbons via Fischer–Tropsch synthesis and upgraded to liquid fuels such as gasoline, jet fuel, and diesel. The research in this area is growing fast around the world, particularly on the role of different catalysts and use of Fischer–Tropsch process.

1.4.2 Hydrogen Production

One main specification of quality and clean fuel is its high H/C (low C/H) ratio. Fuels with high H/C produce less carbon and have higher heating values. Heavy oil and residues have low H/C (high carbon content), and they are converted into lighter fuels through cracking processes. Heavy oils are also rich in heteroatoms such as sulfur, nitrogen, metals, and other impurities, which can be removed through hydroprocessing unit in a refinery. Upgrading of residues and heavy oil requires hydrogen.

Most of the hydrogen required by the refineries is produced by steam methane reforming (SMR), in which methane reacts with steam to produce hydrogen and CO_2 and a high temperature (700–1,000 °C) is used. This process not only directly contributes to carbon emission but also requires large quantities of heat. For this reason, research groups around the world are intensifying their efforts to produce more clean and blue hydrogen to reduce carbon emission. The main obstacle in these efforts is the cost. Partial Oxidation (POX) is an alternative method that can produce syngas from oxygen and hydrocarbons such as methane, as shown in Figure 1.11. In the POX method, pure oxygen is used (less than required stochiometric for full oxidation), followed by water–gas shift reaction (DOE, 2024).

$$CH_4 + \frac{1}{2}O_2 \rightarrow CO + 2H_2 \text{ (+heat)}$$

$$CO + H_2O \rightarrow CO_2 + H_2 \text{ (+small amount of heat)}$$

Use of nickel-based catalysts (such as Ni/CeO_2) for the POX process can reduce required temperature and consequently heat requirement. POX requires

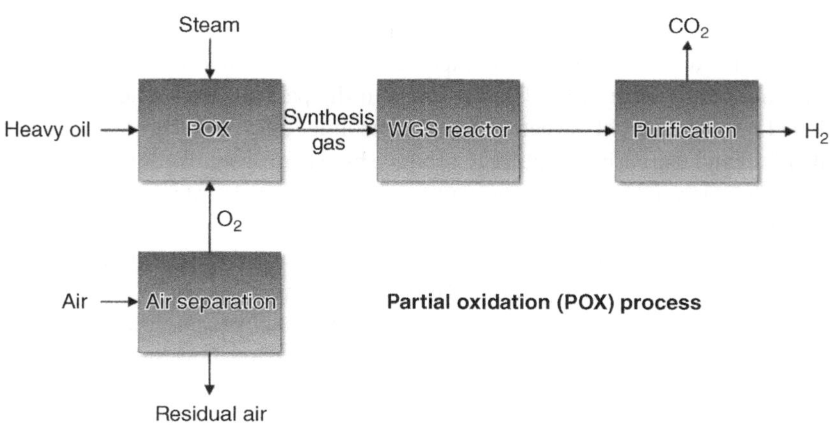

Figure 1.11 Methane production through POX method. *Source*: Li et al. (2024)/MDPI/CC BY 4.0.

the use of pure oxygen for efficient oxidation. The reactions involved are (Amiri, 2024):

$$\text{Air} \rightarrow \text{Air Separation Unit (ASU)} \rightarrow O_2$$

$$O_2 + \text{Heavy Oil} + \text{Steam} \rightarrow \text{Partial Oxidation (POX)} \rightarrow \text{Syngas}$$

Most recently Li et al. (2023) suggested decarbonizing fossil fuels by recovering and sequestering solid carbon instead of gaseous CO_2. In this context, methane pyrolysis splits CH_4 directly into its components, i.e., hydrogen and carbon. Unlike other technologies involving fossil fuels, the major benefit of methane pyrolysis is the production of CO_2-free hydrogen (Pelucchi et al., 2024). In this method, solid carbon is produced, which has many applications in the industry. This method is attractive but not yet commercialized. Production of hydrogen from renewable sources is discussed in Chapter 9.

1.4.3 Next Generation Refineries

Globally, fuel demand continues to grow slowly with compound annual growth rates (CAGRs) from 2018 to 2028 of <1% for both gasoline and diesel. Large-scale modern refineries will be built in regions where demand for fuel exists, and they will shift the global supply and demand balance. These changing dynamics will put downward pressure on gasoline and diesel cracks in oversupplied markets and will drive down margins for refiners unable to adjust their product slate. New refineries also provide opportunities to reduce GHG emissions in the original design rather than post-hoc process modifications.

In 2022, the Houston Chronicle reported that the $5.6 billion project proposed by El Campo-based Prairie Energy Partners, a subsidiary of Southern Rock Energy Partners, to build a new 250,000-barrel-per-day state-of-the-art refinery near Victoria, Texas. This project is expected to be completed in 2026 and is the first refinery to be built in the United States in the past 40 years. The refinery is designed to process light, sweet crude into gasoline, ultra-low sulfur diesel, and bunker fuel and use carbon capture to produce blue hydrogen. In addition, renewable power generated by a 150-megawatt-per-hour peaking plant would be used in conjunction with recycled wastewater to make green hydrogen. It is expected to reduce the CO_2, SO_x, NO_x, volatile organic compounds, and GHG emissions by at least 95% in comparison to traditional Gulf Coast refineries.

Modern refineries are based on the UltraFuels philosophy for simple design that focuses on the upgrading of three liquid products: (1) Ultra Low Sulfur Gasoline (with Octane number 87 regular and 93 premium gasoline), (2) Ultra Low Sulfur Diesel Fuel, and (3) Very Low Sulfur Fuel Oil. Hydrogen is required internally, and minimum blending is required. A schematic of these simple design refineries is shown in Figure 1.12. These are also discussed in Chapter 7.

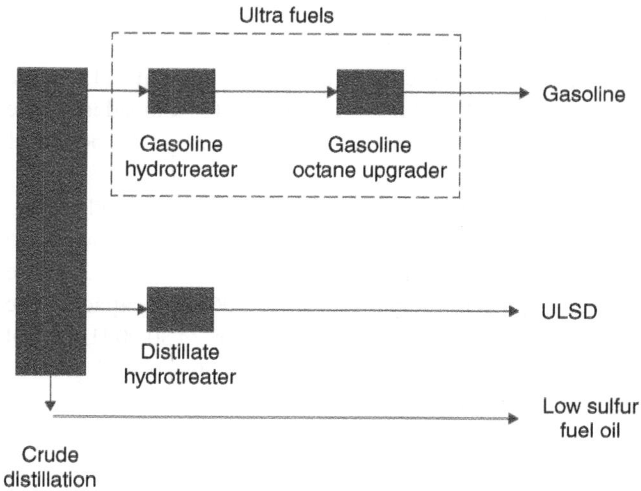

Figure 1.12 Simple design with focus on light liquid products. *Source*: Ultrafuels (2024)/ Get UltraFuels.

Carp and Couth (2021) of Honeywell UOP discussed the changing market from fuels to petrochemicals. In mature markets like the United States, refineries are revisiting their business models, asset base, and technology options. This shift is due to regional and global changes in both fuel specifications and demand. It is anticipated that fuel demand will continue to soften and that petrochemical demand growth will remain strong throughout the world. The refinery of the future will have the flexibility to meet rapidly changing market needs, will be integrated into petrochemicals to unlock new value, and will also contain the connected technology required to optimize performance every day. Today, U.S. refineries have access to a host of new technologies that create an alternative path to profitability. They can unlock new value from the heavy end of crude, as well as manage molecules efficiently to produce immediately marketable products or their own intermediates for petrochemicals. The strategy utilizes the latest bottom-of-barrel, maximum conversion hydrocracking, and aromatics technologies (Carp and Couth, 2021).

1.4.4 Cost Factors and Refinery Capital Cost

Economics is an important element in the sustainable operation of a refinery. The process design and operation must be economically viable, or it will not proceed. Alternative designs must be compared technically, environmentally, and economically. The two main factors in new refinery economics are the cost of the feedstock (crude oil price) and the capital cost to build the refinery. Projection of crude oil price based on 2007 prices for three growth cases is shown in Figure 1.13. According to the reference price, the projected value in 2022 is about \$115, while the actual average price of crude oil in 2022 was about \$101. Considering the impact of the pandemic on the energy market, this prediction for the period of 15 years is quite reasonable.

As discussed in Chapter 7, oil refining is a capital-intensive business. Planning, designing, permitting, and building a new medium-sized refinery is a five to seven year process with a cost of billions of dollars, as shown in Table 1.3. These investment costs are given for typical refinery of about 200,000 bbl/d capacity. The value of α is based on prices before pandemic with a benchmark of 30,000 USD/bbl of crude oil processed. The refinery cost is directly related to the degree of conversion of heavy compounds in residues and heavy oil into lighter compounds

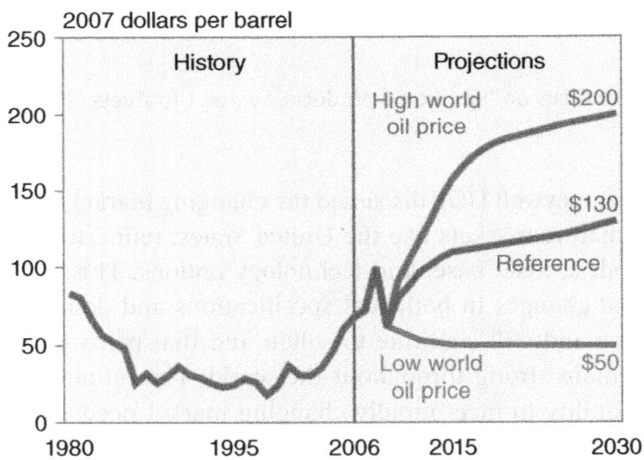

Figure 1.13 World oil prices (USD) in three economic growth cases. *Source*: Adapted from IEA (2007).

Table 1.3 Investment, energy, and hydrogen consumption according to refinery complexity.

Type of refinery	Investment (billion \$)	Energy consumption (wt% of crude)	Hydrogen consumption
Simple	α (≈5)	≈4	β
Classical conversion	3α (≈15)	≈6	2β
Deep conversion	5α (≈25)	≈10	5β

Source: ASTM MNL58/ASTM International.

and fuels. The figures in the above table are not absolute, as the hydrogen and therefore the value of β will depend on operating conditions, which in turn depend on factors such as the nature of the crude oil and the intended sulfur content of products. Further discussion on the refining investment is given in Chapter 14. This chapter is intended to provide a strategic response to the risk and uncertainty facing the petrochemical refining industry today. It builds upon a tripod of industry values "Safety, Environment, Profit" and outlines the leading edge of quantitative methodologies to help management decide the way forward.

1.4.5 Life Cycle Analysis

It can be misleading to analyze GHG reductions from a technology in isolation because providing the feed and handling the products of that technology can have their own GHG impacts. Life cycle analysis (LCA) attempts to account for all the impacts of technology or process modifications so that alternatives can be compared on an equal basis. Life cycle impact assessments (LCIA) convert Life Cycle Impact (LCI) data into potential environmental impacts such as global warming potential (GWP), water footprint, energy use, and air quality and toxicity. Biomass-derived syngas often exhibits lower GWP compared to fossil-based syngas, though emissions from fertilizer use can offset some advantages. Interpreting the LCA through hotspot analysis, scenario analysis, or sensitive analysis helps identify environmental hotspots and evaluate potential improvements. For example, cultivation and syngas production often dominate environmental impacts, particularly in water and energy use; exploring alternatives, such as using biochar as a catalyst or implementing carbon capture, can enhance sustainability. These issues are discussed in Chapter 13.

References

Amiri, J. (2024). *Hydrogen Research*. Fyfe. https://fyfe.com.au/sectors/hydrogen/ (accessed 8 April 2024).

API (2024). *Where are the Pipelines?* American Petroleum Institute. https://www.api.org/oil-and-natural-gas/wells-to-consumer/transporting-oil-natural-gas/pipeline/where-are-the-pipelines/ (accessed 8 April 2024).

BASF (2022). *SABIC and Linde Start Construction of the World's First Demonstration Plant for Large-scale Electrically Heated Steam Cracker Furnaces*. BASF. https://www.basf.com/us/en/who-we-are/sustainability/whats-new/sustainability-news/2022/basf-sabic-and-linde-start-construction-of-the-worlds-first-demonstration-plant-for-large-scale-electrically-heated-steam-cracker-furnaces (accessed 8 August 2024).

Beloe, W. (2019). *Herodotus*. Cambridge University Press; Original translation from W. H. Colburn and R. Bentley, *Herodotus*, London, 1830.

C2ES (2008). *Addressing Emissions from Coal Use in Power Generation*. Center for Climate and Energy Solutions. https://www.c2es.org/document/addressing-emissions-from-coal-use-in-power-generation/ (accessed 8 April 2024).

CAP (2024a). *Canadian Oil and Gas Production*. Canadian Association of Petroleum Producers. https://www.capp.ca/wp-content/uploads/2023/12/Canadian-Oil-and-Gas-Production.pdf (accessed 10 June 2024).

CAP (2024b). *GHG Emissions*. Canadian Association of Petroleum Producers. https://www.capp.ca/wp-content/uploads/2024/03/GHG-Emissions-v032724.pdf (accessed 10 June 2024).

Carp, S. and Couth, K. (2021). *Advances in Crude and Chemicals*. Honeywell UOP. https://www.digitalrefining.com/article/1002610/advances-in-crude-to-chemicals (accessed 10 June 2024).

Century, J.R. (2008). Tar sands: key geologic risks and opportunities. *The Leading Edge* 27 (9): 1202–1204.

da Silva, M.W. (2022). *Crude Oil Refining: A Simplified Approach*. CRC Press.

DNV (2024). Energy transition outlook 2024. A global and regional forecast to 2050. https://www.dnv.com/energy-transition-outlook/download/?utm_source=googlecpc&utm_medium=search&utm_campaign=eto_2024_download&gad_source=1&gclid=EAIaIQobChMIqojAit68jAMVidXCBB0p7Ap8EAAYASAAEgLxnPD_BwE (accessed 20 March 2025).

DOE (2024). "Hydrogen Production: Natural Gas Reforming" US Department of Energy, https://www.energy.gov/eere/fuelcells/hydrogen-production-natural-gas-reforming accessed 8 April 2024.

EIA (2023). *U.S. Oil and Natural Gas Wells by Production Rate*. U.S. Energy Information Administration. https://www.eia.gov/petroleum/wells/ (accessed 8 April 2024).

EIA (2024). *Top US Refineries Operable Capacity*. Energy Information Administration, U.S. Energy Department. www.eia.gov/energyexplained/oil-and-petroleum-products/refining-crude-oil-refinery-rankings.php (accessed 20 March 2025).

EIA (2025a). *Jet Fuel Made Up a Record Share of U.S. Refinery Output in 2024*. U.S. Energy Information Administration (accessed 24 March 2025). www.linkedin.com/company/u-s-energy-information-administration/posts/?feedView=all

EIA (2025b). *U.S. Ethane Production, Consumption and Exports Set New Records in 2024*. Energy Information Administration (EIA). (accessed 20 March 2025). www.linkedin.com/pulse/us-ethane-production-consumption-exports-h1mve/

Elbersen, W. (2005). Biorefineries for the chemical industry–a dutch point of view. In: *Biorefineries-Industrial Processes and Products* (ed. B. Kamm, P.R. Gruber, and M. Kamm). Wiley. https://doi.org/10.1002/9783527619849.ch4.

Energy Institute (2024). Statistical Review of World Energy, 73rd Ed. https://www.energyinst.org/statistical-review/home (accessed 14 May 2025).

GEM (2024). *Global Oil Infrastructure Tracker*. Global Energy Monitor. https://globalenergymonitor.org/projects/global-oil-infrastructure-tracker/ (accessed 8 April 2024).

Government of Canada (2023). Greenhouse gas reporting program (GHGRP) – facility greenhouse gas (GHG) data. https://open.canada.ca/data/en/dataset/a8ba14b7-7f23-462a-bdbb-83b0ef629823 (accessed 8 April 2024).

Hasell, J. and Roser, M. (2019). How do we know the history of extreme poverty? https://ourworldindata.org/extreme-history-methods (accessed 8 April 2024).

IEA (2007). *World Energy Outlook 2007*. International Energy Agency. www.iea.org/reports/world-energy-outlook-2007 (accessed 10 August 2024).

IEA (2023). Emissions from Oil and Gas Operations in Net Zero Transitions A World Energy Outlook Special Report on the Oil and Gas Industry and COP28, International Energy Agency, June 2023. https://iea.blob.core.windows.net/assets/2f65984e-73ee-40ba-a4d5-bb2e2c94cecb/EmissionsfromOilandGasOperationinNetZeroTransitions.pdf

IEA (2024). *Gas Flaring*. International Energy Agency. https://www.iea.org/energy-system/fossil-fuels/gas-flaring (accessed 10 June 2024)

Kiehl, J.T. and Trenberth, K.E. (1997). Earth's annual global mean energy budget. *Bulletin of the American Meteorological Society* 78 (2): 197–208. https://doi.org/10.1175/1520-0477(1997)078<0197:EAGMEB>2.0.CO;2.

Li, F., Liu, D., Sun, K. et al. (2024). Towards a future hydrogen supply chain: a review of technologies and challenges. *Sustainability* 16: 1890. https://doi.org//10.3390/su16051890.

Li, T., Long, J., Du, W. et al. (2023). Three pathways towards elimination of CO_2 emissions from industrial plants that use hydrocarbon fuels. *Journal of Cleaner Production* 391: 136159.

Madugula, A.C.S., Sachde, D., Hovorka, S.D. et al. (2021). Estimation of CO_2 emissions from petroleum refineries based on the total operable capacity for carbon capture applications. *Chemical Engineering Journal Advances* 8: 100162.

Pelucchi, S., Mocellin, P., and Galli, F. (2024). Technoeconomic Framework of Natural Gas Pyrolysis as an Alternative to Flaring and Venting in the Oil Extraction Sector, AIChE Annual Meeting, October 27–31, San Diego, CA, USA.

Rana, M.S., Ancheyta, J., Riazi, M.R. et al. (2013). Chapter 33: Future directions in petroleum and natural gas refining. In: *MNL58: Petroleum Refining and Natural Gas Processing* (ed. M.R. Riazi, S. Eser, S.S. Agrawal et al.), 769–800. ASTM International.

Riazi, M.R. (2005). *MNL50: Characterization and Properties of Petroleum Fractions*. ASTM International.

Roser, M. and Ritchie, H. (2023). How has world population growth changed over time? https://ourworldindata.org/population-growth-over-time (accessed 8 April 2024)

Schmidt, G.A., Ruedy, R., Miller, R.L. et al. (2010). Attribution of the present-day total greenhouse effect. *Journal of Geophysical Research* 115 (D20). https://doi.org/10.1029/2010JD014287.

Statista (2024a). Demand for crude oil worldwide from 2005 to 2022, with a forecast for 2023. https://www.statista.com/statistics/271823/global-crude-oil-demand/ (accessed 8 April 2024).

Statista (2024b). Number of active petroleum refineries in the United States in selected years from 1982 to 2023. https://www.statista.com/statistics/1447051/operative-oil-refineries-in-the-us/ (accessed 8 April 2024).

Statista (2024c). Number of land and offshore oil rigs worldwide at the end of each year from 2010 to 2023. https://www.statista.com/statistics/1128408/number-of-global-oil-rigs-by-type/ (accessed 8 April 2024).

Statista (2024d). Number of crude oil tankers worldwide as of April 2020 by type. https://www.statista.com/statistics/468405/global-oil-tanker-fleet-by-type/ (accessed 8 April 2024).

Statista (2024e). Number of operational and planned crude oil refineries worldwide as of 2023. https://www.statista.com/statistics/1445314/number-of-oil-refineries-worldwide/ (accessed 8 April 2024).

U.N. (2015). *The Paris Agreement, 2015*. United Nations. https://unfccc.int/process-and-meetings/the-paris-agreement?gclid=CjwKCAiAiP2tBhBXEiwACslfnkS7jxlEyt5in69yh1V4AeBp8XZvM8a04UnCGR56R6XHpexyzOQP1xoC1S4QAvD_BwE (accessed 11 April 2024).

U.S. Geological Survey (2003). Heavy oil and natural bitumen: strategic petroleum resources. https://pubs.usgs.gov/fs/fs070-03/fs070-03.pdf (accessed June 2024).

Ultrafuels (2024). UltraFuels. https://www.getultrafuels.com (accessed 16 January 2025).

Wikipedia (2024). Energy density. https://en.wikipedia.org/wiki/Energy_density (accessed 8 April 2024).

Wood Mackenzie (2020). Oil and gas market. https://www.woodmac.com/store/industry-sector/oil-and-gas-markets/ (accessed 10 December 2023).

Xu, C. and Bell-Hammer, L. (2022). Global oil and gas reserves increase in 2022. *Oil & Gas Journal*. (5 December 2022).

Part I

Cleaner Petroleum Production

2

Carbon Capture, Storage, and Enhanced Oil Recovery

Eduardo Manrique

Citation Oil & Gas Corp., Lead Reservoir Engineer, Reservoir Department, Houston, Texas

2.1 Introduction

Oil and gas continue to be key contributors to the global energy demand. However, using fossil fuels has also been associated with increased greenhouse gas (GHG) emissions (i.e., CO_2 and CH_4). Considering the relevance of CO_2 emissions on climate change, in December 2015, "The Paris Agreement" was adopted by 196 countries and entered into force in late 2016 (UN, 2024). This multilateral agreement on climate change seeks to limit the increase of global temperatures to 1.5 °C by the end of this century by reducing GHG emissions, impacting the future of oil and gas operations.

Almost four years later, the Paris Agreement was followed by another major global event, the COVID-19 pandemic (4Q 2019–2020). The latter disrupted human and industrial activity, reducing CO_2 emissions drastically (Davis et al., 2022; Le Quéré et al., 2020). The emission reduction registered during the pandemic accelerated energy transition and global decarbonization plans, or "Net zero or carbon neutrality scenarios" (IPPC, 2018). Although there is an open debate about whether "Net Zero" is achievable (Fankhauser et al., 2022; Farouq Ali and Soliman, 2022, 2023), there is a global consensus to tackle CO_2 emissions in decades to come.

The proposed solutions to reduce CO_2 emissions from fossil fuels include the accelerated adoption of renewable energy (i.e., solar, wind, green hydrogen), low-carbon solutions, and energy efficiency processes. However, regardless of the possible scenarios to meet future energy demand, energy outlooks predict the need

Cleaner Petroleum Production and Refining Technologies, First Edition, Edited by M. R. Riazi and Harvey W. Yarranton.

for oil and gas in the energy mix by 2050. In the worst-case scenario, 50% of today's global consumption of fossil fuels is forecasted (EIA, 2023; IEA, 2023; OPEC, 2023). Therefore, despite the associated uncertainties, there is a global impetus to use carbon capture, utilization, and storage (CCUS) technologies to reduce the carbon footprint of hard-to-abate industries. These CO_2-intensive industries would be hard to shift entirely away from fossil fuels as an energy source.

This chapter provides a comprehensive review and will serve as a guide to CCUS, a broad topic with multiple areas requiring detailed evaluation specific to each project or region. This chapter is structured to provide a general overview of the main steps of CCUS in Section 2.2, focusing on geologic storage in Section 2.3 and Enhanced Oil Recovery (EOR) in Section 2.4. These two utilization approaches are commercially available and key in removing large CO_2 volumes currently emitted into the atmosphere. The impact of global decarbonization and energy efficiency goals on thermal EOR (tEOR), CO_2-EOR projects, and polymer flooding (PF) are also reviewed in Section 2.4. Finally, Section 2.5 discusses the technical, non-technical, and economic challenges of CCUS implementation.

2.2 CCUS: Main Steps

CCUS is one key strategy for reducing CO_2 emissions and has been widely documented by international organizations and in the literature (Dziejarski et al., 2023; ETC, 2022; IEA, 2024; Liu et al., 2023; OGCI, 2024). Therefore, it will not be described in detail in this chapter. CCUS fundamentals consist of capturing CO_2 emissions, mainly from anthropogenic sources, for re-utilization or geological storage instead of venting into the atmosphere. The main steps can be summarized as follows:

- **Carbon Capture:** CO_2 emissions are captured from different industrial sources (i.e., coal or gas power plants, refineries, cement or steel plants). Capture is the most energy-intensive and costly step, especially in diluted CO_2 streams (7–16%) such as refineries and cement plants. In other words, the higher the CO_2 concentration from the industrial source (i.e., petrochemical plants), the lower the cost of CO_2 capture (Dziejarski et al., 2023; Hughes and Zoelle, 2022; Mahasenan and Brown, 2005; Rubin et al., 2012). Generally, capture technologies can deliver CO_2 streams with 95% to 99.9% (food grade) wt concentrations at different pressures. Depending on the CO_2 utilization, additional CO_2 dehydration and compression will be required.
- **CO_2 Compression and Transportation:** CO_2 can be transported via trucks, rail, ship, and pipelines and must meet composition standards with CO_2 purity higher than 95% (NPC, 2021). CO_2 is typically transported

supercritically to optimize transportation volumes and costs. This study will use the United States as a reference case for being the largest operator of CO_2 pipelines worldwide. In the United States, most CO_2 pipelines (ANSI Class 900) typically operate between 1,200 and 2,200 psig, with some pipelines having a maximum operating pressure of 2,500–2,800 psig. For over five decades, the United States has safely operated CO_2 pipelines, reaching the end of 2022 with a pipeline network of more than 5,300 miles (PHMSA, 2024a).

- **CO_2 Storage:** The storage of CO_2 in geologic formations represents one of the options that can contribute to reducing anthropogenic emissions due to the large storage capacity available. Although geologic formations have been proposed to store CO_2 (i.e., coal seams, basalt, shale reservoirs, and saline aquifers), oil and gas reservoirs are the most studied. The experience gained with CO_2 Enhanced Oil Recovery projects (CO_2-EOR) over the last five decades consists of a solid experience that can be extrapolated to other geologic formations (DNV, 2012; Dziejarski et al., 2023; Mwenketishi et al., 2023; NETL, 2024a). Some of the most relevant CO_2 injection in saline aquifer experiences will be summarized later in this section. CO_2-EOR in the United States and overseas will be presented later in this chapter.
- **CO_2 Utilization:** This chapter focuses on geological storage and EOR as a permanent solution to lessen CO_2 emissions. Nevertheless, CO_2 utilization is a broader area under investigation to find new possible applications. Some examples include its conversion into chemical products (e.g., low-molecular-weight carboxylic acids, urea, polymers), fuels (e.g., methanol, synfuels), mineralization, or new construction materials (Saravanan et al., 2021; Valluri et al., 2022; Yousaf et al., 2022). Most recently, formate solutions produced from the electrochemical reduction of CO_2 have been proposed as a possible alternative for EOR and storage of a modified CO_2 molecule (New Options in Figure 2.1). However, this research is in the early stages of evaluation (Oyenowo et al., 2024).

Figure 2.1 shows a simplified schematic of the CCUS process. The first main difference is the CO_2 content in the exhaust gas streams. Emitters with

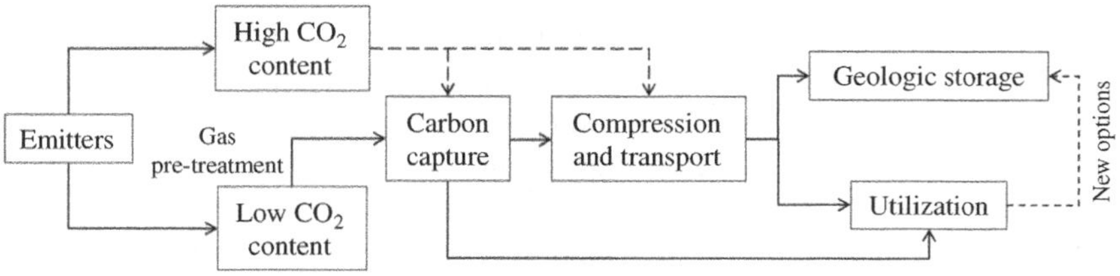

Figure 2.1 Simplified CCUS process.

high-purity CO_2 are characteristic of petrochemical plants (e.g., ammonia, ethanol, and ethylene oxide), natural gas processing Fisher–Tropsch coal to liquids, and others. In contrast, emitters with low-purity CO_2 include power plants, refining processes, and cement and steel industries, which generally require higher energy consumption, capital, and operating expenses (Hughes and Zoelle, 2022; Rubin et al., 2012). Bui et al. (2018) described the work (kilo Joules per mol of CO_2 – kJ/mol CO_2) associated with capturing CO_2 as a function of its concentration from the source of the emissions. This comprehensive review used an alternate approach to describe the dependency of CO_2 concentration of different sources with the energy intensity required to generate CO_2 streams of different purity. The impact of CO_2 concentration in the emission source on project economics will be described later in this chapter.

Depending on its final use (e.g., EOR, geologic storage, or development of new products), high-purity CO_2 may require minor gas treatments and dehydration before compression and transport. However, low-purity CO_2 generally must undergo gas pre-treatment (industry or process-dependent) and carbon capture. CO_2 compression and transport will also depend on its final use. As mentioned earlier, pipelines represent the main transportation of CO_2, generally operating between 1,200 and 2,200 psig, as described in the following sections. CO_2 compression and transportation for other uses may vary depending on the final utilization. Although it has yet to be fully addressed in the literature, new options for EOR or geologic storage may include using chemical products and storing CO_2 via transformed products. However, these feasible options are out of the scope of this chapter.

2.3 Geologic CO_2 Storage

The experience of CO_2 storage not associated with EOR projects is limited to a few commercial-scale applications. CO_2 storage in the Brazilian pre-salt is a good example, starting injection in 2010 (Pereira Nunes et al., 2024 and references therein). More recently, CO_2 storage initiatives in the North Sea (The Northern Lights Project) and Texas Gulf Coast validate the significance of CO_2 storage as part of the strategy to reduce carbon emissions in years to come (Equinor, 2024; ExxonMobil, 2024). This section will summarize four illustrative cases documented in the literature with mixed results (Table 2.1).

Sleipner has been the pioneer in storing substantial amounts of CO_2 for over two decades. The CO_2 source comes from the gas separation (Amine scrubbers) produced from the Sleipner Vest field and injected (0.85–1 million tonnes of CO_2 per annum – Mt CO_2/yr) into the Utsira sand, a regional aquifer in the Norwegian continental shelf. All the operations are performed offshore, requiring a CO_2

Table 2.1 Example of CO_2 storage projects in aquifers.

Field, country	Start and status	CO_2 source	References
Sleipner, Norway	1996 (Active)	Gas processing plant (anthropogenic)	Furre et al. (2017), Hauber (2023)
In-Salah, Algeria	2004 (Finished)	Gas processing plant (anthropogenic)	Mathieson et al. (2011), Ringrose et al. (2013)
The Quest CSS Facility, Canada	2015 (Active)	Hydrogen (H_2) production (anthropogenic)	Alberta Department of Energy (2023), IEAGHG (2019)
Gorgon, Australia	2019 (Active)	Gas processing plant (anthropogenic)	Robertson and Mousavian (2022), Trupp et al. (2021)

pipeline of approximately 12.5 km. It is estimated that this project has stored more than 20 Mt CO_2 since it began. The final storage capacity is still to be defined due to existing uncertainties associated with numerical predictions (Furre et al., 2017; Hauber, 2023). Simulation studies predict that Utsira can increase the CO_2 storage capacity by considering water production to reduce the aquifer pressure and extend CO_2 injection until the cap seal rock pressure safety limit is reached (Zhang et al., 2022a).

The In-Salah CCS project started in 2004. The main goal of this Joint Industrial Project was to reduce the CO_2 content of the produced gas to meet sales specifications. The stripped CO_2 from the gas produced in different gas fields was planned to be stored in a bottom sandstone aquifer of a gas-producing formation at the Krechba gas field in Algeria. The CO_2 capture was based on the MEA amine process before dehydration, compression, and transportation to the storage site. The original plan was estimated to store about 17 Mt CO_2 during the project's lifetime. However, the project had to be suspended in 2011 due to the over-pressure of the aquifer and the potential risks of losing the integrity of the seal rock. When the project was suspended, it was estimated that a total of 3.8 Mt CO_2 was successfully stored in the Krebcha formation (Hauber, 2023; Mathieson et al., 2011; Ringrose et al., 2013).

The Quest CCS project started commercial operations during the second half of 2015. This project is based on the CO_2 capture of three H_2 manufacturing units at the Edmonton Scotford heavy oil (HO) upgrader. The CO_2 is separated from the produced gas stream using amine solvent technology, then dehydrated, compressed, transported ($\approx$65 km pipeline), and injected into a saline formation (Basal Cambrian Sands) for permanent storage. By the end of 2022, the project has not reported major operational issues, and more than 7 MtCO_2 have been stored. CO_2 injectivity has been defined as excellent despite a few injectivity impairments due to suspected secondary mineral phase precipitation (e.g., halite).

Gorgon, located on Barrow Island, is one of the largest natural gas projects in the world. The development plan for these gas resources considered capturing and injecting the separated CO_2 in the Dupuy Formation, a deeper sandstone saline reservoir. CO_2 is separated from natural gas produced using commercial amine scrubber technology (methyl diethanolamine or MDEA). The estimated plan was to inject up to 4 $MtCO_2$/yr (approximately four times larger than Sleipner) during the project's life. However, like many projects in the world, Gorgon is facing technical and economic challenges that need to be revised to meet the scheduled targets (Flett et al., 2009; Marshall, 2022; Robertson and Mousavian, 2022; Trupp et al., 2021).

Unlike the Sleipner and Quest facilities as of today, In-Salah and Gorgon, CCUS projects are good examples of the potential challenges of CCUS projects in non-hydrocarbon-producing formations, where the volume of information (e.g., reservoir compartmentalization and pay continuity) and historical data (e.g., injectivity, pressures) is generally limited.

2.4 Enhanced Oil Recovery

Energy efficiency and gas flaring reduction represent the short-term solutions to mitigate carbon footprint as part of the energy transition plans in the oil and gas industry (Darrold et al., 2024; ElMasry and Shakirov, 2024; Jansen, 2019; Rambech et al., 2016; Schopp et al., 2022; Talaei et al., 2020). This strategy has also been incorporated into oil and gas recovery processes (Al Rahbi et al., 2022; Angga et al., 2022; Farajzadeh et al., 2022).

EOR methods have also adopted energy efficiency and carbon footprint strategies to develop mature assets sustainably. The literature has widely documented EOR definitions and processes, generally classified as Chemical, Gas, and Thermal methods (Alvarado and Manrique, 2010a; Babadagli, 2020; Manrique et al., 2007). However, only a few have been implemented commercially worldwide. The latter excludes natural gas injection as a secondary recovery (pressure maintenance) or in a Water-Alternating-Gas (WAG) scheme. Despite the multiple EOR applications at the pilot and commercial scales, the main objective of this chapter is to discuss the impact of tEOR and CO_2-EOR because decarbonization trends are expected to impact most of these recovery methods and energy transition plans proposed globally. However, at the end of this section, a brief description of the role of PF in reducing the carbon footprint of conventional waterfloods will also be presented.

2.4.1 Thermal EOR and Carbon Capture

Most global EOR production comes from thermal methods, especially cyclic (steam stimulation) and continuous steam injection (steam flooding) in heavy (HO) and extra-heavy oil (XHO) reservoirs. In situ combustion (ISC) has also been tested at pilot and commercial scales, but this recovery process has not significantly impacted worldwide production (Alvarado and Manrique, 2010b). Although steam injection is under commercial application in different countries with vast resources of HO and XHO reservoirs, Canada and the United States, specifically California, are two examples of regions with steam injection projects for several decades. For example, Kern River Field, California, has been producing for more than 120 years, and steam flooding started in 1964 (Bursell, 1970; Popa et al., 2022). In 2020, this field produced approximately 45 KBOPD (California Department of Conservation, 2023), which validated the efficiency of this recovery process.

Figure 2.2 shows the history of oil production by steam injection in California (2010–2020) and Canada (2010–2022). In the case of California (Figure 2.2a), the daily production reported from steam injection (cyclic steam and steam flooding) represents the total production of the five largest fields (Belridge South, Cymric, Kern River, Lost Hills, and Midway Sunset) under tEOR (California Department of Conservation, 2015, 2020, 2023). From 2010 to 2020, these five fields produced about 1.1 billion barrels of oil. The production history of Canadian steam injection projects (Figure 2.2b) comes from cyclic and SAGD (Steam Assisted Gravity Drainage), annually reported by the Alberta Energy Regulator (AER), previously the Energy Resources Conservation Board or ERCB (Alberta Energy Regulator, 2024a). It is worth mentioning that the cumulative oil production of Canadian

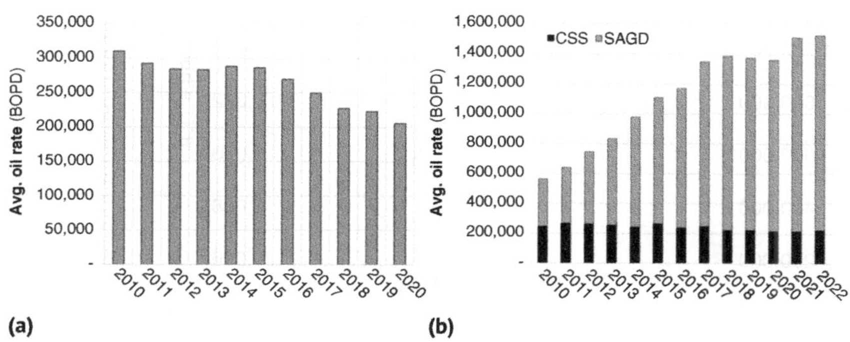

Figure 2.2 Daily oil production from steam injection in (a) California's largest heavy oil (HO) fields and (b) Canadian oil sands (Alberta Energy Regulator, 2024a; California Department of Conservation, 2015, 2020, and 2023).

steam injection projects is approximately 5.3 billion barrels of oil over the period reported (2010–2022).

In 2022, the last report available, the total production of Canadian steam injection came from three cyclic steam stimulation (CSS) projects (Cold Lake, Peace River, and Primrose and Wolf Lake) and 23 commercial SAGD projects, having Christina Lake, Firebag, Foster Creek, and Surmont as the largest oil sand producers (Alberta Energy Regulator, 2024b).

Despite steam injection's high oil recovery efficiency, this recovery process is under scrutiny due to its high energy intensity and carbon footprint. Given the increasing trend of oil production by steam injection in Canada, especially commercial SAGD floods, it will be used as a case study to evaluate the reported carbon intensity of "in situ" oil sand development (oil sand mining was not considered in this study). Figure 2.3 shows the average total production and the carbon or emission intensity (tons of CO_2 equivalent per barrel – tCO_2e/bbl) of oil sands reported from steam injection from 2011 to 2022 (Alberta Government, 2024; Alberta Energy Regulator, 2024a).

According to official information reported by the Alberta Government and AER, oil production is increasing, and the tons of CO_2 emitted per barrel of oil produced are decreasing. In 2011, the average emission intensity was estimated at 0.0778 tCO_2e/bbl for a total oil production of 0.64 million bbls/day. By 2022, the carbon intensity was reduced by approximately 16% (0.0657 tCO_2e/bbl), producing

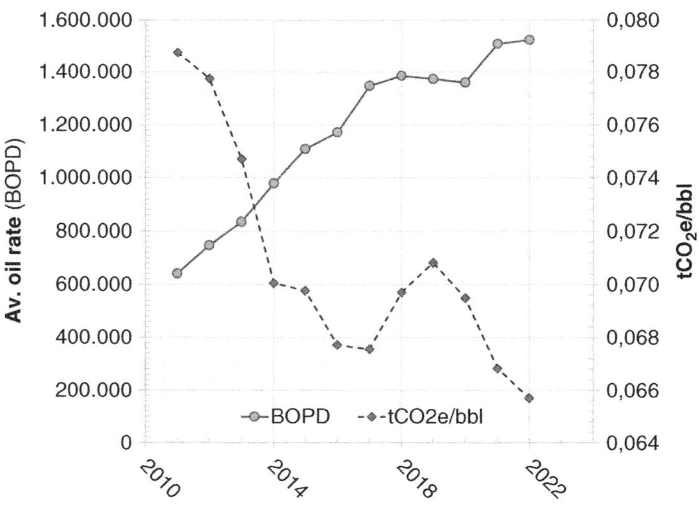

Figure 2.3 Daily oil production and emission intensity (tCO_2e/bbl) reported from oil sand steam (in situ) projects (Alberta Government, 2024; Alberta Energy Regulator, 2024a).

an average rate of 1.52 million bbls/day. These positive trends suggest the incorporation of energy efficiency and emission mitigation strategies pointing in the right direction for the sustainable development of oil sand resources in Canada as projected by the energy transition plans of the Canadian government and globally.

To put these volumes in perspective, omitting the net carbon release into the atmosphere, for an average emission intensity of 0.0707 tCO_2e/bbl from 2011 to 2022 (Alberta Government, 2024) and an oil rate of 1 million bbls/day, represents approximately 25.8 $MtCO_2e$/yr. This mass of CO_2 represents 64.5% of the estimated CO_2 stored (40 $MtCO_2$) in the Weyburn CO_2-EOR project since its start in 2000 (Whitecap Resources Inc., 2024), and the estimated emissions of eight to nine refineries of 500 KBOPD of refining capacity based on a linear correlation of total operable capacity (bbl/day) vs. CO_2 emissions (metric ton/day) for 128 U.S. refineries reported by Saketh Madugula et al. (2021). However, the tCO_2e/bbl may differ with oil refining operations in different world regions.

However, there is a debate about whether the reported Canadian carbon intensity reduction is accurate or whether it is fast enough to meet the CO_2 emission goals by 2050 (Hannouf et al., 2021; Fu and Legge, 2023; Liggio et al., 2019; Liggio and Li, 2023). Regardless of how the information documented in the literature is interpreted, this chapter aims to summarize the status, pros, and cons of the proposed decarbonization strategies by the international community. Figure 2.4 summarizes some of the approaches evaluated and ongoing efforts to increase energy efficiency and reduce the carbon footprint of steam injection projects.

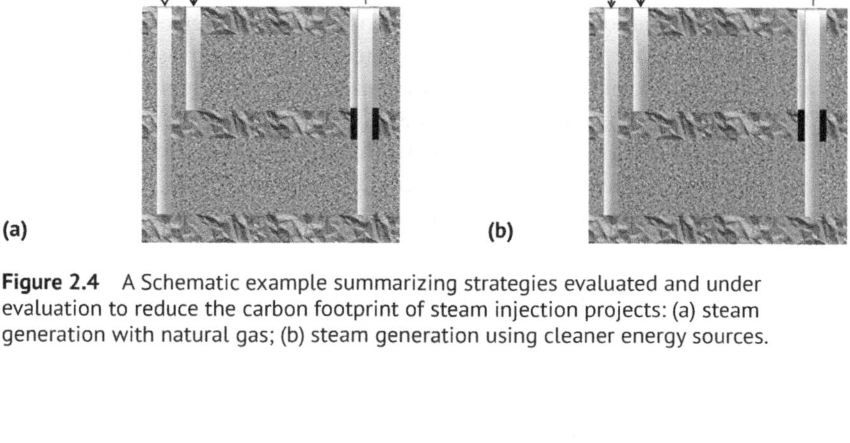

Figure 2.4 A Schematic example summarizing strategies evaluated and under evaluation to reduce the carbon footprint of steam injection projects: (a) steam generation with natural gas; (b) steam generation using cleaner energy sources.

The proposed scenarios to reduce the carbon emission intensity of tEOR projects can be summarized in two main strategies:

- Continue the development of HO and XHO by steam injection using gas-fueled steam generators, integrating technologies (e.g., hybrid steam injection and CCUS), and considering a closed-cycle approach (Figure 2.4a). Some of the examples of hybrid technologies with ongoing efforts reporting energy efficiency and carbon footprint benefits include CSS with foams, flue gases, or nitrogen (Han et al., 2020; Liu et al., 2024; Perez et al., 2023b; Sun et al., 2013; Tang et al., 2022). Generally, gas sources are portable nitrogen generation units or flue gas from steam generators. However, the main objective of CO_2 Capture is to integrate steam floods and SAGD projects with CO_2-EOR or storage in deeper or nearby formations, optimizing CO_2 transportation costs, as will be described later in this chapter. CCUS technologies are commercially available, and the following section will summarize decades of experience. Canada has proven decades of experience in CO_2-EOR or storage. Joffre Viking and Weyburn CO_2-EOR/storage (Pyo et al., 2003; Whitecap Resources Inc., 2024; Whittaker et al., 2011) are well documented in the literature. More recently, the CO_2 capture of an oil sand upgrader, transportation, and storage (Quest CSS Project) represents an important milestone in proving sustainable options to develop HO and XHO resources (Canadian Energy Centre, 2022; IEAGHG, 2019). The experience gained in this project will also contribute to consolidating CCUS strategies in the refining sector.
- Development of HO and XHO by steam methods using cleaner energy sources (e.g., solar, nuclear) for steam generation (Figure 2.4b). Coalinga Field in California (Sandler et al., 2012) and Amal West Field in Oman (Palmer and O'Donnell, 2014) are examples of HO fields that report using solar steam injection. Whether this approach is technically and economically viable is out of the scope of this study. However, some limitations include but are not limited to intermittent steam quality, Direct Normal Irradiation, or DNI (location dependent), and availability of large areas (may include deforestation), among others. Although both projects were not commercially expanded in other regions, a renewed interest in reexploring this approach is likely. The use of nuclear energy for steam generation has been proposed but has not been commercialized for SAGD or tEOR in general (Bersak and Kadak, 2007; Finan and Kadak, 2010). In this case, the debate should be whether nuclear energy should replace the use of fossil fuels. However, this is also out of the scope of the chapter.

In addition to the possible options to develop HO and XHO resources described above, several initiatives evaluated at the laboratory and numerical simulation scale explore new ways to increase energy efficiency and seek the use of flue gases generated from steam generation (Chang, 2020). Recently, the integration of tEOR with the Organic Rankine Cycle (ORC) for electricity generation (Aikman, 2022; Bashti et al., 2023; Rodriguez et al., 2023) and downhole steam generation based on mist water and air co-injection has also been proposed (Perez et al., 2023a). Despite multiple research efforts, steam injection methods will continue dominating tEOR. However, tEOR must improve energy efficiency and reduce the carbon footprint at competitive costs. Although CCUS has been proposed as one of the options to meet these goals, tEOR may face multiple challenges in years to come.

2.4.2 CO_2 Enhanced Oil Recovery

CCUS technologies are commercially available and represent one of the key options for reducing CO_2 emissions from industrial sources or gas and oil operations. Captured and compressed CO_2 can be safely stored in geologic formations, especially oil and gas reservoirs where these fluids have been confined for geologic periods. Although CO_2-EOR has been proven an effective way to increase oil recovery and, at the same time, reduce CO_2 emissions (closed-cycle projects) for decades, opponent sectors claim that this will extend the dependence on fossil fuels, negatively impacting global decarbonization goals (ETC, 2022). Regardless of the vision of CO_2-EOR and storage, countries have recognized that energy security will prevail over the energy transition with increased energy demand (DNV, 2023; Rabbi et al., 2022; Yergin, 2024). Therefore, all options available to meet decarbonization goals must be considered to secure energy supply with the lowest impact on the environment possible.

CO_2-EOR has been successfully evaluated for over six decades worldwide (Brock and Bryan, 1989; Martin and Taber, 1992). In the United States, initial projects relied more on natural sources of CO_2 (Allis et al., 2001), and as the pipeline network was developed, the number of projects has continuously increased since the 2000s (Advanced Resource International, 2021; Alvarado and Manrique, 2010b; Kuuskraa and Wallace, 2014; Manrique et al., 2007).

Table 2.2 summarizes a short list of representative CO_2-EOR projects using natural and anthropogenic sources since the early 1970s. As of 2020, the United States reported 142 projects with an average oil production of 273,000 BOPD. The total supply of these CO_2 projects was estimated at 1.6 Bcf/d, where approximately 19% (0.3 Bcf/d) came from anthropogenic sources. However, this represents approximately a 47% decrease in CO_2 supply compared to 2019 (Advanced Resource International, 2021). The latter confirms the complexity of energy transition and decarbonization goals, making it difficult to predict the demand for anthropogenic CO_2 for future EOR or storage in the United States.

Regardless of the current trends of CO_2-EOR projects in the United States, there is no doubt about the commercial experience available to implement CCUS

Table 2.2 Example of commercial U.S. CO_2-EOR projects.

Field (state)	Start (status)	CO_2 source	References
Scurry Area Canyon Reef Operators Committee (SACROC) Unit, (TX)	1972 (Active)	Gas processing plants and natural sources	Brock and Bryan (1989), Ghahfarokhi et al. (2016), Kalteyer (2020), Kindermorgan (2024)
East Vacuum Grayburg San Andres Unit (EVGSAU), (NM)	1985 (Active)	Natural sources	Brock and Bryan (1989), Moffitt et al. (2015)
Rangely Weber, (CO)	1986 (Active)	Gas processing plant (anthropogenic)	Brock and Bryan (1989), Cooper (2016), Masoner et al. (1996)
The Goldsmith-Landreth San Andres Unit (GLSAU), (TX)	2009 (Active)	Natural sources	Kalitsun et al. (2023), Kindermorgan (2024)
Beaver Creek, (WY)	2008 (Active)	Gas processing plant (Anthropogenic)	Peterson et al. (2012), Robertson (2022)
Petra Nova, (TX)	2017 (Active)	Coal Fired Power Plant (Anthropogenic)	EIA (2017), Jacobs (2023), Jenkins (2015), U.S. DOE (2020)

projects in the next decades. However, despite the vast experience of CO_2-EOR, the current approach cannot be extrapolated elsewhere. Building the CO_2 pipeline infrastructure will require years (PHMSA, 2024a), if not decades, to fully materialize either CO_2-EOR or CO_2 storage projects. The required CO_2 infrastructure will require important capital investments and especially public acceptance, as will be described later in this chapter.

Regarding CO_2-EOR opportunities in the United States, developing commercial projects in reasonable short time frames is possible. Beaver Creek represents a good example (Table 2.2). Beaver Creek screening studies started during the mid-2000s, and the CO_2-EOR project started in July 2008. In addition to surface and well facilities, this project only required building a 47-mile pipeline interconnected to Exxon's CO_2 distribution pipeline in Wyoming. The source of CO_2 is from the Shute Creek gas processing plant also in Wyoming (Peterson et al., 2012; Robertson, 2022).

In contrast, Petra Nova (Table 2.2) represents a billion-dollar investment that has faced multiple challenges. This project aimed to evaluate a post-combustion carbon capture technology in a coal-fired W.A. Parish power plant. This project was designed to capture 90% of the CO_2 of the flue gas stream using a semi-commercial solvent technology (≈1.4 million metric tons per year). The captured CO_2 is compressed, dried, and transported in an 80-mile pipeline to West Ranch Field (TX) for EOR and storage. However, due to the pandemic and low oil prices, among other difficulties, the project was suspended in May 2020. Petra Nova restarted approximately three years later. Project economics are expected to improve due to the current oil price environment, the increase of the 45Q tax credits from US\$27.61/tCO_2 (in 2023) to US\$60/tCO_2 for EOR projects established in the recently approved Inflation Reduction Act of 2022 (Congressional Research Service, 2023a), and good oil response observed in West Ranch Field. The impact of the 45Q tax credit on project economics will be discussed later in this chapter.

Regarding the international experience of CO_2-EOR, there are well-documented projects, some listed in Table 2.3. Bati Raman represents one of the few CO_2-EORs in a carbonate-HO reservoir with 450–1,000 cp oil viscosity at reservoir temperature (150 °F/65 °C). The CO_2 source was the Dodan field, a natural CO_2 reservoir located 55 miles away from the field. This case study is an excellent example of pushing the limits of technology in a viscous oil complex vuggy carbonate formation (Özgür, 2019; Sahin et al., 2012). The remaining CO_2 floods listed in Table 2.3 are considered relevant to validate the existent experience in CO_2-EOR using natural and anthropogenic sources. It is also important to remark that these projects were implemented before the complex requirements of claiming the newly proposed tax credits.

Table 2.3 Example of commercial international CO_2-EOR projects.

Field, country	Start (status)	CO_2 source	References
Bati Raman, Turkey	1986 (Active)	Natural source	Özgür (2019), Sahin et al. (2012)
Forest Reserve, Trinidad and Tobago	1974 (Finished)	Ammonia plant (Anthropogenic)	Mohammed-Singh and Singhal (2004), Butcher et al. (2014)
Oropouche, Trinidad and Tobago	1990 (Finished)		
Buracica, Brazil	1991 (Finished)	Fertilizer plant (Anthropogenic)	Estublier et al. (2011), Medina-Ketzer et al. (2014)
Joffre Viking, Canada	1991 (Active)	Ethylene plant (Anthropogenic)	Stephenson et al. (1993), Whitecap Resources Inc. (2024)
Weyburn, Canada	2000 (Active)	Coal gasification (Anthropogenic)	Whittaker et al. (2011), Whitecap Resources Inc. (2024)

2.4.3 Polymer Flooding

Beyond steam injection and CO_2-EOR, PF is the only chemical method contributing to global EOR production. Despite its relatively straightforward implementation and low costs, the number of commercial projects has not increased compared to the many pilots documented in the literature (Manrique et al., 2017; Standnes and Skjevrak, 2014). Canada represents the best example of commercial polymer floods in viscous oil reservoirs (1,500–8,000 cp range). Oil production peaked between 2014 and 2015 at approximately 74.5 KBOPD, decreasing to an average of 50 KBOPD in 2022 (Alberta Energy Regulator, 2024b). Most recently, Argentina has reported a growing trend in oil production from polymer floods from medium to light oil reservoirs. Average daily production was reported at approximately 11.1 and 14.6 KBOPD during 2022 and 2023, respectively (Gobierno de Argentina, 2024).

In some instances, PF may replace steam injection in viscous oil reservoirs (<8,000 cp) if reservoir permeabilities are high enough to sustain high injection rates. However, this strategy will lead to a significant decrease in recoverable oil that can impact oil reserves and, hence, the value of HO assets. The argument to consider this change in HO reservoir development strategies can be associated with the reduction of emissions generated during steam generation, the liberation of natural gas (used for steam generation) to gas or blue hydrogen (H_2 production coupled with CCUS) markets, and potential incomes from carbon tax credits due to the production of HO with a lower carbon footprint. If the benefits can offset the cost of reserve loss, this could be a strategy to avoid HO reservoirs becoming stranded assets. However, this will be country-dependent (e.g., legislation and regulations) and beyond the scope of the chapter.

The benefits of PF in conventional oil reservoirs lie in its potential to reduce water handling, leading to a reduced carbon footprint compared to water flooding (Dupuis et al., 2021; Farajzadeh et al., 2022; Ghosh et al., 2023; Shirazi et al., 2024; Skauge et al., 2024). This relatively recent concept represents how the industry rapidly incorporates carbon footprint estimations of existing technologies, incorporating new requirements to contribute to decarbonization strategies aligned with the proposed energy transition.

2.5 CCUS: Challenges and General Economics

Based on the status of the technology, there is a vast experience with CO_2-EOR using anthropogenic sources to support the future of CCUS globally. However, countries and industries will face multiple challenges that must be overcome to make CCUS a viable option. For simplification purposes, CCUS challenges can be classified into five groups, some of which are summarized as follows but are not limited to:

- Technical: Storability, CO_2 injection strategy (EOR or depleted oil and gas reservoirs vs. saline aquifers), geologic storage, and CO_2 capture technologies.
- Regulatory: Legal and regulatory frameworks and tax incentives.
- Social: Public acceptance, CO_2 pipelines (private vs. public lands), and assessing safety concerns.
- Legal Liability: Permanent monitoring to mitigate risks of potential leaks due to contention loss or natural events (i.e., seismic events).
- Economics: Capital investments, cost of capture technologies, costs of CO_2-EOR/storage, and long-term CO_2 monitoring.

This section will summarize the most relevant challenges in fully adopting CCUS globally. It is important to remark that all CCUS challenges listed above can be interrelated and impact project economics, as will be described later.

2.5.1 Technical Challenges

2.5.1.1 CO_2 Capture

CO_2 capture represents one of the most technical and economic challenges CCUS projects face worldwide. This section will summarize the status of the CO_2 capture technologies and their commercial status, highlighting key ongoing projects. At the end of this chapter, some examples of CO_2 capture costs and their impact on project economics will be presented.

Fundamentally, carbon capture consists of separating CO_2 from industrial processes and the exhaust or reformed gases from stationary sources with pre-, post-, or oxy-combustion operations and has been widely documented in the literature. As of today, most of the commercial applications of CO_2 capture used in CO_2-EOR or CO_2 storage are from industrial separation processes and post-combustion processes (Bajpai et al., 2022; Dziejarski et al., 2023; Freeman and Bhown, 2011; Hong, 2022; IEAGHG, 2014; Madejski et al., 2022; NETL, 2024b):

- **Industrial Separation Processes:** Natural gas purification or separation (to meet BTU and commercialization requirements) and petrochemical (e.g., ethanol and fertilizers) production represent the largest anthropogenic CO_2 sources used for CO_2-EOR and storage globally (Tables 2.1–2.3). These CO_2 sources are preferred because they either generate high concentrations of CO_2 streams (>85%) or can be separated at lower costs, as will be described later in this chapter. Although the low-cost hydrogen production from natural gas, also called grey H_2 (CO_2 byproduct is vented to the atmosphere), has not been widely used for CO_2-EOR or CO_2 storage, blue H_2 (separate the

CO_2 byproduct and use it for EOR or storage) also represents another option under consideration (Global CCS Institute, 2021a; Incer-Valverde et al., 2023; Massarweh et al., 2023; U.S. DOE, 2024; Yañez et al., 2020).

- **Post-combustion:** In this case, the CO_2 is separated and captured from the flue gases generated during the combustion of fossil fuels (e.g., coal, fuel oil, and natural gas-fired power plants), which generally produce low CO_2 concentration streams (<15%) at low partial pressures. Despite the existence of commercial technologies to capture CO_2 from post-combustion processes, the main technical and economic challenges are the high energy consumption and costs (CAPEX and OPEX), respectively. Petra Nova (Table 2.2) is one example of an active post-combustion CO_2 capture from a coal-fired power plant used for CO_2-EOR and storage in the United States.

CO_2 capture from pre- and oxy-combustion or other processes (e.g., DAC or Direct Air Capture) has not significantly contributed to CO_2-EOR and storage (NETL, 2024b). One exemption of CO_2 capture from a pre-combustion process is the CO_2-EOR and storage at Weyburn, Canada (Table 2.2). The CO_2 source comes from a coal gasification plant in North Dakota and is transported in a 205-mile pipeline to Canada (Dakota Gasification Company, 2024). Recently, a Colombian case study evaluated the gasification of heavy refinery residues as a strategy for H_2 production and CO_2-EOR in nearby fields (Yañez et al., 2022). However, this approach is associated with large capital investment and high energy intensity (additional emissions to account for). Therefore, its applicability will be limited and region-specific (e.g., countries with large hydropower and other renewable installed capacity).

The technology selected for CO_2 capture and separation systems will depend on the source and composition of industrial emissions. The technologies available can include chemical and physical processes such as adsorption and membranes, among others (Figure 2.5). Although the technologies shown in Figure 2.5 are not exhaustive, the chapter will only focus on the available commercial technologies. CO_2 capture technologies are also widely documented in the literature and commonly categorize their commerciality in terms of the Technology Readiness Level or TRL, being TRL 9 for normal commercial service to TRL 3 or lower for technologies at the Research and Development level (Bajpai et al., 2022; Dziejarski et al., 2023; Freeman and Bhown, 2011; Global CCS Institute, 2021, 2022; Hong, 2022; Takht-Ravanchi and Sahebdelfar, 2014; U.S. GAO, 2022).

It is well known that the high cost and energy consumption are the most important challenges of conventional CO_2 capture technologies and have been the subject of multiple studies over the past two decades (Dziejarski et al., 2023; Figueroa et al., 2008; Global CCS Institute, 2021b; Hong, 2022; MacDowell et al., 2010). However, the commercialization of new proposed technologies has not

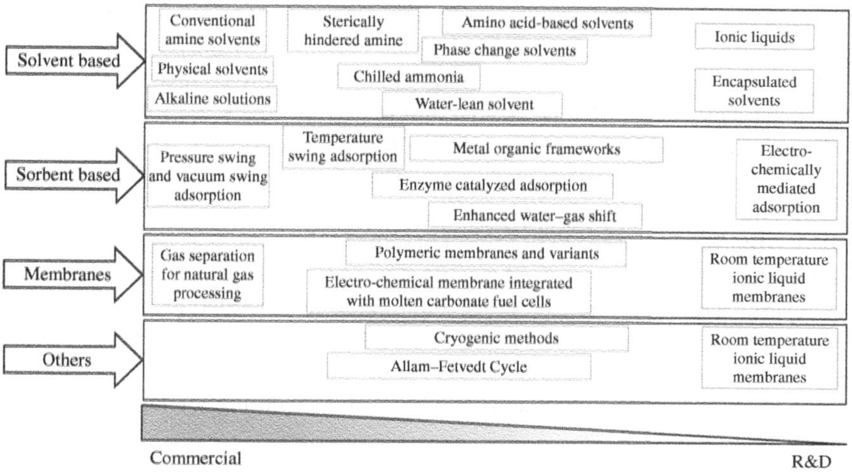

Figure 2.5 Summary of CO_2 capture technologies and their estimated level of commerciality. Adapted from NETL (2024b).

been achieved, and as of today, the same commercial technologies are responsible for most of the CO_2 captured worldwide. In an industrial assessment of 92 post-combustion with CO_2 capture (PCC) projects a decade ago, 57% used absorption (solvent-based) technologies, followed by 15% of capture systems based on adsorption methods. The remaining 29% was distributed between membranes, mineralization, or biofixation (Freeman and Bhown, 2011). Ten years later, the same or improved commercial CO_2 capture technologies dominate the CCS market (Dziejarski et al., 2023; Global CCS Institute, 2021b; Hong, 2022; U.S. GAO, 2022).

Today, most CO_2-EOR and CO_2 storage in saline aquifers use anthropogenic CO_2 from gas processing plants, petrochemical production (e.g., fertilizers, ethanol, methanol), H_2 production, and, to a lesser extent, power generation. In these processes, conventional amine-based solvents (e.g., MEA – monoethanolamine, DEA – diethanolamine, TEA – triethanolamine, MDEA – methyl diethanolamine) capture the CO_2 (Figure 2.5). Some examples of projects reporting the use of CO_2 capture technology using conventional amine solvents include Sleipner, In-Salah, Quest, and Gorgon (Table 2.1), Snøhvit (Norway), and Boundary Dam (Canada). The latter is from a power generation plant using the captured CO_2 for EOR. Although strong inorganic alkaline solutions (e.g., NaOH – sodium hydroxide and KOH – potassium hydroxide) or weak alkaline solutions such as aqueous ammonia ($NH_3.H_2O$) or potassium carbonate (K_2CO_3), also known as the Benfield process, are also commercially available, they have not been frequently used for CO_2-EOR/storage applications.

Among the commercial physical solvents mostly used for CO_2 capture technologies, Rectisol (methanol) and Selexol (dimethyl ethers or polyethylene glycol) are the most common technologies implemented. Examples include the Great Plains Synfuels Plant in North Dakota and the Shute Creek gas treating facility in Wyoming. The Great Plains Synfuels Plant produces synthetic natural gas from coal gasification, and the CO_2 captured is based on the Rectisol process (NETL, 2024c). The CO_2 generated in this process was dehydrated, compressed, and transported to the Weyburn CO_2-EOR project in Canada since the year 2000 (Table 2.3). The Shute Creek gas treating facility started operations in 1986. Initially, the CO_2 was separated using the Selexol process and later incorporated proprietary cryogenic distillation (Controlled Freeze Zone or CFZ™) technology (Tengku-Hassan et al., 2022). The CO_2 captured from this facility has been supplying CO_2 for multiple fields in Colorado (e.g., Rangely Weber) and Wyoming for more than four decades (Robertson, 2022).

Some examples of projects using different CO_2 capture technologies include Petra Nova and Petrobras Santos Basin Pre-Salt oil fields. Petra Nova (Table 2.2) uses a solvent-based CO_2 capture system based on a sterically hindered amine absorption technology. These systems are generally based on a mixture of solvents (e.g., amino alcohol or ethers, piperidine alcohols, and piperazine compounds). Although this technology is commercially available for CO_2 extraction from natural gas combustion processes, Petra Nova represents the largest post-combustion (Coal-fired power plant) application of this technology (Dooley et al., 2009; Dziejarski et al., 2023; Garcia et al., 2022; Global CCS Institute, 2021b). Regarding the Brazilian Pre-Salt CSS project, the CO_2 capture system is based on a commercial polymeric membrane (Figure 2.5). CO_2 is separated from produced natural gas, compressed, and reinjected offshore for CO_2-EOR and storage. As of 2022, the project has reinjected 40.8 million tons of CO_2 (Dziejarski et al., 2023; Garcia et al., 2022; Global CCS Institute, 2021b; Godoi and dos Santos Matai, 2021; Petrobras, 2023; Pizarro and Branco, 2023).

As is well documented, commercial and promising new CO_2 capture technologies have their advantages and disadvantages (Bajpai et al., 2022; Dziejarski et al., 2023; Hong, 2022; Takht-Ravanchi and Sahebdelfar, 2014). However, the commercialization of new CO_2 capture technologies has not significantly progressed during the last two decades (Dooley et al., 2009; Figueroa et al., 2008; Global CCS Institute, 2021b, 2022). Although important progress has been made in developing cryogenic methods, metal–organic frameworks (MOFs), and polymeric membranes (Jana and Modi, 2024; Mahajan and Lahtinen, 2022; Shen et al., 2022), the same commercial technologies have the largest market share today. Therefore, the promised lower CO_2 capture costs of proposed technologies might not be feasible in the near future, especially because most of the current sanctioned projects must rely on existing proven technologies rather than technologies that have not been tested under continuous operations for a representative time. The latter will negatively impact project economics, making it difficult to justify future investments in CCUS projects, understanding that CCUS projects will require approximately 5–10 years from their visualization to their final implementation.

Generally, geological assessment and validation may take three to five years for CO_2 storage projects in aquifers or saline formations (Buchmiller, 2023; DNV, 2012, 2017; ISO, 2017, 2019). These times are shorter for CO_2-EOR or CO_2 storage in depleted oil and gas reservoirs with enough reservoir characterization, injection, and production history. CO_2 sources and capture technologies are commonly assessed in parallel with geologic/reservoir assessment. However, once the geological storage site is validated or certified and the project is financially approved, the engineering studies will also require three to five years from the sanctioning to the construction and final implementation of the project. This will depend on the region and operators (private or state oil companies). At this point, the CO_2 capture technology must be already defined. This is important because already sanctioned CCUS projects are expected to rely mainly on current commercial technologies to meet decarbonization goals by 2030 and beyond. This impact on ongoing efforts to develop new CO_2 capture technologies is still being determined but will provide uncertainties for decision-makers.

In summary, sanctioning projects will be required to meet short-term CO_2 reduction targets, if any. In addition to significant capital investments, these projects will require detailed engineering studies and lengthy technical and regulatory approval. Additionally, given the number of commercial technologies and providers available today, it is possible that the technology providers need help meeting the demand, causing delays in project construction and implementation.

2.5.1.2 CO_2 Injection Strategy and Geologic Storage

Based on the reservoir characterization, historical injection, and production performance, CO_2-EOR, depleted, or mature oil and gas reservoirs are considered the preferred targets for CO_2 injection because they are structures that proved to safely trap hydrocarbons for geologic periods. This strategy also reduces uncertainties related to injectivity, hydrocarbon pore volumes (HCPV), original oil/gas in place (OOIP/OGIP), and frac gradients, leading to a better predictability of CO_2 storability compared to CO_2 storage in saline aquifers. An additional advantage of CO_2-EOR, or its injection in depleted or mature oil and gas reservoirs, is the availability of existing wells and facilities to help reduce the capital expenditure (CAPEX) of CCUS projects. However, fields with a large number of wells can also be seen as a disadvantage due to possible sources for CO_2 leaks during the life span of the projects.

For the scenario of CO_2 storage in saline aquifers, all the know-how developed in CO_2-EOR confirms the availability of technologies to transfer the experiences and lessons learned over five decades, summarized in Section 2.4.2 of this chapter. However, the description of aquifers is generally limited compared to oil and gas reservoirs. Aquifers will not always have the number of wells to develop a detailed reservoir characterization, bringing some uncertainties regarding well injectivity, cement stability, and the presence of fracture systems and their connectivity (especially onshore and with no seismic data available), among others. As of today, Sleipner CO_2 storage represents a solid case to support the viability of CO_2 storage in saline aquifers (Furre et al., 2017; Hauber, 2023). However, the In-Salah and Gorgon projects (Table 2.3) also demonstrate the challenges CO_2 storage in saline aquifers can face during their execution. Therefore, despite the existing experience in CO_2-EOR, every CO_2 storage project will have its characteristics. Hence, project management must include detailed monitoring programs and contingency plans to assess unexpected responses during the implementation and execution.

Regarding the storability of CO_2, multiple assessments confirm the existence of enough storage capacity. Therefore, storage capacity should not be considered a limiting factor in implementing CCUS even though storage volumetric calculations are general estimates with some level of uncertainties due to the lack of data availability, especially in saline aquifers (Bachu et al., 2007; IEA, 2021; Kearns et al., 2017; USGS, 2013). Kearns et al. (2017) estimated that the worldwide CO_2 storage capacity ranges from 8,000 to 55,000 gigatons (Gt). Specifically, the total CO_2 storage capacity in the United States was estimated at 2,400 and 3,700 Gt (USGS, 2013). However, in the case of CO_2-EOR, the incremental oil recoveries in the United States were estimated at 25–32 billion barrels with a potential CO_2 retention between 7.4 and 9.5 Gt (USGS, 2022). To provide some perspective on these volumes, it has been estimated that only 197 Mt of CO_2 were stored in 20 CCS (Carbon Capture and Storage) between 1996 and 2020 (Zhang et al., 2022b). On the other hand, the CO_2-EOR project at Weyburn has reported that an estimated 40 Mt of CO_2 has been stored since the project started in 2000 (Whitecap Resources Inc., 2024).

It is worth mentioning that the International Organization for Standardization (ISO) and Det Norske Veritas (DNV) provide a detailed description of the requirements and recommendations to certify the safe storage of CO_2 from the implementation to the closure of the geologic formations based on the ISO standard 27914 (DNV, 2012, 2017; ISO, 2017). Similar standard guidelines exist for implementing CO_2-EOR (Buchmiller, 2023; ISO, 2019). Existing ISO standards are periodically updated or replaced (ISO, 2024a, b). Therefore, it is important to review possible changes as new information from ongoing projects is incorporated into these standards. An additional difference between the use of anthropogenic CO_2

for CO_2-EOR and CO_2 storage can be the type of well required for each application. This distinction applies to U.S. projects where CO_2-EOR can be implemented with Class II while CO_2 storage requires Class VI wells. This underground injection control (UIC) regulation is defined to mitigate the risks of endangering underground human consumable water during the injection of fluids, in this case, carbon dioxide (U.S. EPA, 2024). The recommendations and requirements, including permits and approvals by regulatory authorities, to implement CO_2-EOR and storage can be technically and economically challenging, especially in countries without oil and gas activities. Finally, the standards available for CO_2-EOR and storage are also meant to guide regulators globally.

2.5.2 Legal and Regulatory Framework

Undoubtedly, CCUS represents one of the key strategies for reducing CO_2 emissions (Dziejarski et al., 2023; ETC, 2022; IEA, 2024; Liu et al., 2023; OGCI, 2024). However, to achieve the CO_2 emission reduction goals established by "The Paris Agreement," countries must have the proper legal and regulatory framework. This will provide the required domain for companies to adopt and invest in lower-emission and higher-efficiency technologies required to reduce CO_2 and GHG emissions. It will also support possible disputes or interpretations of legally binding local legislation and international agreements (IEA, 2022; Voigt, 2023).

As described in this chapter, there is widespread experience using anthropogenic CO_2 for EOR and storage applications worldwide. However, few countries already have the legal and regulatory framework to deploy these complex and capital-intensive projects successfully. The growing trend in CO_2-EOR and storage projects does not mean countries will meet their CO_2 emissions reduction goals. Recently, it was reported that more than 20 national and subnational jurisdictions had established the legal and regulatory framework for CCUS (IEA, 2022). This number is overwhelmingly low compared to the 196 countries that ratified "The Paris Agreement" (UN, 2024), even though most countries will not have the capabilities to implement CO_2-EOR or storage projects, especially countries without established oil and gas operations. Although the IEA members (31) and association (11) countries might not represent a significant number of countries to claim that there are not enough countries with the legal and regulatory framework for CCUS, this data suggests that the lack thereof can represent an important challenge to meet emission reduction goals globally.

The legal and regulatory framework for international CCUS projects has been summarized and continuously updated (IEA, 2022). This information is a great resource for countries developing proper legislation seeking to adopt CCUS strategies to meet CO_2 emission reduction goals. However, regardless of the experience of CCUS projects worldwide (projects listed in Tables 2.1–2.3 and references therein), this chapter will only consider the Norwegian and especially the U.S. experiences. In the case of Norway, the CO_2 regulation associated with oil and gas operations dates to the early 1990s (Norwegian Offshore Directorate, 2024a). Having this regulation in place definitively contributed to the successful implementation of Sleipner and Snøhvit CO_2 storage projects and becoming one of the most relevant CO_2 storage hubs in the North Sea (Buchmiller, 2023; Hauber, 2023; Norwegian Offshore Directorate, 2024b, c).

The legal and regulatory framework in the United States started earlier, mainly driven by the use of natural sources of CO_2 for EOR projects since the early 1970s (Table 2.2), not specifically for CCUS. However, CO_2-EOR triggered the legislation for CO_2 pipelines since the Hazardous Liquid Pipeline Act 1979. Since then, the oil production of CO_2-EOR has been supported by more than 5,300 miles safely operated over the past five decades (PHMSA, 2024a, b), and it is expected to support a large fraction of CCUS projects in years to come. Drilling and completion have also been regulated, defining well types. Most CO_2-EOR projects are operated using Class II wells. However, CO_2 geologic storage projects must use Class VI wells requiring stricter permits. They must meet safety regulations and protect underground drinking water sources, among other requirements (U.S. EPA, 2024 and references therein).

CO_2 transportation and well construction legislation have also been combined with EOR tax incentives since the 1990s to promote domestic oil production, and the incentives are implemented depending on the oil price (Buckner, 1994; U.S. IRS, 2024). Most recently, Section 45Q was enacted (2008) to incentivize investment in CO_2-EOR and CO_2 geologic storage. These benefits have been extended to promote the use of cleaner energies and technologies over the years until the most recent update sanctioned as the Inflation Reduction Act. Under this new legislation, the 45Q tax credits were increased from US\$27.61/tCO_2 (in 2023) to US\$60/tCO_2 for EOR projects and from US\$40.89/tCO_2 (in 2023) to US\$85/tCO_2 for CO_2 storage projects. Another important benefit of this updated regulation is that the annual CO_2 capture requirements were significantly reduced for power plants and different facilities (Congressional Research Service, 2023a). Despite the different legislations, it is worth noticing that the new 45Q tax credit for CO_2 storage projects was slightly below the Norwegian Carbon Credit ($\approx$US\$91/tCO_2) but higher than several European countries during the 1Q of 2023 (Mengden, 2023).

The cases of Norway and the United States summarized above are good examples of countries developing and updating their legal and regulatory frameworks related to CO_2 emissions and CCUS over decades. However, although the experience gained in Norway, the United States, and other regions with well-documented CO_2-EOR or storage projects (e.g., Tables 2.1 and 2.2), it cannot be assumed that the existing legal and regulatory framework can be easily implemented worldwide. The legal and regulatory framework is country-specific, and it may take years to consolidate the required legislation to deploy CO_2-EOR or CO_2 storage projects. Until this happens, the industry will be limited to committing significant capital investments associated with these projects. These challenges are expected to be similar if countries and governments decide to meet the emission reduction goals by considering alternatives to meet their energy demands (e.g., Hydrogen, Gas to Liquids), where the legislation available globally is also limited. However, these topics are out of the scope of the chapter.

2.5.3 Social Challenges

EOR technical assessments sometimes overlook non-technical or soft issues (Alvarado and Manrique, 2010a). Social challenges are one of them. When engineers evaluate CO_2-EOR and CO_2 storage projects, they may involuntarily omit that high-pressure CO_2 pipelines will cross public lands and communities (kilometers from the source to the sink), and storage sites might be beneath local communities and farms.

The HSE (Health-Safety-Environment) plans, project details, and their benefits must be regularly divulged within the communities involved, and all possible concerns must be assessed rigorously and promptly. This openness is important, especially in countries without experience in CO_2 capture and storage projects (e.g., noise, energy consumption, massive infrastructure, employment needs, and demands), and safety concerns (e.g., high-pressure pipelines, possible blow-outs, toxicity) are some of the topics that may arise during the planning and implementation of projects. It is typically more challenging in onshore than offshore environments, although the public perception might be similar.

To demonstrate the impact of social influence on current efforts in CCUS projects, recent cases of CO_2 pipeline permits and safety concerns of CO_2 storage will be used as examples. CO_2 pipelines are critical to meet the CO_2 emissions reduction targets via CO_2-EOR and storage. Despite a low frequency of CO_2 pipeline accidents (CCS Facts, 2024; PHMSA, 2024c; Vitali et al., 2022; Xi et al., 2023), CO_2 ruptures and associated incidents can happen (CCS Facts, 2024; U.S. Department of Transportation, 2022). Such incidents can become one of the main challenges, if not the top one, for future CO_2 pipeline permits in the United States and overseas. Some examples of social challenges associated with recent CO_2 pipeline permits and safety concerns of CO_2 storage in Australia (Quinn, 2022) and the United States (Congressional Research Service, 2023b; George, 2024) are summarized as follows:

- Glencore, one of the largest mining companies in the world, proposed a CO_2 storage pilot demonstration that was recently rejected in Queensland, Australia. The decision came after pressure from farmers and local groups arguing that this project could affect the Great Artesian Basin groundwater resources. This project was planned to capture the CO_2 generated at the Millmerran coal-fired power plant, compress it at supercritical conditions, transport it by truck, and inject the CO_2 in the Precipice aquifer near Moonie. The pilot was considering injecting up to 110,000 tCO_2/yr over three years. After rejecting this project, revisiting the Australian existing legislation for CCUS is likely (Argus, 2024; Morris, 2024).

- In early 2021, Summit Carbon Solutions submitted a proposal to build a 2,000-mile CO_2 pipeline across five states in the United States (from Iowa and Nebraska to North Dakota). The North Dakota regulator denied the application two and a half years later. Soon after, the South Dakota regulator denied the pipeline route approval, arguing that the petition did not provide enough support to prove that the pipeline would not harm the environment and the communities along the states. However, despite the permit rejection, the company claimed that 73% of South Dakota landowners voluntarily agreed to provide access to their lands to build the pipeline. It is expected that Summit will refile its application. The main goal of this proposal is to gather CO_2 from approximately 60 facilities based on fermentation processes of biorefineries (e.g., ethanol plants) across Iowa, Nebraska, Minnesota, and South and North Dakota and transport it for geologic storage in North Dakota (Congressional Research Service, 2023b; Pipeline & Gas Journal, 2023; Summit Carbon Solutions, 2024). In June 2024, the Iowa Utilities Board granted Summit Carbon permission to continue with Iowa's 688-mile CO_2 pipeline section. The approval was conditional on compliance with the construction standards, maintenance protocols, and insurance policy. The operator must meet current regulations to begin project construction, which is expected to operate in 2026 (Holcomb, 2024).
- In March 2021, Navigator CO_2 Ventures ("Navigator") announced the plan to build a 1,300-mile CO_2 pipeline known as the Heartland Greenway project. Although this project shared some routes of Summit Carbon Solutions (described above), this CO_2 pipeline network also involved five U.S. states (Illinois, Iowa, Minnesota, Nebraska, and South Dakota). The initial intent was not to interconnect with the Summit pipeline system. In September 2023, the South Dakota Public Utilities Commission (PUC) denied the pipeline permit. After this resolution, Navigator announced the cancellation of the project. This CCUS project was also based on CO_2 capture from ethanol and fertilizer plants across the states involved, which were to be transported and permanently stored at site locations in Illinois. The final scope of this project was estimated to store approximately 15 million tCO_2/yr at full capacity. However, safety concerns (CO_2 pipeline and CO_2 storage) and landowners lack of support are some of the reasons claimed to cancel this

initiative (Congressional Research Service, 2023b; Lydersen, 2023; Navigator, 2022; Pipeline & Gas Journal, 2023).

- Wolf Carbon Solution proposed the construction of a 280-mile CO_2 pipeline between Illinois and Iowa, the United States. This project (Mt. Simon Hub) was announced in January 2022, and almost two years later, the Illinois Commerce Commission denied the permit for its construction (Illinois Commerce Commission, 2024). The company withdrew the application and is expected to reapply in 2024, while the status of the Iowa permit was unknown by the time this chapter was written. However, fractions of the pipeline must be completed due to previous agreements with an ethanol producer in Iowa. As in the previous two project proposals (Summit and Navigator), the safety concerns of local communities and farmers are the main reason for the denial of permits. This CCUS project also proposes the transport of supercritical CO_2 (up to 12 million tCO_2/yr) captured from ethanol and cogeneration facilities for its geologic storage in Illinois (Adams, 2023; Congressional Research Service, 2023b; Wolf Carbon Solutions, 2024).
- Tallgrass Energy ("Tallgrass") announced the Trailblazer Project (May 2022). This project consists of the conversion of an existing 392-mile natural gas pipeline built in the 1980s into a CO_2 pipeline to transport the CO_2 capture from ethanol facilities at Archer Daniels Midland (ADM) corn-processing complex in Nebraska to the state of eastern Wyoming in the United States. This project was approved on October 2023 by the Federal Energy Regulatory Commission (FERC). When the final abandonment of the natural gas is completed, FERC will no longer have the regulatory authority over this pipeline and will be transferred to the Pipeline and Hazardous Material Safety Administration under the U.S. Department of Transportation (PHMSA, 2024a). The project is scheduled to be in service in 2024, and a key aspect of this approval was the agreement reached with the local communities. The deal struck by the operators and the community, including environmental groups, was a US$1.1 million to Nebraska Communities and committed to have first responders along the CO_2 pipeline route (Congressional Research Service, 2023b; Dunker, 2024; Tallgrass, 2024).

The examples summarized above, except for the case of the Tallgrass project (pipeline repurpose without requiring major permit requirements), clearly show the impact social influence can have on proposed CCUS projects regardless of the existing legal and regulatory framework. Therefore, given the unpredictable nature of regulatory and public perceptions of CCUS projects, investment commitments toward CO_2 emission reduction initiatives involving CO_2-EOR and CO_2 storage will be delayed.

From the technical and economic perspective, the number of ethanol and fertilizer plants operating in the Iowa, Nebraska, and South Dakota corridor represents a great advantage for potential low-cost CO_2 capture due to its high concentrations (Section 2.5.1) and the experience operating CO_2 projects for more than five decades in the United States. However, the latter demonstrates the importance of assessing community needs and concerns well before projects are proposed to avoid delays and even cancellations that can impact CO_2 reduction emissions goals committed by governments and companies worldwide. Finally, regardless of the experience and safety track of operating CO_2 projects in a given region, there will need to be more evidence to convince other governments and communities about the benefits and safety of CCUS projects. All project requirements will be country and region-specific, and extrapolation of experiences and regulations during the planning of CCUS projects must be avoided.

2.5.4 CCUS: Basic Costs and Economics

As described earlier, countries (governments) and industries will face multiple challenges in making CCUS a viable option to meet decarbonization goals agreed upon after ratifying the Paris Agreement. However, given the complexity and substantial number of scenarios that may apply to a CCUS project in a particular region, this section will provide relevant information to guide preliminary assessments and feasibility studies that will contribute to recognizing the magnitude of carbon capture, transportation, and utilization for EOR or permanent geologic storage projects. However, the variables presented can also be useful for other applications for CO_2 utilization.

It is important to note that CCUS projects are capital-intensive endeavors. Therefore, the first major hurdle to overcome is financing these initiatives. This is particularly challenging in countries with weak economies, where the ability to increase debt and provide the necessary incentives may be limited or even unaffordable. Figure 2.6 provides an example of the capital investments required for carbon capture (with or without storage) projects, as reported by the U.S. National Energy Technology Laboratory carbon and storage database (NETL, 2024b).

The NETL database includes 417 worldwide projects, of which approximately 55% report total project costs. However, the cases consolidated in Figure 2.6 are for projects (149) reporting the capital investment in US dollars regardless of status (e.g., active, completed, planning, or terminated). Notably, 74.5% of the projects reported in the database range from million-dollar small projects to large-scale pilot or demonstration projects with costs below US\$0.7 billion. From the selected sample, only 24.5% are projects with capital investments equal to or higher than US\$1 billion. Petra Nova (Table 2.2) is an example of a US billion-dollar investment project capturing CO_2 using ammonia-based solvents from a

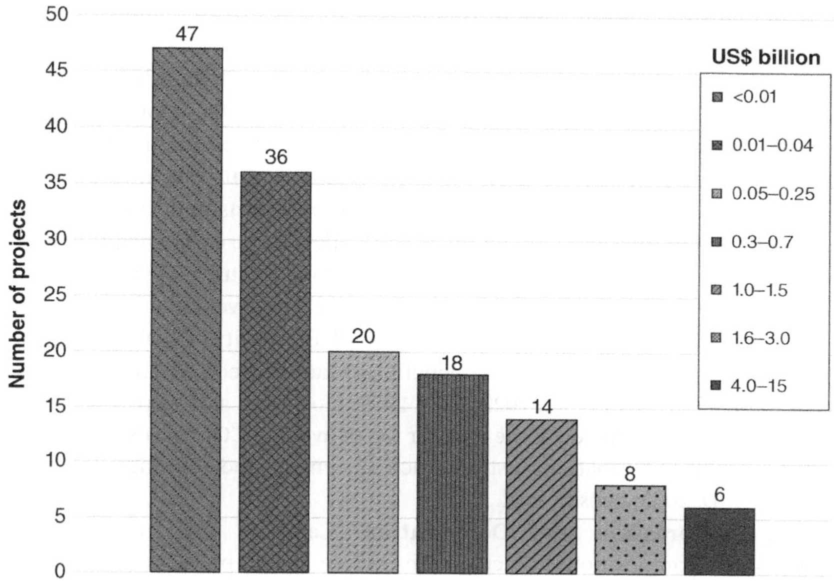

Figure 2.6 Example of the capital cost of carbon capture and storage projects. *Source:* Adapted from NETL (2024b).

240 MW coal-fired power plant for CO_2-EOR (EIA, 2017; Jacobs, 2023; Jenkins, 2015; NETL, 2024b; U.S. DOE, 2020). To provide some perspective on this level of investment, a project like Petra Nova can represent 4 to 7% for major oil and gas companies with annual investments between US$15 and 25 billion (Kimani, 2023). However, it can represent more than 10% of companies investing US$10 billion or less per year (e.g., small national oil companies and independents). The latter represents a big challenge for oil and gas and other industries with large emissions (e.g., cement, steel).

Regarding the itemized CCUS cost, carbon capture, transportation, and utilization (EOR or geologic storage) costs have been widely addressed in the literature (Alberta Department of Energy, 2023; Dziejarski et al., 2023; ETC, 2022; Global CCS Institute, 2021b; Hong, 2022; IEAGHG, 2014; NETL, 2010; U.S. GAO, 2022). Despite the CCUS cost available, the information widely varies due to the different applications and types of projects. The main CCUS costs are generally divided into the following segments:

- **Carbon Capture:** Based on current commercial technologies, this process can account for up to 80% of the total costs of the CCUS projects. Carbon capture technologies were briefly described previously (Section 2.5.1). Generally, capture costs for high-concentration CO_2 streams are cheaper than industries emitting highly diluted gas streams that can require gas pre-treatments. Carbon capture technologies typically deliver CO_2 streams that require dehydration and compression for transportation, and these costs are frequently not included in this segment. Finally, carbon capture technologies can be energy-intensive, and the capture cost does not always include the associated energy penalty. In other words, capture technologies will require additional energy that will add additional costs and CO_2 emissions depending on the source of energy used to operate the carbon capture technology. Generally, validating the costs of carbon capture technologies can provide misleading information regarding the improvements claimed. For this reason, the use of the Levelized Cost of Electricity (LCOE) is another option proposed to compare the benefits, or lack thereof, of carbon capture technologies (Baker et al., 2022; ETC, 2022; Hochman and Appasamy, 2024; IEAGHG, 2014). However, the cost per ton of avoided CO_2 has also been used for other hard-to-abate industries, such as cement and refining (Bui et al., 2018; IISD, 2023; Leeson et al., 2017).

- **CO_2 Compression and Dehydration:** Captured CO_2 must require minimum compositional specifications before its compression for transportation (Bielka et al., 2023; European Commission, 2017; Hong, 2022; IPPC, 2024). In some instances, these costs are included in CO_2 transportation costs. Therefore, the cost allocation needs to be validated before economic evaluations.
- **CO_2 Transportation:** Most of the worldwide experience for CO_2-EOR and storage has been developed based on CO_2 pipeline transportation in onshore projects and offshore to a lesser extent. Costs depend highly on pipeline distances, diameter, and flow rates (Smith et al., 2021). However, CO_2 transportation by ship, truck, and rail has also been used for different CO_2 utilization (e.g., food industry, EOR, etc.). In offshore environments, it is possible to foresee the combination of CO_2 transportation via ships and pipelines.
- **CO_2 Utilization:** Despite the multiple uses of CO_2, this chapter only considers the injection of CO_2 in geologic formations (EOR and storage) as described in Sections 2.3 and 2.4. These costs will depend, among other things, on whether the project is onshore or offshore, reservoir depth, and the number and type (Class II or VI) of wells.
- **Monitoring and Verification:** The costs in this area are not fully validated, especially for long-term geologic storage. This lack of validation represents a challenge that has not been properly addressed. Regarding the costs, it will also depend on the type of project. For example, the use of 4D seismic can be considered in offshore environments. However, it may be limited to onshore projects. This choice can represent a significant difference in costs. Reservoir depth and number of wells are other factors impacting these costs.

Monitoring and verification costs will also vary by region according to the specific requirements of the existing regulations. The type of monitoring strategy, tools (e.g., logging), gas and water analysis (e.g., isotopic composition), and frequency will dictate the costs of this segment of a particular CCUS project.

Table 2.4 summarizes the costs documented in the literature reviewed. It also includes the estimated average cost of CO_2-EOR and CO_2 storage projects, which can be compared against the possible incomes of these projects.

As can be seen, the different segments of CCUS show a wide range of costs due to the multiple variables and conditions of these projects. Direct Air Carbon Capture (DAC or DACC) was not included because the expected volume captured of a highly diluted CO_2 stream is not considered attractive enough to be used for CO_2-EOR or storage. Additionally, reported capture costs range from US$140 to 340 /t$CO_2$ (ETC, 2022; Hong, 2022), making these projects uneconomic, as will be summarized in the example shown in Figure 2.7.

The estimated costs of different segments of CCUS presented in Table 2.4 assume a total cost of US$100/t$CO_2$ for capture, dehydration, and compression. Other costs are considered reasonable averages. However, a high cost for monitoring and verification was assumed because the estimates reported in the literature are considered optimistic, and stricter programs should be expected in the

Table 2.4 Estimated costs for different segments of CCUS projects.

Segment	Type	Cost range (US/tCO_2$)	Avg. costs[c] (US/tCO_2$)
Capture Technologies	High CO_2 concentration (70–100%)[a]	10–35	100
	Low CO_2 concentration (5–45%)[b]	50–160	
Compression and Dehydration	—	13–25	
Pipeline Transportation	Onshore	5–40	20
	Offshore	10–60	
Utilization	EOR and storage	10–30	20
	Storage	3–20	12
Monitoring and Verification	—	3–10	10

a. Natural gas processing, bioethanol, fertilizers, and similar.
b. Cement, iron, steel, and power generation and refining (excludes DAC).
c. Average cost used for cost vs. potential incomes presented in Figure 2.7.

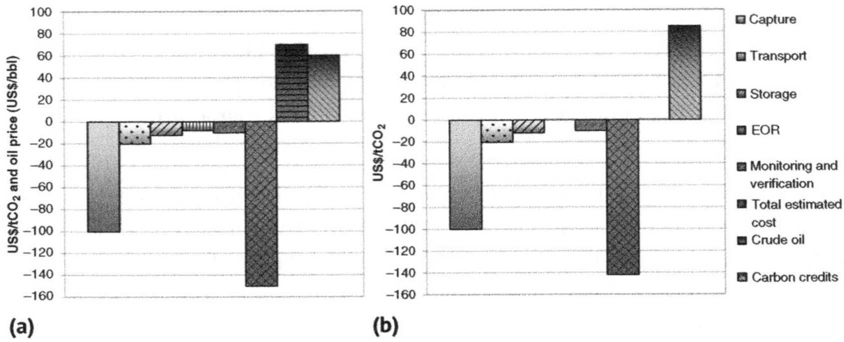

Figure 2.7 Example of costs and possible incomes of CCUS for CO_2-EOR (a) and storage (b) projects.

future. These costs will be compared with the two possible incomes to offset the costs of CCUS. An average oil price of US$70/bbl and the 45Q tax credits recently announced in the Inflation Reduction Act of 2022 represent the carbon credits used in the scenarios presented in Figure 2.7. The 45Q tax credit for CO_2-EOR and CO_2 storage are US$60/t$CO_2$ and US$85/t$CO_2$, respectively (Congressional Research Service, 2023a).

Based on these assumptions, Figure 2.7 compares a possible carbon capture scenario from a low CO_2 concentration stream and uses it for CO_2-EOR (Figure 2.7a) and CO_2 storage (Figure 2.7b). Although this simple example does not represent a comprehensive economic evaluation, neither of the cases can offset the costs of CCUS, estimated at US$150/t$CO_2$ for CO_2-EOR and US$142/t$CO_2$ for CO_2 storage. These results have not changed much since 2010 (Alvarado and Manrique, 2010b), especially because of the limited progress in developing lower-cost CO_2 capture technologies despite the recent increase in carbon tax benefits (e.g., the 45Q tax credit) to promote implementing CCUS projects.

However, these scenarios can improve project economics for carbon capture from high CO_2 concentrations (e.g., natural gas processing, bioethanol, fertilizers), where costs are much lower (Table 2.4). By reducing carbon capture and compression in half (US$50/t$CO_2$), CO_2-EOR and storage suggest the possibility of profitable projects and potentially for some CO_2 storage applications. This cost reduction explains why many active or proposed projects in the United States and abroad use these CO_2 sources (see examples in Tables 2.1–2.3). Project economics can be further improved by the development of Hubs, as suggested in different states in the United States, to gather and transport CO_2 from multiple facilities (e.g., ethanol plants) for geologic storage. However, these projects may face social challenges that delay or suspend their implementation (see examples in Section 2.3).

It is worth mentioning that the total cost of US\$100/tCO_2 is within the range of CO_2 capture, transport, and storage cost reported (≈US\$103/tCO_2) in the Quest CCS project (Table 2.1) assuming a rate of exchange of CAN\$0.75 per US\$1 (Alberta Department of Energy, 2023). However, as discussed in this chapter, CCUS cost, operational expenditures (OPEX), monitoring, and verification are specific to every project (e.g., CO_2 source and its distance to the injection site, type of reservoir, number of wells, reservoir depth, etc.) and region (e.g., regulatory framework, social aspects, carbon tax benefits, etc.).

Based on current uncertainties associated with CCUS in several regions, the preferred approach is to start with CO_2-EOR in countries with oil and gas operations (Canadian Energy Centre, 2022). This strategy will accelerate project implementation in well-known reservoirs and provide opportunities to optimize (e.g., well availability) and offset (e.g., oil revenues) CCUS costs. As these projects evolve, the know-how developed will support further expansion of CO_2-EOR and storage. At the same time, countries and companies can distribute capital investments over time as the number of projects increases. Colombia is one recent example evaluating CCUS strategies combined with CO_2-EOR (Manrique et al., 2022; Yañez et al., 2020). A recent evaluation identified the production of blue hydrogen by gasifying heavy refinery residues integrated with CO_2-EOR in fields relatively close to the refinery. Despite the encouraging results preliminarily estimated, project economics showed negative Net Present Values, and the viability of this project will rely on carbon credits and potential international carbon markets (Yañez et al., 2022). The latter is a clear example of the challenges CCUS projects will face in years to come.

2.6 Closing Remarks

The role of CCUS will be critical for mitigating GHG emissions. Therefore, the oil and gas industry will be critical in reducing CO_2 emissions through EOR and storage projects. The experience gained over five decades of safely operating CO_2-EOR and storage projects provides a good opportunity to contribute to decarbonization goals agreed upon by the international community.

Regarding oil recovery processes, the first approach may focus on reducing the carbon footprint (increasing energy efficiency) of thermal EOR projects, especially steam injection methods. This reduction can be achieved by using CO_2 or flue gas combined with steam injection (e.g., hybrid methods) or capturing and disposing of the CO_2 generated from steam generation for CO_2-EOR or storage in deeper intervals or nearby fields. The use of solar steam cannot be ruled out. However, based on previous experiences (e.g., California and Oman), it did not meet technical and commercial expectations. If these approaches are unprofitable, significant resources may become stranded assets. Polymer flooding can be considered an alternate strategy for HO reservoirs, preferably with oil viscosities below 8,000 cp at reservoir temperatures based on the Canadian experience. However, lower oil recovery factors and ultimate recoverable reserves should be expected.

Despite the global consensus and set targets for CO_2 emission reduction in the following decades, the current number of new and planned projects does not suggest that the proposed goals are achievable. Several "non-technical" factors were not properly envisioned. Some of them include:

- Energy security will prevail over energy transition. Governments are responsible for providing energy at reasonable costs. Global hydrocarbon consumption and exploration efforts, especially in natural gas, suggest that countries are not ready to reduce fossil fuel consumption in the short term but can start an energy transition and decarbonization plans at a slower pace.
- CCUS projects are capital-intensive, and large capital investments from the government and private sectors are unlikely without a well-established regulatory framework and tax incentives. This large capital barrier is especially true for the large emitters responsible for the upfront capital (CO_2 capture and energy penalty) of CCUS projects, regardless of the potential benefits of reducing CO_2 emissions. Otherwise, the alternate solutions and the required infrastructure (e.g., H_2, nuclear, and renewables) to meet the energy demand are neither low-cost nor feasible in every region. These options are not exempt from government incentives that can be unaffordable in countries with weak economies.
- Project timeline represents another challenge of CCUS projects. These projects can take five to eight years, from the technical evaluation to the final construction and operation. The implementation time will depend on multiple factors (e.g., CO_2 source, source–sink distance, regulatory framework, permits, etc.). Therefore, projects without approvals for commissioning the construction of CCUS facilities will make CO_2 emission reduction goals extremely difficult in the next decade.
- Social challenges have been underestimated. This chapter presented several examples of how difficult the evaluation of CO_2 pipeline proposals can be, requiring extensive review processes by the regulatory authorities and negotiations with local communities. The examples provided have led to project cancellations (e.g., Glencore, Australia, and Navigator and Wolf Carbon Solutions in the United States) or lengthy and partial approvals, such as the proposal submitted by Summit Carbon Solutions, which is estimated to take five years from the project submittal (2021) to start its operation in 2026, assuming all safety and regulatory conditions are met.

- Regions without experience or activity in oil and gas may also impose some limitations (e.g., project timelines). In practice, CO_2 capture and storage can be implemented in any region. However, not having the industry support (e.g., drilling, completion, reservoir characterization and monitoring, operation, maintenance, etc.) will require establishing the industrial support associated with the experience required for CCUS projects. Achieving this will not only require important resources and time, but it can also delay decarbonization goals. Capturing and transporting supercritical CO_2 by ship overseas seems improbable or may apply to limited regions (e.g., the North Sea).

As described in this chapter, implementing CCUS projects can be technically challenging and complex. Therefore, when evaluating CCUS, multidisciplinary teams must address multiple activities, preferably simultaneously. Some of these activities include but are not limited to:

- Assess CO_2 sources, composition, and volumetrics regionally or countrywide over time. If available, start with high CO_2 concentration stream sources (e.g., natural gas processing, bioethanol, fertilizers, and similar). This strategy can reduce capture costs and project implementation times.
- Identify possible CO_2 sinks (the closer to the CO_2 capture, the better). If existing gas or oil pipelines are operating, the probability of getting permits for CO_2 pipelines might be higher using the same routes. If not, assess possible routes and trigger a team to assess possible social challenges, address their needs, and promptly respond to their concerns. Do not take CO_2 pipeline permits for granted.
- Once the CO_2 sinks are identified, the decision will depend on whether the use of CO_2 will be for EOR and storage or strictly storage. Based on reservoir characterization and uncertainties, CO_2-EOR or CO_2 storage in depleted oil or gas reservoirs should be the best option compared to CO_2 storage in aquifers, which typically lack static and dynamic information. Project economics also favors CO_2-EOR, and the perception that this approach supports fossil fuel use should not be the driver but the decarbonization and be part of global energy transition plans. Additionally, starting with CO_2-EOR will be easier/faster and contribute to developing the learning curve and offsetting CCUS costs.
- Laboratory and numerical prediction studies. Numerical models are generally available for oil and gas fields but not aquifers. Hence, starting numerical predictions with available (black oil) models can help to frame the scope of the CO_2 injection based on current or past injection rates, number of injectors required to dispose of the volumes of CO_2 captured, breakthrough times, CO_2 recycling and reinjection (e.g., sizing facilities), incremental oil or CO_2 storage (e.g., in depleted reservoirs), and other variables key to estimate

project economics. Numerical simulation and overall project uncertainties can be reduced as more data becomes available (e.g., compositional PVT, CO_2 and crude oil compatibility, minimum miscibility pressure tests, core-flood displacement, CO_2 solubility in reservoir brine, possible CO_2 expansion "Joule-Thompson" effects of CO_2 on wellbore integrity, well tests, updated geologic models, etc.). Regardless of these efforts, the results must be considered a guideline because deviation from the numerical approach should be expected once the project is implemented. Hence, given the nature of each geologic site, projects must include contingency plans.
- By understanding previous steps, preliminary economics can be used to estimate possible economic scenarios, and the required tax incentives can be estimated. Suppose the proposed projects plan to claim tax benefits from national or international carbon markets. In that case, it is important to verify if the selected geologic site meets all the requirements (e.g., site certification) and if the required investments for monitoring and verification are included in project economics. Do not assume tax incentives are given. Most countries have not developed or cannot afford the required regulatory framework.

Finally, every reservoir or CO_2 sink is specific and will have its unique properties. An important underestimated challenge is that the CO_2 captured and stored must be verified for long or geologic periods. The latter can limit the potential of CO_2-EOR and storage to support the expected decarbonization goals worldwide. Thus, gas-to-liquid (GTL) technologies may provide alternate solutions to support the reduction of CO_2 emissions in decades to come, as will be described in the following chapters of this book.

References

Adams, A. (2023). Iowa-Illinois carbon dioxide pipeline application. https://capitolnewsillinois.com/news/iowa-illinois-carbon-dioxide-pipeline-application-withdrawn (accessed 2 June 2024).

Advanced Resource International (2021). U.S. CO_2 enhanced oil recovery survey 2021 update. https://www.eoriwyoming.org/projects-resources/publications/eori-library/co2-eor-survey-update-2021 (accessed 24 March 2024).

Aikman, M.J. (2022). *Clean energy from oil: a process to generate low cost, low carbon electricity from mature and depleted oil fields*. Paper presented at the ADIPEC, Abu Dhabi, UAE. https://doi.org/10.2118/210940-MS.

Al Rahbi, M.S., Al Sidairi, N.A., Al Ghafri, A.M. et al. (2022). *Game changer in subsurface flare reduction*. Paper presented at the ADIPEC, Abu Dhabi, UAE. https://doi.org/10.2118/210873-MS.

Alberta Department of Energy (2023). Quest carbon capture and storage project: annual report, 2022. https://open.alberta.ca/publications/quest-carbon-capture-and-storage-project-annual-report-2022 (accessed 17 May 2024).

Alberta Energy Regulator (2024a). ST-53 annual Alberta in situ oil sands summary reports. https://www.aer.ca/providing-information/data-and-reports/statistical-reports/st98/crude-bitumen/well-activity (accessed 1 January 2024).

Alberta Energy Regulator (2024b). ST-53 2022 Alberta in situ oil sands summary. https://www.aer.ca/providing-information/data-and-reports/statistical-reports/st98/crude-bitumen/well-activity (accessed 1 January 2024).

Alberta Government (2024). Alberta oil sands greenhouse gas emission intensity analysis. https://open.alberta.ca/opendata/alberta-oil-sands-greenhouse-gas-emission-intensity-analysis (accessed 16 March 2024).

Allis, R., Chidsey, T., Gwynn, W. et al. (2001). Natural CO_2 reservoirs on the Colorado Plateau and Southern Rocky Mountains: candidates for CO_2 sequestration. https://geology.utah.gov/docs/emp/co2sequest/pdf/reservoirs.pdf (accessed 24 March 2024).

Alvarado, V. and Manrique, E. (2010a). *Enhanced Oil Recovery: Field Planning and Development Strategies*. Gulf Professional Publishing: Elsevier.

Alvarado, V. and Manrique, E. (2010b). Enhanced oil recovery: an update review. *Energies* 3 (9): 1529–1575. https://doi.org/10.3390/en3091529.

Angga, I.G., Bellout, M., Kristoffersen, B.S. et al. (2022). Effect of CO_2 tax on energy use in oil production: waterflooding optimization under different emission costs. *SN Applied Sciences* 4: 313. https://doi.org/10.1007/s42452-022-05197-4.

Argus (2024). Queensland to review CCS after rejecting Glencore plan. https://www.argusmedia.com/en/news-and-insights/latest-market-news/2572227-queensland-to-review-ccs-after-rejecting-glencore-plan (accessed 28 May 2024).

Babadagli, T. (2020). Philosophy of EOR. *Journal of Petroleum Science and Engineering* 188: 106930. https://doi.org/10.1016/j.petrol.2020.106930.

Bachu, S., Bonijoly, D., Bradshaw, J. et al. (2007). CO_2 storage capacity estimation: methodology and gaps. *International Journal of Greenhouse Gas Control* 1 (4): 430–443. https://doi.org/10.1016/S1750-5836(07)00086-2.

Bajpai, S., Shreyash, N., Singh, S. et al. (2022). Opportunities, challenges and the way ahead for carbon capture, utilization and sequestration (CCUS) by the hydrocarbon industry: towards a sustainable future. *Energy Reports* 8: 15595–15616. https://doi.org/10.1016/j.egyr.2022.11.023.

Baker, R., Alizadeh Sahraei, O., Dal-Cin, M.M. et al. (2022). A technology development matrix for carbon capture: technology status and R&D gap assessment. *Frontiers in Energy Research* 10. https://doi.org/10.3389/fenrg.2022.908658.

Bashti, S., Sadeghi, A., McCoy, S. et al. (2023). *Could the post-SAGD heat recovery supply the direct air CO_2 capture (DAC) energy in a net negative carbon emission* *environment*? Paper presented at the SPE Canadian Energy Technology Conference and Exhibition, Calgary, Canada. https://doi.org/10.2118/212816-MS.

Bersak, A.F. and Kadak, A.C. (2007). Integration of nuclear energy with oil sands projects for reduced greenhouse gas emissions and natural gas consumption. https://www.ourenergypolicy.org/wp-content/uploads/2012/03/MITWhitePaper.pdf (accessed 17 March 2024).

Bielka, P., Kuczyński, S., and Nagy, S. (2023). CO_2 compression and dehydration for transport and geologic storage. *Energies* 16 (4): 1804. https://doi.org/10.3390/en16041804.

Brock, W.R. and Bryan, L.A. (1989). *Summary results of CO_2 EOR field tests, 1972-1987*. Paper presented at the Low Permeability Reservoirs Symposium Low Permeability Reservoirs Symposium, Denver, Colorado. https://doi.org/10.2118/18977-MS.

Buchmiller, D. (2023). Applying ISO standards to geologic storage and EOR projects - IEAGHG technical report: 2022-11. https://ieaghg.org/files/Webinar_IEAGHG_Technical_Report_2022-11_rev0.pdf (accessed 5 May 2024).

Buckner, L.A. (1994). *Tax incentive for enhanced oil recovery projects*. Paper presented at the SPE/DOE Improved Oil Recovery Symposium, Tulsa, Oklahoma. https://doi.org/10.2118/27781-MS.

Bui, M., Adjiman, C.S., Bardow, A. et al. (2018). Carbon capture and storage (CCS): the way forward. *Energy & Environmental Science* 11 (5): 1062–1176. https://pubs.rsc.org/en/content/articlelanding/2018/ee/c7ee02342a (accessed 16 October 2024).

Bursell, C.G. (1970). Steam displacement Kern River field. *Journal of Petroleum Technology* 22: 1225–1231. https://doi.org/10.2118/2738-PA.

Butcher, S.A., Isaac, J., and Frontin-De Peaza, C. (2014). *Reintroduction of carbon dioxide supply from central to south Trinidad via pipeline – perspectives and possibilities for using present system*. Paper presented at the SPE Energy Resources Conference, Port of Spain, Trinidad and Tobago. https://doi.org/10.2118/SPE-169941-MS.

California Department of Conservation (2015). 2014 preliminary report of California oil & gas production statistics. https://www.conservation.ca.gov/calgem/pubs_stats/annual_reports/Pages/annual_reports.aspx (accessed 18 December 2023).

California Department of Conservation (2020). 2019 annual report of state oil & gas supervisor. https://www.conservation.ca.gov/calgem/pubs_stats/annual_reports/Pages/annual_reports.aspx (accessed 18 December 2023).

California Department of Conservation (2023). California state oil & gas supervisor. annual report 2020. https://www.conservation.ca.gov/calgem/pubs_stats/annual_reports/Pages/annual_reports.aspx (accessed 18 December 2023).

Canadian Energy Centre (2022). Carbon capture, utilization and storage (CCUS) and the energy transition: assessing the future use of CCUS in Canada's oil and gas sector. https://www.canadianenergycentre.ca/wp-content/uploads/2022/07/CEC-Research-Brief-23-CCUS-V4-June-30-2022.pdf (accessed 17 March 2024).

Carbon Capture and Storage (CCS Facts) (2024). CO_2 pipelines are the sewage system for the fossil fuel industry funded by the public. https://www.carboncapturefacts.org/pipeline-health-and-safety (accessed 1 June 2024).

Chang, J. (2020). *Flue gas waste heat utilization for enhanced oil production – water hot-gas injection for Lloydminster reservoirs.* Paper presented at the SPE Canada Heavy Oil Conference, Virtual. https://doi.org/10.2118/199949-MS.

Congressional Research Service (2023a). The section 45Q tax credit for carbon sequestration. https://crsreports.congress.gov/product/pdf/IF/IF11455 (accessed 10 April 2024).

Congressional Research Service (2023b). Siting challenges for carbon dioxide (CO_2) pipelines. https://crsreports.congress.gov/product/pdf/IN/IN12269 (accessed 1 June 2024).

Cooper, J. (2016). *Rangely weber sand unit WAG management optimization.* Paper presented at the CO_2 Conference, Midland, Texas. https://www.co2conference.net/wp-content/uploads/2016/12/2-Chevron-Final-2016-Midland-CO2-Forum-WAG-Optimization.pdf (accessed 26 March 2024).

Dakota Gasification Company (2024). CO_2 capture and Storage. https://www.dakotagas.com/about-us/CO2-capture-and-storage/index (accessed 11 May 2024).

Darrold, L.K., Ong, K., and Prabagar, M. (2024). *Prioritisation of impactful decarbonisation programmes across existing assets (OTC-34763).* Paper presented at the Offshore Technology Conference Asia, Kuala Lumpur, Malaysia. https://doi.org/10.4043/34763-MS.

Davis, S.J., Liu, Z., Deng, Z. et al. (2022). Emissions rebound from the COVID-19 pandemic. *Nature Climate Change* 12: 410–417. https://doi.org/10.1038/s41558-022-01332-6.

Det Norske Veritas – DNV (2012). Geological storage of carbon dioxide – recommended practice (DNV-RP-J203). https://www.dnv.com/oilgas/download/dnv-rp-j203-geological-storage-of-carbon-dioxide (accessed 18 May 2024).

Det Norske Veritas – DNV (2017). Certification of sites and projects for geological storage of carbon dioxide - DNVGL-SE-0473. https://www.dnv.com/oilgas/download/dnv-se-0473-certification-of-sites-and-projects-for-geological-storage-of-carbon-dioxide/ (accessed 18 May 2024).

Det Norske Veritas – DNV (2023). Trilemma and transition: the momentum to break barriers. https://www.dnv.com/news/energy-security-is-top-priority-in-the-energy-trilemma-for-2023-240553/ (accessed 24 March 2024).

Dooley, J.J., Davidson, C.L., and Dahowski, R.T. (2009). An assessment of the commercial availability of carbon dioxide capture and storage technologies as of June 2009. https://www.pnnl.gov/main/publications/external/technical_reports/PNNL-18520.pdf (accessed 10 March 2024).

Dunker, C. (2024). Carbon pipeline developer, environmental group strike deal to benefit Nebraska landowners. https://theindependent.com/carbon-pipeline-developer-environmental-group-strike-deal-to-benefit-nebraska-landowners/article_d1c75e5e-fc71-11ee-873e-5b6b5c1210b1.html (accessed 2 June 2024).

Dupuis, G., Dupuis, G., Al-Khoury, P. et al. (2021). Using polymer EOR to reduce carbon intensity while increasing oil recovery. *EAGE IOR* V2021: 1–20. https://doi.org/10.3997/2214-4609.202133048.

Dziejarski, B., Krzyżyńska, R., and Andersson, K. (2023). Current status of carbon capture, utilization, and storage technologies in the global economy: a survey of technical assessment. *Fuel* 342. https://doi.org/10.1016/j.fuel.2023.127776.

ElMasry, H. and Shakirov, A. (2024). *Global perspective in gas flaring*. Paper presented at the Offshore Technology Conference Asia, Kuala Lumpur, Malaysia. https://doi.org/10.4043/34859-MS.

Energy Transition Commission – ETC (2022). Carbon capture, utilisation and storage in the energy transition: vital but limited. https://www.energy-transitions.org/publications/carbon-capture-use-storage-vital-but-limited/#download-form (accessed 27 January 2024).

Equinor (2024). The northern lights project. https://www.equinor.com/energy/northern-lights (accessed 20 October 2024).

Estublier, A., Dino, R., Schinelli, M.C. et al. (2011). CO_2 injection in Buracica – long-term performance assessment. *Energy Procedia* 4: 4028–4035. https://doi.org/10.1016/j.egypro.2011.02.344.

European Commission (2017). Requirements for safe and reliable CO_2 transportation pipeline (SARCO2) – final report. https://data.europa.eu/doi/10.2777/205449 (accessed 16 June 2024).

ExxonMobil (2024). ExxonMobil secures the largest CO_2 offshore storage site in the U.S. https://corporate.exxonmobil.com/news/news-releases/2024/10102024_exxonmobil-secures-largest-carbon-offshore-storage-site-in-the-us (accessed 20 October 2024).

Fankhauser, S., Smith, S.M., Allen, M. et al. (2022). The meaning of net zero and how to get it right. *Nature Climate Change* 12: 15–21. https://doi.org/10.1038/s41558-021-01245-w.

Farajzadeh, R., Glasbergen, G., Karpan, V. et al. (2022). Improved oil recovery techniques and their role in energy efficiency and reducing CO_2 footprint of oil production. *Journal of Cleaner Production* 39. https://doi.org/10.1016/j.jclepro.2022.133308.

Farouq Ali, S.M. and Soliman, M.Y. (2022). *Relevance of petroleum engineering under the present uncertainties*. Paper presented at the SPE Annual Technical Conference and Exhibition, Houston, Texas. https://doi.org/10.2118/210310-MS.

Farouq Ali, S.M. and Soliman, M.Y. (2023). *Limitations and fallacies of carbon capture and storage (CCS) and impact on oil and gas production*. Paper presented at the

SPE Annual Technical Conference and Exhibition, San Antonio, Texas. https://doi.org/10.2118/214950-MS.

Figueroa, J.D., Fout, T., Plasynski, S. et al. (2008). Advances in CO_2 capture technology - The U.S. Department of Energy's carbon sequestration program. *International Journal of Greenhouse Gas Control* 2 (1): 9–20. https://doi.org/10.1016/S1750-5836(07)00094-1.

Finan, A.E. and Kadak, A.C. (2010). Integration of nuclear energy into oil sands projects. *ASME Journal of Engineering Gas Turbines Power* 132 (4). https://doi.org/10.1115/1.3098421.

Flett, M., Brantjes, J., Gurton, R. et al. (2009). Subsurface development of CO_2 disposal for the Gorgon project. *Energy Procedia* 1 (1): 3031–3038. https://doi.org/10.1016/j.egypro.2009.02.081.

Freeman, B.C. and Bhown, A.S. (2011). Assessment of the technology readiness of post-combustion CO_2 capture technologies. *Energy Procedia* 4: 1791–1796. https://doi.org/10.1016/j.egypro.2011.02.055.

Fu, L. and Legge, A.H. (2023). Uncertainty and bias in Liggio et al. (2019) on CO_2 emissions from oil sands operations. *Nature Communications* 14 (1): 5406. https://doi.org/10.1038/s41467-023-40818-5.

Furre, A.K., Eiken, O., Alnes, H. et al. (2017). 20 Years of monitoring CO_2 injection at sleipner. *Energy Procedia* 114: 3916–3926. https://doi.org/10.1016/j.egypro.2017.03.1523.

Garcia, J.A., Villen-Guzman, M., Rodriguez-Maroto, J.M. et al. (2022). Technical analysis of CO_2 capture pathways and technologies. *Journal of Environmental Chemical Engineering* 10 (5). https://doi.org/10.1016/j.jece.2022.108470.

George, V. (2024) Illinois rejects carbon pipeline permit for one earth energy. https://carbonherald.com/illinois-rejects-carbon-pipeline-permit-for-one-earth-energy/ (accessed 28 May 2024).

Ghahfarokhi, R.B, Pennell, S., Matson, M. et al. (2016). *Overview of CO_2 injection and WAG sensitivity in SACROC*. Paper presented at the SPE Improved Oil Recovery Conference, Tulsa, Oklahoma. https://doi.org/10.2118/179569-MS.

Ghosh, P., Wilton, R.R., Bowers, A. et al. (2023). Reinjection of produced (Sheared) polymer in a Canadian viscous oil reservoir: considerations to improve economics and reduce carbon footprint. *SPE Journal* 28: 251–267. https://doi.org/10.2118/209347-PA.

Global CCS Institute (2021a). Blue hydrogen. https://www.globalccsinstitute.com/wp-content/uploads/2021/04/Circular-Carbon-Economy-series-Blue-Hydrogen.pdf (accessed 11 May 2024).

Global CCS Institute (2021b). Technology readiness and costs of CCS. https://www.globalccsinstitute.com/wp-content/uploads/2021/03/Technology-Readiness-and-Costs-for-CCS-2021-1.pdf (accessed 27 April 2024).

Global CCS Institute (2022). State of the art: CCS technologies 2022. https://www.globalccsinstitute.com/wp-content/uploads/2022/05/State-of-the-Art-CCS-Technologies-2022.pdf (accessed 17 May 2024).

Gobierno de Argentina (2024). Producción de petróleo y gas desde 2019 (Sesco Web). https://www.argentina.gob.ar/economia/energia/hidrocarburos/produccion-de-petroleo-y-gas (accessed 26 April 2024).

Godoi, J.M.A. and dos Santos Matai, P.H.L. (2021). Enhanced oil recovery with carbon dioxide geosequestration: first steps at pre-salt in Brazil. *Journal of Petroleum Exploration and Production* 11 (3): 1429–1441. https://doi.org/10.1007/s13202-021-01102-8.

Han, X., Zhong, L., Tang, X. et al. (2020). *Ten-year pilot test of steam and flue gas stimulation for offshore heavy oilfield*. Paper presented at the Offshore Technology Conference Asia, Kuala Lumpur, Malaysia. https://doi.org/10.4043/30188-MS.

Hannouf, M., Assefa, G., and Gates, I. (2021). Carbon intensity threshold for Canadian oil sands industry using planetary boundaries: is a sustainable carbon-negative industry possible? *Renewable and Sustainable Energy Reviews* 151. https://doi.org/10.1016/j.rser.2021.111529.

Hauber, G. (2023). Norway's Sleipner and Snøhvit CCS: industry models or cautionary tales? https://ieefa.org/media/3828/download (accessed 13 April 2024).

Hochman, G. and Appasamy, V. (2024). The case for carbon capture and storage technologies. *Environments* 11: 52. https://doi.org/10.3390/environments11030052.

Holcomb, M. (2024). 688-Mile summit carbon pipeline project approved for construction by Iowa regulators. *Pipeline and Gas Journal*. https://pgjonline.com/news/2024/june/688-mile-summit-carbon-pipeline-project-approved-for-construction-by-iowa-regulators?oly_enc_id=0806I9485478J4U (accessed 26 June 2024).

Hong, W.Y. (2022). A techno-economic review on carbon capture, utilisation and storage systems for achieving a net-zero CO_2 emissions future. *Carbon Capture Science & Technology* 3. https://doi.org/10.1016/j.ccst.2022.100044.

Hughes, S. and Zoelle, A. (2022). *Cost of Capturing CO_2 from Industrial Sources*. National Energy Technology Laboratory (NETL). https://www.osti.gov/servlets/purl/1887586 (accessed 10 March 2024).

IEAGHG (2014). Assessment of emerging CO_2 capture technologies and their potential to reduce costs - 2014/TR4. https://ieaghg.org/publications/2014-tr4-emerging-capture-technologies/ (accessed 11 May 2024).

IEAGHG (2019). The shell quest carbon capture and storage project, 2019/04. https://ieaghg.org/publications/technical-reports/reports-list/9-technical-reports/949-2019-04-the-shell-quest-carbon-capture-and-storage-project (accessed 17 March 2024).

Illinois Commerce Commission (2024). Public comments for 23-0475. https://www.icc.illinois.gov/docket/P2023-0475/public-comments (accessed 28 May 2024).

Incer-Valverde, J., Korayem, A., Tsatsaronis, G. et al. (2023). "Colors" of hydrogen: definitions and carbon intensity. *Energy Conversion and Management* 291. https://doi.org/10.1016/j.enconman.2023.117294.

International Energy Agency - IEA (2021). The world has vast capacity to store CO_2: net zero means we'll need it. https://www.iea.org/commentaries/the-world-has-vast-capacity-to-store-co2-net-zero-means-we-ll-need-it (accessed 3 May 2024).

International Energy Agency - IEA (2022). Legal and regulatory framework for CCUS: an IEA CCUS handbook. https://www.iea.org/reports/legal-and-regulatory-frameworks-for-ccus (accessed 26 May 2024).

International Energy Agency - IEA (2023). World energy outlook 2023. https://www.iea.org/reports/world-energy-outlook-2023 (accessed 10 February 2024).

International Energy Agency - IEA (2024). Carbon capture, utilisation and storage. https://www.iea.org/energy-system/carbon-capture-utilisation-and-storage (accessed 10 March 2024).

International Institute for Sustainable Development – IISD (2023). Why the cost of carbon capture and storage remains persistently high. https://www.iisd.org/articles/deep-dive/why-carbon-capture-storage-cost-remains-high (accessed 20 October 2024).

International Organization for Standardization - ISO (2017). Carbon dioxide capture, transportation, and geological storage: geologic storage - ISO 27914:2017. https://www.iso.org/obp/ui/en/#iso:std:iso:27914:ed-1:v1:en (accessed 5 May 2024).

International Organization for Standardization - ISO (2019). Carbon dioxide capture, transportation, and geological storage: carbon dioxide storage using enhanced oil recovery (CO_2-EOR) - ISO 27916:2019. https://www.iso.org/obp/ui/en/#iso:std:iso:27916:ed-1:v1:en (accessed 5 May 2024).

International Organization for Standardization - ISO (2024a). Carbon dioxide capture, transportation, and geological storage: geologic storage - ISO/CD 27914. https://www.iso.org/standard/84578.html (accessed 5 May 2024).

International Organization for Standardization - ISO (2024b). Carbon dioxide capture, transportation, and geological storage: carbon dioxide storage using enhanced oil recovery (CO_2-EOR) – ISO/AWI 27916. https://www.iso.org/standard/89142.html (accessed 5 May 2024).

IPPC (2018). *IPPC Special Report on Global Warming of 1.5°C* (ed. V. Masson-Delmotte, P. Zhai, H.-O. Pörtner et al.) (SR15). Intergovernmental Panel on Climate Change. https://www.ipcc.ch/site/assets/uploads/sites/2/2019/06/SR15_Full_Report_High_Res.pdf (accessed 27 February 2024).

IPPC (2024). Transport of CO_2. https://archive.ipcc.ch/pdf/special-reports/srccs/srccs_chapter4.pdf (accessed 16 June 2024).

Jacobs, T. (2023). A new day begins for shuttered Petra Nova CCUS. https://jpt.spe.org/a-new-day-begins-for-shuttered-petra-nova-ccus (accessed 28 March 2024).

Jana, A. and Modi, A. (2024). Recent progress on functional polymeric membranes for CO_2 separation from flue gases: a review. *Carbon Capture Science & Technology* 11. https://doi.org/10.1016/j.ccst.2024.100204.

Jansen, E. (2019). *Refineries of the future - the refinery as an ENERGY HUB*. Paper presented at the SPE Kuwait Oil & Gas Show and Conference, Mishref, Kuwait. https://doi.org/10.2118/198184-MS.

Jenkins, J. (2015). Financing mega-scale energy projects; a case study of the Petra Nova carbon capture project. https://www.paulsoninstitute.org/wp-content/uploads/2015/10/CS-Petra-Nova-EN.pdf (accessed 9 April 2024).

Kalitsun, V., Fazelipour, W., Thiele, M. et al. (2023). *Application of streamline-based WAG injection rate management to improve oil recovery of GLSAU CO_2 flood*. Paper presented at the SPE Annual Technical Conference and Exhibition, San Antonio, Texas. https://doi.org/10.2118/214828-MS.

Kalteyer, J. (2020). *A case study of SACROC CO_2 flooding in marginal pay regions: improving asset performance*. Paper presented at the SPE Improved Oil Recovery Conference, Virtual. https://doi.org/10.2118/200460-MS.

Kearns, J., Teletzke, G., Palmer, J. et al. (2017). Developing a consistent database for regional geologic CO_2 storage capacity worldwide. *Energy Procedia* 114: 4697–4709. https://doi.org/10.1016/j.egypro.2017.03.1603.

Kimani, A. (2023). Upstream spending to rise to $485 billion in 2023. https://oilprice.com/Energy/Crude-Oil/Upstream-Spending-To-Rise-To-485-Billion-In-2023.html (accessed 14 June 2024).

Kindermorgan (2024). Carbon dioxide (CO_2) operations. https://www.kindermorgan.com/Operations/CO2/Index (accessed 26 March 2024).

Kuuskraa, V. and Wallace, M. (2014). CO_2-EOR set for growth as new CO_2 supplies emerge. *Oil & Gas Journal* 112 (4): 66–76. https://www.adv-res.com/pdf/CO2-EOR-set-for-growth-as-new-CO2-supplies-emerge.pdf (accessed 25 March 2024).

Le Quéré, C., Jackson, R.B., Jones, M.W. et al. (2020). Temporary reduction in daily global CO_2 emissions during the COVID-19 forced confinement. *Nature Climate Change* 10: 647–653. https://doi.org/10.1038/s41558-020-0797-x.

Leeson, D., Mac Dowell, N., Shah, N. et al. (2017). A techno-economic analysis and systematic review of carbon capture and storage (CCS) applied to the iron and steel, cement, oil refining and pulp and paper industries, as well as other high purity sources. *International Journal of Greenhouse Gas Control* 61: 71–84. https://doi.org/10.1016/j.ijggc.2017.03.020.

Liggio, J. and Li, S.M. (2023). Reply to: uncertainty and bias in Liggio et al. (2019) on CO_2 emissions from oil sands operations. *Nature Communications* 14: 5407. https://doi.org/10.1038/s41467-023-40819-4.

Liggio, J., Li, S.M., Staebler, R.M. et al. (2019). Measured Canadian oil sands CO_2 emissions are higher than estimates made using internationally recommended methods. *Nature Communications* 10: 1863. https://doi.org/10.1038/s41467-019-09714-9.

Liu, D., Liu, Z., Pan, G. et al. (2024). *Pilot tests of steam flooding after cyclic steam-N_2-CO_2 stimulation in Bohai offshore with large well spacing*. Paper presented at the Offshore Technology Conference Asia, Kuala Lumpur, Malaysia. https://doi.org/10.4043/34911-MS.

Liu, E., Lu, X., and Wang, D. (2023). A systematic review of carbon capture, utilization and storage: status, progress and challenges. *Energies* 16 (6): 2865. https://doi.org/10.3390/en16062865.

Lydersen, K. (2023). Navigator CO_2 pipeline is canceled, but Illinois opponents say the fight isn't over. https://energynews.us/2023/10/20/navigator-co2-pipeline-is-canceled-but-illinois-opponents-say-the-fight-isnt-over/ (accessed 28 May 2024).

MacDowell, N., Florin, N., Buchard, A. et al. (2010). An overview of CO_2 capture technologies. *Energy & Environmental Science* 3 (11): 1645–1669. http://dx.doi.org/10.1039/C004106H.

Madejski, P., Chmiel, K., Subramanian, N. et al. (2022). Methods and techniques for CO_2 capture: review of potential solutions and applications in modern energy technologies. *Energies* 15 (887). https://doi.org/10.3390/en15030887.

Mahajan, S. and Lahtinen, M. (2022). Recent progress in metal-organic frameworks (MOFs) for CO_2 capture at different pressures. *Journal of Environmental Chemical Engineering* 10 (6). https://doi.org/10.1016/j.jece.2022.108930.

Mahasenan, N. and Brown, D.R. (2005). Beyond the big picture: characterization of CO_2-laden streams and implications for capture technologies. *Proceedings of the 7th International Conference on Greenhouse Gas Control Technologies*, Vancouver, Canada (5 September 2004). Elsevier Science Ltd. https://doi.org/10.1016/B978-008044704-9/50218-4.

Manrique, E., Ahmadi, M., and Samani, S. (2017). Historical and recent observations in polymer floods: an update review. *CT&F - Ciencia, Tecnología Y Futuro* 6 (5): 17–48. https://doi.org/10.29047/01225383.72.

Manrique, E., Maya G., Peñuela, J. et al. (2022). *Evaluation of CO_2-EOR and storage opportunities from industrial sources and thermal EOR projects*. Presentation at Petrophase 2022, Bucaramanga, Colombia. https://petrophase2022.org/wp-content/uploads/2022/06/memories_PetroPhase_2022.pdf (accessed 22 June 2024).

Manrique, E., Muci, V., and Gurfinkel, M. (2007). EOR field experiences in carbonate reservoirs (SPE-100063-PA). *SPE Reservoir Evaluation & Engineering* 10 (6): 667–686. https://doi.org/10.2118/100063-PA.

Marshall, J.P. (2022). A social exploration of the west Australian gorgon gas, carbon capture and storage project. *Clean Technologies* 4: 67–90. https://doi.org/10.3390/cleantechnol4010006.

Martin, D.F. and Taber, J.J. (1992). Carbon dioxide flooding. *Journal of Petroleum Technology* 44: 396–400. https://doi.org/10.2118/23564-PA.

Masoner, L.O., Abidi, H.R., and Hild, G.P. (1996). *Diagnosing CO_2 flood performance using actual performance data*. Paper presented at the SPE/DOE Improved Oil Recovery Symposium, Tulsa, Oklahoma. https://doi.org/10.2118/35363-MS.

Massarweh, O., Al-Khuzaei, M., Al-Shafi, M. et al. (2023). Blue hydrogen production from natural gas reservoirs: a review of application and feasibility. *Journal of CO_2 Utilization* 70. https://doi.org/10.1016/j.jcou.2023.102438.

Mathieson, A., Midgely, J., Wright, I. et al. (2011). In Salah CO_2 storage JIP: CO_2 sequestration monitoring and verification technologies applied at Krechba, Algeria. *Energy Procedia* 4: 3596–3603. https://doi.org/10.1016/j.egypro.2011.02.289.

Medina-Ketzer, J.M., Machado, C.X., Rockett, G.C. et al. (2014). Brazilian Atlas of CO_2 capture and geological storage. https://www.globalccsinstitute.com/archive/hub/publications/190903/brazilian-atlas-co2-capture-geological-storage.pdf (accessed 28 March 2024).

Mengden, A. (2023). Carbon taxes in Europe. https://taxfoundation.org/data/all/eu/carbon-taxes-in-europe-2023/ (accessed 27 May 2024).

Moffitt, P., Pecore, D., Trees, M. et al. (2015). *East vacuum grayburg San Andres unit, 30 years of CO_2 flooding: accomplishments, challenges and opportunities*. Paper presented at the SPE Annual Technical Conference and Exhibition, Houston, Texas. https://doi.org/10.2118/175000-MS.

Mohammed-Singh, L.J. and Singhal, A.K. (2004). *Lessons from Trinidad's CO_2 immiscible pilot projects 1973-2003*. Paper presented at the SPE/DOE Symposium on Improved Oil Recovery, Tulsa, Oklahoma. https://doi.org/10.2118/89364-MS.

Morris, N. (2024). AgForce, National Farmers' Federation oppose Glencore's Great Artesian Basin carbon storage trial. https://www.abc.net.au/news/2024-03-07/agforce-nff-oppose-glencore-great-artesian-basin-ccs-trial/103501062 (accessed 28 May 2024).

Mwenketishi, G.T., Benkreira, H., and Rahmanian, N. (2023). A comprehensive review on carbon dioxide sequestration methods. *Energies* 16: 7971. https://doi.org/10.3390/en16247971.

National Energy Technology Laboratory (NETL) (2010). Carbon dioxide enhanced oil recovery. https://www.netl.doe.gov/sites/default/files/netl-file/CO2_EOR_Primer.pdf (accessed 16 June 2024).

National Energy Technology Laboratory (NETL) (2024a). Carbon storage FAQs. https://netl.doe.gov/carbon-management/carbon-storage/faqs/carbon-storage-faqs (accessed 9 March 2024).

National Energy Technology Laboratory (NETL) (2024b). Carbon capture and storage database. https://netl.doe.gov/carbon-management/carbon-storage/worldwide-ccs-database (accessed 11 May 2024).

National Energy Technology Laboratory (NETL) (2024c). Great plains synfuels plant. https://netl.doe.gov/research/Coal/energy-systems/gasification/gasifipedia/great-plains (accessed 17 May 2024).

National Petroleum Council (NPC) (2021). A roadmap to at-scale deployment of carbon capture, use, and storage: chapter six – CO_2 transport. https://www.energy.gov/sites/default/files/2022-10/CCUS-Chap_6-030521.pdf (accessed 9 March 2024).

Navigator (2022). The Heartland Greenway: a navigator project. https://www.midwestccus.org/wp-content/uploads/Anderson_MRCI2022_NavigatorCO2-presentation.pdf (accessed 2 June 2024).

Norwegian Offshore Directorate (2024a). Regulations relating to measurement of petroleum for fiscal purposes and for calculation pd CO_2-tax (the measurement regulations). https://www.sodir.no/en/regulations/regulations/measurement-of-petroleum-for-fiscal-purposes-and-for-calculation-of-co2-tax/ (accessed 26 May 2024).

Norwegian Offshore Directorate (2024b). Announcement of acreage for CO_2 storage on the Norwegian continental shelf. https://www.sodir.no/en/whats-new/news/general-news/2024/announcement-of-acreage-for-co2-storage-on-the-norwegian-continental-shelf/ (accessed 26 May 2024).

Norwegian Offshore Directorate (2024c). Eight companies have applied for CO_2 storage acreage. https://www.sodir.no/en/whats-new/news/general-news/2024/eight-companies-have-applied-for-co2-storage-acreage/ (accessed 26 May 2024).

Oil and Gas Climate Initiative (OGCI) (2024). Carbon capture, utilization and storage. https://www.ogci.com/ccus (accessed 10 March 2024).

Organization of the Petroleum Exporting Countries - OPEC (2023). World oil outlook 2045. https://woo.opec.org/index.php (accessed 11 February 2024).

Oyenowo, O.P., Wang H., Mirzaei-Paiaman, A. et al. (2024). Geochemical impact of high-concentration formate solution injection on rock wettability for enhanced oil recovery and geologic carbon storage. *Energy & Fuels* 38: 6138–6155. https://doi.org/10.1021/acs.energyfuels.3c05081.

Özgür, E. (2019). The role of enhanced oil recovery in the upstream petroleum sector, a case study from Turkey. *Bulletin Mineral Research and Exploration* 158: 291–297. https://dergipark.org.tr/tr/download/article-file/508563 (accessed 25 March 2024).

Palmer, D. and O'Donnell, J. (2014) *Construction, operations and performance of the first enclosed trough solar steam generation pilot for EOR applications*. Paper presented at the SPE EOR Conference at Oil and Gas West Asia, Muscat, Oman. https://doi.org/10.2118/169745-MS.

Pereira Nunes, J.P., Seabra, G.S., and de Sousa, L.C. (2024). A review of CO_2-injection projects in the Brazilian pre-salt — storage capacity and geomechanical constraints. *International Journal of Greenhouse Gas Control* 137. https://doi.org/10.1016/j.ijggc.2024.104232.

Perez, R., Guerrero, J., and Rodriguez, H. (2023a). In-situ steam generation using mist water-air injection as enhanced oil recovery and energy efficiency process: kinetic modeling and numerical simulation approach. *SPE Reservoir Evaluation & Engineering* 26: 365–381. https://doi.org/10.2118/212853-PA.

Perez, R.A., Rodriguez, H.A., Romero, J.E. et al. (2023b). *Incorporating hybrid technology of CSS + foam in heavy oil field development plans: practical experiences and lessons learned*. Paper presented at the SPE Latin American and Caribbean Petroleum Engineering Conference, Port of Spain, Trinidad and Tobago. https://doi.org/10.2118/213199-MS.

Peterson, C.A., Pearson, E.J., Chodur, V.T. et al. (2012). *Beaver Creek Madison CO_2 enhanced oil recovery project case history; Riverton, Wyoming*. Paper presented at the SPE Improved Oil Recovery Symposium, Tulsa, Oklahoma. https://doi.org/10.2118/152862-MS.

Petrobras (2023). Pre-salt is heading towards decarbonization. https://noticias.petrobras.com.br/en/business/pre-salt-is-heading-towards-decarbonization-06-09-2023/ (accessed 19 May 2024).

Pipeline & Gas Journal (2023). South Dakota denies permit for summit carbon solutions 495-mile CO_2 pipeline segment. https://pgjonline.com/news/2023/september/south-dakota-denies-permit-for-summit-carbon-solutions-495-mile-co2-pipeline-segment?oly_enc_id=080%E2%80%A6 (accessed 28 May 2024).

Pipeline and Hazardous Material Safety Administration (PHMSA) (2024a). Annual report mileage for hazardous liquid or carbon dioxide systems. https://www.phmsa.dot.gov/data-and-statistics/pipeline/annual-report-mileage-hazardous-liquid-or-carbon-dioxide-systems (accessed 9 March 2024).

Pipeline and Hazardous Material Safety Administration (PHMSA) (2024b). Hazardous liquid pipeline safety act of 1979. https://www.phmsa.dot.gov/working-phmsa/state-programs/hazardous-liquid-pipeline-safety-act-1979 (accessed 26 May 2024).

Pipeline and Hazardous Material Safety Administration (PHMSA) (2024c). Pipeline incident 20 years trend. https://www.phmsa.dot.gov/data-and-statistics/pipeline/pipeline-incident-20-year-trends (accessed 1 June 2024).

Pizarro, J.O.S. and Branco, C.C.M. (2023). A future pathway for CO_2 geological storage (CCS) in the offshore pre-salt of Santos basin. Paper presented at the Offshore Technology Conference Brazil, Rio de Janeiro, Brazil. https://doi.org/10.4043/32770-MS.

Popa, A., Lolley, C., Munoz, J. et al. (2022). *Successful reservoir management and optimization of mature steamflood projects using artificial intelligence*. Paper presented at the SPE Annual Technical Conference and Exhibition, Houston, TX. https://doi.org/10.2118/209988-MS.

Pyo, K., Damian-Diaz, N., Powell, M. et al. (2003). *CO_2 flooding in Joffre Viking Pool*. Paper presented at the Canadian International Petroleum Conference, Calgary, Alberta. https://doi.org/10.2118/2003-109.

Quinn, M. (2022). Carbon capture and storage; a review of Australian projects. *The APPEA Journal* 62: 334–341. https://doi.org/10.1071/AJ21161.

Rabbi, M.F., Popp, J., Máté, D. et al. (2022). Energy security and energy transition to achieve carbon neutrality. *Energies* 15: 8126. https://doi.org/10.3390/en15218126.

Rambech, E.L., Vandenbussche, V., Thylander, E.K. et al. (2016). *Energy efficiency: collecting experiences from 20 years of efforts on the Norwegian continental shelf.* Paper presented at the SPE International Conference and Exhibition on Health, Safety, Security, Environment, and Social Resp., Stavanger, Norway. https://doi.org/10.2118/179282-MS.

Ringrose, P.S., Mathieson, A.S., Wright, I.W. et al. (2013). The In Salah CO_2 storage project: lessons learned and knowledge transfer. *Energy Procedia* 37: 6226–6236. https://doi.org/10.1016/j.egypro.2013.06.551.

Robertson, B. and Mousavian, M. (2022). *Gorgon Carbon Capture and Storage: The Sting in the Tail.* Institute for Energy Economics and Financial Analysis. https://ieefa.org/sites/default/files/2022-04/Gorgon-Carbon-Capture-and-Storage-The-Sting-in-the-Tail_April-2022.pdf (accessed 14 April 2024).

Robertson, E.P. (2022). *CO_2 Enhanced Oil Recovery in Wyoming: 2020 Update.* EORI Bulletin, #2202/2022. https://www.eoriwyoming.org/projects-resources/publications/geology/co2-eor-in-wyoming-update-2020 (accessed 28 March 2024).

Rodriguez, H.A., Perez, R.A., Guerrero, J.E. et al. (2023). *An innovative combination of thermal enhanced oil recovery process with organic Rankine cycle to improve energy efficiency.* Paper presented at the SPE Latin American and Caribbean Petroleum Engineering Conference, Port of Spain, Trinidad and Tobago. https://doi.org/10.2118/213144-MS.

Rubin, E.S., Mantripragada, H., Marks, A. et al. (2012). The outlook for improved carbon capture technology. *Progress in Energy and Combustion Science* 38 (5): 630–671. https://doi.org/10.1016/j.pecs.2012.03.003.

Sahin, S., Kalfa, U., Celebioglu, D. et al. (2012). *A quarter century of progress in the application of CO_2 immiscible EOR project in Bati Raman Heavy Oil Field in Turkey.* Paper presented at the SPE Heavy Oil Conference Canada, Calgary. https://doi.org/10.2118/157865-MS.

Saketh Madugula, A.D., Sachde, D., Hovorka, S.D. et al. (2021). Estimation of CO_2 emissions from petroleum refineries based on the total operable capacity for carbon capture applications. *Chemical Engineering Journal Advances* 8: 100162. https://doi.org/10.1016/j.ceja.2021.100162.

Sandler, J., Fowler, G., Cheng, K. et al. (2012). *Solar-generated steam for oil recovery: reservoir simulation, economic analysis, and life cycle assessment.* Paper presented at the SPE Western Regional Meeting, Bakersfield, California. https://doi.org/10.2118/153806-MS.

Saravanan, A., Senthil Kumar, P., Vo, D.V. et al. (2021). A comprehensive review on different approaches for CO_2 utilization and conversion pathways. *Chemical Engineering Science* 236: 116515. https://doi.org/10.1016/j.ces.2021.116515.

Schopp, U., Herbertson, J., Perez Bajo, L. et al. (2022). *Flaring management guidance for the oil and gas industry*. Paper presented at the SPE Annual Technical Conference and Exhibition, Houston, Texas. https://doi.org/10.2118/210484-MS.

Shen, M., Tong, L., Yin, S. et al. (2022). Cryogenic technology progress for CO_2 capture under carbon neutrality goals: a review. *Separation and Purification Technology* 299. https://doi.org/10.1016/j.seppur.2022.121734.

Shirazi, M., Mahani, H., Tamsilian, Y. et al. (2024). Full life cycle review of water-based CEOR methods from pre-injection to post-production. *Fuel* 356. https://doi.org/10.1016/j.fuel.2023.129574.

Skauge, T., Skauge, A., Lugo, N. et al. (2024). *How polymer flooding reduces CO_2 emissions and energy consumption – an exergy return on exergy investment case study*. Paper presented at the SPE Improved Oil Recovery Conference, Tulsa, Oklahoma. https://doi.org/10.2118/218231-MS.

Smith, E., Morris, J., Kheshgi, H. et al. (2021). The cost of CO_2 transport and storage in global integrated assessment modeling. *International Journal of Greenhouse Gas Control* 109. https://doi.org/10.1016/j.ijggc.2021.103367.

Standnes, D.C. and Skjevrak, I. (2014). Literature review of implemented polymer field projects. *Journal of Petroleum Science and Engineering* 122: 761–775. https://doi.org/10.1016/j.petrol.2014.08.024.

Stephenson, D.J., Graham, A.G., and Luhning, R.W. (1993). Mobility control experience in the Joffre Viking miscible CO_2 flood. *SPE Reservoir Engineering* 8: 183–188. https://doi.org/10.2118/23598-PA.

Summit Carbon Solutions (2024). Project footprint. https://summitcarbonsolutions.com/project-footprint/ (accessed 2 June 2024).

Sun, Y., Zhao, L., Lin, T. et al. (2013). *Case study: thermal enhance Bohai offshore heavy oil recovery by co-stimulation of steam and gases*. SPE Heavy Oil Conference-Canada, Calgary, Alberta, Canada. https://doi.org/10.2118/165410-MS.

Takht-Ravanchi, M. and Sahebdelfar, S. (2014). Carbon dioxide capture and utilization in petrochemical industry: potentials and challenges. *Applied Petrochemical Research* 4: 63–77. https://doi.org/10.1007/s13203-014-0050-5.

Talaei, A., Oni, A.O., Ahiduzzaman, M. et al. (2020). Assessment of the impacts of process-level energy efficiency improvement on greenhouse gas mitigation potential in the petroleum refining sector. *Energy* 191. https://doi.org/10.1016/j.energy.2019.116243.

Tallgrass (2024). Tallgrass to capture and sequester CO_2 emissions from ADM corn processing complex in Nebraska. https://tallgrass.com/newsroom/press-releases/tallgrass-to-capture-and-sequester-co2-emissions-from-adm-corn-processing-complex-in-nebraska (accessed 2 June 2024).

Tang, X., Yu, C., Bai, Y. et al. (2022). *Field trial of nitrogen-assisted cyclic steam stimulation in post-CHOPS wells, case study in Sudan*. Paper presented at the SPE

Thermal Well Integrity and Production Symposium, Banff, Alberta, Canada. https://doi.org/10.2118/212147-MS.

Tengku-Hassan, T.N.A., Shariff, A.M., Mohd Pauzi, M.M. et al. (2022). Insights on cryogenic distillation technology for simultaneous CO_2 and H_2S removal for sour gas fields. *Molecules* 27: 1424. https://doi.org/10.3390/molecules27041424.

Trupp, M., Ryan, S., Mendoza, I.B. et al. (2021). Developing the world's largest CO_2 Injection System – a history of the Gorgon Carbon Dioxide Injection System. *Proceedings of the 15th Greenhouse Gas Control Technologies Conference* (15–18 March 2021). https://ssrn.com/abstract=3815492. (accessed 14 April 2024).

U.S. Department of Energy – (U.S. DOE) (2020). Petra Nova - W.A. Parish post-combustion CO_2 capture and sequestration demonstration project: final scientific/technical report. https://www.osti.gov/servlets/purl/1608572 (accessed 9 April 2024).

U.S. Department of Energy – (U.S. DOE) (2024). Hydrogen production: natural gas reforming. https://www.energy.gov/eere/fuelcells/hydrogen-production-natural-gas-reforming (accessed 11 May 2024).

U.S. Department of Transportation (2022). Failure investigation report - Denbury gulf coast pipelines, LLC - pipeline rupture. https://www.phmsa.dot.gov/sites/phmsa.dot.gov/files/2022-05/Failure%20Investigation%20Report%20-%20Denbury%20Gulf%20Coast%20Pipeline.pdf (accessed 1 June 2024).

U.S. Energy Information Administration – EIA (2017). Petra Nova is one of two carbon capture and sequestration power plants in the world. https://www.eia.gov/todayinenergy/detail.php?id=33552 (accessed 27 March 2024).

U.S. Energy Information Administration – EIA (2023). Annual energy outlook. https://www.eia.gov/outlooks/aeo/pdf/AEO2023_Narrative.pdf (accessed 10 February 2024).

U.S. Environmental Protection Agency - U.S. EPA (2024). Underground injection control: well classes. https://www.epa.gov/uic/underground-injection-control-well-classes (accessed 5 May 2024).

U.S. Geologic Survey - USGS (2013). National assessment of geologic carbon dioxide storage resources – results. https://pubs.usgs.gov/circ/1386/pdf/circular1386.pdf (accessed 3 May 2024).

U.S. Geologic Survey - USGS (2022). National assessment of carbon dioxide enhanced oil recovery and associated carbon dioxide retention resources – results. https://pubs.usgs.gov/circ/1489/cir1489.pdf (accessed 3 May 2024).

U.S. Government Accountability Office (U.S. GAO) (2022). Decarbonization: technology assessment. https://www.gao.gov/assets/730/723198.pdf (accessed 27 April 2024).

U.S. Internal Revenue Service - U.S. IRS (2024). Update for Form 8830 (Rev. March 2022) per Notice 2023-57. https://www.irs.gov/forms-pubs/update-for-form-8830-rev-march-2022-per-notice-2023-57 (accessed 26 May 2024).

United Nations - UN (2024). The Paris Agreement. https://unfccc.int/process-and-meetings/the-paris-agreement?gclid=CjwKCAiAiP2tBhBXEiwACslfnkS7jxlEyt5in69yh1V4AeBp8XZvM8a04UnCGR56R6XHpexyzOQP1xoC1S4QAvD_BwE (accessed 10 February 2024).

Valluri, S., Claremboux, V., and Kawatra, S. (2022). Opportunities and challenges in CO_2 utilization. *Journal of Environmental Sciences* 113: 322–344. https://doi.org/10.1016/j.jes.2021.05.043.

Vitali, M., Zuliani, C., Corvaro, F. et al. (2022). Statistical analysis of incidents on onshore CO_2 pipelines based on PHMSA database. *Journal of Loss Prevention in the Process Industries* 77. https://doi.org/10.1016/j.jlp.2022.104799.

Voigt, C. (2023). The power of the Paris Agreement in international climate litigation. *Review of European, Comparative & International Environmental Law* 32 (2): 237–249. https://doi.org/10.1111/reel.12514.

Whitecap Resources Inc. (2024). CO_2 sequestration - Weyburn unit: total sequestered to date. https://www.wcap.ca/sustainability/co2-sequestration (accessed 17 March 2024).

Whittaker, S., Rostron, B., Hawkes, C. et al. (2011). A decade of CO_2 injection into depleting oil fields: monitoring and research activities of the IEA GHG Weyburn-midale CO_2 monitoring and storage project. *Energy Procedia* 4: 6069–6076. https://doi.org/10.1016/j.egypro.2011.02.612.

Wolf Carbon Solutions (2024). Mt. Simon hub. https://wolfcarbonsolutions.com/mt-simon-hub/ (accessed 2 June 2024).

Xi, D., Lu, H., Fu, Y. et al. (2023). Carbon dioxide pipelines: a statistical analysis of historical accidents. *Journal of Loss Prevention in the Process Industries* 84. https://doi.org/10.1016/j.jlp.2023.105129.

Yañez, E., Castillo, E., Manrique, E. et al. (2022). An integrated approach to assess CCUS potential in a decarbonisation strategy for the oil industry: a case study. *Proceedings of the 16th Greenhouse Gas Control Technologies Conference (GHGT-16)* (23–24 October 2022). http://dx.doi.org/10.2139/ssrn.4288099 (accessed 11 May 2024).

Yañez, E., Ramirez, A., Nuñez-Lopez, V. et al. (2020). Exploring the potential of carbon capture and storage-enhanced oil recovery as a mitigation strategy in the Colombian oil industry. *International Journal of Greenhouse Gas Control* 94. https://doi.org/10.1016/j.ijggc.2019.102938.

Yergin, D. (2024). The return of energy security. https://www.spglobal.com/en/research-insights/featured/special-editorial/look-forward/the-return-of-energy-security (accessed 24 March 2024).

Yousaf, M., Zaman, M., Mahmood, A. et al. (2022). Carbon dioxide utilization: a critical review from a multiscale perspective. *Energy Science & Engineering* 10 (12). https://doi.org/10.1002/ese3.1303.

Zhang, K., Lau, H.C., and Zhangxin, C. (2022a). Extension of CO_2 storage life in the Sleipner CCS project by reservoir pressure management. *Journal of Natural Gas Science and Engineering* 118. https://doi.org/10.1016/j.jngse.2022.104814.

Zhang, Y., Jackson, C., and Krevor, S. (2022b). An estimate of the amount of geological CO_2 storage over the period of 1996–2020. *Environmental Science & Technology Letters* 9 (8): 693–698. https://doi.org/10.1021/acs.estlett.2c00296.

3

Fugitive Methane Emissions

Stuart Riddick

Department of Science, Engineering and Aviation, University of the Highlands and Islands, Perth, Perthshire, UK

This chapter focuses on methods to detect and mitigate fugitive methane emissions. Section 3.1 describes emission types from oil/gas production and transport, including vented, fugitive, and maintenance emissions. Sections 3.2 and 3.3 introduce methane and micrometeorology, respectively. Section 3.4 explains the methods used to quantify emissions, Section 3.5 describes technologies used to quantify methane, and Section 3.6 explains how technologies and methods can be applied to create full system solution. Section 3.7 describes the current challenges in fugitive methane quantification and the aims for future work.

3.1 Methane Oil and Gas Emission Sources

On an oil or gas production site, methane is emitted from equipment almost all the time. Some of the emissions are routine and permitted (vented), some are uncontrolled and unexpected (fugitives), and some are the result of maintenance.

Venting of gas is built into some pieces of equipment's design for safety reasons. For example, pneumatic devices on a wellhead use produced gas to open and close valves and vent to the atmosphere. Gas is emitted from pneumatic devices almost continuously, and, although individual emissions are small, cumulatively these can comprise up to one-third of emissions at production sites, depending on-site equipment and configuration. Similarly, gas is emitted almost continuously from rod packing vents on reciprocating compressors while they are operating normally; as rod packing ages, it degrades and emissions are no longer considered vented emissions. Also, glycol dehydrators are used to remove water, which can

Cleaner Petroleum Production and Refining Technologies, First Edition, Edited by M. R. Riazi and Harvey W. Yarranton.

cause corrosion in the pipes, from produced natural gas. When glycol absorbs water, it can also absorb small amounts of methane, which are released to the atmosphere when the glycol is regenerated. Maintenance on production facilities often requires individual pieces of equipment, usually compressors, to be isolated and gas removed from the system. This is typically described as a "blowdown" event where a relatively large amount of the gas is vented to the atmosphere over a very short time. Typically, vented emissions are relatively well quantified, as gas is released by design and volumes are known.

Fugitive emissions are quite different from vented emissions. Fugitives are the result of an upset condition and can come from any part of the system. As gas is at pressure, any gap in tubing, valves, or vents can result in a fugitive emission. Fugitives have generally been observed to be smaller than vented emissions, but are considered a safety, economic, and environmental hazard as they serve no engineering purpose. Fugitive emissions are typically only detected during leak detection and repair (LDAR) surveys, and the probability of detection is dependent on instrumentation used and skill of the operator. The frequency of LDAR surveys depends on national, state, and company policy.

Combustion emissions are the result of unburnt gas used as fuel escaping from the system. This is typically known as gas slip and is emitted from flares, compressor engines, and heaters used in separators to help split produced gas from liquids. The destruction efficiency of flares is assumed to be 98% and was first quantified using measurements made downwind of flares in Tulsa, Oklahoma, USA in 1982 and Detroit, Michigan, USA in 2010 (Ewing et al., 2010; McDaniel, 1983). Both experiments were conducted in wind speeds less than 2 m/s. Several studies have since reported that this destruction efficiency may be an overestimation, and efficiency may be affected by environmental conditions (wind speed) and the condition of the flare (Johnson and Kostiuk, 2002; Riddick and Mauzerall, 2023). A 2022 aircraft-based study reported average observed flaring destruction removal efficiencies across the Eagle Ford, Bakken, and Permian basins of 95.2% (95% CI: 94.3–95.9%) and a total effective destruction removal efficiency of 91.1% (95% CI: 90.2–91.8%), when unlit flares are accounted for (Plant et al., 2022).

Typical fugitive and vented emissions used by the Environmental Protection Agency (EPA) to generate facility-level emission estimates were based on measurements made at 334 facilities across the United States (American Petroleum Institute, 2004; Campbell et al., 1996; GRI and EPA, 1996; US EPA, 2023) and range from the relatively small, 29 g CH_4/h per wellhead, to the much larger, 1.8 kg CH_4/h per tank (Table 3.1). Recent studies (Barkley et al., 2023; Riddick et al., 2023; Varon et al., 2023) suggest that these emission factors may underestimate emission by up to a factor of five, as the measurement of 334 facilities is unlikely to have observed the long-tail right-skewed emission distribution common to oil and gas emissions distributions (Plant et al., 2022; Vaughn et al., 2017; Zavala-Araiza et al., 2015, 2017). Zavala-Araiza et al. (2017) used a Monte Carlo approach coupled with measurement

Table 3.1 Emission factors for typical vented and fugitive emissions from production equipment on well pads using EPA emission factors (US EPA, 2023) and accounting for the long-tail distribution Zavala-Araiza et al. (2017).

Source	EPA emission factors (g CH_4 equip/h)	Accounting for long-tail distribution (g CH_4 equip/h)
Intermittent bleed pneumatic controller	72	530
Chemical injection pumps	173	180
Compressors	80	100
Dehydrators	0.09	0.3
Leaks	13	150
Tanks	0.4	73

to estimate the effect of large, infrequent emissions on the average emission, showing that emission factors for pneumatic controllers, leaks, and condensate tanks are likely an underestimate (Table 3.1).

As described above, fugitive emissions are uncontrolled, can occur at any point on a facility, and if left undiscovered can emit enough methane to the atmosphere to be a safety, economic, and environmental hazard. Finding and repairing fugitive emissions has become a priority for many oil and gas operators, and several LDAR methods and instrumentation have been developed to increase the probability of leak detection. The following sections will describe technologies used in LDAR and how measurements can be used to quantify leak size.

3.2 Methane

In general, the amount of gas emitted by a source cannot be directly quantified. Instead, gas concentrations are typically measured, and combined with a flow rate or environmental conditions, the emission rate can be inferred. In this section, we will discuss currently available instrumentation that can be used to quantify methane concentrations. Firstly, justification is provided on the importance of measuring methane, and a definition is given to clarify the term concentration as compared to mixing ratios.

3.2.1 Why Quantify Methane?

Methane is a colorless gas and typically produced during the breakdown of organic matter. It is the primary component of natural gas and comprises between 60 and 98% of natural gas, depending on where the gas has been extracted. For example, natural gas extracted in the Permian basin, Texas is referred to as wet gas, as it has a larger component of heavier hydrocarbons (66% methane, 13% ethane, 10% propane, 5% butane, and 3% other heavier hydrocarbons) (Howard et al., 2015), while gas extracted from the Arkoma Basin in Arkansas is dry (98% methane, 1% ethane, and 1% other heavier hydrocarbons) (Speight, 2020).

Methane gas is a powerful greenhouse gas, with 25 times stronger greenhouse warming potential than carbon dioxide over 100 years and 84 times over 20 years. Increases in atmospheric methane strongly suggest that anthropogenic methane emissions are contributing to a changing climate (700 ppb in 1850 to 1,900 ppb in 2022; IPCC, 2006, 2014, 2022). Therefore, quantifying methane emission rates from human activities, such as oil and gas production, have become a priority in order to better understand how emissions can be mitigated and reduced.

3.2.2 Methane Emission Rates

Methane emission rates from a point source are typically quantified as either the mass or volume flow per unit time. Mass flow rates are standard, while volume flow rates depend on temperature and pressure and calculated using the ideal gas law. Academia (e.g., kilograms per hour, kg/h), industry (e.g., standard cubic feet per hour, scfh), and regulators tend to use different standard units, and online tools are available for simple conversion between units (Highwood, 2024). As mentioned above, emission rates are not directly measured but can be calculated from methane concentrations.

3.2.3 Mixing Ratios Versus Concentration

One critical point of clarification is the difference between the concentration and mixing ratio of a gas, which are often used synonymously. The volume mixing, also called the molar mixing ratio, of a gas is the number of moles of the gas divided by the number of moles of air. Due to its abundance in the atmosphere, atmospheric methane mixing ratios are typically reported in parts per million by volume (ppmv), but larger amounts of methane can be reported as a percentage (%).

The concentration of a gas is the mass of the gas per unit volume ($[X]$, g/m^3). Often, an instrument will claim to measure the concentration but report the mixing ratio, which can lead to some confusion. In most cases, the inaccuracy of using "concentration" to describe a "mixing ratio" is trivial; however, it is often a point of note during the peer review process where precision is important. If required, the methane mixing ratio can be converted to concentration using the Ideal Gas Law. Instrumentation that detects the amount of methane in a sample of air will typically measure the volume mixing ratio (ppmv), but manufacturers often report these results as "concentrations" and use the less-precise unit of parts per million (ppm). Henceforth, we will keep to peer-reviewed standards and use the terms volume mixing ratio, concentration, and units of ppmv, as described above.

The quantification of the volume mixing ratio of methane in a sample of air can be achieved using a range of instruments and detectors, which vary in price, accuracy, precision, and portability. Higher cost instruments tend to be more accurate and precise but tend to have larger power requirements and require stable environmental conditions. Lower cost instruments can be deployed in most environments, but data can require a great deal of quality assurance/quality control (QA/QC), and data are only useable in a limited set of conditions. The following section describes the instrumentation currently available to quantify the methane volumetric mixing ratio in air. In each case, the basic principles of operation are described as well as typical instrumentation specifications (lowest detectable methane volume mixing ratio, precision, accuracy, drift, cost, and practical considerations).

3.2.4 Methane Absorption Spectra

Historically, measuring the abundance of methane in an air sample has relied on either the thermal conductivity (TC) or the combustive properties of the gas mixture (as described in the following section). This has allowed for the quantification of methane mixing ratios from ~0.1 to 100%, but methods are not sensitive enough to quantify parts per million levels of methane typically observed 10s of meters downwind of relatively large sources (kilograms of methane per hour). To address this, many new quantification methods have been developed that use the absorption properties of methane gas in the infrared range to quantify mixing ratio in a sample of air. Such instrumentation measures the methane gas' absorption of radiation at wavelengths of 1.7, 2.3, 3.3, and 7.6 μm. These instruments are able to quantify methane mixing ratios to parts per billion.

3.3 Micrometeorology

Several quantification approaches use measurements made downwind to infer the emission rate of a source. Typically, they rely on some understanding of how the gas disperses, both horizontally and vertically, into the ambient flow of air. Horizontal forces are caused by the rate at which air particles move parallel to the surface, i.e., the horizontal wind speed. Vertical forces are caused by the difference in temperature between the near-surface atmosphere, where solar heating can cause the ground to be hotter than the atmosphere, resulting in vertical movement of air, i.e., convective turbulence, and aerodynamic obstructions that physically block the movement of air and force particles to move over the obstruction, i.e., mechanical turbulence. Plume-based methods rely on a statistical representativeness of the plume in given atmospheric conditions and present the plume as an

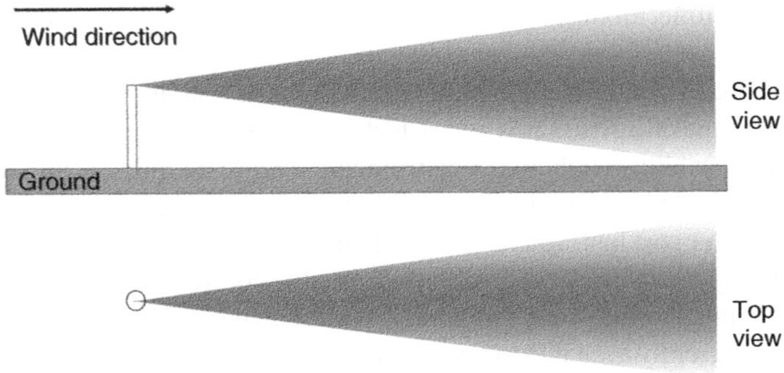

Figure 3.1 Time-averaged three-dimensional cone of a plume of gas forming from the vertical and horizontal dispersal of gas.

idealistic, time-averaged three-dimensional cone with a higher concentration of gas toward the center (Figure 3.1).

3.3.1 Atmospheric Stability

The size and shape of the cone vary depending on the atmospheric conditions (Seinfeld and Pandis, 2016). In neutral conditions (i.e., winds greater than 3 m/s), horizontal forces are greater than vertical forces, and the cone is narrow (Figure 3.2a). As wind speed decreases, the shape of the plume depends on the temperature difference between the ground and the atmosphere. During periods of high solar insolation, i.e., during sunny days, the ground is at a higher temperature than the near-surface air, and air near the surface is heated, becomes more buoyant, and travels upward. These are unstable atmospheric conditions and result in tall plumes, and the gas disperses quickly into the atmosphere (Figure 3.2b). During periods of low solar insolation or at night, the ground is typically at a lower temperature than the air, and upward movement of air is very small. Any gas emitted is trapped close to the ground (Figure 3.2c). The atmospheric stability can either be (1) estimated using the wind speed and solar insolation, (2) calculated from wind speeds at different heights, or (3) directly measured.

The Monin–Obukhov length (L, m) is the quantitative way of describing the atmospheric stability and is a measure of the height above the surface at which buoyancy forces dominate the turbulent forces (Seinfeld and Pandis, 2016). The Monin–Obukhov length is calculated from the density of air $(\rho, \mathrm{kg/m^3})$, the specific heat capacity of air at constant pressure $(\widehat{c_p}, \mathrm{J/kg/K})$, the air temperature at z = 0 (T_0, K), the von Karman constant (k, 0.4), the acceleration due to gravity $(g, \mathrm{m/s^2})$ and the sensible heat flux which measures the heat transfer by convection $(H, \mathrm{J/m^2/s})$ (Equation 3.1). The sensible heat flux $(H, \mathrm{W/m^2})$ can be

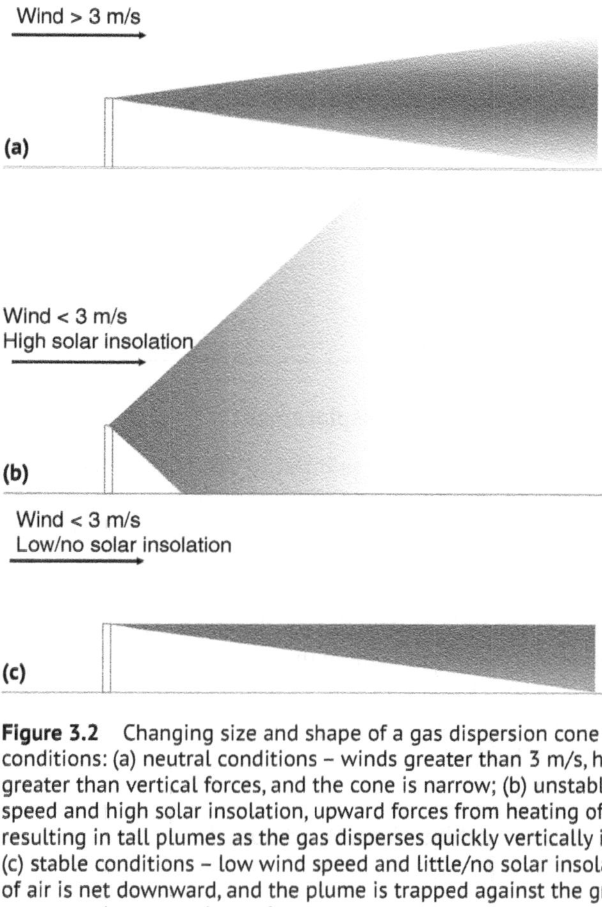

Figure 3.2 Changing size and shape of a gas dispersion cone in varying atmospheric conditions: (a) neutral conditions – winds greater than 3 m/s, horizontal forces are greater than vertical forces, and the cone is narrow; (b) unstable conditions – low wind speed and high solar insolation, upward forces from heating of the ground are higher, resulting in tall plumes as the gas disperses quickly vertically into the atmosphere; and (c) stable conditions – low wind speed and little/no solar insolation, vertical movement of air is net downward, and the plume is trapped against the ground, resulting in high concentrations near the surface.

calculated from the density of air, specific heat capacity of air, the air temperature (T_a, K), absolute temperature of air at $z = 0$ (T_0, K), wind speed (u, m/s) and the transfer coefficient for sensible heat flux (CH, 1×10^{-3}).

$$L = -\frac{\rho \widehat{c_p} T_0 u_*^3}{kgH} \tag{3.1}$$

The Monin–Obukhov length can also be converted to the more qualitative scale, the Pasquill–Gifford Stability Class (PGSC). The PGSCs run alphabetically from A (very unstable) through D (neutral) to G (very stable) (Table 3.2). While the PGSCs are less definite, they are often used to classify the atmospheric stability without having to take a measurement.

Table 3.2 Range of Monin–Obukhov values (L) for qualitative atmospheric stability conditions.

Values of L (m)	Stability conditions	Pasquill–Gifford stability class
$-100 < L < 0$	Very unstable	A
$-10^5 < L < -100$	Unstable	B/C
$L < -10^5$ or $L > 10^5$	Neutral	D
$100 < L < 10^5$	Stable	E/F
$0 < L < 100$	Very stable	G

3.3.2 Roughness Length and Zero-plane Displacement

Two other aerodynamic properties often used as input to dispersion models are the roughness length (z_0, m) and the zero-plane displacement (z_d, m) (Seinfeld and Pandis, 2016). The roughness length is an aerodynamic property of the surface and derived from either wind profile data or can be taken from published values of similar surfaces (Table 3.3). As a rule of thumb, z_0 can be taken as one-tenth of the height of the aerodynamic obstruction in the fetch. The zero-plane displacement is the height above the ground where the momentum of the air due to mechanical turbulence is zero. This is typically measured using a sonic anemometer, but as a rule of thumb z_d could be taken as two-thirds the height of the aerodynamic obstruction in the fetch.

3.3.3 Estimating Atmospheric Stability from Wind Speed and Solar Insolation

A lookup table can be used to estimate the PGSC from wind speed and irradiance (Table 3.4). In creating this lookup table, strong insolation is defined as sunny midday in midsummer in England, and slight insolation as similar conditions in midwinter (Pasquill and Smith, 1983). Quantitatively, strong sunlight is irradiance greater than 1 kW/m^2, moderate sunlight is irradiance between 0.5 and 1 kW/m^2 and slight sunlight is irradiance less than 0.5 kW/m^2.

3.3.4 Calculating Aerodynamic Properties from Wind Speeds at Different Heights

Roughness height (z_0, m), friction velocity (u^*, m/s), and Monin–Obukhov length (L, m) can be calculated from a wind profile, i.e., wind speeds measured at different heights (Figure 3.3a) (Seinfeld and Pandis, 2016). The roughness length

Table 3.3 Description of common aerodynamic obstructions and associated roughness lengths.

Roughness length (m)	Description of aerodynamic obstructions
0.0002	Open water or tidal flat
0.005	Featureless land or ice
0.03	Flat terrain (grass or very low vegetation)
0.1	Low crops
0.25	Open landscape
0.5	Landscape with bushes or small trees
1	Mature forest or low-rise built-up area
Over 2	City center or large forest with clearings

Table 3.4 Lookup table to estimate the Pasquill–Gifford stability class from wind speed and irradiance.

	Pasquill–Gifford stability class				
Wind speed (m/s)	Strong sunlight	Moderate sunlight	Slight sunlight	Night or overcast	Night clear
2	A	A	B		
3	B	B	C	E	F
4	B	C	C	D	E
5	C	C	D	D	D
6	C	D	D	D	D

is calculated as the exponential of the intercept, with the natural logs of wind measurement heights plotted against wind speeds (Figure 3.3b). The friction velocity, calculated as the gradient of the natural logs of wind measurement heights plotted against wind speeds multiplied by the von Karman constant ($k = 0.41$), can then be used to calculate the Monin–Obukhov length (Equation 3.1).

3.3.5 Directly Measuring Atmospheric Stability

Micrometeorological data, used in atmospheric dispersion approaches, can also be measured directly using a sonic anemometer. The sonic anemometer uses ultrasonic transducers and receivers to measure the wind velocity in three orthogonal

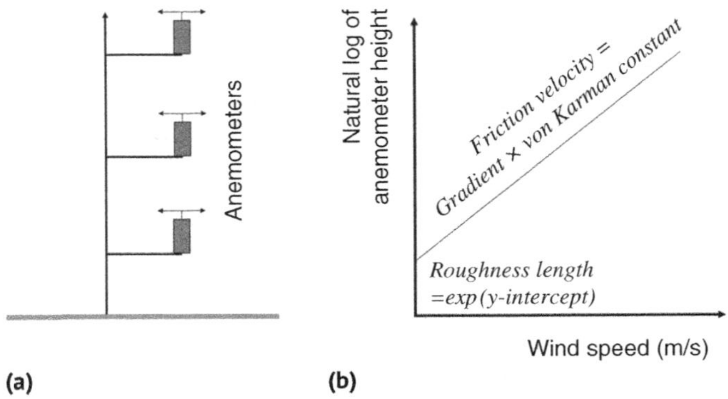

Figure 3.3 (a) Aerodynamic properties of a surface calculated from a measured wind profile, and (b) the roughness height (z_0, m), friction velocity (u^*, m/s), and Monin–Obukhov length (L, m) from the natural logs of wind measurement heights plotted against wind speeds.

directions (u, v, and w vectors). The wind speed in a particular direction is measured using the time it takes for a sound wave to pass between the transducer pair. The friction velocity is directly calculated from the orthogonal wind speeds and the Monin–Obukhov length (L, m) calculated from u^* (Equation 3.1).

3.4 Methane Emission Quantification Methodologies

Methane concentrations in air can be measured using a variety of methods and technologies (this will be described in Section 3.5); however, the concentration tells you very little about the size of the emission. Emission rates can only be calculated using the measured concentration and other parameters in a model/algorithm, depending on the method used. Section 3.4.1 describes some methods commonly used to quantify methane emissions from oil and gas infrastructure.

3.4.1 Chamber Methods

Chamber methods are perhaps the most simplistic for quantifying the emission rate of a source. As the name suggests, the source is enclosed within a measurement chamber and the emission calculated. The main shortcoming of a chamber method is that the emission source needs to be completely enclosed within the chamber; therefore, this makes measuring a large source, such as a compressor,

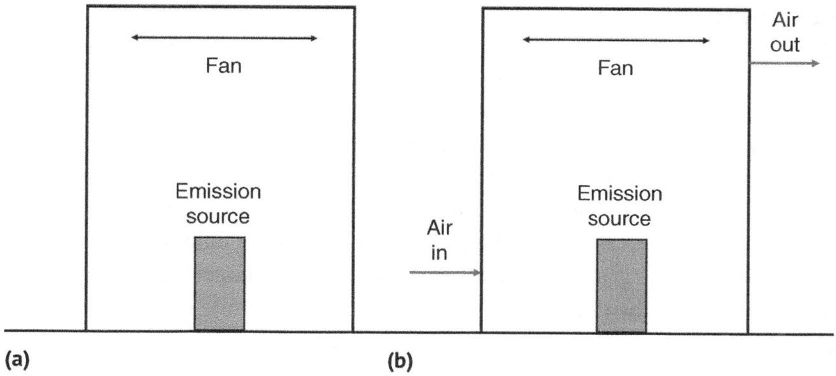

Figure 3.4 Schematics of the (a) static and (b) dynamic chamber methods for emission quantification.

very difficult. The main advantage of using a chamber to quantify the emission is that it is relatively inexpensive, as methane accumulates to high concentrations (percent methane) and low-cost sensors can be used. Two chamber methods are described here: the static chamber and dynamic chamber (Figure 3.4). A third method, the Hi-Flow sampler, is described in Section 3.3.2, as it uses a bag instead of a chamber, but this approach works in much the same way as the dynamic chamber.

3.4.1.1 Static Chambers

A static chamber comprises a rigid chamber sealing the source within a fixed volume of air (V, m^3) (Figure 3.4a). As the source emits, the methane concentration increases over time, and the emission rate (Q, g/s) can be calculated from the rate of change of methane concentration (C, g /m^3) within the chamber over time (t, s). In this calculation (Equation 3.2), the volumetric mixing ratio must be converted to concentration (Collier et al., 2014; Pihlatie et al., 2013).

$$Q = \frac{dC}{dt}V \tag{3.2}$$

Technology used to measure the rate of change of concentration within the chamber can be inexpensive, e.g., integrated infrared (INIR) sensors; however, care should be taken to avoid heated or combustion devices, as the enclosed gas can become explosive. Another shortcoming of this method is that natural gas at production sites is unprocessed and can include aromatic hydrocarbons and possibly BTEX compounds, which puts the operator at risk when the chamber is removed (Riddick et al., 2022b).

3.4.1.2 Dynamic Chambers

Unlike static chambers, dynamic chambers rely on dilution to keep the gas collecting to explosive concentrations. Dynamic chambers comprise a container enclosing the air around an emission source and a flow of air being passed through the chamber (Figure 3.4b). If the emission rate is constant, the gas emitted from the source will become well mixed inside the container, and a steady-state concentration will be reached over time (Aneja et al., 2006). The emission rate of the source (Q, g/s) is calculated (Equation 3.3) from the methane concentration at steady state $\left(C_{\text{eq}},\ \text{g/m}^3\right)$, the background methane concentration in the air being pumped into the chamber $\left(C_b,\ \text{g/m}^3\right)$, and the flow rate of air through the chamber $\left(q,\ \text{m}^3/\text{s}\right)$.

$$Q = \left(C_{\text{eq}} - C_b\right)q \tag{3.3}$$

The methane concentration is typically calculated from a volumetric mixing ratio measured by a methane detector with a range of thousands of ppm to % methane. If the lower or upper measurement limit of the sensor is not reached or exceeded, the flow rate of the pump can be decreased/increased to change the steady-state concentration. As such, dynamic chambers are relatively inexpensive to run (INIR sensors cost ~$500) and avoid the health and safety risks of the static chamber, as concentrations within the chamber can be controlled. Dynamic chambers are typically used to quantify emissions from relatively small sources, ~10 g/h, using relatively small pump rates, ~10 l/min, and have been used extensively to quantify emissions from abandoned oil and gas wells (Riddick et al., 2022b, 2023).

3.4.2 Component-level Measurements

Some quantification methods measure emissions from individual sources at the component level and sum over the whole site to generate facility-wide emissions estimates. The advantage of this approach is that relatively inexpensive methods can be used to detect, estimate, and quantify emissions (as described in the following text), and training required to make measurements is minimal. The shortcomings of component-level measurements are that they are very time-consuming, and sources may be missed if they are hard to reach or the emissions are intermittent. These difficulties infer that component-level emission estimates are typically biased low and likely an underestimate of the actual emissions.

3.4.2.1 Bubble Method

The first component-level measurement approach has been added here to highlight how simple quantification methods can be. The "bubble" method is almost mythical, as few operators can remember the specifics of the approach but vaguely recall its use in leak detection and quantification, specifically related to estimating emissions from natural gas pipelines. The method follows that a soap solution, comprising a specific concentration of soap, is applied to the suspected emission source, and the emission rate is determined from the rate and size of the bubbles formed. As the formation of bubbles is affected by atmospheric pressure and wind velocity, the method (bubble formation) was calibrated using controlled gas release of natural gas by pipeline operators. Unfortunately, the calibration and soap solution concentration data are proprietary information and were never released to the public. However, the "bubble" method remains a viable, inexpensive method of estimating emission from a point source and could be used to make component-level emissions inventories.

3.4.2.2 EPA Method 21

The U.S. EPA's "Method 21 – Determination of volatile organic compound leaks" is an approach designed to determine if a piece of process equipment is leaking volatile organic compounds (VOCs) and is not designed to generate methane-centric emissions estimates (EPA, 2017). Process equipment includes the following on production equipment: valves, flanges and other connections, pumps and compressors, pressure relief devices, process drains, open-ended valves, pump and compressor seal system degassing vents, accumulator vessel vents, agitator seals, and access door seals. Method 21 does not specifically recommend a VOC detector for use in Method 21, but the instrument must be an intrinsically safe, pump-driven, non-methane-specific VOC monitoring instrument with a scale readable to $\pm 2.5\%$. As such, Method 21 is simply a leak detection protocol and not a method to directly measure the mass emission rate from individual sources. However, as Method 21 is industry standard in the United States, it would be remiss not to include a description here.

3.4.2.3 Hi-Flow

Hi-Flow samplers are commercially available instruments that work in the same way as the dynamic chamber and output an emission rate. Hi-Flow samplers comprise a bag attached to a methane sensor via tubing (Vaughn et al., 2022) carrying a flow of air from the emission point at a high flow rate. The bag is placed over the emission source to ensure gas is not lost, and air is drawn at a relatively high flow rate, ~250 l/min, passed over the methane sensor, and the emission rate is calculated by the instrument using Equation 3.3. As the flow of air is large, Hi-Flow samplers are typically used to quantify larger emissions, between 50 g CH_4/h and 9 kg CH_4/h to an accuracy of $\pm 10\%$, and typically used in leak detection and quantification surveys on oil and production sites (Riddick et al., 2023). Typically, Hi-Flow samplers are used in conjunction with optical gas imaging (OGI) cameras or U.S. EPA's Method 21 to quantify the emission rate when the leak has been found.

3.4.3 Gaussian Plume Approach

A Gaussian Plume approach can be used to generate an idealized, time-averaged representation of the gas concentration at a point in x, y, z space downwind of a point source emission event (Seinfeld and Pandis, 2016). Here, idealized is used to describe an emission event where the emission rate, wind speed, wind direction, and atmospheric stability are constant and there are no aerodynamic obstructions between the emission source and the detector. For most emission sources, it is unlikely that all of these variables are constant; however, this can be overcome by averaging each variable over time to minimize the effects of turbulence over very small time scales but preserving any variation in concentration, which is caused by environmental or atmospheric change (Hensen and Scharff, 2001).

The concentration of methane $(X, \mathrm{g/m^3})$ at a distance downwind of the source (x, m), the lateral distance of the detector from the center line of the plume (y, m), and at a height above ground level (z, m) can be calculated. To calculate the concentration, the emission rate of the gas $(Q, \mathrm{g/s})$, the wind speed $(u, \mathrm{m/s})$ the height of the source (h_s, m), the height of the boundary layer (h, m), and the PGSC must also be known (Equation 3.4). The PGSC is not used directly in the calculation, instead it is used to determine the standard deviations of the lateral (σ_y, m) and vertical (σ_z, m) mixing ratio distributions (Pasquill, 1962) using a lookup table published through the U.S. EPA (US EPA, 1995). The assumptions are that (1) the source is emitting CH_4 at a constant rate; (2) the mass of methane is conserved when reflected at the surface of the ground or the top of the boundary layer; (3) wind speed and vertical eddy diffusivity are constant with time; (4) there is uniform vertical mixing; (5) terrain is relatively flat between source and detector; (6) the distance between the source and detector is less than 100 m; and (7) the emission is a point source.

$$X(x,y,z) = \frac{Q}{2\pi u \sigma_y \sigma_z} e^{\frac{-y^2}{2\sigma_y^2}} \left(e^{\frac{-(z-h_s)^2}{2\sigma_z^2}} + e^{\frac{-(z+h_s)^2}{2\sigma_z^2}} + e^{\frac{-(z-2h+h_s)^2}{2\sigma_z^2}} + e^{\frac{-(z+2h-h_s)^2}{2\sigma_z^2}} + e^{\frac{-(z-2h-h_s)^2}{2\sigma_z^2}} \right) \tag{3.4}$$

Typically, the Gaussian Plume approach is used to infer an emission rate using a downwind concentration. This approach has been used extensively when quantifying emissions from oil and gas infrastructure, and it is the most used approach in leak detection and quantification (LDAQ) technology (Caulton et al., 2018; Riddick et al., 2019). The major strengths of the Gaussian Plume approach are

(1) the equation is relatively straightforward (i.e., easy to populate), (2) the equation always gives an answer even when an assumption (listed 1–7) may have been violated, and (3) the emissions can be calculated in real-time with a relatively low-technology approach. However, this approach can result in the calculated downwind concentration either being overestimated, e.g., if the source emission is intermittent, or underestimated, e.g., if the source is near an aerodynamic obstacle that causes gas to pool around it. Another shortcoming of the Gaussian Plume approach is that the EPA's lookup table for the standard deviations of the lateral mixing ratio distribution only includes values for source-to-detector distances greater than 100 m (Riddick et al., 2022b). This can conceivably be overcome, and the equation can be used for distance less than 100 m if the detector is placed directly downwind of the source and the *y*-component is zero (Riddick et al., 2023). Despite these limitations, the Gaussian Plume approach is very versatile and has been used to estimate emissions from a variety of sources, including landfills. Although a landfill is not a point source, it can be considered a point source when observed from a long distance or as a collection of individual point sources within a defined area.

3.4.4 Lagrangian Stochastic

While the Gaussian Plume approach relies on pre-calculated algorithms to represent air movement in a given atmospheric condition, a Lagrangian stochastic (LS) approach models the random movement (or walk) of an air parcel from a source as it is affected by horizontal and vertical forces acting on it (Flesch et al., 1995). Typically, a Lagrangian dispersion model calculates the emission rate of a source from the downwind methane concentration and meteorological or micrometeorological data. Unlike the Gaussian approach, Lagrangian models will acknowledge when a solution cannot be calculated, i.e., if the wind is not blowing from the source to the detector, providing a level of QC on the output data.

Forward modeling, i.e., modeling the release of gas from a source and determining if it will reach a downwind detector, requires many calculated particle trajectories to be calculated to provide a statistically acceptable calculated emission (Flesch et al., 2004, 2005, 2009). As a result, forward modeling approaches are computationally inefficient and take a long time to run. Running the Lagrangian model in a backward mode, i.e., modeling air from a detector to the source, is more computationally efficient and faster. The backward Lagrangian model calculates an emission rate from a source by modeling the movement of thousands of parcels of methane gas and by generating a methane concentration $(X_{sim}, g/m^3)$ to emission rate $(Q_{sim}, g/s)$ ratio for a given set of atmospheric condition input by the user. The emission rate $(Q, g/m^2/s)$ is then calculated from this simulated ratio, the measured methane concentration (X_m, g/m^3), and the background methane concentration (X_b, g/m^3) (Equation 3.5).

$$Q = \frac{X_m - X_b}{\left(\frac{X_{\text{sim}}}{Q_{\text{sim}}}\right)} \tag{3.5}$$

The open-source software package WindTrax (www.thunderbeachscientific.com) uses a backward Lagrangian dispersion model to calculate emission rate from point source emissions typical to those found on oil and gas operations. WindTrax is versatile and only requires data on the location of the source, the methane concentration downwind of the source, and a wind speed and direction to calculate the emission rate. Used in this data-limited way, the model applies default settings to make assumptions on stability, roughness length, and atmospheric temperature. More data can be added to the model runs, including PGSC, Monin–Obukhov length, air temperature, air pressure, altitude, and roughness length. It is assumed that with greater data input, the uncertainty in emission quantification decreases; however, to what extent is currently unclear (Denmead, 2008).

Ideal methane concentration sampling periods are between 5 and 120 minutes to adequately preserve real changes to CH_4 emission caused by changing environmental or atmospheric factors. The key limitations of WindTrax are that it assumes: (1) the emission rate of the source is constant over time; (2) atmospheric conditions are constant over time; (3) wind direction remains constant; (4) the fetch between the source and detector has a roughness length of less than 15 cm (similar to a wheat field); and (5) the distance between the source and detector is less than 1 km (Flesch et al., 2005, 2009; Laubach et al., 2008; Sommer et al., 2005).

3.4.5 Tracer Flux

Unlike the dispersion approaches (Gaussian and Lagrangian), the tracer flux method uses the controlled release of a second "tracer" gas (Q_t, g/m^2/s) to infer the emission rate of a source (Lamb et al., 1995; Vaughn et al., 2017). Concentrations of methane (X_m, g/m^3) and the tracer gas (X_t, g/m^3) are measured downwind of a source, and the methane emission rate (Q, g/m^2/s) is calculated (Equation 3.6) (Lamb et al., 1995). More specifically, at least one tracer gas is released at the same location as the methane source, and a transect perpendicular to the wind direction is traversed and the gas concentrations measured. The unknown methane emission rate is calculated from the ratio of the methane and tracer integrated downwind plumes. To improve the accuracy of the quantified emission rate and to ensure the correct plume is being measured, two tracer gases can be used with the gas releases bookending the methane emission source (Subramanian et al., 2015).

$$Q = Q_t \cdot \frac{\int_{y_{\min}(p)}^{y_{\max}(p)} \left[X_{\text{plume}}(p) - X_{\text{bgnd}}(p)\right] \, dy_p}{\int_{y_{\min}(t)}^{y_{\max}(t)} \left[X_{\text{plume}}(t) - X_{\text{bgnd}}(t)\right] \, dy_t} \tag{3.6}$$

The main advantage of the tracer flux method is that it does not require any measurements of the atmospheric stability or account for dispersion of gas over aerodynamic obstacles; the method assumes that any effects of dispersion will be similar between the methane and tracer gas and result in similar dispersion patterns downwind. Therefore, the principal assumptions of the tracer flux approach are that (1) the tracer and methane have the same dispersion properties and (2) the source does not emit any of the tracer gas (Lamb et al., 1995). As such, nitrous oxide and acetylene are often used as tracers in quantifying methane emissions from oil and gas operations. The uncertainty of quantification using the tracer flux method varies with wind conditions, but controlled release experiments have shown it to be in the region of $\pm 25\%$ (Allen, 2016; Fredenslund et al., 2019; Lamb et al., 1995).

3.4.6 Mass Balance Approaches

Similar to the tracer flux method, the mass balance does not need to consider micrometeorology. This makes both methods easier to parameterize than the atmospheric dispersion approaches (Cambaliza et al., 2014; Heimburger et al., 2017). The mass balance approach uses the difference in the methane concentrations integrated over the plume width (y, m) and height (z, m) upwind (X_b, g/m^3) and downwind (X_e, g/m^3) concentrations from a source to infer the emission rate (Q, g/s; Equation 3.7) (Denmead, 2008). Specifically used in Equation 3.7 are the wind speed at heights (u_z, m/s) between the ground and above the plume and the downwind methane concentrations measured in two-dimensional plane ($X_{e,y,z}$, kg/m^3) perpendicular to the wind direction at a constant distance downwind. In most cases the upwind, background concentration ($X_{b,y,z}$, kg/m^3) will be constant with height and laterally (Conley et al., 2017).

$$Q = \int_0^z \int_{-y}^{y} u_z \left(X_{e,y,z} - X_{b,y,z}\right) \, dy \, dz \tag{3.7}$$

Practically, a mass balance measurement can either be undertaken using a point sensor to measure the concentration laterally across the plume or by using line-averaged sensors (open-path TDLAS) to measure the integrated concentration

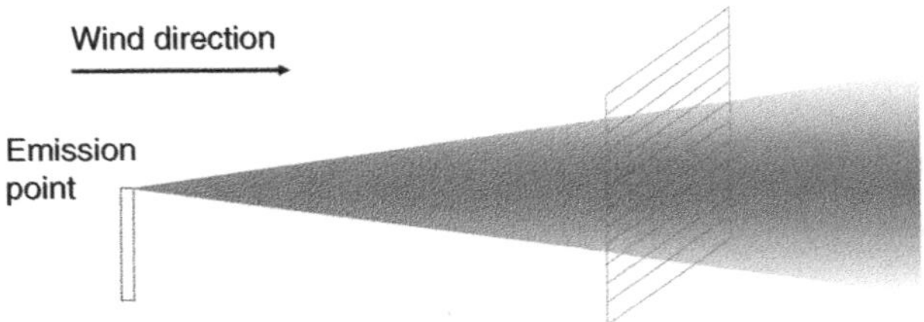

Figure 3.5 Schematic of, a mass balance measurement, which can either be done by using a point sensor to measure the concentration laterally across the plume or by using line-averaged sensors (open-path TDLAS) to measure the integrated concentration across the plume.

across the plume (Figure 3.5) (Conley et al., 2017; Denmead, 2008). In both cases, the transects of the plume must be measured from the ground to above the plume, as should the wind speed, which will increase logarithmically with height and is affected by the roughness length of obstacles in the fetch.

As mentioned above, the key advantage of the mass balance approach is that fewer assumptions are made about the atmospheric stability, which results in lower uncertainty in the calculated emission rate. However, in many practical cases the wind speed is approximated to a single value at a single height, which could result in significant uncertainty, especially when calculating emissions from oil and gas infrastructure (Schwietzke et al., 2019). One approach to overcome aerodynamic obstructions is to measure far enough downwind to allow the plume to reform. With the development of low-weight trace gas methane analyzers, the mass balance approach is being used to quantify whole-site emissions from oil and gas facilities. It also has the flexibility to be used to target individual pieces of equipment by isolating the emission point and making near-source measurements.

3.4.7 Plume Imaging: Aircraft and Satellite Methods

Aircraft- and satellite-based observations use a common quantification approach, albeit at different scales and with different detection limits. At this time, aircraft-based observations (fixed-wing) can quantify emissions down to 5 kg CH_4/h in favorable conditions (Kunkel et al., 2023), while satellite-based platforms have a quantification limit of ~100 kg CH_4/h (Cooper et al., 2022). These are changing with developments in technology and retrieval algorithms. Typically, both systems can use infrared spectrometers to detect methane plumes from a source. The spectrometer data are then used to calculate the source emission rate from the measured column-averaged methane mixing ratio and the wind speed. The algorithms used to quantify emission are typically proprietary to solution providers.

Aircraft observations are often used with context cameras to quantify emissions from individual pieces of equipment on oil and gas production facilities, while satellite images have coarser spatial resolution but can be used to quantify total facility emissions (Duren et al., 2019).

Even though limits of detection are higher than the other methods, satellite quantification of methane emissions is attractive to oil and gas operators as it is relatively low effort, with hardware and analysis conducted by the satellite operator. At present, there are two main satellite types used for this application: earth observation satellites that scan broad areas (e.g., Sentinel-2 and Landsat 8) and targeted systems that can be routed to specific sites (e.g., GHGSat). Currently, the limit of detection for wide-area satellites is estimated at 1,400 kg CH_4/h $\pm$ 50%, while targeted systems are estimated at 200 kg CH_4/h $\pm$ 13% (Sherwin et al., 2023).

3.5 Measuring Methane Concentrations

As noted above, methane emission rates cannot be directly quantified and instead are calculated from the measured methane concentration in air. All quantification methods presented in Section 3.4 require the measurement of a methane concentration (or methane mixing ratio). To obtain these data, several sensors/instruments have been developed, all of which have detection limits associated with the technology used. In the following section, an outline of how nine methane sensing platforms work is discussed along with key references. The nine methane sensors are catalytic combustion (CC), TC, OGI, INIR, metal oxide (MOX), TDLAS, LiDAR, cavity ring down spectrometers (CRDS), and Airborne Visible/Infrared Spectrometer (AVIRIS). This list is not exhaustive but presents the technologies in most commonly used to quantify emissions from oil and gas infrastructure at the present time. It should also be noted that instrument development is a fast-evolving space, and specifications of instrumentation deployed at oil and gas facilities are constantly evolving.

3.5.1 CC Sensors

CC, or pellistor bead, sensors are used to detect the presence of combustible gases in air and can be used to quantify the abundance of methane. These devices were first developed by Alan Baker in the 1960s to sense combustible gas in coal mines to improve safety (Baker, 1963). Combustion occurs when a combustible gas mixed with air comes into contact with a bead coated in a catalytic surface, which is heated to ~500 °C (Jones et al., 1987). Heat given off increases the temperature of the catalytic surface, as a function of the amount of combustible gas, and reduces its resistance accordingly (Figure 3.6). Used in a Wheatstone bridge

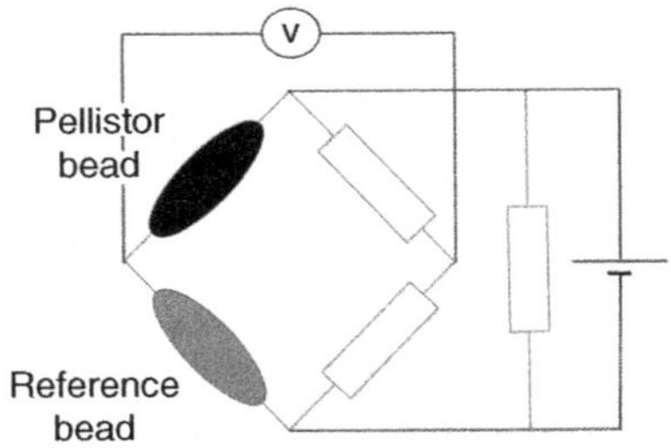

Figure 3.6 Schematic of a Wheatstone bridge circuit using resistance change in the pellistor bead to infer the abundance of a combustible gas.

circuit, the resistance of the catalyst can be taken as a proxy for the mixing ratio of gas once the system has been calibrated (Moseley, 1997). Note that the CC sensor cannot distinguish between methane and other combustible gases. CC sensors can detect near-background concentrations (~10 ppm) and have an upper detection threshold of ~5% methane (50,000 ppmv) as they require enough oxygen for the gas to become flammable (Jones et al., 1987). Given the detection range of these instruments, they are typically used for quantifying emissions using chamber-type methods and were used to measure lower emission rates, down to 57 g CH_4/h, in the Bacharach Hi-Flow sampler (Bacharach Inc., 2015).

3.5.2 TC Sensors

TC sensors comprise two beads, which are heated to 500 °C; one bead is exposed to the gas/air mixture (detector bead), while the second bead is sealed in a chamber containing clean air (reference bead) (Daynes, 1933; Gardner et al., 2023). When the detector bead is exposed to an air/gas mixture with a TC significantly different to ambient air, the heat lost from the bead will increase and result in a change in resistance. The resistance change is determined through a comparison to the reference bead (Szulczyński and Gębicki, 2017). The mixing ratio of the gas can then be inferred from the difference between detector and reference bead resistances. TC sensors to measure methane mixing ratios above the lower explosive limit (5–100% CH_4) (Gardner et al., 2023). TC sensors were collocated with CC sensors in the Bacharach Hi-Fow sampler and used to measure higher emission rates, i.e., measurements were automatically switched to the TC sensor when the mixing ratio detected by the CC sensor in the Bacharach Hi-Flow exceeded 50,000 ppm (Bacharach Inc., 2015).

3.5.3 Integrated Infrared Sensors

INIR sensors typically measure the absorption of infrared radiation (usually 3.3 μm) as it passes through a sample of air to infer the volumetric mixing ratio of methane in a sample of air (Vaughn et al., 2022). Often sensors will use a second, non-methane wavelength of infrared radiation, as a reference to compensate for signal loss through temperature and humidity (Figure 3.7). The lower detection limit for methane is 100 ppm with a resolution of 10 ppm and time taken for the instrument to record 90% of full-scale gas concentration (T_{90}) of 20 seconds (Jayarathne et al., 2022). As many hydrocarbons absorb wavelengths of 3.3 μm, this is not a methane-specific sensor, but these sensors can measure from 100 ppm to 100% methane and are intrinsically safe and have been used to replace the dual TC and CC sensors in new iterations of Hi-Flow samplers (Vaughn et al., 2022).

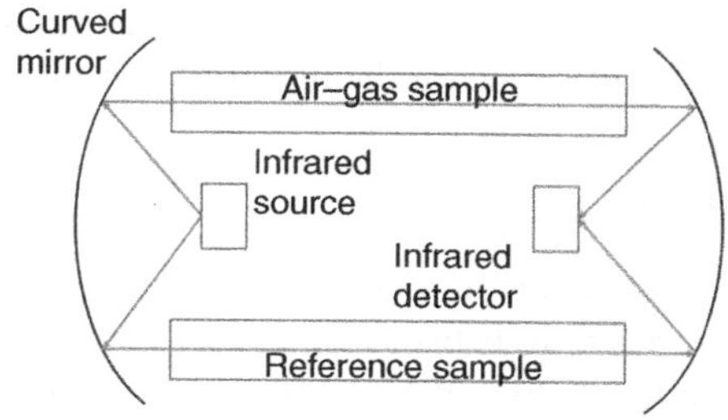

Figure 3.7 Schematic of an integrated infrared (INIR) sensor.

3.5.4 MOX Sensors

MOX sensors comprise a heated semiconductor-based MOX sensing strip that changes resistance when exposed to gas (Kohl, 1990). Tin oxide strips doped with noble metals, e.g., platinum or palladium, are often used to detect methane (Haridas and Gupta, 2013; Xue et al., 2019). As the concentration of methane in air increases, more methane is taken up, decreasing the resistance of the MOX strip. Therefore, if the voltage is measured across the MOX, the amount of methane in the air can be inferred by the change in resistance (Kohl, 1990).

MOX sensors have shown that they can be calibrated and report representative methane mixing ratios between background (2 ppm) and 1,000 ppm (Eugster and Kling, 2012; Riddick et al., 2020; Shah et al., 2023); however, they do have some shortcomings (Furuta et al., 2022). Firstly, MOX sensors are highly sensitive to environmental conditions and must be corrected for relative humidity and temperature (Eugster and Kling, 2012; Figaro, 2005, 2020; Shah et al., 2023). Some MOX sensors work poorly in relative humidity less than 35% (Eugster and Kling, 2012; Shah et al., 2023), but this can be corrected for using machine learning algorithms (Andrews et al., 2023). Secondly, each MOX sensor's response is slightly different, which means that every sensor's output must be individually calibrated (Collier-Oxandale et al., 2018; Riddick et al., 2022a). Instrument calibration usually requires the instrument's response to be tested against a gas of known concentration; however, as MOX sensors do not work in dry gas (as typical of calibration gases supplied by a gas cylinder), alternative methods must be used to calibrate the sensor, such as co-location with high-precision instruments (Shah et al., 2023). Thirdly, even though MOX sensors are relatively easy to deploy, the heater required for operation makes the power needs of the sensor package too high to be supplied by a battery for long-term measurement (>200 mW) (Riddick et al., 2022a). The option of short-duration measurements is not an option either, as it takes time for the heater to warm up.

3.5.5 OGI Cameras

OGI cameras are video cameras that use a cooled detector that filters for radiation of wavelengths between 3.2 and 3.4 μm, and can therefore be used to observe methane plumes (TELEDYNE FLIR, 2024). As mentioned above, methane and other hydrocarbons absorb radiation of these wavelengths. When the difference between the IR radiation absorbed by a gas plume and the IR emitted by the ambient surroundings is high enough, the plume can be seen on the video camera's screen (Burr et al., 2006; Burr and Hengartner, 2006; Gålfalk et al., 2016). The definition of the image in the screen is determined by the specifications of the OGI camera, the environmental conditions, the difference between the temperature of the gas plume and ambient (i.e., the greater the difference, the better), and the experience level of the camera operator (Ravikumar et al., 2017; Zimmerle et al., 2020).

3.5.6 Tunable Diode Laser Absorption Spectrometers

Tunable diode laser absorption spectrometers (TDLAS) typically comprise a laser tuned to the absorption wavelength of methane, a cavity in which the air sample is contained, and a detector used to quantify the amount of radiation absorbed by the methane gas in the sample (Figure 3.8) (Schiff et al., 1994). Following the Beer–Lambert Law, the methane concentration in a sample of air

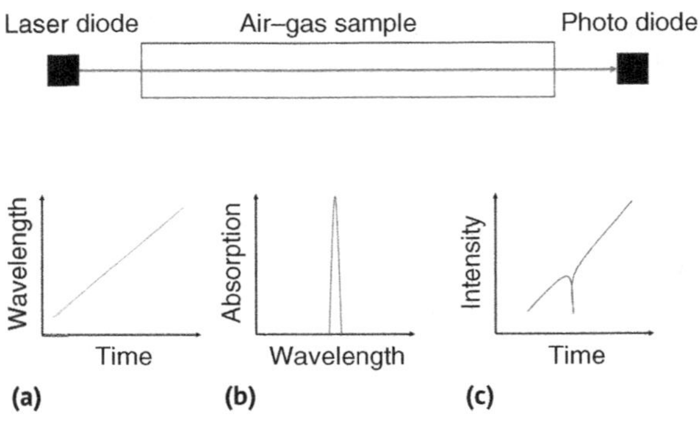

Figure 3.8 Schematic of a closed-path tunable diode laser absorption spectrometer (TDLAS). TDLAS typically comprises a laser emitting radiation over a range of wavelengths, a cavity in which the air sample is contained, and a detector used to quantify the amount of radiation absorbed by the methane gas in the sample. (a) Laser diode wavelength scan, (b) absorption by the methane, and (c) laser intensity detected.

can be calculated from the change in the laser's intensity as it passes through the sample and a measure of how much radiation methane absorbs at a given temperature and pressure. Therefore, the volumetric mixing ratio of methane integrated over the laser length in the cavity can be calculated from the cell temperature, cell pressure, and the length of the cavity. The Beer–Lambert Law assumes a constant laser intensity (i.e., with no fluctuation in intensity over the tunable wavelengths), and laser fluctuations can result in a reduction in measurement precision. Detection limits and precision can be increased by increasing the path length of the laser. This is typically achieved by using multipath cells where a mirror reflects the laser multiple times along the cavity (Allen, 1998). TDLAS systems can either be closed path, where the air-gas sample is pumped into a measurement cavity of fixed length (Schiff et al., 1994) (Figure 3.8), or open path, where the laser beam is directed onto a reflective surface a distance away (maximum ~50 m) and the methane between the laser and screen is quantified (Iseki et al., 2000; Well et al., 2005).

The strengths of TDLAS instruments are that they have a relatively low minimum detection limit (as low as 1 ppb), a high signal-to-noise ratio, can measure in near-real time, and gases can be targeted and therefore assayed with a high degree of specificity (Hodgkinson and Tatam, 2013). The shortcomings of the TDLAS are that only one gas can be observed by an instrument at a time, the absorption within the cavity can be affected by dust and/or particulates, and lasers require significant power to maintain operation temperatures.

3.5.7 Laser Imaging, Detection, and Ranging

Laser imaging, detection, and ranging (LiDAR) uses much the same technology as an open-path TDLAS (Iseki et al., 2000), but is often used while the laser is pointed downward from an aircraft (Hunter and Thorpe, 2017). Coupled with downward-pointing optical camera (to provide visual context) and a topography-detecting camera (to provide pathlength and reflectivity data), methane emissions can be observed by aircraft-mounted LiDAR travelling at 200 m above the ground (Johnson et al., 2021). The major advantage of LiDAR-based measurements is that they can observe many sites during a relatively short period of time; however, LiDAR currently has an emission detection threshold in the range of kilograms of methane per hour depending on the environmental conditions during measurement (Kunkel et al., 2023).

3.5.8 Cavity Ring Down Spectrometers

CRDS use the change in laser intensity as it is reflected within a cavity to quantify the methane within a cell (O'Keefe and Deacon, 1988). The intensity of the laser

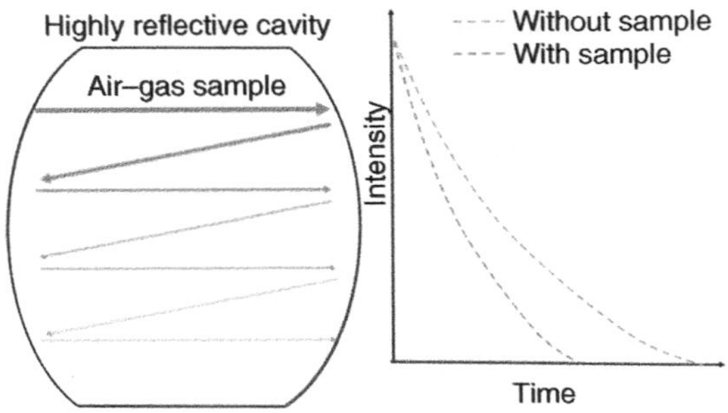

Figure 3.9 The change in intensity of a laser pulse within the highly reflective cavity of a cavity ring down spectrometer (CRDS).

pulse decreases as it passes multiple times through the methane-air sample, and the concentration of the methane in the air sample can be determined as a measure of the rate of change of laser intensity over time (Figure 3.9) (Berden and Engeln, 2009). CRDS instruments are capable of making sub-ppb methane mixing ratio measurements.

CRDS instruments are more precise than TDLAS instruments, as they are not affected by laser intensity fluctuations and are more sensitive, as the path length of the laser is much longer (due to the pulse being reflected multiple times within the cavity) (Hodgkinson and Tatam, 2013). However, the time taken for a single measurement with CRDS is much longer than the TDLAS, and the instrument cost is higher due to the requirements of the highly reflective cavity.

3.5.9 Airborne Visible/Infrared Spectrometers

Infrared imaging spectrometers, such as the AVIRIS used by NASA, comprise downward-looking cameras tuned to measure light intensity in 224 contiguous spectral channels (bands) between wavelengths of 0.38 and 2.51 μm (Thompson et al., 2015). The AVIRIS can be mounted on either an aircraft or a satellite, and the methane enhancement can be retrieved from the AVIRIS data (Cusworth et al., 2019; Thorpe et al., 2014, 2017). As quantification relies on reflected solar radiance, measurements can only take place when there is enough light (observations are better toward midday), when there is low cloud cover, and over a surface that will reflect infrared radiation, and AVIRIS needs to adapt their methods for methane retrieval over water as compared to land, as water absorbs infrared wavelengths (Ayasse et al., 2019, 2022).

3.6 Full System Solutions

Sections 3.4 and 3.5 present instrumentation and quantification methods, respectively. For each quantification method, the methane mixing ratio must be converted to an emission rate using an algorithm or modeling approach. From the specifications it is clear that not all instruments can be used with each quantification method, and the two must be paired accordingly. The methane mixing ratios measured using chamber methods over oil and gas infrastructure (abandoned oil and gas wells or emissions from subsurface leaks) are typically in the range of thousands of ppm and can quickly exceed percent methane. This means that CC, TC, INIR, molecular property spectrometer, and MOX sensors can be used. Optical devices inside chamber quickly become saturated, and upper detection limits are exceeded. As with chamber methods, equipment-based quantification is conducted very close to the emission source, and sensors with upper detection thresholds less than 1,000 ppm are of little use.

Sensors required for any of the plume-based measurements (Gaussian Plume or LS) must be able to detect low ppm mixing ratios. This means that MOX, TDLAS, and CRDS can be used to measure methane mixing ratios to parameterize algorithms and models. With increased precision and accuracy comes less uncertainty in emission quantification. Anything that has a lower detection limit greater than 10 ppmv cannot conceivably be used for quantification. The LS method, e.g., the WindTrax modeling environment, has the extra utility to it can also use line-averaged concentrations, as measured by an open-path TDLAS, for example, to quantify emissions.

As with plume dispersion methods, tracer flux requires the measurements from at least two trace gas analyzers, the target gas (typically methane) and the tracer gas (e.g., acetylene, nitrous oxide, or sulfur hexafluoride), but the instruments need to be able to respond fast enough to measure changes in the plume while travelling at speed. It is not currently clear if a MOX sensor could respond fast enough to measure the mixing ratios as the sensor passes through the plume while mounted on a vehicle. Similarly, mass-balance measurements typically measure near-background mixing ratios while moving at speed, and it is unlikely that MOX sensors can respond fast enough to these changes.

3.7 Challenges and Needs for Future Work

3.7.1 Quantifying Fugitives

The development of systems to detect and quantify fugitive methane has been undertaken by a number of commercial companies. This is a fast-evolving space with the development and refining of sensors coupled to analytics that can be used to quantify the emissions in a relatively turbulent aerodynamic landscape. Current solutions are improving on localization and have been shown to quantify emissions to a factor of two during controlled release experiments (Bell et al., 2023).

3.7.1.1 Why Fugitives Are So Difficult to Detect and Quantify – The Simplest Emission Scenario

Figure 3.10 shows a typical configuration of the well pad, consisting of wellheads (width: 0.1 m × depth: 2 m × height: 2 m), separators (1 × 4 × 2), condensate/water tanks (4 × 4 × 4), combustors (1 × 1 × 4) and compressors (5 × 10 × 5). If we take, for example, the simplest emission scenario, e.g., a single fugitive emission with no confounding vented emissions. Due to the size of the average emission (Table 3.5), taken from the MEET model and measurement (Allen, 2016; Allen et al., 2022; GRI and EPA, 1996), only certain methods could be used to quantify the emissions. Table 3.5 provides a summary of the suitability of methods to quantify an "average" methane emission from fugitives, and these methods and their challenges are discussed.

3.7.1.2 Chamber Methods

Chamber methods, including the Hi-Flow, can quantify emissions from most sources on a production site once the chamber/bag has been placed over the emission point (Table 3.5). The major shortcoming of chamber methods is that they take so long to physically move around the 1,200 possible emission points.

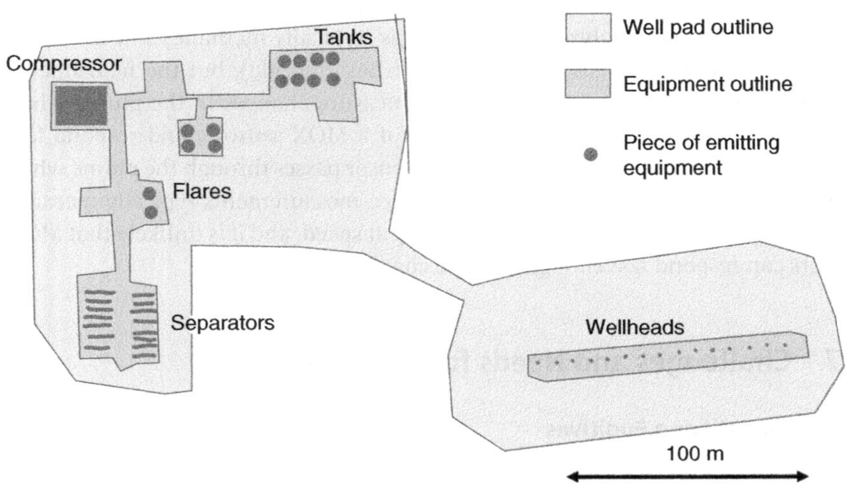

Figure 3.10 Typical configuration of an oil and gas production site in the Denver-Julesburg Basin, CO.

Table 3.5 Suitability of methods to quantify an "average" methane emission from fugitive types. Q refers to the average emission rate in kg CH_4/h. Methods are as follows: C – Chamber; HF – Hi-Flow; GP – Gaussian Plume; LS – Lagrangian stochastic; TF – Tracer flux; MB – Mass balance; A – Aircraft; and S – Satellite. A check mark represents emissions that lie in the method's range of quantification; a cross mark indicates emissions that lie outside the method's range of quantification; and grey fill represents emissions that are challenging for the method to physically measure. This assumes that the method used is combined with a suitable instrument for measuring the methane mixing ratio, as described in Section 3.6.

Equipment	Description	Q	C	HF	GP	Ls	TF	MB	A	S
Compressor	Blowdown vent leak	0.021	✓	×	×	×	×	×	×	×
	Component leak	0.070	✓	✓	✓	✓	×	×	×	×
	Rod packing leak	9.1	×	✓	✓	✓	✓	✓	✓	×
Flare	Flared gas malfunction	0.032	✓	×	✓	✓	×	×	×	×
	Unflared gas from flare	0.30	✓	✓	✓	✓	✓	✓	×	×
Separator	Separator component leak	0.059	✓	✓	✓	✓	×	×	×	×
Tank	Component leak	0.034	✓	×	✓	✓	×	×	×	×
	Vent leak	0.47	✓	✓	✓	✓	✓	✓	×	×
	Hatch leak	0.057	✓	✓	✓	✓	×	×	×	×
Wellhead	Component leak	0.12	✓	✓	✓	✓	✓	✓	×	×

Screening, using an OGI camera or by U.S. EPA Method 21, can speed this up, but there is a potential for missing an emission source. Additionally, measuring emissions from pieces of equipment with moving parts or hot surfaces, such as compressors or combustors, is inherently dangerous and impractical.

3.7.1.3 Downwind Methods

Hypothetically, downwind methods, such as Gaussian Plume and LS, can be used to quantify all fugitive emissions from sources on a production site (Table 3.5). The instrumentation required is portable enough to be placed downwind of an emission source, and the emission can be calculated in near real-time. The ADED program run in the "ideal" conditions at the Methane Emission Technology Evaluation Center (METEC) in Fort Collins, Colorado, highlights the difficulties that can be encountered when quantifying emissions using downwind methods. Most survey methods tested by the ADED program use continuous fence line monitors, typically MOX sensors, located ~10 m from the well pad and infer emissions using a Gaussian Plume approach with meteorological data from a rotating cup anemometer.

As the emission source is continuous and from a point source, the downwind plume *should* be relatively Gaussian, and emission *could* be inferred using the Gaussian equation (Equation 3.4). However, given the results of the ADED campaign, it is clear that some of the assumptions made by Gaussian approach are being violated. The obvious issue with the Gaussian approach is that the air is being disturbed as it moves over oil and gas infrastructure, principally condensate and water tanks. To overcome this, methane concentrations can be measured farther downwind; however, the increased distance means that measured concentrations are much lower and require more expensive instrumentation (e.g., TDLAS or CRDS). This means that even the simplest "ideal" scenario cannot be resolved by downwind methods.

3.7.1.4 Other Methods

Average emission rates of most fugitive emission sources are too small to be detected by either aircraft or satellite, and many are undetectable to either Tracer Flux or Mass Balance (Table 3.5). However, this may not be a problem, as these approaches usually attempt to identify only larger emissions, and most methods (excluding satellites) can detect emissions of kilograms of methane per hour. Therefore, as none of these methods have the micrometeorological limitations of the downwind methods, routine surveys using these methods should be able to detect, locate, and quantify an "ideal" fugitive source.

3.7.1.5 Why Fugitives Are So Difficult to Detect and Quantify – the Real World

"Non-ideal" emission sources in the real world are more difficult to quantify than "ideal" sources. Most emissions from oil and gas production facilities are not fugitive; vented and combustion emissions (as described in Section 3.1) contribute more to the total emission. The key difficulty here being – can the fugitive emissions be detected from the background noise of the other legitimate emission sources? For very large fugitives, e.g., tens of kilograms per hour, it is likely that most methods (apart from satellites) can detect these emissions from most pieces of equipment (e.g., wellheads, separators, and tanks). However, detecting large fugitive emissions from compressors, i.e., when rod packing is damaged, could be confiscated by the combustion emissions from the driver exhaust. Even at site level with access, it is difficult to distinguish between vented, fugitive, and combustion emissions from compressors.

Adding to the complexity of the problem, emissions (fugitive, vented, and combustion) are not constant and are only released as a result of the mechanical processes as various vessels on the site are pressurized. Take the example of an open thief hatch on a condensate tank, which has popped open as a result of too much pressure in the tank. After the initial large methane release, emissions from the hatch will be much lower as the well is brought to pressure (i.e., when the well is closed), and methane will only be released as it volatilizes from any liquid in the tank. In low temperatures, this emission rate could be very low and likely undetectable by any survey method. As the valves are released and the wellhead opened, gas, oil, and water are separated, and the condensate enters the tank. The pressure drop between the separator and tank likely results in gas to be released (typically called flash gas), which will result in hatch emissions. Even if a tank hatch is open, it is only likely that a survey method will detect the emission if the observation happens at the same time as the emission. This is likely to be detected by longer surveys, OGI, or U.S. EPA Method 21, but becomes less likely to be detected as the observation period becomes shorter (Tracer Flux – 30 minutes; Mass Balance – 5 minutes; Aircraft – 30 seconds; and Satellite – 8 seconds). The applicability of the downwind dispersion methods depends on the emission behavior, as one of the assumptions of the methods is that the emission is continuous.

3.7.2 How to Reduce Fugitives

Considering multiple emission sources and variability in emissions makes detecting fugitive emission a non-trivial problem, aside from detecting the really big emissions (100s of kilograms of methane per hour), which would likely be detected by satellite. Developments in machine learning could be leveraged to train observation platforms to distinguish between vented emissions, with relatively known emission time signals, and from fugitives, which are typically always on or as a result of vessel pressurization.

Another approach being taken is the reduction in vented emissions. Many operators are changing from using traditional gas-driven pneumatics to air-driven ones or those powered by electricity. Vented gas emissions from compressors, e.g., compressor blowdowns or rod packing, could be captured and routed to the export lines, or gas-driven compressors could be replaced with those that are electrically driven. There is also a case for the centralization of all processing activities, where mixed phase produced is piped from the production site to a central site, thus removing all separators and tanks on-site.

Regulatory change has also resulted in reductions in basin-wide emissions from the Denver-Julesburg Basin in Colorado. In 2008 and 2012, estimated methane loss normalized against production at 4% (Pétron et al., 2012, 2014); this reduced to 2.1% in 2015 (Peischl et al., 2018) and 0.3% in 2020 (Riddick et al., 2022c) despite a 10-fold increase in production. Between 2008 and 2020, the Colorado Department of Public Health and Environment (CDPHE) implemented regulations to reduce methane emissions, which included 2013, adopting LDAR requirements; 2014, establishing regulatory requirements for oil and gas methane; 2017–2023, changes to the Air Quality Control Commission Rulemaking, Regulation 7, that introduced requirements for compressors, pneumatics, equipment leaks, and fugitive emissions; and 2018, flowline testing and guidelines for venting of storage tanks.

3.7.3 Future of Methane Measurements

In summary, technology and approaches exist that can detect, quantify, and result in reduction of emissions from oil and gas production. However, the greatest limitation to the efficacy of fugitive emission reduction is the comprehensive understanding of the emission sources and being able to distinguish fugitives from non-fugitives. Currently, no technological approach has been able to resolve this problem, other than the time-consuming measurements of individual components. In the future, it may be possible to use a bottom-up approach, such as the MAES model, to generate reasonable uncertainty bounds of emission and compare these to measurements made by the top-down approaches described above. An LDAR team could be alerted if the top-down exceeds the bottom-up emissions by a statistically significant amount. This approach would also benefit from measurements made by multiple technologies to reduce the uncertainty bounds of the top-down estimates.

Future methane detection and quantification platforms include both area flux and point source mapping satellites. Area flux mapping platforms observe the total region emissions with platform-dependent pixel sizes ranging from 0.1 to 10 km, while point source imagers quantify emissions on a much smaller scale, less than 60 m (Jacob et al., 2022). All area mappers report column dry methane mixing ratio, and data are open source (Jacob et al., 2022); these include MethaneSAT, launched by EDF in March 2024, which has a pixel size of 130 × 400 m (Rohrschneider et al., 2021); GOSAT-GW by JAXA/MOE/MIES, planned to launch in 2024, with a pixel size of between 1 × 1 km and 10 × 10 km (NIES, 2024); and Sentinel-5, launched in 2020 by the European Space Agency, with a pixel size of 7.5 × 7.5 km (Fletcher, 2020). Point source platforms include GHGSat and Carbon Mapper. By 2023, GHGSat Inc. has released 12 satellites with pixel sizes down to 25 × 25 m, and images can be purchased (GHGSat, 2024; Jervis et al., 2021), while Carbon Mapper, launching in 2024, has a pixel size of 30 × 30 m, and data products are open source (Carbon Mapper, 2024; Duren, 2021). Satellites have the advantage over other localization and quantification methods, as they can survey large areas very quickly. This is a rapidly developing branch of science, and efforts are being directed toward controlled releases to validate the emission estimates made by satellites (Dowd et al., 2024; Nesser et al., 2024; Rouet-Leduc and Hulbert, 2024; Sherwin et al., 2023, 2024; Vollrath et al., 2024).

References

Allen, D.T. (2016). Emissions from oil and gas operations in the United States and their air quality implications. *Journal of the Air & Waste Management Association* 66: 549–575. https://doi.org/10.1080/10962247.2016.1171263.

Allen, D.T., Cardoso-Saldaña, F.J., Kimura, Y. et al. (2022). A methane emission estimation tool (MEET) for predictions of emissions from upstream oil and gas well sites with fine scale temporal and spatial resolution: model structure and applications. *Science of The Total Environment* 829: 154277. https://doi.org/10.1016/j.scitotenv.2022.154277.

Allen, M.G. (1998). Diode laser absorption sensors for gas-dynamic and combustion flows. *Measurement Science and Technology* 9: 545–562. https://doi.org/10.1088/0957-0233/9/4/001.

American Petroleum Institute (2004). *Compendium of Greenhouse Gas Emissions Estimation Methodologies for the Oil and Gas Industry*. Washington, DC: API.

Andrews, B., Chakrabarti, A., Dauphin, M. et al. (2023). Application of machine learning for calibrating gas sensors for methane emissions monitoring. *Sensors* 23: 9898. https://doi.org/10.3390/s23249898.

Aneja, V.P., Blunden, J., Claiborn, C.S. et al. (2006). Dynamic chamber system to measure gaseous compounds emissions and atmospheric-biospheric interactions. In: *Environmental Simulation Chambers: Application to Atmospheric Chemical Processes, Nato Science Series: IV: Earth and Environmental Science* (ed. I. Barnes and K.J. Rudzinski), 97–109. Dordrecht: Kluwer Academic Publishers. https://doi.org/10.1007/1-4020-4232-9_7.

Ayasse, A.K., Dennison, P.E., Foote, M. et al. (2019). Methane mapping with future satellite imaging spectrometers. *Remote Sensing* 11: 3054. https://doi.org/10.3390/rs11243054.

Ayasse, A.K., Thorpe, A.K., Cusworth, D.H. et al. (2022). Methane remote sensing and emission quantification of offshore shallow water oil and gas platforms in the Gulf of Mexico. *Environmental Research Letters* 17: 084039. https://doi.org/10.1088/1748-9326/ac8566.

Bacharach Inc. (2015). HI FLOW® Sampler for natural gas leak rate measurement: instruction 0055-9017 operation and maintenance. https://www.mybacharach.com/wp-content/uploads/2015/08/0055-9017-Rev-7.pdf (accessed 24 May 2024).

Baker, A. (1963). Apparatus for detecting combustible gases having an electrically conductive member enveloped in a refractory material. U.S. Patent 3,092,799. 3.

Barkley, Z., Davis, K., Miles, N. et al. (2023). Quantification of oil and gas methane emissions in the Delaware and Marcellus basins using a network of continuous tower-based measurements. *Atmospheric Chemistry and Physics* 23: 6127–6144. https://doi.org/10.5194/acp-23-6127-2023.

Bell, C., Ilonze, C., Duggan, A. et al. (2023). Performance of continuous emission monitoring solutions under a single-blind controlled testing protocol. *Environmental Science & Technology* 57: 5794–5805. https://doi.org/10.1021/acs.est.2c09235.

Berden, G. and Engeln, R. (eds.) (2009). *Cavity Ring-Down Spectroscopy: Techniques and Applications*, 1e. Wiley. https://doi.org/10.1002/9781444308259.

Burr, T. and Hengartner, N. (2006). Overview of physical models and statistical approaches for weak gaseous plume detection using passive infrared hyperspectral imagery. *Sensors* 6: 1721–1750. https://doi.org/10.3390/s6121721.

Burr, T., Foy, B., Fry, H. et al. (2006). Characterizing clutter in the context of detecting weak gaseous plumes in hyperspectral imagery. *Sensors* 6: 1587–1615. https://doi.org/10.3390/s6111587.

Cambaliza, M.O.L., Shepson, P.B., Caulton, D.R. et al. (2014). Assessment of uncertainties of an aircraft-based mass balance approach for quantifying urban greenhouse gas emissions. *Atmospheric Chemistry and Physics* 14: 9029–9050. https://doi.org/10.5194/acp-14-9029-2014.

Campbell, L.M., Campbell, M.V., and Epperson, D.L. (1996). Methane Emissions from the Natural Gas Industry, Volume 2: Technical Report, Final Report, GRI-94/0257.1 and EPA-600/R-96-080b. Gas Research Institute and U.S. Environmental Protection Agency.

Carbon Mapper (2024). Carbon mapping. https://carbonmapper.org/

Caulton, D.R., Li, Q., Bou-Zeid, E. et al. (2018). Quantifying uncertainties from mobile-laboratory-derived emissions of well pads using inverse Gaussian methods. *Atmospheric Chemistry and Physics* 18: 15145–15168. https://doi.org/10.5194/acp-18-15145-2018.

Collier, S.M., Ruark, M.D., Oates, L.G. et al. (2014). Measurement of greenhouse gas flux from agricultural soils using static chambers. *Journal of Visualized Experiments* 90: 52110. https://doi.org/10.3791/52110.

Collier-Oxandale, A., Casey, J.G., Piedrahita, R. et al. (2018). Assessing a low-cost methane sensor quantification system for use in complex rural and urban environments. *Atmospheric Measurement Techniques* 11: 3569–3594. https://doi.org/10.5194/amt-11-3569-2018.

Conley, S., Faloona, I., Mehrotra, S. et al. (2017). Application of Gauss's theorem to quantify localized surface emissions from airborne measurements of wind and trace gases. *Atmospheric Measurement Techniques* 10: 3345–3358. https://doi.org/10.5194/amt-10-3345-2017.

Cooper, J., Dubey, L., and Hawkes, A. (2022). Methane detection and quantification in the upstream oil and gas sector: the role of satellites in emissions detection, reconciling and reporting. *Environmental Science Atmospheres* 2: 9–23. https://doi.org/10.1039/D1EA00046B.

Cusworth, D.H., Jacob, D.J., Varon, D.J. et al. (2019). Potential of next-generation imaging spectrometers to detect and quantify methane point sources from space. *Atmospheric Measurement Techniques* 12: 5655–5668. https://doi.org/10.5194/amt-12-5655-2019.

Daynes, H.A. (1933). *Gas Analysis by Measurement of Thermal Conductivity.* Cambridge, England: University Press.

Denmead, O.T. (2008). Approaches to measuring fluxes of methane and nitrous oxide between landscapes and the atmosphere. *Plant Soil* 309: 5–24. https://doi.org/10.1007/s11104-008-9599-z.

Dowd, E., Manning, A.J., Orth-Lashley, B. et al. (2024). First validation of high-resolution satellite-derived methane emissions from an active gas leak in the UK. *Atmospheric Measurement Techniques* 17: 1599–1615. https://doi.org/10.5194/amt-17-1599-2024.

Duren, R.M. (2021). Carbon Mapper: on-orbit performance predictions and airborne prototyping, American Geophysical Union Fall Meeting. New Orleans, 13–17 December 2021, A53F-05. https: //agu.confex.com/agu/fm21/meetingapp.cgi/Paper/985888 (accessed 23 July 2022).

Duren, R.M., Thorpe, A.K., Foster, K.T. et al. (2019). California's methane super-emitters. *Nature* 575: 180–184. https://doi.org/10.1038/s41586-019-1720-3.

EPA (2017). Environmental Protection Agency - Method 21 - determination of volatile organic compound leaks. https://www.epa.gov/sites/default/files/2017-08/documents/method_21.pdf (accessed 24 May 2024).

Eugster, W. and Kling, G.W. (2012). Performance of a low-cost methane sensor for ambient concentration measurements in preliminary studies. *Atmospheric Measurement Techniques* 5: 1925–1934. https://doi.org/10.5194/amt-5-1925-2012.

Ewing, B., Roesler, D., and Evans, S. (2010). Performance Test of a Steam-Assisted Elevated Flare with Passive FTIR - Detroit. Final Report; Marathon Petroleum Company, LP & Clean Air Engineering: Detroit, MI, USA.

Figaro (2005). TGS 2600 – for the detection of air contaminants, On- line product data sheet. https://www.figarosensor.com/product/docs/TGS2600B00%20(0913).pdf (accessed 24 May 2024).

Figaro (2020). Production information. TGS 2611 - for the detection of Methane. https://www.figarosensor.com/product/docs/TGS%202611C00(1013).pdf (accessed 26 August 2021).

Flesch, T., Wilson, J., Harper, L. et al. (2005). Estimating gas emissions from a farm with an inverse-dispersion technique. *Atmospheric Environment* 39: 4863–4874. https://doi.org/10.1016/j.atmosenv.2005.04.032.

Flesch, T.K., Harper, L.A., Powell, J.M. et al. (2009). Inverse-dispersion calculation of ammonia emissions from Wisconsin dairy farms. *Transactions of the American*

Society of Agricultural and Biological Engineers 52: 253–265. https://doi.org/10.13031/2013.25946.

Flesch, T.K., Wilson, J.D., and Yee, E. (1995). Backward-time Lagrangian stochastic dispersion models and their application to estimate gaseous emissions. *Journal of Applied Meteorology and Climatology* 34: 1320–1332. https://doi.org/10.1175/1520-0450(1995)034<1320:BTLSDM>2.0.CO;2.

Flesch, T.K., Wilson, J.D., Harper, L.A. et al. (2004). Deducing ground-to-air emissions from observed trace gas concentrations: a field trial. *Journal of Applied Meteorology and Climatology* 43: 487–502. https://doi.org/10.1175/1520-0450(2004)043<0487:DGEFOT>2.0.CO;2.

Fletcher, K. (2020). *Sentinel-5 ESA's Polar-Orbiting Atmospheric Composition Mission in Support of Copernicus Services*. Noordwijk, the Netherland: ESA Communications.

Fredenslund, A.M., Rees-White, T.C., Beaven, R.P. et al. (2019). Validation and error assessment of the mobile tracer gas dispersion method for measurement of fugitive emissions from area sources. *Waste Management* 83: 68–78. https://doi.org/10.1016/j.wasman.2018.10.036.

Furuta, D., Sayahi, T., Li, J. et al. (2022). Characterization of inexpensive metal oxide sensor performance for trace methane detection. *Atmospheric Measurement Techniques* 15: 5117–5128. https://doi.org/10.5194/amt-15-5117-2022.

Gålfalk, M., Olofsson, G., Crill, P. et al. (2016). Making methane visible. *Nature Climate Change* 6: 426–430. https://doi.org/10.1038/nclimate2877.

Gardner, E.L.W., Gardner, J.W., and Udrea, F. (2023). Micromachined thermal gas sensors—a review. *Sensors* 23: 681. https://doi.org/10.3390/s23020681.

GHGSat (2024). Methane emissions monitoring. https://www.ghgsat.com/en/ (accessed 24 May 2024).

GRI and EPA (1996). Harrison, M.R., Shires, T.M., Wessels, J.K., Cowgill, R.M. Methane Emissions from the Natural Gas Industry, Volumes 1–15, Final Report, GRI-94/0257 and EPA-600/R-96- 080, Gas Research Institute and US Environmental Protection Agency, June 1996.

Haridas, D. and Gupta, V. (2013). Study of collective efforts of catalytic activity and photoactivation to enhance room temperature response of SnO_2 thin film sensor for methane. *Sensors and Actuators B: Chemical* 182: 741–746. https://doi.org/10.1016/j.snb.2013.03.100.

Heimburger, A.M.F., Harvey, R.M., Shepson, P.B. et al. (2017). Assessing the optimized precision of the aircraft mass balance method for measurement of urban greenhouse gas emission rates through averaging. *Elementa: Science of the Anthropocene* 5: 26. https://doi.org/10.1525/elementa.134.

Hensen, A. and Scharff, H. (2001). Methane emission estimates from landfills obtained with dynamic plume measurements. *Water, Air and Soil Pollution: Focus* 1: 455–464. https://doi.org/10.1023/A:1013162129012.

Highwood (2024). Highwood Emissions Management. How Much Gas is That?!

Hodgkinson, J. and Tatam, R.P. (2013). Optical gas sensing: a review. *Measurement Science and Technology* 24: 012004. https://doi.org/10.1088/0957-0233/24/1/012004.

Howard, T., Ferrara, T.W., and Townsend-Small, A. (2015). Sensor transition failure in the high flow sampler: implications for methane emission inventories of natural gas infrastructure. *Journal of the Air & Waste Management Association* 65: 856–862. https://doi.org/10.1080/10962247.2015.1025925.

Hunter, D. and Thorpe, M.J. (2017). Gas mapping LiDAR aerial verification program final report. Alberta Upstream Petroleum Research Fund Project.

IPCC (2006). 2006 IPCC guidelines for national greenhouse gas inventories, Volume 5, waste [WWW Document]. https://www.ipcc-nggip.iges.or.jp/public/2006gl/vol5.html (accessed 25 October 2022).

IPCC (2014). *Climate Change 2013 - The Physical Science Basis: Working Group I Contribution to the Fifth Assessment Report of the Intergovernmental Panel on Climate Change*. Cambridge: Cambridge University Press. https://doi.org/10.1017/CBO9781107415324.

IPCC (2022). Climate change 2022: impacts, adaptation and vulnerability. In: *Contribution of Working Group II to the Sixth Assessment Report of the Intergovernmental Panel on Climate Change* (ed. H.-O. Pörtner, D.C. Roberts, M. Tignor et al.). Cambridge, UK and New York, NY, USA: Cambridge University Press.

Iseki, T., Tai, H., and Kimura, K. (2000). A portable remote methane sensor using a tunable diode laser. *Measurement Science and Technology* 11: 594–602. https://doi.org/10.1088/0957-0233/11/6/302.

Jacob, D.J., Varon, D.J., Cusworth, D.H. et al. (2022). Quantifying methane emissions from the global scale down to point sources using satellite observations of atmospheric methane. *Atmospheric Chemistry and Physics* 22: 9617–9646. https://doi.org/10.5194/acp-22-9617-2022.

Jayarathne, N.J.R.R., Smits, K.M., Riddick, S.N. et al. (2022). Understanding mid-to large underground leaks from buried pipelines as affected by soil and atmospheric conditions – field scale experimental study. https://doi.org/10.18738/T8/MRX54H.

Jervis, D., McKeever, J., Durak, B.O.A. et al. (2021). The GHGSat-D imaging spectrometer. *Atmospheric Measurement Techniques* 14: 2127–2140. https://doi.org/10.5194/amt-14-2127-2021.

Johnson, M.R. and Kostiuk, L.W. (2002). A parametric model for the efficiency of a flare in crosswind. *Proceedings of the Combustion Institute* 29: 1943–1950. https://doi.org/10.1016/S1540-7489(02)80236-X.

Johnson, M.R., Tyner, D.R., and Szekeres, A.J. (2021). Blinded evaluation of airborne methane source detection using Bridger Photonics LiDAR. *Remote Sensing of Environment* 259: 112418. https://doi.org/10.1016/j.rse.2021.112418.

Jones, T.A., Moseley, P.T., and Tofield, B.C. (1987). *Atomic Energy Research Establishment* (Harwell, E. (UKAEA)). The Chemistry of Solid State Gas Sensors.

Kohl, D. (1990). The role of noble metals in the chemistry of solid-state gas sensors. *Sensors and Actuators B: Chemical* 1: 158–165. https://doi.org/10.1016/0925-4005(90)80193-4.

Kunkel, W.M., Carre-Burritt, A.E., Aivazian, G.S. et al. (2023). Extension of methane emission rate distribution for Permian basin oil and gas production infrastructure by aerial LiDAR. *Environmental Science & Technology* 57: 12234–12241. https://doi.org/10.1021/acs.est.3c00229.

Lamb, B.K., McManus, J.B., Shorter, J.H. et al. (1995). Development of atmospheric tracer methods to measure methane emissions from natural gas facilities and urban areas. *Environmental Science & Technology* 29: 1468–1479. https://doi.org/10.1021/es00006a007.

Laubach, J., Kelliher, F.M., Knight, T.W. et al. (2008). Methane emissions from beef cattle - a comparison of paddock- and animal-scale measurements. *Australian Journal of Experimental Agriculture* 48: 132. https://doi.org/10.1071/EA07256.

McDaniel, M. (1983). *Flare Efficiency Study*. Washington, D.C.: U.S. Environmental Protection Agency. EPA/600/2-83/052 (NTIS PB83261644).

Moseley, P.T. (1997). Solid state gas sensors. *Measurement Science and Technology* 8: 223–237. https://doi.org/10.1088/0957-0233/8/3/003.

Nesser, H., Jacob, D.J., Maasakkers, J.D. et al. (2024). High-resolution US methane emissions inferred from an inversion of 2019 TROPOMI satellite data: contributions from individual states, urban areas, and landfills. *Atmospheric Chemistry and Physics* 24: 5069–5091. https://doi.org/10.5194/acp-24-5069-2024.

NIES (2024). Global observing satellite for greenhouse gases and water cycle. GOSAT-GW mission: background, aims, and mission requirements. [WWW Document]. https://gosat-gw.nies.go.jp/ (accessed 23 August 2023).

O'Keefe, A. and Deacon, D.A.G. (1988). Cavity ring-down optical spectrometer for absorption measurements using pulsed laser sources. *Review of Scientific Instruments* 59: 2544–2551. https://doi.org/10.1063/1.1139895.

Pasquill, F. (1962). Atmospheric diffusion. By F. Pasquill. London (Van Nostrand Co.), 1962. Pp. xii, 297; 60s. *Quarterly Journal of the Royal Meteorological Society* 88: 202–203. https://doi.org/10.1002/qj.49708837622.

Pasquill, F. and Smith, F.B. (1983). *Atmospheric Diffusion*, 3e. Chichester: Ellis Horwood, (John Wiley & Sons).

Peischl, J., Eilerman, S.J., Neuman, J.A. et al. (2018). Quantifying methane and ethane emissions to the atmosphere from central and western U.S. oil and natural gas production regions. *Journal of Geophysical Research: Atmospheres* 123: 7725–7740. https://doi.org/10.1029/2018JD028622.

Pétron, G., Frost, G., Miller, B.R. et al. (2012). Hydrocarbon emissions characterization in the Colorado Front Range: a pilot study: Colorado Front Range emissions study. *Journal of Geophysical Research: Atmospheres* 117: n/a-n/a. https://doi.org/10.1029/2011JD016360.

Pétron, G., Karion, A., Sweeney, C. et al. (2014). A new look at methane and nonmethane hydrocarbon emissions from oil and natural gas operations in the Colorado Denver-Julesburg Basin. *Journal of Geophysical Research: Atmospheres* 119: 6836–6852. https://doi.org/10.1002/2013JD021272.

Pihlatie, M.K., Christiansen, J.R., Aaltonen, H. et al. (2013). Comparison of static chambers to measure CH_4 emissions from soils. *Agricultural and Forest Meteorology* 171–172: 124–136. https://doi.org/10.1016/j.agrformet.2012.11.008.

Plant, G., Kort, E.A., Brandt, A.R. et al. (2022). Inefficient and unlit natural gas flares both emit large quantities of methane. *Science* 377: 1566–1571. https://doi.org/10.1126/science.abq0385.

Ravikumar, A.P., Wang, J., and Brandt, A.R. (2017). Are optical gas imaging technologies effective for methane leak detection? *Environmental Science & Technology* 51: 718–724. https://doi.org/10.1021/acs.est.6b03906.

Riddick, S.N. and Mauzerall, D.L. (2023). Likely substantial underestimation of reported methane emissions from United Kingdom upstream oil and gas activities. *Energy & Environmental Science* 16: 295–304. https://doi.org/10.1039/D2EE03072A.

Riddick, S.N., Ancona, R., Cheptonui, F. et al. (2022a). A cautionary report of calculating methane emissions using low-cost fence-line sensors. *Elementa: Science of the Anthropocene* 10: 00021. https://doi.org/10.1525/elementa.2022.00021.

Riddick, S.N., Ancona, R., Mbua, M. et al. (2022b). A quantitative comparison of methods used to measure smaller methane emissions typically observed from superannuated oil and gas infrastructure. *Atmospheric Measurement Techniques* 15: 6285–6296. https://doi.org/10.5194/amt-15-6285-2022.

Riddick, S.N., Cheptonui, F., Yuan, K. et al. (2022c). Estimating regional methane emission factors from energy and agricultural sector sources using a portable measurement system: case study of the Denver–Julesburg basin. *Sensors* 22: 7410. https://doi.org/10.3390/s22197410.

Riddick, S.N., Mauzerall, D.L., Celia, M. et al. (2019). Methane emissions from oil and gas platforms in the North Sea. *Atmospheric Chemistry and Physics* 19: 9787–9796. https://doi.org/10.5194/acp-19-9787-2019.

Riddick, S.N., Mauzerall, D.L., Celia, M. et al. (2020). The calibration and deployment of a low-cost methane sensor. *Atmospheric Environment* 230: 117440. https://doi.org/10.1016/j.atmosenv.2020.117440.

Riddick, S.N., Mbua, M., Riddick, J.C. et al. (2023). Uncertainty quantification of methods used to measure methane emissions of 1 g CH_4 h^{-1}. *Sensors* 23: 9246. https://doi.org/10.3390/s23229246.

Rohrschneider, R.R., Wofsy, S.C., Franklin, J.E. et al. (2021). The MethaneSAT mission. *35th Annual Small Satellite Conference Proceedings*, virtual, 7–12 August 2021, SSC21-II-05, 1–7.

Rouet-Leduc, B. and Hulbert, C. (2024). Automatic detection of methane emissions in multispectral satellite imagery using a vision transformer. *Nature Communications* 15: 3801. https://doi.org/10.1038/s41467-024-47754-y.

Schiff, H.I., Mackay, G.I., and Bechara, J. (1994). The use of tunable diode laser absorption spectroscopy for atmospheric measurements. *Research on Chemical Intermediates* 20: 525–556. https://doi.org/10.1163/156856794X00441.

Schwietzke, S., Harrison, M., Lauderdale, T. et al. (2019). Aerially guided leak detection and repair: a pilot field study for evaluating the potential of methane emission detection and cost-effectiveness. *Journal of the Air & Waste Management Association* 69: 71–88. https://doi.org/10.1080/10962247.2018.1515123.

Seinfeld, J.H. and Pandis, S.N. (eds.) (2016). *Atmospheric Chemistry and Physics: From Air Pollution to Climate Change*, 3e. Hoboken, New Jersey: John Wiley & Sons, Inc.

Shah, A., Laurent, O., and Lienhardt, L. (2023). Characterising the methane gas and environmental response of the Figaro Taguchi Gas Sensor (TGS) 2611-E00. *Atmospheric Measurement Techniques* 16: 3391–3419. https://doi.org/10.5194/amt-16-3391-2023.

Sherwin, E.D., El Abbadi, S.H., Burdeau, P.M. et al. (2024). Single-blind test of nine methane-sensing satellite systems from three continents. *Atmospheric Measurement Techniques* 17: 765–782. https://doi.org/10.5194/amt-17-765-2024.

Sherwin, E.D., Rutherford, J.S., Chen, Y. et al. (2023). Single-blind validation of space-based point-source detection and quantification of onshore methane emissions. *Scientific Reports* 13: 3836. https://doi.org/10.1038/s41598-023-30761-2.

Sommer, S.G., McGinn, S.M., and Flesch, T.K. (2005). Simple use of the backwards Lagrangian stochastic dispersion technique for measuring ammonia emission from small field-plots. *European Journal of Agronomy* 23: 1–7. https://doi.org/10.1016/j.eja.2004.09.001.

Speight, J. (2020). Processing gas from tight formations. In: *Shale Oil and Gas Production Processes*, 451–518. Elsevier. https://doi.org/10.1016/B978-0-12-813315-6.00008-7.

Subramanian, R., Williams, L.L., Vaughn, T.L. et al. (2015). Methane emissions from natural gas compressor stations in the transmission and storage sector: measurements and comparisons with the EPA greenhouse gas reporting program protocol. *Environmental Science & Technology* 49: 3252–3261. https://doi.org/10.1021/es5060258.

Szulczyński, B. and Gębicki, J. (2017). Currently commercially available chemical sensors employed for detection of volatile organic compounds in outdoor and indoor air. *Environments* 4: 21.

TELEDYNE FLIR (2024). Insights from the Field: How OGI Cameras Improve Gas Leak Detection and Environmental Health. https://www.flir.com/discover/instruments/gas-detection/insights-from-the-field-how-ogi-cameras-improve-gas-leak-detection-and-environmental-health/(accessed 24 May 2024)

Thompson, D.R., Leifer, I., Bovensmann, H. et al. (2015). Real-time remote detection and measurement for airborne imaging spectroscopy: a case study with methane. *Atmospheric Measurement Techniques* 8: 4383–4397. https://doi.org/10.5194/amt-8-4383-2015.

Thorpe, A.K., Frankenberg, C., and Roberts, D.A. (2014). Retrieval techniques for airborne imaging of methane concentrations using high spatial and moderate spectral resolution: application to AVIRIS. *Atmospheric Measurement Techniques* 7: 491–506. https://doi.org/10.5194/amt-7-491-2014.

Thorpe, A.K., Frankenberg, C., Thompson, D.R. et al. (2017). Airborne DOAS retrievals of methane, carbon dioxide, and water vapor concentrations at high spatial resolution: application to AVIRIS-NG. *Atmospheric Measurement Techniques* 10: 3833–3850. https://doi.org/10.5194/amt-10-3833-2017.

US EPA (1995). *Industrial Source Complex (ISC3) Dispersion Model.* User's Guide. EPA 454/B 95 003a (vol. I) and EPA 454/B 95 003b (vol. II). Research Triangle Park, NC: U.S. Environmental Protection Agency.

US EPA (2023). United States Environmental Protection Agency. Natural Gas and Petroleum Systems in the GHG Inventory: Additional Information on the 1990-2021 GHG Inventory (published April 2023). Last Accessed 4th January 2024. [WWW Document]. https://www.epa.gov/ghgemissions/natural-gas-and-petroleum-systems-ghg-inventory-additional-information-1990-2021-ghg (accessed 23 August 2023).

Varon, D.J., Jacob, D.J., Hmiel, B. et al. (2023). Continuous weekly monitoring of methane emissions from the Permian Basin by inversion of TROPOMI satellite observations. *Atmospheric Chemistry and Physics* 23: 7503–7520. https://doi.org/10.5194/acp-23-7503-2023.

Vaughn, T.L., Bell, C.S., Yacovitch, T.I. et al. (2017). Comparing facility-level methane emission rate estimates at natural gas gathering and boosting stations. *Elementa Science of the Anthropocene* 5: 71. https://doi.org/10.1525/elementa.257.

Vaughn, T.L., Ross, C., Zimmerle, D.J. et al. (2022). Open-source high flow sampler for natural gas leak quantification [WWW Document]. California Air Resources Board. https://energy.colostate.edu/wp-content/uploads/sites/28/2022/08/FACF_High_Flow_Final_Report_ada.pdf (accessed 26 October 2022).

Vollrath, C., Hugenholtz, C.H., and Barchyn, T.E. (2024). Onshore methane emissions measurements from the oil and gas industry: a scoping review. *Environmental Research Communications* 6: 032001. https://doi.org/10.1088/2515-7620/ad3129.

Well, B.V., Murray, S., Hodgkinson, J. et al. (2005). An open-path, hand-held laser system for the detection of methane gas. *Journal of Optics A: Pure and Applied Optics* 7: S420–S424. https://doi.org/10.1088/1464-4258/7/6/025.

Xue, D., Wang, P., Zhang, Z. et al. (2019). Enhanced methane sensing property of flower-like SnO_2 doped by Pt nanoparticles: a combined experimental and first-principle study. *Sensors and Actuators B: Chemical* 296: 126710. https://doi.org/10.1016/j.snb.2019.126710.

Zavala-Araiza, D., Alvarez, R.A., Lyon, D.R. et al. (2017). Super-emitters in natural gas infrastructure are caused by abnormal process conditions. *Nature Communications* 8: 14012. https://doi.org/10.1038/ncomms14012.

Zavala-Araiza, D., Lyon, D.R., Alvarez, R.A. et al. (2015). Reconciling divergent estimates of oil and gas methane emissions. *Proceedings of the National Academy of Sciences* 201522126. https://doi.org/10.1073/pnas.1522126112.

Zimmerle, D., Vaughn, T., Bell, C. et al. (2020). Detection limits of optical gas imaging for natural gas leak detection in realistic controlled conditions. *Environmental Science & Technology* 54: 11506–11514. https://doi.org/10.1021/acs.est.0c01285.

4

Technologies to Reduce GHG Emissions from Heavy Oil Production

Dr. Bruce S. Carey

Peters & Co. Limited, Research, Calgary, Alberta, Canada

4.1 Introduction

Heavy oils are produced worldwide by various in situ recovery processes, with the process selected being based generally on the oil viscosity and the reservoir characteristics. Definitions of "heavy oil" and "bitumen" vary within the industry, but for reference herein, we will broadly define them as summarized in Table 4.1.

Heavy oils (100–10,000 mPa·s) can be produced by primary recovery means or by water injection or polymer injection to provide a more favorable mobility ratio with the resident heavy oil. Bitumens, with their higher viscosities (>10,000 mPa·s, often over 1 million mPa·s), and some high-viscosity heavy oils are produced primarily using thermal methods, in which steam is injected into the reservoir to reduce the viscosity of the oil and allow it to be produced at economic rates. All of these processes require energy, whether mechanical, chemical, or thermal,

Table 4.1 General properties defining heavy oils and bitumen.

Oil	Viscosity, mPa·s	Gravity, °API
Heavy	100–10,000	10–22
Extra-heavy	100–10,000	<10
Bitumen	>10,000	<10

Source: Adapted from Hein (2016).

Cleaner Petroleum Production and Refining Technologies, First Edition, Edited by M. R. Riazi and Harvey W. Yarranton.

to recover the hydrocarbon resource, and thus have a broad range of associated Greenhouse Gas (GHG) emissions. Thermal methods in particular have GHG emission intensities that are orders of magnitude higher than other recovery methods because they burn methane, and thus emit carbon dioxide (CO_2), to generate steam. Consequently, thermal methods, and new technologies that have been developed to improve or replace them, will be the primary focus of this chapter.

The chapter is intended to provide historical perspectives, the current status (as of year-end 2024), and the outlooks for new technologies to reduce GHG emissions during the production of heavy oil resources. There have been recent published literature reviews on this subject that span the range of development from the conceptual to the commercial. (Bayestehparvin et al., 2019; Lee and Babadagli, 2021; Harding, 2024). This chapter builds upon and extends those works by focusing primarily on more-detailed analyses of only those recovery technologies that have been field-tested. It is not intended as a compendium of all recovery technologies that have ever been proposed, but rather as a basis to appreciate the realistic impact and challenges that these technologies may have in the next decade.

Review of the literature from the past 30 years indicates that the heavy oil/bitumen industry in Alberta Canada has been a global focal point for the development and field testing of next-generation, lower-GHG technologies for heavy oil recovery. Therefore, the Alberta experience provides the basis for much of the discussion and examples in the chapter.

The chapter is organized as follows:

- Section 4.1 gives a brief overview of worldwide heavy oil and bitumen resources and production and introduces the basis for the use of thermal methods for recovering heavy oils.
- Section 4.2 provides an overview of the commercial thermal methods currently used, examining their key performance metrics, linking these metrics to environmental performance indicators, and highlighting the technical and practical challenges that are facing the industry.
- Section 4.3 presents details of the extensive range of solvent-based technologies that have the potential to significantly reduce associated GHG intensities compared to current processes. Aspects of the physical principles, examples of field tests that have been conducted, and the field test results are presented.
- Section 4.4 presents several other recovery processes and technologies associated with heavy oil production. Some of these processes have been tested in the past but have been shelved or reduced to low levels of use and are included here for completeness. Others are in current use by the industry and are helping to maintain downward pressure on GHG intensities by optimizing production performance. The section wraps up with a presentation of

evolving technologies that are longer-term, but have the potential to almost completely eliminate or sequester associated emissions.
- Section 4.5 provides a table of all of the technologies examined in this chapter and summarizes their published or estimated GHG intensity reduction levels.

The chapter concludes with a summary of key takeaways and an appendix presenting the timeline of all solvent-based field tests that have been conducted.

4.1.1 Heavy Oil Worldwide Resources and Production

Heavy oil and bitumen accumulations comprise a substantial resource worldwide. As summarized in Table 4.2, one estimate of the combined volumes of these two resources totals 134 billion barrels, with about half of the heavy oil being located in Venezuela and about half of the bitumen found in Alberta, Canada.

Recent production levels of heavy oils/bitumens are presented in Table 4.3. The Canadian rate includes approximately 240,000 m^3/d from oil sands mining in Alberta, which is discussed in a separate chapter. For reference, world oil production in 2019 was 15.1 m^3/d (BP, 2020). Viscous oils thus accounted for about 10% of this total.

4.1.2 The Need to Reduce Viscosity

Many heavy oils have sufficiently low viscosities that they can be produced at economic rates via primary production, i.e., using only the reservoir pressure to drive the oil to producing wells. This condition is generally true for the oil in the Orinoco Belt in Venezuela, in which the viscosities of the heavy oils at reservoir temperatures generally range from 2,000 to 7,000 mPa·s (Martinez, 1987).

Table 4.2 Estimated remaining technically recoverable heavy oil and bitumen reserves.

Region	Heavy oil (10^6 m^3)	% of total	Region	Bitumen (10^6 m^3)	% of total
Venezuela	42.2	55	Alberta, Canada	26.2	46
United States	12.5	16	United States	12.1	21
Middle East	12.4	16	Africa	6.8	12
Asia	4.7	6	Asia	6.8	12
Other areas	4.5	6	Other areas	5.4	9
TOTAL	**76.3**	**100**	**TOTAL**	**57.3**	**100**

Source: Adapted from Hein (2016).

Table 4.3 Heavy oil/bitumen production rates for top producers in Q2 2019.

Country	Production rates (10^3 m^3/d)
Canada	508
Mexico	144
Iraq	134
China	128
Iran	103
Venezuela	102
Brazil	84
USA	78
Colombia	69
Oman	67
TOTAL	**1,417**

Source: Adapted from Passalacqua and Strack (2020).

Some other heavy oils with higher viscosities can be produced using secondary methods, in which injected water or viscosified water (using polymers) is used to push the viscous oil to producing wells. However, most bitumen and extra heavy reservoirs have viscosities that are too high (100,000 to over 1,000,000 mPa·s at reservoir temperatures) to provide economic production rates, unless heat is used to reduce their viscosity in situ.

Heat can significantly reduce the viscosity of heavy oils, as shown in Figure 4.1. The methods that have been most used historically to heat the oil have injected steam at 200–300 °C. These methods burn natural gas (essentially methane) to generate steam and consequently are energy intensive and generate significant GHG emissions.

4.1.3 Notes on GHG Emissions

4.1.3.1 Scope 1, 2, and 3 Carbon Emissions

In considering the development of technologies to reduce GHG emissions from the recovery of heavy oils, or any other hydrocarbons, it is important to maintain a view of the broader context of GHG emissions associated with these processes. Specifically, industry and governments in the United States and Canada categorize emissions into three types or "scopes" as defined in Table 4.4. In this table, and the remainder of this chapter, the term "CO_2e" denotes the CO_2 aggregated

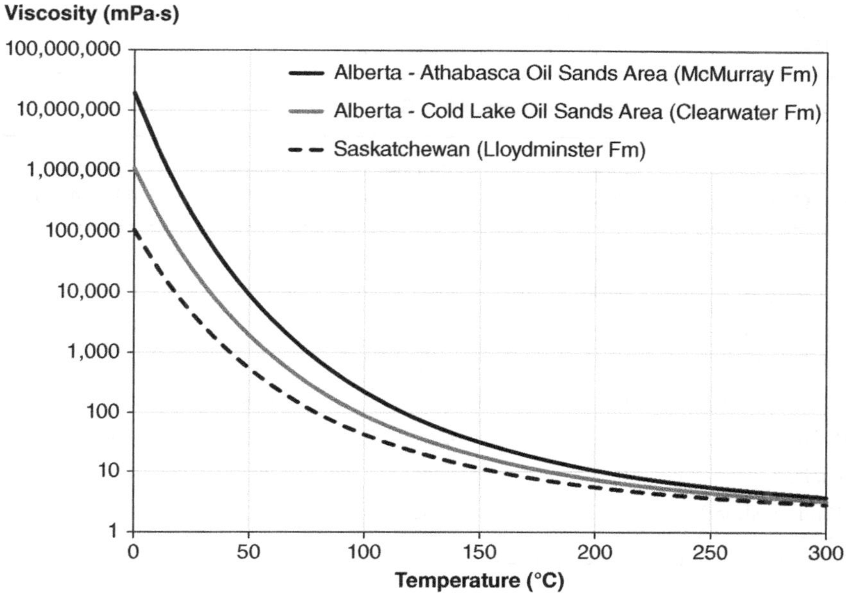

Figure 4.1 Impact of temperature on heavy oils and bitumens is shown using the Svrcek and Mehrotra (1988) equation. The curves have been generated using the one-parameter viscosity equation; parameters were computed to fit the viscosities at 15 °C (McMurray = 10^6 mPa·s; Clearwater = 2×10^5 mPa·s; and Lloydminster = 20×10^3 mPa·s). *Source*: Svrcek-Mehrotra (1988)/U.S. Department of Energy/Public Domain.

Table 4.4 General definitions of GHG emission scope categories for facilities.

Scope	General definition	Example (for heavy oil production)
1	Direct emissions from sources controlled or owned at the facility	CO_2e generated by the burning of natural gas to generate steam for recovery of the heavy oil
2	Indirect emissions energy brought into the facility	CO_2e generated in providing the electricity that has been purchased for the heavy oil operation
3	All other indirect emissions associated the facility	CO_2e generated as automotive emissions by the gasoline produced from the heavy oil

Source: Adapted from Business Development Bank of Canada.

equivalent of all compounds emitted that contribute to GHG effects. Methane, for example, has a much higher GHG impact than CO_2, i.e., a higher CO_2e value. Most discussions in literature focus on Scope 1 emissions, due to their direct attribution to the specific project, company, or industry being discussed.

Table 4.5 Typical values and ranges of direct and life-cycle emissions for energy sources.

	Emissions (tonne-CO_2e/GJ)			
Energy sources	**Direct (Scope 1)**	**Life-cyle (Minimum)**	**Life-cyle (Median)**	**Life-cyle (Maximum)**
Coal	0.228	0.222	0.246	0.273
Natural gas	0.111	0.123	0.147	0.195
Geothermal	0	0.0018	0.0114	0.0237
Solar (utility)	0	0.0054	0.0144	0.0540
Solar concentrated)	0	0.0026	0.0081	0.019
Hydro	0	0.0003	0.0072	0.66
Nuclear	0	0.0011	0.0036	0.033
Wind (onshore)	0	0.0021	0.0033	0.0168

Source: Adapted from Schlömer et al. (2014).

4.1.3.2 Carbon Emissions from Electrical Power Generation

Electrical power purchased from local utility grids is a component of any oilfield operation. For heavy oil, purchased energy can be used to replace or offset direct emissions, such as using electricity versus steam to heat heavy oil in situ. It is important to understand the implications of this offset to the overall environment, since it may be less of a "net positive" when full life-cycle (for Scopes 1 and 2) is considered. An example is using electricity from coal-fired power plants versus using natural gas in steam generators.

Estimates for CO_2 generation from power plants can vary significantly depending on the age, size, and type of facility. Table 4.5 provides a useful comparison of the typical emission intensities from different types of power plants. For "green" energy sources, although there are no direct emissions (Scope 1), there are small levels of life-cycle emissions that arise primarily from infrastructure and supply chain emissions.

4.2 Current Commercial Thermal Methods

Since the 1960s, there have been three steam-based recovery processes used commercially around the world to recover extra-heavy oils and bitumens: cyclic steam stimulation (CSS), steam flooding, and steam-assisted gravity drainage (SAGD).

4.2.1 Process Descriptions

4.2.1.1 Cyclic Steam Stimulation

The CSS process, also referred to as the "Huff and Puff" process, is based around single wells that can be either vertical or horizontal and are used to both inject steam and produce fluids. A schematic of a vertical well process is shown in Figure 4.2. In this process, specified volumes of steam are injected into the reservoir. In ultrahigh viscosity bitumen reservoirs such as in Alberta, steam in the early cycles must be injected at high, fracturing pressures for it to enter the reservoir and contact bitumen. In lower-viscosity heavy oil reservoirs, such as in California and elsewhere, the steam can be injected at lower pressures. The injected steam heats the formation and reduces the viscosity of the bitumen. After a specified volume of steam has been injected, the well is shut-in and left to "soak" for a period of time. The well is then put on production; the pressure in the reservoir from the injected steam drives the heated bitumen back to the well to be produced. When the pressure is depleted and production rates have fallen to low levels, a new cycle of injection/production is begun. In general, each subsequent cycle requires more steam than the previous cycle. The process is continued until the expected performance of the next cycle is projected to be uneconomic, i.e., insufficient oil is produced during the cycle to offset the cost of steam and associated operating costs.

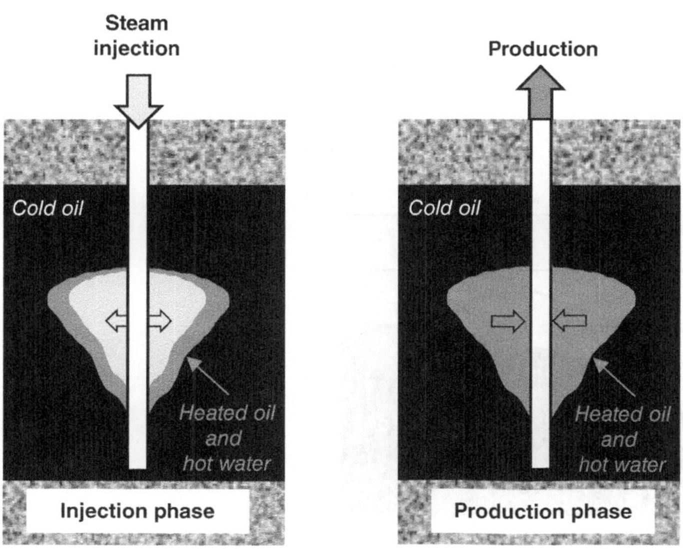

Figure 4.2 Schematic of cyclic steam stimulation (CSS).

There are currently only three CSS projects in the Alberta oil sands at Imperial Cold Lake, CNRL Primrose/Wolf Lake, and CNRL Peace River. There is no expectation that any new CSS projects will be proposed in the province because of the high-pressure nature of the process required in the Alberta oil sands.

4.2.1.2 Steam Flooding

This is a multi-well process, in which steam is continuously injected into one well to heat and displace the oil to an adjacent well, as shown in Figure 4.3. In some reservoirs worldwide (e.g., the Duri field in Sumatra, Indonesia), the viscosities of heavy oils can be sufficiently low that steam flooding can be used as the initial depletion process. In Alberta, steam flooding is also used as a follow-up process to CSS once the oil between adjacent wells has been sufficiently heated to make the bitumen mobile. Like CSS, vertical wells or horizontal wells can be used for this process.

In Alberta, Canada, the CSS and steam flooding processes are used in three projects at Cold Lake (Imperial Oil), Primrose (CNRL), and Peace River (CNRL). Production from CSS and steam flooding at these three projects averaged over 35,000 m^3/d in 2023. This comprised approximately 14% of the 247,000 m^3/d total production by thermal processes in Canada (Alberta Energy Regulator, 2023). For comparison, production in California CSS and steam flood projects totaled 35,000 m^3/d in 2020 (California Energy Commission, 2020).

4.2.1.3 Steam-assisted Gravity Drainage

SAGD is also a multi-well process, which uses pairs of horizontal wells. As shown in Figure 4.4, a horizontal producer (P) is drilled near the base of the pay zone.

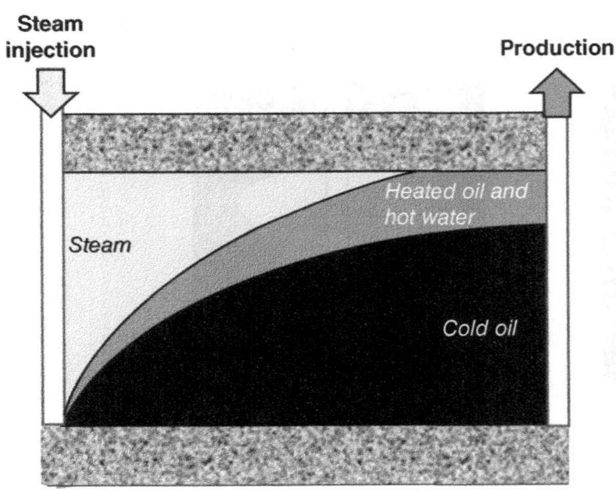

Figure 4.3 Schematic of a steam flood.

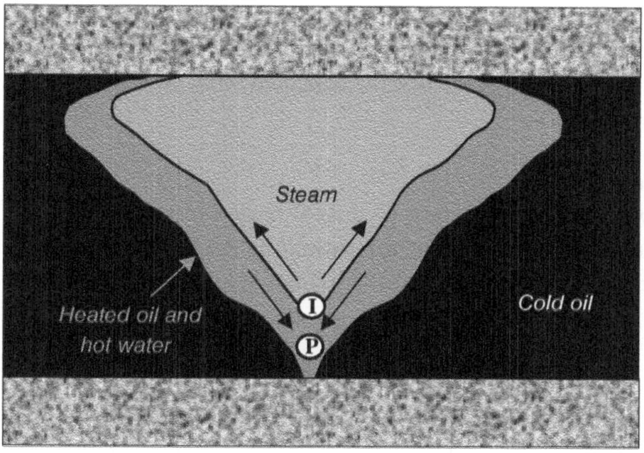

Figure 4.4 Schematic of steam-assisted gravity drainage (SAGD).

The corresponding horizontal injector (I) is drilled about 5 m directly above the producer. Steam is first circulated within the wells to heat the formation between the wells and initiate communication. Production is then begun, and steam injection is continued to create a steam chamber above the wells. The heated bitumen on the edge of the chamber drains under the influence of gravity to the producer. Because gravity is a weak driving force, the vertical gravity drainage rate is low. To generate a sufficient rate to make the process economic, the horizontal wells used can be quite long, with some wells now over 1,600 m in length.

The 23 commercial thermal projects in Alberta accounted for 211,000 m^3/d in 2023 (Alberta Energy Regulator, 2023). There is a thermal industry in Saskatchewan, in which the reservoirs and projects are much smaller than those in Alberta. Yearly production from the thermal projects in Saskatchewan totaled about 10% of Alberta's production.

4.2.2 Key Performance Indicators for Thermal Methods

There are two key performance indicators for thermal processes: the Steam-Oil Ratio (SOR) and well productivity (m^3/d/well). The SOR is expressed as the volume of liquid water used to generate the steam, termed "Cold Water Equivalent" or CWE. The SOR provides an indication of

- the thermal efficiency of the thermal process – how much steam is required to recover a barrel of oil, either on a monthly basis or on a cumulative basis (CSOR),

- the economics of the project – how much natural gas must be purchased (to generate steam) per barrel of oil produced, as well as how much capital must be spent for steam generation to produce the expected daily oil rate from the project.

4.2.2.1 Steam-oil Ratio

The SOR for a SAGD well pair is a function of several reservoir and operational parameters. An excellent example of a model for estimating the CSOR for a SAGD well pair at a specific time is provided by the following equation (Edmunds and Peterson, 2007):

$$\mathrm{CSOR} = \frac{T_{\mathrm{steam}} - T_{\mathrm{res,i}}}{H_{\mathrm{lv}}\,\phi(S_{\mathrm{o,i}} - S_{\mathrm{o,f}})} \times \frac{C_{\mathrm{vr}} + \sqrt{k_{\mathrm{t}} C_{\mathrm{vo}}\, t}}{h\eta_{\mathrm{s}}} \tag{4.1}$$

where:

C_{vo} = the volumetric heat capacity of the overburden
C_{vr} = the volumetric heat capacity of the initial reservoir
$S_{\mathrm{o,i}}$ = the initial oil saturation
$S_{\mathrm{o,f}}$ = the final oil saturation in steam chamber
T_{steam} = the temperature of injected steam
$T_{\mathrm{res,i}}$ = the initial reservoir temperature
ϕ = the reservoir porosity
h = the reservoir pay thickness
H_{lv} = the latent heat of steam condensation
η_{s} = the effective sweep efficiency
k_{t} = the thermal conductivity of the overburden
t = time

Inspection of the terms in this equation shows that for changes in individual parameters (with all others being held constant), the CSOR decreases as

- porosity and/or initial oil saturation increase (relatively more oil to be produced),
- steam temperature is decreased (less heat is used, but this will also have a negative impact on rate),
- pay thickness is increased (relatively more of the injected heat is used in the pay relative to being lost to the overburden).

Note that oil viscosity is not explicitly in the equation. It is implicit in the equation through the time required to reach the specified effective sweep efficiency, which is dependent on the rate of drainage around the steam chamber, as discussed in the following text.

4.2.2.2 Well Productivity

In the case of SAGD, well productivity refers to the oil drainage rate. The original equation for the drainage rate (m^3/day per meter of well length) from SAGD is given by Butler (1997):

$$q = \sqrt{\frac{2\phi \Delta S_o kg\alpha h}{m\nu_s}} \tag{4.2}$$

where:

- q = the drainage rate per meter of wellbore length
- α = the thermal diffusivity of the reservoir
- ϕ = the porosity of the reservoir
- k = the permeability of the reservoir
- g = the gravitational constant
- h = the height of the steam chamber
- ΔS_o = the difference in oil saturations in the steam chamber and in the reservoir
- ν_s = the kinematic oil viscosity at the steam temperature
- m = the constant relating the dimensionless viscosity to the dimensionless temperature and has a value between 3 and 4

Since the initial development of the SAGD process and drainage rate equation, reservoir simulations and observations from over two decades of field operations have indicated that the factors in Equation 4.2 are appropriate, but the expectation of the square-root factor has been relaxed to allow for variable exponents for each factor to be between 0.5 and 1. An example of the impact of these factors on the drainage (production) rate is shown in the simulation results shown in Figure 4.5.

Inspection of the graph shows that increasing the permeability-thickness product (K × h) from 37.5 to 175 (a factor of 5) increases the corresponding drainage rates for each temperature by an average of 3.75 times, equivalent to an exponent of 0.75. This supports the assumption that an exponent between 0.5 and 1.0 is reasonable.

Figure 4.5 highlights the trade-offs that are involved in selecting the steaming temperature. As noted earlier, reducing the steaming temperature reduces the SOR, thus making the process more thermally efficient. However, this action can substantially reduce the corresponding drainage (production) rate. In other words, it is likely not economically viable to reduce the steam temperature substantially in order to reduce GHG intensity.

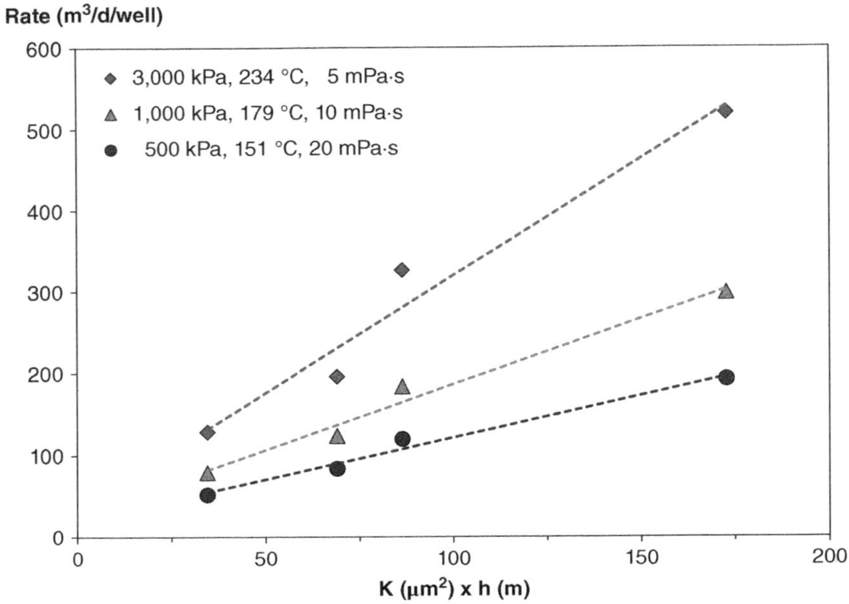

Figure 4.5 Examples of impacts of permeability, pay thickness, and temperature/viscosity on simulated SAGD drainage rates. *Source*: Adapted from Edmunds and Chhina (2001).

4.2.3 The GHG Problems/Challenges with Current Thermal Methods

From an environmental perspective, there are three high-level problems associated with existing thermal methods:

- The burning of natural gas to make steam generates substantial GHGs.
- The loss of a significant fraction of the injected heat to the reservoir and surrounding rock.
- Stagnant industry SORs (and therefore stagnant GHG intensity).

Each problem/challenge is discussed in the following sections.

4.2.3.1 Burning of Natural Gas to Generate Steam

Steam is generated by the combustion of natural gas, resulting in significant volumes of CO_2. In 2022, 725,000 m^3 of steam per day were generated by the Alberta thermal industry (Alberta Energy Regulator, 2022) requiring approximately 52 million m^3 of gas per day. This volume is equivalent to using 71 m^3 of natural gas per cubic meter of steam generated. Producing this amount of steam generated 41 million tonnes of CO_2e for the year 2021 (Government of Canada, 2024). CO_2e includes the total CO_2 equivalent for all volumes of GHGs generated, primarily CO_2 and methane (CH_4). The corresponding "GHG Intensity" (weight of CO_2e generated per cubic meter of hydrocarbon produced) was approximately 47 kg (0.47 tonnes) of CO_2e for every cubic meter of bitumen produced. For comparison, the average GHG intensity for the thermal industry in California in 2019 was 73 kg (0.73 tonnes) of CO_2e for every cubic meter of bitumen produced (California Energy Commission, 2020). The basis for this higher intensity will be discussed shortly.

Since the preponderance of CO_2e generated for all thermal projects results from the burning of natural gas, it is not surprising that there is an excellent correlation between the GHG intensity of a project and its SOR as shown in Figure 4.6. The three SAGD projects shown span the range of SORs in the industry (data from geoSCOUT, 2024; and the Government of Canada, 2024). This correlation, for either individual projects or the industry as a whole, suggests that a reasonable "rule of thumb" is

$$\text{Annual GHG Intensity (tonnes-}CO_2e/m^3) \sim 0.157 \times \text{Annual SOR} \tag{4.3}$$

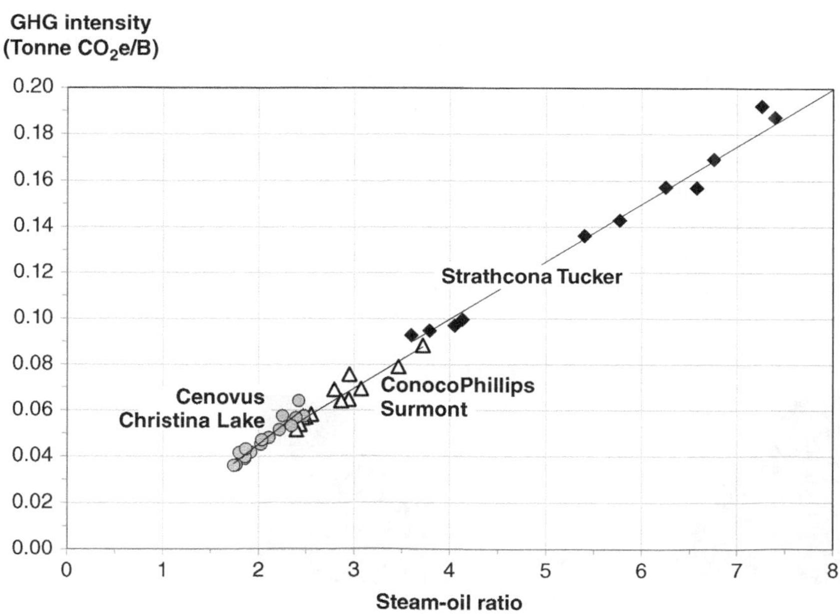

Figure 4.6 Correlation of annual GHG intensities with steam-oil ratios for three SAGD projects, 2012–2022. *Source*: Data from geoSCOUT and Government of Canada.

The overall SOR for the Alberta industry in 2023 was 3.0, which is consistent with the calculated industry GHG intensity of 0.47 tonnes CO_2 e/m^3 noted earlier. Likewise, the higher GHG intensity (0.73 tonnes CO_2 e/m^3) calculated for the California thermal industry in 2019 is consistent with the higher SORs in the more mature thermal projects in California (California Energy Commission, 2020). Reductions in SORs are thus required for GHG intensities to be reduced.

4.2.3.2 Heat Losses

Although the drawings shown earlier for the thermal processes in Figures 4.2–4.4 indicate the general locations of the injected steam and heated fluid within the reservoir pay zone, they do not show the amount of "unproductive" heat that is used to heat the rock in the reservoir, as well as the heat that is lost through conduction to the overburden, as shown in Figure 4.7.

In fact, calculations from the analytical model shown in Equation 4.1 to estimate the cumulative SOR indicate that the amount of heat lost to the overlying or underlying strata can be substantial, as shown in Figure 4.8 (model and data from Edmunds and Peterson, 2007, and industry data from geoSCOUT, 2024). The model results show that for the range of reservoir pay thickness and oil saturations tested:

- If no heat is lost outside of the reservoir, recovery of 1 m^3 of oil would require between 1 and 1.5 m^3 of steam.
- The relative amount of heat lost outside of the reservoir to the overlying and underlying strata can be several multiples of the heat that is actually used to recover the oil. This effect is compounded as the reservoir becomes thinner.

For this reason, the general economic cutoff for pay thickness in the SAGD sector of the thermal industry is currently 15 m. As the pay gets thicker, the relative effect diminishes.

Figure 4.8 shows that some SAGD projects are quite efficient, exhibiting cumulative steam-oil ratios (CSORs) as low as 2. But even in such cases, comparison

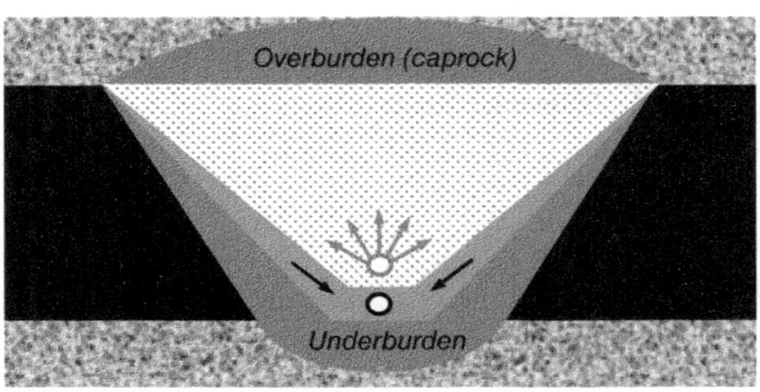

Figure 4.7 Schematic showing heat loss around a SAGD steam chamber.

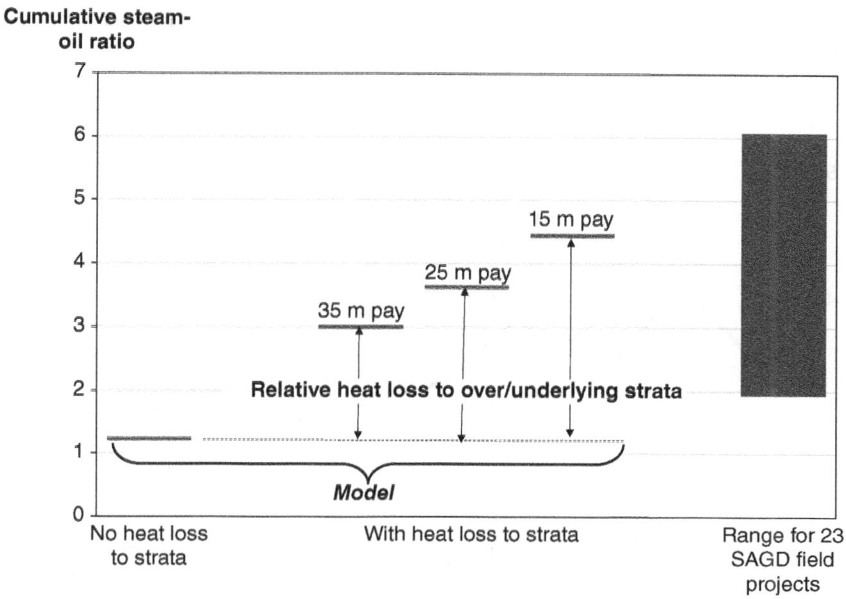

Figure 4.8 Model results showing theoretical impact of pay thickness on SAGD CSOR due to heat loss outside of pay. Model parameters used in Equation 4.1 include $\phi = 0.34$, $S_o = 0.725$ and $t = 10$ years. Ranges of cumulative steam-oil ratios (CSORs) for thermal projects are shown for comparison.

with the model results suggests that only about half of the injected heat (SOR = 1 in the ideal case) is being used to recover oil, with the rest being lost. Although the lost heat does not contribute to the oil recovery, it is still a result of burning natural gas and thus contributes to the overall GHG intensity of a project.

Finally, there is a substantial volume of heavy oil resources that exists in pay zones that are thinner than 15 m. As a result, bitumen in these resources cannot be economically produced by SAGD. Horizontal well CSS has been used in thinner pays, even down to 6 m. (Canadian Natural Resources, 2023), but as noted previously, the industry is unlikely to propose new CSS projects.

4.2.3.3 Stagnant Industry SORs

The annual SOR history of the thermal industry in Alberta is shown in Figure 4.9.

The takeaways from this chart are as follows:

- **Industry:** The annual SOR of the thermal industry has been flat at about 3.0 since 2014.
- **CSS:** The CSS sector currently requires about twice as much steam to recover a barrel of bitumen as the SAGD sector (Annual CSS SOR = 5.0 versus

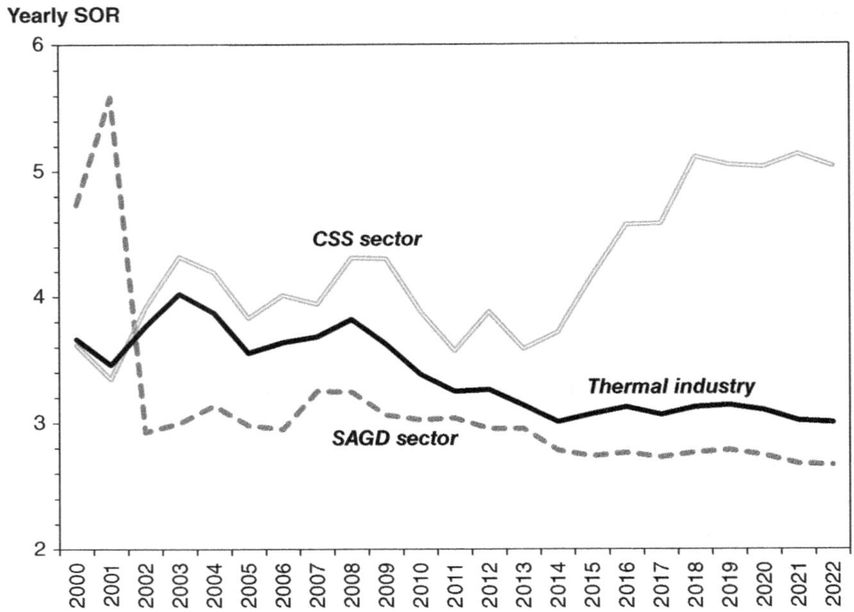

Figure 4.9 Annual steam-oil ratios in Alberta since 2000. Data from geoSCOUT (2024).

Annual SAGD SOR = 2.6). This difference is primarily due to the age of the three CSS projects, as well as the lower-quality resources in which they are being conducted, which have lower permeabilities, porosities, and bitumen saturations.

- **SAGD:** The SAGD sector of the industry has continually reduced its annual SOR from 2008, but it appears to also be approaching a minimum value.

As shown previously, the GHG intensities of projects, and thus the industry as a whole, are essentially proportional to the SOR. The industry can point to the SOR history and claim that from 2009 to 2014, the overall GHG intensity and SOR were reduced by over 15% due to the start-ups of lower-SOR SAGD projects and technological improvements, such as the emerging technologies that are discussed in a subsequent section. But since then, due to the aging of the projects, the overall metrics have remained essentially flat, with only a 2% drop in each metric from 2017 to 2021, which is the last year for which emission data are currently available from the government (Government of Canada, 2024). In addition to any impact of the COVID-19 pandemic on operations, it is also possible that some of the technologies implemented since 2014 (such as non-condensable gas [NCG] injection and flow control devices (FCDs), discussed subsequently) have successfully prevented the industry SORs from rising in this period.

4.2.4 Concluding Comments – Current Recovery Processes

Production of high-viscosity heavy oils, and bitumen in particular, requires a significant reduction of the viscosity of the hydrocarbon to allow it to be produced at economic rates. To achieve this reduction in viscosity, current methods inject steam to heat the viscous oil. These methods have averaged over 250,000 m^3/day of oil per day in Alberta in 2024, with about another 24,000 m^3 per day in Saskatchewan, and 35,000 m^3/d in California.

Steam is generated by burning natural gas, which generates significant quantities of CO_2. Steam injection is not thermally efficient; 50% or more of the injected heat is used to heat the reservoir rock and is lost to the underlying/overlying strata. The thermal industry may have reached a minimum GHG intensity (tonnes CO_2e generated per barrel of oil produced), and further improvement will require new technologies.

The remainder of this chapter presents a review of the methods that have been developed and field-tested to fully or partially decouple heavy oil recovery from the use of steam, thus providing substantial improvements in energy efficiency and environmental performance.

4.3 New Solvent-based Recovery Processes with Improved Environmental Metrics

4.3.1 Introduction and Overview

As noted in the prior section, the key to being able to produce heavy oil is to reduce its viscosity, and the use of steam to heat the oil has been the basis of the industry to date. However, there is another approach that has been under development for almost 30 years – the use of solvents. This section is intended to provide a basic understanding of the use of solvents to recover bitumen and heavy oil. It includes commentary on the key aspects and considerations in their use. It is not intended as a comprehensive technical document on the technology, as the field is data-rich with observations and recommendations based on the specific conditions being studied (bitumen type, use of dead or live oil, temperature, concentration, mode of operation, use of pure alkanes or blends of refinery products, etc.).

4.3.1.1 Solvent Design Considerations in Pilot Tests

Solvent projects conducted by the industry can be categorized according to three attributes: well configuration, solvent type, and solvent concentration.

4.3.1.1.1 Well/Drive Configuration

Solvent processes can be applied in the same well type (horizontal or vertical) and operation type as used in current commercial projects (cyclic or SAGD).

4.3.1.1.2 Solvent Type

Solvents fall into one of three categories:

- Light, low molecular weight (MW) alkanes – ethane, propane, or butane.
- Heavier, higher MW hydrocarbons – pentanes + diluent, condensate, naphtha.
- Other compounds – lower-carbon compounds such as methane, CO_2, or synthetic solvents.

To date, the industry has been evenly divided in its choice of light versus heavy hydrocarbon solvents. The other options for solvents currently remain in the laboratory investigation stage.

4.3.1.1.3 Solvent Concentration

An idealized example shown in Figure 4.10 illustrates some of the concepts involved in the use of solvents to partially or fully replace steam. In this representation, consider a SAGD (steam-only) process that could be conducted at 210 °C. At this temperature, the viscosity of the heated bitumen will be reduced from about 1,000,000 mPa·s at the initial reservoir temperature to 10 mPa·s at the steam

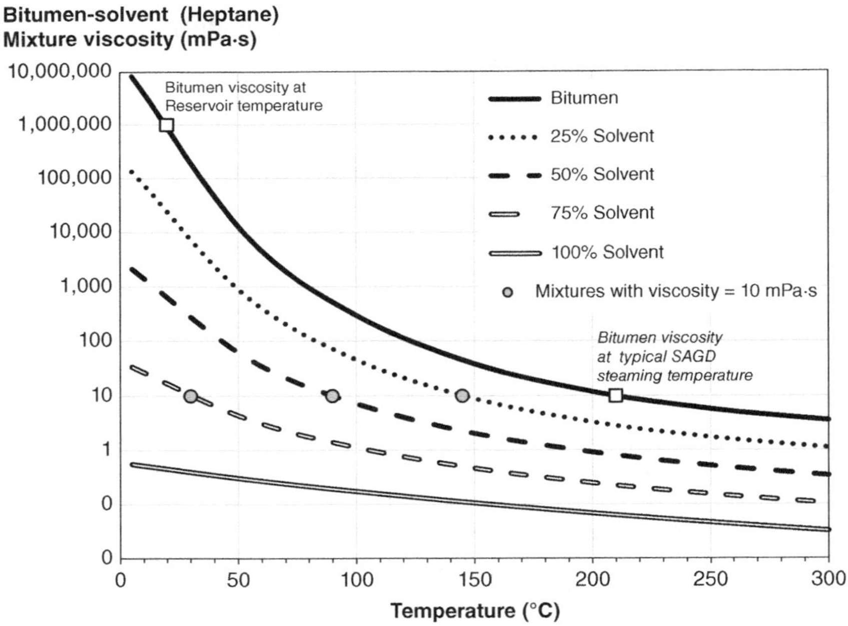

Figure 4.10 Model of temperature and solvent (heptane) blending on viscosity with McMurray bitumen. Viscosity curves were generated using the following mixing model for heptane with Athabasca bitumen: $\ln[\ln(\mu_{mix} + 1)] = w_S \ln[\ln(\mu_S + 1)] + w_B \ln[\ln(\mu_B + 1)]$, where w = weight fraction and μ = viscosity.

temperature. As shown by the shaded circles, this same viscosity can be achieved by bitumen/solvent (heptane) blends of 75/25 (at 145 °C), 50/50 (at 90 °C), or 25/75 (at 30 °C). These conditions represent substantial reductions in the amount of heat energy required to be injected and in the level of corresponding emissions.

Since solvents are relatively expensive, the economic viability of all solvent-based processes requires high levels of solvent recovery for reuse, regardless of whether a light or heavy solvent is used. As discussed subsequently, solvent recovery has been a major concern to address in progressing solvent processes to commercialization.

Table 4.6 summarizes that most field tests (22/33) to date have been conducted using low solvent concentrations with steam, with several (9/32) using pure solvent and only one that has used a high concentration solvent with steam. These numbers do not include any of the commercial applications that are currently in operation as of year-end 2024.

Key distinctions between the low- and high-concentration steam-solvent approaches are as follows:

- **The degree of field testing conducted to date:** Pilot testing has relied on the use of existing facilities and steaming conditions. The use of lower concentrations of the solvents has reduced testing costs and thus made for more cost-effective piloting. As noted above, to date, there has only been one high-concentration field test.
- **The expected level of impact:** The use of high solvent concentrations offers the prospect of significant reductions in steam usage due to the operation at lower temperatures – as suggested in Figure 4.10 – with corresponding significant improvements in performance and environmental metrics.
- **Economic risk due to solvent costs and potential loss:** There is a high economic risk/exposure inherent in the high-concentration processes due to the high costs of solvents, the higher volumes required for high-concentration processes, and the uncertainty/confidence around the level of solvent recovery that can be assured. Until the lower-concentration processes have been commercially demonstrated, progress in this area will remain slow.

Table 4.6 Summary of field tests of hydrocarbon solvents to date.

Solvent concentration	SAGD configuration	Cyclic configuration
100% (solvent only)	4 field tests using light hydrocarbons	6 field tests using light hydrocarbons
High (50–90%)	1 field test using a light hydrocarbon	No field tests conducted
Low (<25%)	20 field tests with light (9) and heavy (11) hydrocarbons	2 field tests with a heavy hydrocarbon

4.3.1.2 Solvent Impacts on Key Performance Indicators

4.3.1.2.1 Solvent-steam Processes

As noted earlier, the low-concentration steam-solvent approach has been by far the most extensively researched and field-tested by industry. Figure 4.10 illustrated how the addition of solvent to steam can maintain a low bitumen-phase viscosity using less steam, or attain an even lower viscosity if it is used with the same amount of steam as used with SAGD. The addition of solvent to the steam can also reduce heat loss to the overburden, as shown in Figure 4.11.

Thus, the addition of solvent is intended to improve the performance of the two key performance indicators for thermal recovery processes:

- Productivity increases by further reducing the viscosity of the hydrocarbon phase.
- SOR decreases (and thus the GHG intensity) by
  - using less steam per barrel of oil produced,
  - reducing heat that is lost to the overburden by the insulating effect of the gas,
  - reducing the project life (further reducing the total amount of steam used and the total heat loss) due to the higher production rates.

There is a trade-off, however, based on the solvent chosen. Since the earliest technical publications regarding steam-solvent processes, it has been identified that

- the higher carbon-number alkanes (e.g., C6 through C10 or refinery products such as naphtha or diluent) produce the most dramatic improvements in rates (because a higher fraction of these components stays in the liquid phase at higher temperatures, thus reducing the viscosity of the oil phase further),

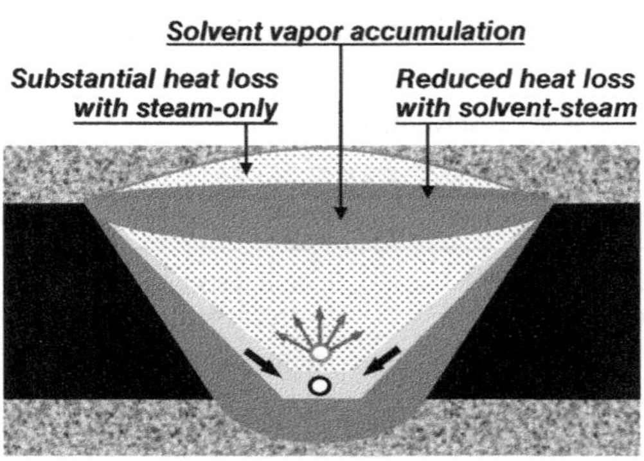

Figure 4.11 Impact of solvent injection on heat loss and drainage rate.

- the lower MW alkanes (propane or butane) provide more improvement in the SOR because they reside more in the gaseous phase and thus provide a better insulating effect at the top of the steam chamber.

This trade-off is shown in the results of a simulation study of the performance of the C3 through C7 alkanes (10% concentration with dead (gas-free) bitumen) compared to SAGD-only, as shown in Figure 4.12.

Features that stand out in this chart are as follows:

- The addition of all solvents reduced the energy (and GHG) intensity at eight years of operation.
- The use of propane actually reduced the hydrocarbon drainage rate and thus the recovery performance after eight years. In theory, if a light solvent is used, and depending on the conditions, the evolution of the vapor phase at the edge of the steam chamber can reduce heat transfer and correspondingly lower the gravity drainage rate.

There are some important qualifications for this chart as follows:

- The results shown are from only one set of conditions used in the full study and are intended primarily to show the directional impact of the solvent selection on the recovery and SOR under these specific conditions.

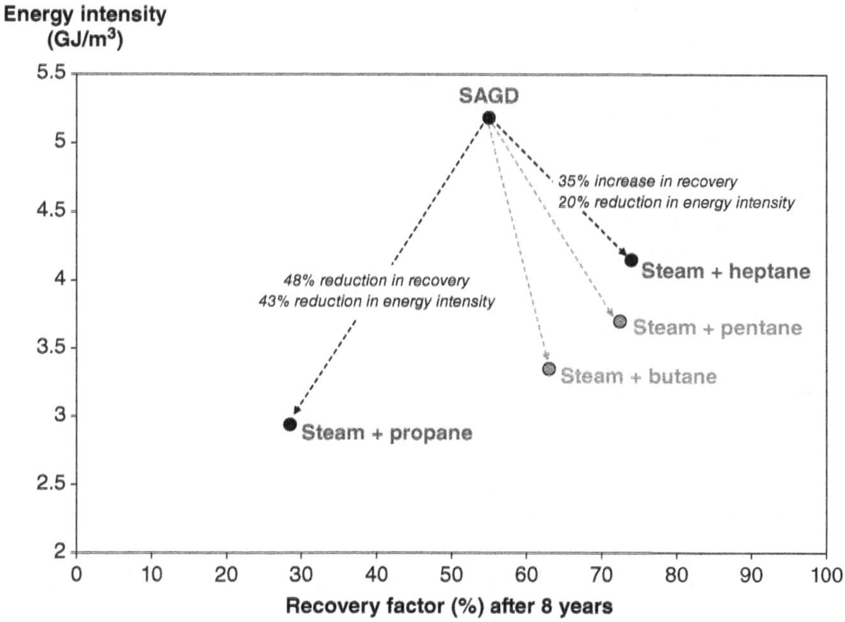

Figure 4.12 Example of the simulated impact of alkane solvent on recovery and energy intensity. *Source*: Adapted from Ardala et al. (2010).

- The study investigated the impacts on both dead (gas-free) and "live" (containing dissolved gas) bitumens from both the Athabasca and Cold Lake oil sands areas. Regardless of the bitumen used, the same directional trends (especially in the impact of propane) were observed, but to a less dramatic level with the live oil.

4.3.1.2.2 Solvent-only Processes

The elimination of steam from the process obviously results in an SOR = 0, but it does not correspondingly eliminate energy requirements or GHG emissions because there will always be associated Scope 2 (and Scope 3) emissions, especially if heat is added at surface or downhole as part of the process design.

The potential impact of solvents on the well productivity metric can be substantial in comparison to purely thermal-based processes. The rate that a lower MW solvent in the gas phase mixes with the bitumen in situ is based on its rate of mass-transfer diffusion into the bitumen, which is in general much slower than the heat-transfer diffusion rate in thermal processes. As discussed in a subsequent section, this has been the primary consideration in the success or failure of gas-phase solvents in field testing. Mass transfer limitations are not restricted to solvent-only processes. For solvent-steam processes that run at low temperatures such as 100 °C, the inherent higher viscosity of the bitumen at this temperature could be expected to begin to impact the productivity due to mass-transfer considerations.

Solvents that are in the liquid form when injected would be expected to benefit from increased mass transfer with the bitumen through convective mixing. But because the application of this type of process has been in a horizontal, cyclic-drive mode (as discussed subsequently), the factors impacting the well productivity will be

- the fraction of the resident bitumen that is contacted by the injected solvent, since the solvent will "finger" through the bitumen and contact only a fraction of the total volume,
- the pressure differential between the pressurized reservoir during injection and the bottom hole pressure during production.

4.3.1.3 Practical Considerations in the Selection of the Solvent

Figure 4.12 illustrated some of the trade-offs in performance that industry has had to consider in choosing an optimal solvent for their operations and resources. An expanded list of the trade-off considerations that operators have had to make in practice is presented in Table 4.7. This list relates to both solvent-steam and solvent-only processes.

Specific projects will be discussed in a subsequent section, but the selection of the type of solvent used to date may be more a reflection differing corporate philosophies in weighing the costs and risks in decision making. In particular, the economics of solvent-based processes are highly sensitive to the level of the

Table 4.7 Practical considerations in the selection of solvents.

Factor	Preferred solvent
Unit cost Lower molecular weight (MW) hydrocarbon solvents are less expensive (propane and butane << condensate)	Lower MW
Production and recovery of solvent Gas phase may implicitly be expected to have higher recovery/ recycle rate	Lower MW
New facilities costs Steam generation facilities are the most expensive component of the capital budget – lower MW solvents have more impact on reducing SOR and thus reduce capex more for a specified production nameplate	Lower MW
Potential for in situ upgrading Lower MW solvents may partially precipitate asphaltenes in situ and produce an upgraded hydrocarbon	Lower MW (as long as no pore plugging by precipitated asphaltenes)
Use of existing infrastructure Most existing projects already have diluent blending facilities for handling and pipelining bitumen. Cost estimates for retrofitting existing facilities to accommodate light solvents (propane, butane) have been deemed by some industry companies to render this option uneconomic. For some processes that use lower MW solvents, it is critical to remove methane from the captured/recycled solvent gas before it can be reinjected. As a result, additional facilities are required to reduce methane concentrations to low levels prior to solvent injection.	Higher MW (diluent)
Project life Higher rates from higher-MW solvents will reduce project life and increase project PV	Higher MW

(relatively expensive) solvent recovered and reused. Industry has estimated that over 80–85% of the solvent must be recovered for the project to be economic. Low solvent recovery has historically been perceived as a substantial risk and has tempered enthusiasm for progressing the technology by many in the industry.

4.3.2 Process Descriptions and Key Field Tests: SAGD Configuration

4.3.2.1 Low-concentration Processes

As will be observed throughout this chapter, operators have different names for their processes, but all of the processes essentially consist of injection of steam and low-to-moderate concentrations of the solvents. A comprehensive list of the field pilots for low-concentration solvent-steam processes, along with all other solvent-based pilots, is provided in the appendix. Projects that have provided specific results of importance are listed in Table 4.8.

There are two key technical/economic observations that can be made from these field results:

- The general reduction in SOR has been in the range of 30%.
- Other than the ConocoPhillips pilot, the pilots have shown that high solvent recoveries have been achieved. In its final report, ConocoPhillips noted that "Within the pilot wells, it was determined that a significant amount of injected solvent could remain in the reservoir due to high sub-cool (related to the pressure differential between injector and producer) operation, reservoir heterogeneity, thermal interactions from adjacent wells, and injection of solvent into mature steam chambers, whereas solvent transport away from the "eSAGD" wells could occur following coalescence with adjacent steam chambers" (ConocoPhillips Canada, 2015).

Table 4.8 Key performance indicators for recently completed low-concentration solvent-steam projects. In addition, Suncor completed its ES-SAGD (steam-diluent) pilot at Firebag Pad 109, and MEG wrapped up the field test of its eMVAPEX (steam-butane process) at Christina Lake Pad A2 in 2023, but neither company has publicly released any results on these metrics.

Project	Process	Solvent	% SOR reduction	% rate uplift	% solvent recovery
Cenovus Christina Lake[a]	SAP	Butane	30	20	87
Imperial Cold Lake[b]	SA-SAGD	Condensate	40	40	85
Connacher Great Divide[c]	SAGD+	Light hydrocarbon	26	32	96
ConocoPhillips Surmont[d]	ES-SAGD	NGL mixture	31	21	36
CNRL Kirby South[e]	LHE-SAGD	Butane	45	NR	85
Suncor Firebag[f]	ES-SAGD	Diluent	20	NR	NR

a. Cenovus (2015, 2018).
b. Imperial Oil (2014) (average of two well pairs).
c. Lau et al. (2014) (average of two well pairs).
d. ConocoPhillips (2015) (average of two well pairs).
e. Canadian Natural Resources (2024).
f. Suncor (2024).
NR, not reported.

One example of a successful field test of a low-concentration process is the pilot conducted by Imperial Oil at Cold Lake from 2010 to 2017 (Imperial Oil, 2014). The pilot used

- a pentanes$^+$ diluent as the solvent, in concentrations up to 20% by volume with steam,
- two 750-m well pairs in the Clearwater Formation, spaced at 150 m.

The pilot was not designed to evaluate the recovery factor associated with SA-SAGD relative to SAGD, but rather to assess the following performance indicators:

- Bitumen rate uplift
- SOR reduction
- Solvent recovery.

One of the well pairs (WP1) was operated initially in SAGD mode and transitioned to solvent injection, while the second well pair (WP2) started with solvent injection and later transitioned to SAGD mode. The solvent injection rates, bitumen production rates, and solvent production rates for the two well pairs are shown in Figure 4.13, with the corresponding cumulative SORs shown in

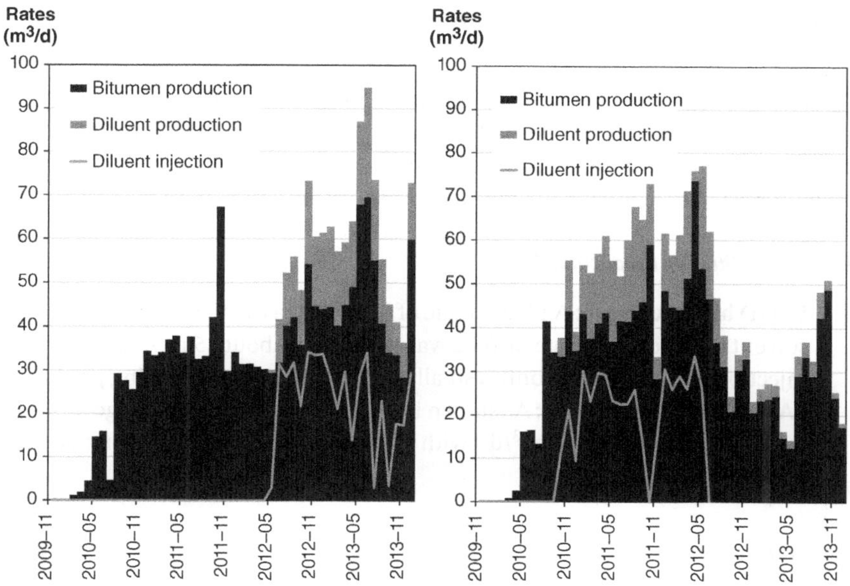

Figure 4.13 Rate performance of the two well pairs (WP1 on left, WP2 on right) at Imperial's Cold Lake SA-SAGD pilot. Solvent production data after 2013 were not available. *Source*: Adapted from Imperial Oil/IETP (2014).

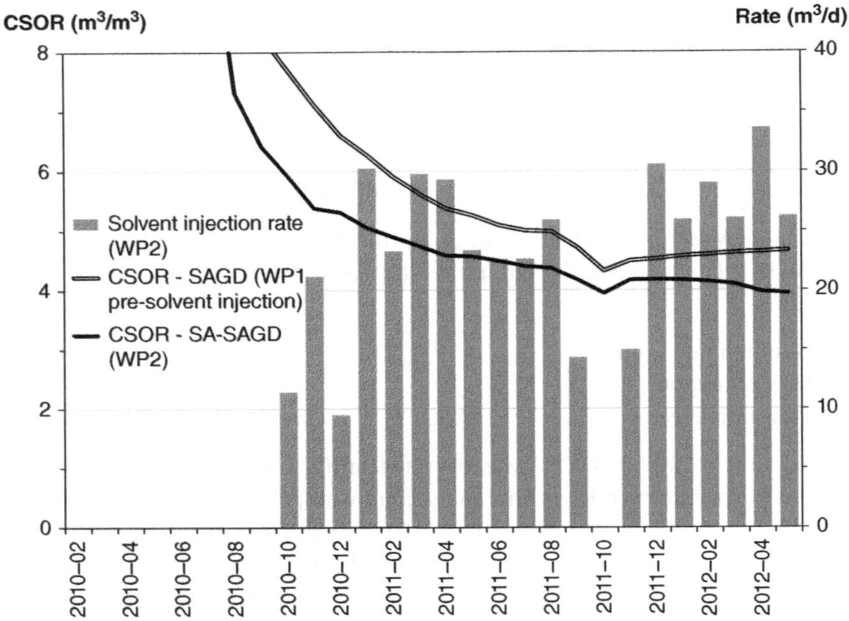

Figure 4.14 CSOR performance of the two well pairs at Imperial's Cold Lake SA-SAGD pilot.

Figure 4.14 and solvent recovery shown in Figure 4.15. Note that determining the level of solvent recovery can be challenging and requires chemical analysis of the injected diluent and both the produced liquid and gas phases.

Figures 4.13–4.15 provide the following observations:

4.3.2.1.1 Bitumen Production Rate

- **SAGD bitumen rate:** A plateau rate of 35 m^3/d was achieved in WP1 prior to solvent injection. The rate in WP2 was reduced to about 25 m^3/d after solvent injection had ceased and only a small amount of solvent was being produced.
- **SA-SAGD bitumen rate:** As shown in both Figure 4.13, the average bitumen rates grew to about 45 m^3/d (with peaks to 50–60 m^3/d) during solvent injection.

4.3.2.1.2 CSORs

- **SAGD:** Prior to solvent injection, WP1 had achieved a stable CSOR of 4.7.
- **SA-SAGD:** By the end of solvent injection, WP2 had achieved a stable CSOR of 3.9.

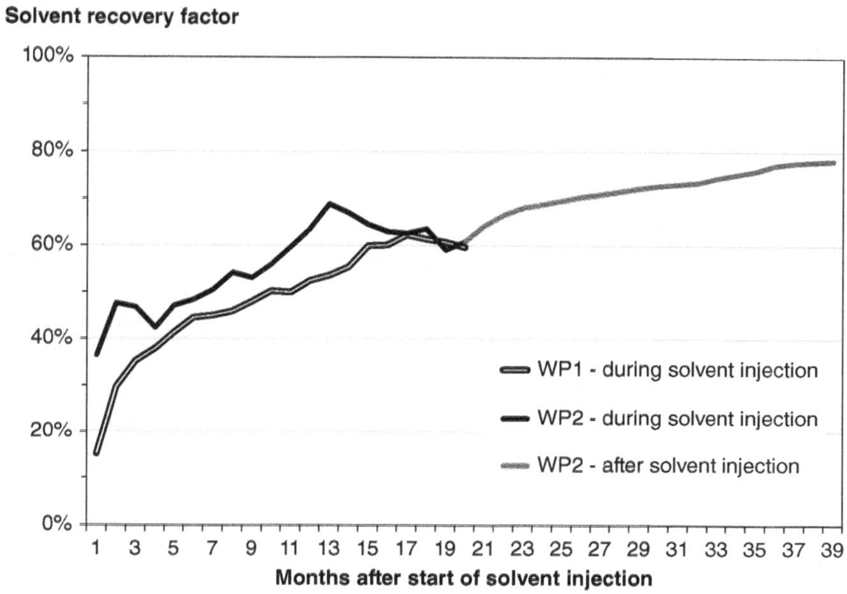

Figure 4.15 Solvent recovery performance of the two well pairs at Imperial's Cold Lake SA-SAGD pilot; data through December 2013.

4.3.2.1.3 Solvent Recovery

- **WP1:** After 20 months of solvent injection – to the end of the reporting period – 60% of the solvent had been recovered (Imperial Oil, 2014).
- **WP2:** After 20 months of solvent injection, 60% of the injected solvent had been recovered; after 39 months of total operation, almost 80% of the solvent had been recovered.

4.3.2.1.4 GHG Intensity Reduction

- **WP1:** Prior to solvent injection, WP1 reached a CSOR of 4.7, which corresponds to an estimated GHG intensity of 0.74 tonnes CO_2e/m^3.
- **WP2:** By the end of solvent injection, WP2 reached a CSOR of 3.9, which corresponds to an estimated GHG intensity of 0.61 tonnes CO_2e/m^3.
- The use of solvent in the process thus reduced the GHG intensity by 17% (0.13 tonnes CO_2e/m^3).

The following conclusions can be made for the SA-SAGD pilot at Cold Lake:

- SA-SAGD was able to increase bitumen production rate by over 25%, from 35 to 45 m^3/d.

- SA-SAGD was able to reduce overall steam usage per barrel of bitumen by 15%.
- The solvent recovery profiles were consistent for both well pairs; ultimate recovery of 80% of the injected was achieved.

The pilot was clearly a technical success and demonstrated that the use of diluent as a solvent could generate performance metrics that could both support an economic commercial application and, in the process, reduce the carbon intensity associated with the production. Imperial subsequently began planning and implementation of a 2,400 m^3/d (15,000 b/d) commercial SA-SAGD project in the Grand Rapids Formation at Cold Lake. Steam injection for this project began in December 2023, and first solvent injection began in mid-year 2024. After four months of production, the project rate exceeded its 15,000 b/d design expectations and operated at an SOR below 2 (refer to Figure 4.14 and Equation 4.3).

4.3.2.1.5 Some Additional Comments on Post-CSS Solvent Injection in Drive Processes

In the prior section that described current thermal processes, it was pointed out that steam flooding is often used as a post-CSS process to extend recovery once further cyclic injection is no longer economically feasible. In the same manner, CSS operators have also tested the post-CSS injection of low concentrations of solvents, both with and without steam in drive processes. Pilot results were as follows:

- In 2014, Shell Canada conducted a pilot of its steam-solvent Vertical Well Steam Drive (VSD) process at Peace River. In this process, steam with 15% solvent (diluent) was injected continuously with steam into a central injector in a five-spot pattern. The surrounding producers had undergone two cycles of CSS. The process was thus a "drive" variant of the cyclic Liquid Addition to Steam for Enhanced Recovery (LASER) process (discussed in the next section). Although Shell indicated that the efficiency of the diluent co-injection in the process was expected to be lower than that of the LASER process, its simulations indicated that the solvent recovery factors would be expected to be much higher, over 80%.

  Injection was conducted for four months in 2014. After one month of injection, the rate of one of the producers had doubled. During the first 10 months of injection, the steam oil ratio for the wells was about 10% lower than during the steam flood. At the time, Shell projected that 86% of the solvent would be recovered at an oil/lost-solvent ratio of 5.2 m^3/m^3. Since that time, the field has changed ownership, and no further activity on this process has been undertaken at Peace River.

- In 2019, Imperial Oil submitted an application for its "Enhanced Late-Life Process" (ELP). ELP is a potential follow-up process to replace steam with solvent for late-life flooding. It uses a light hydrocarbon such as butane as the solvent and operates at a lower temperature and pressure than steam flood. The company expected that the process could contribute 10–15% incremental recovery relative to CSS. The company noted that benefits from the process would be a reduction in GHG intensity by up to 70%, as well as no water usage. In 2023, Imperial indicated that the project had been delayed and requested that the approved confidentiality period be extended by two years to 2026. The status of the pilot at this time has not been presented publicly by Imperial.
- In 2020, CNRL applied to conduct a Light Hydrocarbon Enhance steam flood (LHE-SF) at Primrose (Canadian Natural Resources Limited, 2020). The process is intended to be initiated after approximately five years of steam flooding. It co-injects a light hydrocarbon (propane or butane) with steam at similar operating pressures as the steam flood. The concentration of the proposed light hydrocarbon-to-steam ratio of the process is expected to be higher than in other low-concentration solvent applications and will vary with time. CNRL noted that the LHE-SF process has the potential to
  - significantly lower the steam rate (up to 50%),
  - maintain a similar oil rate as compared to steam flood,
  - result in significantly reduced GHG intensity,
  - decrease the water requirements for steam flood operations.

  The project was granted confidentiality status by the Alberta Energy Regulator through 2025.

4.3.2.2 High-concentration Processes

Although low-concentration processes have been under development and field testing since the early 2000s, interest in higher-concentration processes (50–90%) at moderate temperatures is a more recent development. This recent interest was motivated both by the success of low-concentration pilots and by the potential for achieving the significantly higher levels of GHG reductions offered by the technology. Cenovus and Imperial in particular have been the two operators that have spearheaded field testing of this technology. Historically, these two operators have the most experience with field testing of lower-concentration processes, albeit using different types of solvent, with Cenovus focusing on lighter solvents (propane, butane) and Imperial using heavier, diluent solvents. These two approaches have been carried over into their designs for their field tests of higher-concentration processes. Because this is a leading-edge technology area, all data for their field tests have been proposed to be held as confidential for several years.

The two pilots are summarized in Table 4.9. The Cenovus pilot has been underway for several years, while the Imperial pilot (joint field test with Suncor) is expected to be conducted in 2027.

Cenovus began its pilot at Foster Creek W06 pad with propane injection into one well pair. Prior to all information being held confidential in late-2020, public domain data showed that the process was able to reduce the average SOR from 3.5 to 0.5 m^3/m^3, with hydrocarbon production dropping only by

Table 4.9 Descriptions of high-concentration solvent-steam pilots by Cenovus and Imperial; data from Cenovus, "Foster Creek Thermal Project Amendment to Solvent-Aided Process," 2016; Imperial Oil, "Enhanced Bitumen Recovery Technology Pilot Application," 2019.

Pilot information	Cenovus	Imperial
Field	Foster Creek	Aspen
Pad	W06 (8 well pairs)	Pilot (3 well pairs)
Process name	Solvent Driven Process (SDP)	Enhanced Bitumen Recovery Technology (EBRT)[a]
Solvent	Propane	Gas condensate
Solvent concentration	50–90% (by weight)	90% (by volume)
Steam temperature	80–100 °C	150–200 °C
Pilot/ Process goals	Lower CSOR and lower water treatment costs associated with steam generation	Significant reductions vs. SAGD in • thermal energy (70%) • GHG intensity (60%) • water intensity (90%) • water use (up to 90%) • operating pressures (lower)
Pilot objectives	Quantify the process impact on • oil production rate • Steam-oil ratio (SOR) Quantify recovery of propane	Quantify the process impact on • oil production rate • SOR Quantify impact of pressure
Company expectations	A significant portion of the propane will be produced, although after 3 years of solvent injection about 50% may be retained in the reservoir Incremental oil production	Process could reduce GHG emissions intensities by approximately 60% and result in a reduction of water use by as much as 90% compared to SAGD

(Continued)

Table 4.9 (Continued)

Pilot information	Cenovus	Imperial
Status	Pre-pilot steam injection and production began in 2014, continuing to the end of 2020	In early 2024, Imperial and Suncor announced that the EBRT pilot will be a 50–50 joint investment
	Initially, only one well pair received propane injection	The field trial is expected to start by 2027
		A 3-year confidentiality period has been requested
	All subsequent injection and production information has been held as confidential	

a. The temperature and solvent concentration used in the EBRT process are chosen such that they generate an azeotrope at the operating pressure of the process in the reservoir, i.e., at vapor–liquid equilibrium, the solvent concentrations in the liquid and vapor phases will be the same.

50% (from about 100 to 50 m^3/d), as shown in Figure 4.16. The substantial reduction in SOR corresponds to a reduction in GHG intensity from 0.55 to 0.08 tonne-CO_2e/m^3, a saving of 0.47 tonne-CO_2e/m^3, and an 86% reduction in GHG intensity.

4.3.2.3 Solvent-only Processes

As noted previously, these processes inject 100% solvent either at reservoir temperature (usually 10–15 °C) or at mildly warmed temperatures up to about 60 °C. There are two solvent-only, SAGD-configured processes of note: VAPEX and NSolv.

4.3.2.3.1 VAPEX

The use of solvents in a SAGD configuration was first developed in the 1980s and continued through the 1990s with a series of papers on the VAPEX process by Butler and Mokrys (1991). In this process, a gaseous solvent such as propane or butane is injected into the reservoir via the upper horizontal well at ambient temperature. The solvent diffuses into the bitumen phase, reducing its viscosity and allowing the diluted bitumen to gravity drain to the underlying producing well, as shown previously for the SAGD process in Figure 4.5.

Only a few field tests of the VAPEX process were conducted, as listed in Table 4.10. In general, the tests led to the conclusion that the effective rate of solvent diffusion into the bitumen was too slow to generate economically attractive

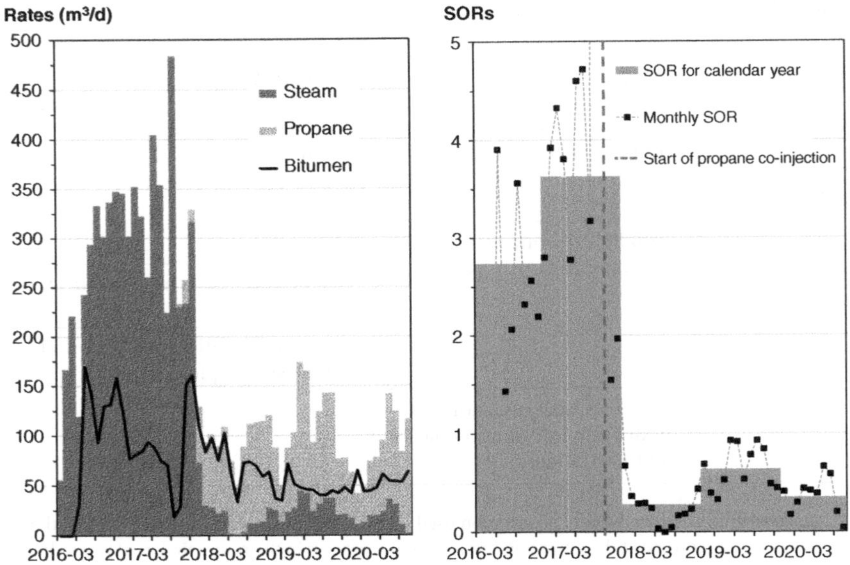

Figure 4.16 Rate (left) and SOR (right) performance of the Cenovus Foster Creek Pad W06 SDP pilot. Data from geoSCOUT (2024).

Table 4.10 Descriptions of solvent-only VAPEX pilots in a SAGD configuration; injection period data from geoSCOUT (2024).

Project	Solvent	Solvent injection period
Encana Foster Creek	Propane	October 2002 to May 2003
Devon Dover	Propane/methane	June 2005 to November 2006
Canadian Natural Primrose	Butane	September 2006 to November 2007

production rates. The field pilot by Devon at Dover (UTF site) is illustrative of the general performance of the process that has been observed. The results shown in Figure 4.17 indicate that the oil production rates achieved during the VAPEX phase were about an order of magnitude lower than had been observed during the initial SAGD phase.

As shown in Figure 4.10, there have been only three other pilots of this type of process in the early 2000s. The results of the DOVAP pilot highlight the fact that the diffusion rate of the solvent into the viscous bitumen at the edge of the chamber is much slower than the rate of heat transfer in SAGD, thus leading to substantially reduced rates. For this reason, industry has essentially terminated

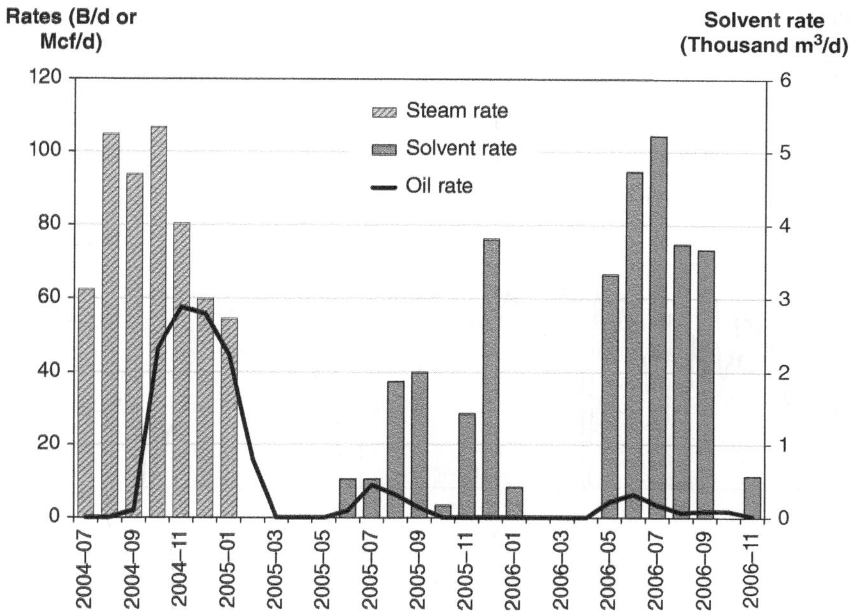

Figure 4.17 Performance of the Devon VAPEX pilot at Dover ("DOVAP"). Data from geoSCOUT.

further field piloting of the isothermal VAPEX process in the Alberta oil sands. As interest in VAPEX began to wane, attention was shifted in the industry to the use of "warmed" solvents (but without using steam).

4.3.2.3.2 Nsolv

In this process, solvent is chosen such that when heated to a moderate temperature, it can be injected into the reservoir as a vapor phase, and when it contacts the bitumen at the cooler reservoir temperature, it condenses to a liquid to mix with the bitumen and dilute it. In this way, solvent mixing with the bitumen can be achieved more rapidly than by gas diffusion through the viscous bitumen.

Proof-of-concept laboratory testing was performed from 2005 to 2007 at the Alberta Research Council and confirmed the potential for the process. The ideal operating temperatures for the process were identified to be between 35 and 75 °C, as this would provide a sufficient increase in the bitumen temperature to significantly reduce the bitumen viscosity (Suncor, 2015). A pilot of the Nsolv process termed the BEST ("Bitumen Extraction Solvent Technology") pilot was initiated by Suncor and Hatch in January 2014 in a 300-m well pair at Suncor's MacKay River/Dover lease. Butane was chosen as the solvent

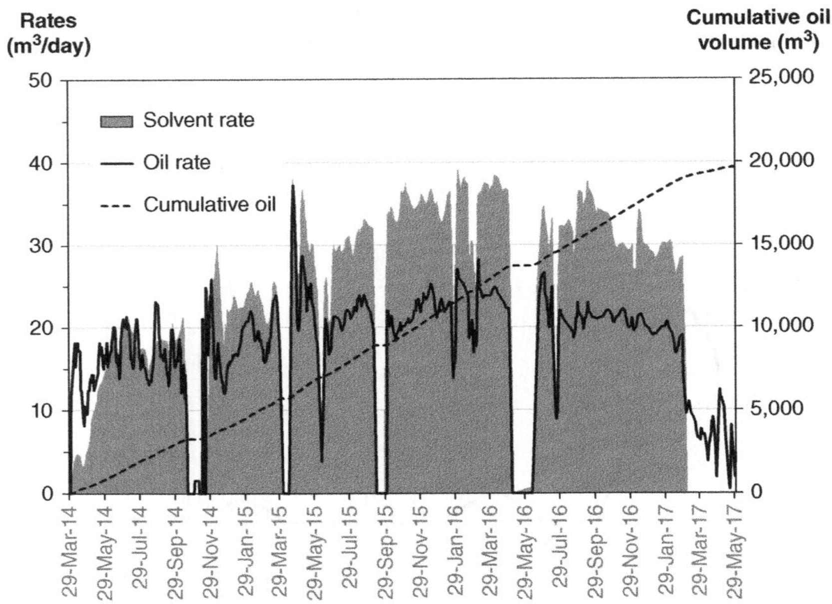

Figure 4.18 Performance of the Suncor/Hatch "BEST" pilot of the Nsolv process at Dover. *Souce*: Adapted from Suncor (2017).

because it exists as a vapor at the selected injection temperature of 60 °C and as a liquid at reservoir temperature (7 °C) and pressure (600 kPa). Solvent injection started in February 2014 and continued until March 2017. Solvent recovery (blowdown) occurred for three months and was discontinued at the end of May 2017. The solvent injection and oil production profiles are shown in Figure 4.18.

Suncor (Suncor, 2017; Suncor 2018) noted that

- 41% of the "recoverable bitumen" was produced.
- by the end of the project, over 96% of the injected solvent had been recycled. Results supported solvent recovery prediction of over 70% for mature well pairs in commercial-scale projects.
- the cumulative solvent–oil ratio at end of the project was 7.6, representing a thermal energy equivalent SOR of 0.71.
- upgrading of the oil from 8 to 14 °API was observed.
- Post-test cores also indicated that
  - the average oil saturation within the solvent chamber was 8%.
  - asphaltene precipitation occurred on a sub-pore scale, as had been observed in laboratory core experiments. As a result, the precipitated

asphaltenes did not result in "pore plugging" that could have substantially reduced flow.

Despite apparently promising results, no further plans for field testing this technology have been publicly announced.

It is important to appreciate that during the reclamation of produced solvent for recycling and further injection, the concentrations of low-molecular-weight gases such as methane in the recycle stream need to be reduced below a minimum value. The presence of these gases raises the dew point of the injected solvent, thus affecting its condensation in the reservoir at cooler temperatures. If produced solvent is to be reused, it must be separated from the methane in the produced gas stream.

4.3.3 Process Descriptions and Key Field Tests: Cyclic Configuration

4.3.3.1 Low-concentration Processes

This area of technology is the most mature of solvent-steam processes. It uses the same principle as in the SAGD configuration; that is, to use solvent to lower the viscosity of the hydrocarbon phase to increase hydrocarbon production rates. However, in this configuration, the solvent is added to the steam in the middle cycles of the CSS process. For equivalent volumes of steam injected, the higher rates achieved from the solvent addition result in

- higher produced oil volumes,
- shorter cycle lengths,
- lower SORs for the cycles (SOR = steam volume injected in cycle/oil volume produced).

Imperial piloted this process, termed "LASER", in the early 2000s in its CSS project at Cold Lake. The pilot test injected solvent with steam (nominally 10% by volume) as a "low-concentration" process into 8 CSS wells, with 19 adjacent CSS wells used as a control group. The performance of the test and control groups during Cycle 7 is shown in Figure 4.19. In this chart, the shaded area represents the range of SORs for CSS that were

- observed in the pilot for the 19 CSS control wells,
- predicted by multivariate analyses of CSS performance for the LASER wells (had they been operated in CSS mode),
- predicted by multivariate analysis of CSS for the 19 CSS control wells.

By the end of the cycle, the LASER wells achieved a cycle SOR of 2.5 compared to an average value of 3.3 (+/−0.2) for CSS, corresponding to a 24% reduction

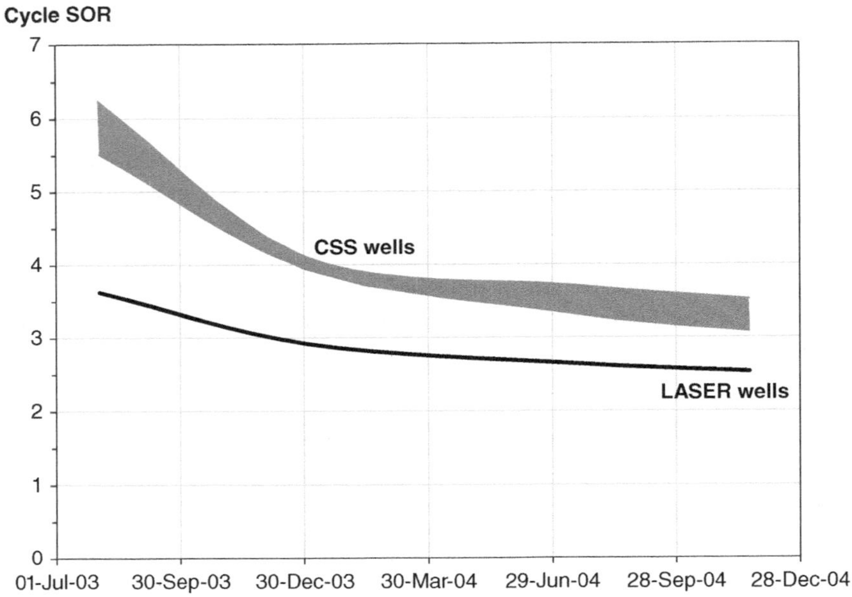

Figure 4.19 Performance of the 8 LASER and 19 CSS wells during the production stage of cycle 7 at Imperial Cold Lake. Data from Leaute' and Carey (2007).

in SOR and GHG intensity (from 0.52 to 0.39 tonne-CO_2e/m^3, a reduction of 0.13 tonne-CO_2e/m^3). By 2010, the process had been sufficiently tested so that it was subsequently put into commercial use on several pads at Cold Lake and has been implemented since then as a standard component of the Imperial's development plan for the resource.

4.3.3.2 High-concentration Processes

There have been no field tests undertaken to advance high-concentration solvent-steam processes in a cyclic configuration.

4.3.3.3 Solvent-only Processes

The general principle of this approach is to inject a low-carbon number solvent (C2–C4) into the reservoir at a pressure that is sufficiently high that the solvent exists in a liquid state and can finger through the bitumen, rapidly mixing with it and diluting it. After a specified volume has been injected, the well is put into production. The drawdown around the well reduces the reservoir pressure, allowing some of the solvent to come out of solution into a vapor phase, providing pressure support to drive the lower-viscosity, diluted bitumen to the well for production. As with CSS, when production has fallen sufficiently, the next cycle of solvent injection is begun. Solvent injection volumes are increased with each cycle to ensure that additional bitumen is contacted.

Depending on the process conditions (temperature, pressure) and solvent used, multiple hydrocarbon phases can form in the reservoir, each containing components of the original bitumen. Much of the literature regarding cyclic solvent injection also discusses the addition of methane to the injected solvent (usually propane) to provide further support to the pressure drive during the production phase of a cycle. Variations of the process have also been proposed as a means to increase recovery in post-primary production operations in lower-viscosity heavy oil reservoirs.

Although there have been many technical papers written on the cyclic injection of solvent as a means to recover bitumen, there has been essentially no field activity to test the process, other than by Imperial Oil. The incentive for this technology was to develop the substantial resources at Cold Lake that exist in zones that have either (1) high saturations with thin pays or (2) thick pays with low bitumen saturations.

The company began to conduct laboratory experiments in 1993, progressing through small-scale proof-of-concept field tests (Imperial Oil Resources, 2010):

- **1996:** Xylene (as a surrogate for lighter hydrocarbon with similar liquid-phase viscosities) was injected into a vertical well to obtain data on injectivity, degree of solvent mixing with the bitumen, and injection pressures.
- **1998:** Ethane was used as a solvent, since its vapor pressure at the reservoir temperature of Cold Lake was equal to the reservoir pressure (3 MPa). The ethane was injected into a 50-m horizontal well. Operational problems complicated interpretation of results.
- **2009:** Propane was injected in a one-cycle test in three deviated wells that had not yet been steamed to gain data on injection and production.

Based on the results of the 2009 pilot, Imperial applied in 2010 for regulatory approval of a substantially expanded pilot testing program. The objective of the pilot was to use three dedicated horizontal wells (200–300 m in length) to evaluate the effectiveness of propane-based solvents over several cycles in terms of key performance indicators for volumetrics, rates, and solvent recovery:

- solvent injectivity,
- the effectiveness of different solvent blends,
- the impact of wellbore length,
- solvent conformance within the reservoir.

In its regulatory application, Imperial specifically noted that

- the technology was intended to target thin or low oil saturation resources in which traditional thermal processes could not be used.

- an incentive to pursue the technology was the significantly. lower level of GHGs associated with the process compared to traditional thermal methods.

The performance of one of the three wells is shown in Figures 4.20 and 4.21. As shown in Figure 4.20, the behavior shown in the cycles of this solvent-based process mimics the trends observed in the CSS process, in that

- in a cycle, the oil rate ramps up and then declines following injection.
- the average calendar date production rate exhibits a maximum in the middle cycles (e.g., cycles 3 or 4).
- the cycle solvent–oil ratio declines to a minimum in a middle cycle and then begins a significant increase.

By the end of 2018, the three wells had completed a total of 13 cycles. The final performance metrics reported are summarized in Table 4.11.

As a result of this pilot, Imperial deemed CSP technology to be commercially viable. Imperial plans to begin commercial deployment by using the technology to develop a thick, but low-quality (6–8 wt% bitumen) resource at Cold Lake. The bitumen recovery of the CSP base process is estimated by Imperial to be between 20 and 50%, depending on a number of factors, such

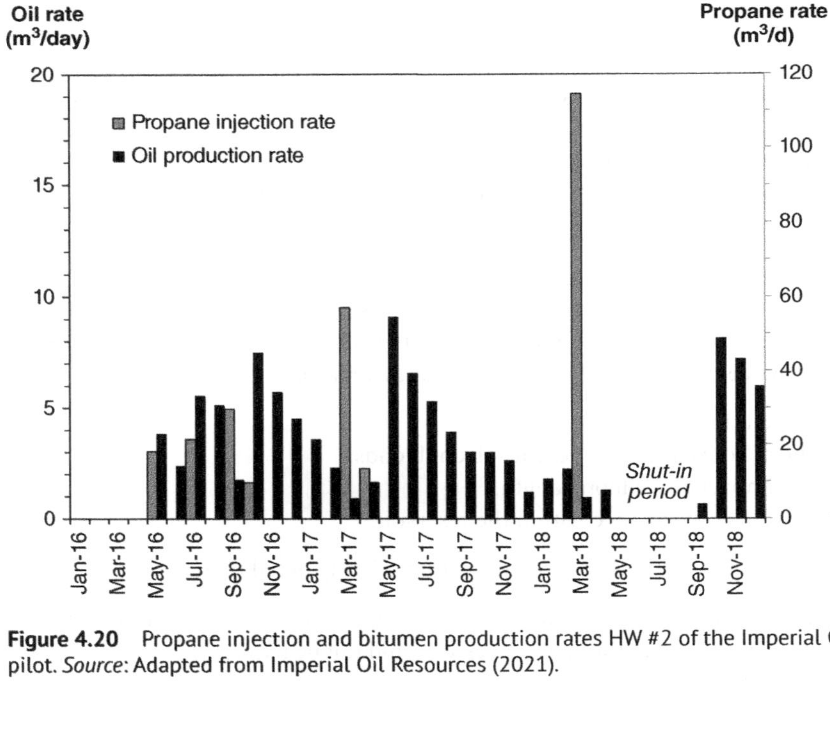

Figure 4.20 Propane injection and bitumen production rates HW #2 of the Imperial CSP pilot. *Source*: Adapted from Imperial Oil Resources (2021).

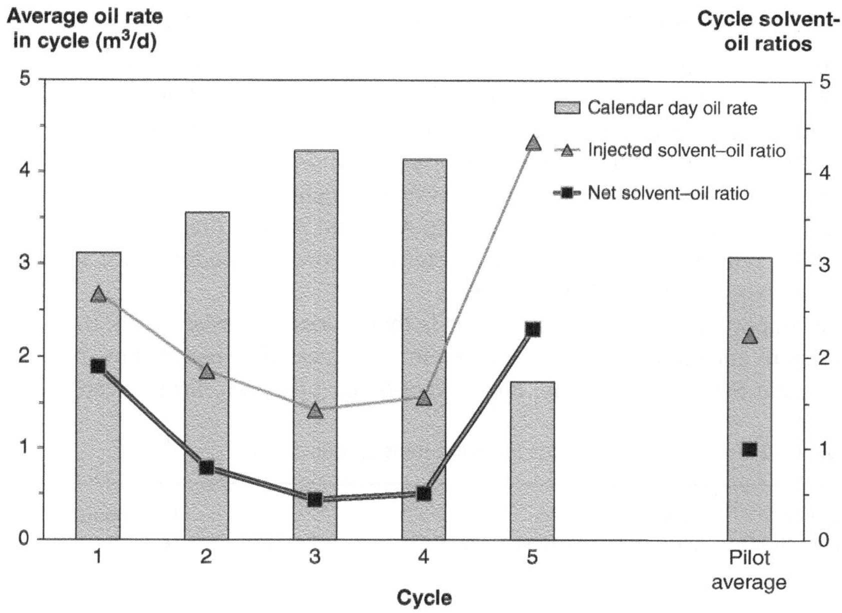

Figure 4.21 Average cycle bitumen rates and solvent–oil ratios in cycles of HW #2 of the Imperial CSP pilot. *Source*: Adapted from Imperial Oil Resources (2021).

Table 4.11 Performance metrics for the Imperial Oil CSP pilot.

Performance metric	Value
Total volume of propane injected	27,074 m^3
Total bitumen production	11,406 m^3
Injected solvent-to-oil ratio	2.4
Total solvent recovery (to end 2018)	55%

Source: Adapted from Imperial Oil Resources (2021).

as the well spacing, resource thickness, and resource quality. In its current online Innovation Brochure (Imperial Oil, 2024), the company indicates that the process can reduce GHG emissions by up to 90% by eliminating the use of steam. In its final report-out in 2018 to the ERA (Imperial Oil Resources, 2018), Imperial estimated that the CSP process could reduce the GHG intensity by almost 80% in comparison to CSS (0.143 tonnes-CO_2e/m^3 vs. 0.671 tonnes-CO_2e/m^3). The final performance metrics for the pilot are shown in Table 4.11.

4.3.4 Expected GHG Reductions for New Recovery Technologies

Table 4.12 lists the information that oil sands producers that have conducted field trials of solvent-based processes have provided on the expected impact on GHGs on their operations. One point that stands out from the table is that the general reduction in GHG intensity expected for a solvent process, whether using low or high concentrations, is greater for the lower MW, lighter solvents such as propane than for heavier solvents such as diluent. A second point is that the expectations for GHG reductions are generally consistent with the field performance and simulation expectations for SOR reductions.

An important point to keep in mind is that unless otherwise noted, the GHG emissions and reductions discussed herein, and as usually presented by the operators, are Scope 1 emissions. For solvent-based processes, Scope 2 emissions

Table 4.12 Industry expectations for GHG intensity reductions.

Solvent concentration	Solvent type	Process name	Operator	GHG intensity reduction
100%	Light	Cyclic Solvent Process[a]	Imperial	Up to **90%**
		NsolvBEST Process[b]	Suncor	**80%** or better
High	Light	Solvent Driven Process[c]	Cenovus	Up to **90%**
	Heavy	Enhanced Bitumen Recovery Technology[a,d]	Imperial	Up to **60%**
Low	Light	Light Hydrocarbon Enhanced SAGD[e]	Canadian Natural	**~45%***
		Solvent Assisted Process[f]	Cenovus	At least **25%***
		eMVAPEX[g]	MEG	As much as **43%**
	Heavy	LASER[a]	Imperial	Up to **25%**
		Solvent-Assisted-SAGD Process[a]	Imperial	Up to **25%**
		Expanding Solvent-SAGD[d]	Suncor	Potential **30%**
		SAGD+[h]	Connacher	**32%***

*Denotes SOR-basis.

a. Imperial Oil (2024).
b. Suncor (2015).
c. Cenovus FCCL Ltd. (2018).
d. Suncor (2024).
e. Canadian Natural (2022).
f. Cenovus Energy Inc. (2021).
g. Emissions Reduction Alberta (2024).
h. Lau et al. (2014).

associated with the volumes of solvent that are not recovered also need to be considered when ultimate comparisons of processes and any associated environmental improvements are made.

4.3.5 Practical Considerations and Challenges for Implementation of New Recovery Processes

From the material presented in this chapter thus far, it might be tempting to conclude that the addition of solvent to steam in all projects could quickly reduce the industry's GHG emissions and intensity by at least 25% in the near term. Unfortunately, the following realities present challenges to this aspiration.

4.3.5.1 Applicability

The economics of solvent processes remain marginal, as attested to by the need to recover such high levels of solvent, estimated to be >80% by industry. Poor economics could preclude implementation of the processes in existing pads that

- have high SORs and low productivities,
- have recovered more than 50% of their ultimate recovery,
- are located next to depleted pads that could act as thief zones for the injected solvent.

A rough estimate of these exclusions corresponds to about 50% of current industry production.

Cenovus has publicly indicated (Cenovus Energy Inc. (2024).) that although the potential remains for solvents at its future standalone projects, it does not plan to use solvents in its existing thermal projects at Foster Creek and Christina Lake. These projects already have such high levels of performance (low SORs and low GHG intensities) that there would appear to be only marginal impacts from implementation.

4.3.5.2 Cost to Retrofit

The cost of retrofitting existing pads for operation with solvents can be high, in particular for projects using light hydrocarbons. As of this writing, MEG (eMVAPEX) has publicly announced that it will not be using solvents in its existing thermal project at Christina Lake, but will await future stand-alone developments (MEG Energy Corp., 2023).

4.3.5.3 Minimal Impact on Emissions

Even where solvents, particularly heavier solvents, can be used in existing projects, such implementations will reduce the GHG intensity of the projects (GHG/m^3 bitumen), but should not be expected to necessarily reduce the GHG emissions (tonnes GHG per year). Solvents lower the SOR of the project and so will produce more oil using the existing steam capacity of the projects. Realistically, if sufficient developable resources remain for the project, operators could be expected to continue to operate using the maximum steaming capacities, since any shut-in of steam would defray the economic present value of the projects. Projects would thus continue to burn the same amount of natural gas until resources were depleted, and the GHG emissions from the projects would remain unchanged.

4.3.5.4 Competition from Other Initiatives

As discussed in the following section, there are two technology initiatives to reduce emissions that are underway that provide potential challenges to need for new recovery process to reduce GHG emissions (notwithstanding the potential for the new recovery processes to improve performance metrics over current methods): Carbon Capture and Storage (CCS) and green power.

4.3.5.4.1 Carbon Capture and Storage

Six operators of the major in situ oil sands projects are currently engaged in the Pathways Alliance to provide CCS of the CO_2 generated by the in situ industry. The CO_2e emissions from the in situ sector exceeded 40 million tonnes in 2022. The Alliance originally expected that the project could sequester approximately 10–12 million tonnes annually by 2030 and as much as 40 million tonnes per year by 2050 (Pathways Alliance, 2023). With a significant level of financial commitment required and with the corresponding level of impact that CCS can have on GHG emissions, in situ operators may be challenged to invest substantial additional funds in solvent projects that offer less impact on both intensities and emissions.

4.3.5.4.2 Green Power

At the same time that the Pathways Alliance is addressing reduction of GHG emissions through capture and storage of generated GHGs, other longer-term initiatives are underway to eliminate emission generation entirely through

- electrical heating using green power,
- development of local green power with Small Modular Reactors (SMRs) using nuclear power.

Both of these approaches are still in their early stages, but in the longer term could make significant contributions. Their success could challenge the use of solvents, especially depending on the future of the Pathways initiative.

4.3.6 Concluding Comments – New Recovery Processes

4.3.6.1 Historical Development

The evolution of solvent-based processes is now approaching three decades of development, testing, and application. Apart from the commercialization of the

LASER process at Cold Lake, it may appear that the technology has been slow to live up to its promises of improving recoveries and reducing GHG intensities on a commercial scale. There is, in fact, a reference to solvent use for heavy oil well stimulation dating back to the 1960s (Jeffries-Harris and Coppel, 1969).

A counterargument would point out that both of the main thermal technologies, CSS and SAGD, did not achieve commercial status until after more than two decades of research and field testing.

- CSS in Canada was first tested in the 1960s but first applied on a commercial scale in the mid-1980s by Imperial Oil at Cold Lake – 2 decades.
- SAGD was invented in the early 1980s, first field-tested at the Underground Test Facility in the mid-1990s, and put into commercial operation in the early 2000s – 2 decades.

Given the additional historical concerns over

- the sensitivity of the process economics to solvent recovery,
- the cost of implementation,

it may not be so surprising that progress for commercialization of solvent technologies has seemed slow. But two trends suggest that progress has been made in maturing the technology to commercial readiness.

4.3.6.1.1 Trend in Number of Technical Publications

Since 1990, the use of solvents to enhance thermal recovery performance has been a prolific field for publications – 391 publications were identified since 1990 with direct relation to the use of solvents for heavy oil recovery. The distribution of the dates of these publications (One Petro, 2024) shows an interesting correspondence with the progress that has been made toward commercialization, as shown in Figure 4.22. The cumulative curve closely resembles the well-known "S-Curve" for technology development and commercialization, showing periods corresponding to a slow start-up, rapid buildup, and final recent tapering-off of academic and industrial publications.

4.3.6.1.2 Trend in Types of Field Tests

The trend in publication shown in Figure 4.22 is built upon by three periods of industry field activity:

- **1990–2004:** Theoretical work, laboratory experiments, and single-well field experiments to establish the proof-of-concept and winnow out lower-potential processes,
- **2005–2017:** Field testing in small to moderate-sized projects to reduce uncertainty and establish key performance indicators, such as the Cenovus SAP

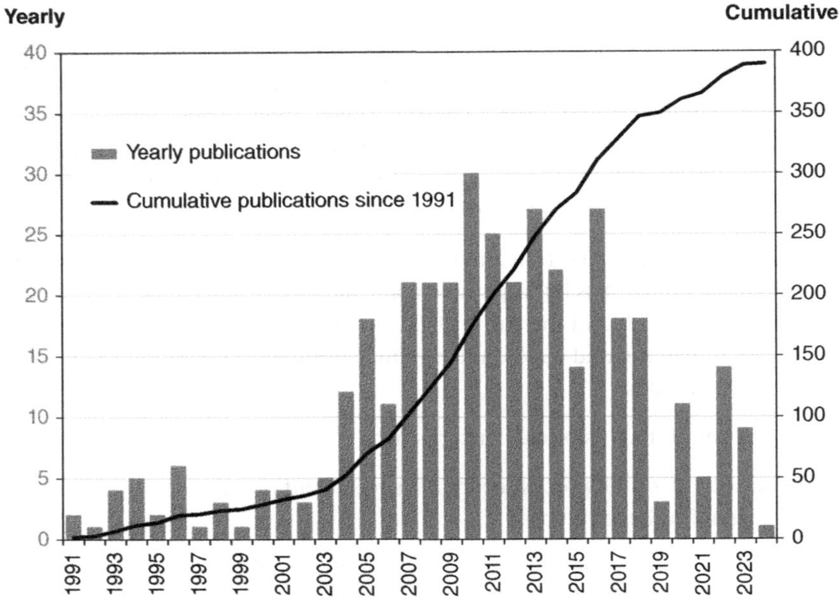

Figure 4.22 Historical distribution of annual publications related to the use of solvents in heavy oil recovery (One Petro, 2024).

pilots at Christina Lake and Foster Creek and the Imperial SA-SAGD field test at Cold Lake,

- **2018–2023:** Field testing on larger scales to establish the basis for commerciality, such as MEG's eMVAPEX pilot at Christina Lake (A02 pad) and Suncor's ES-SAGD pilot at Firebag (Pad 109S).

These field tests have demonstrated that

- the required recovery levels for solvent can be achieved,
- SORs can be substantially reduced,
- rates can be increased with solvent addition to steam.

A qualitative representation of the progress made by solvent-based technologies since 2010 is shown in Figure 4.23. Before 2010, there were no new solvent-based recovery processes being used. Since that time, there have now been four processes that have reached to point of being declared commercial, with one having several years of commercial experience (LASER), two having started or about to start in the current timeframe (SA-SAGD and LHE), and one deemed as commercial but with no implementation date yet released by the company (CSP).

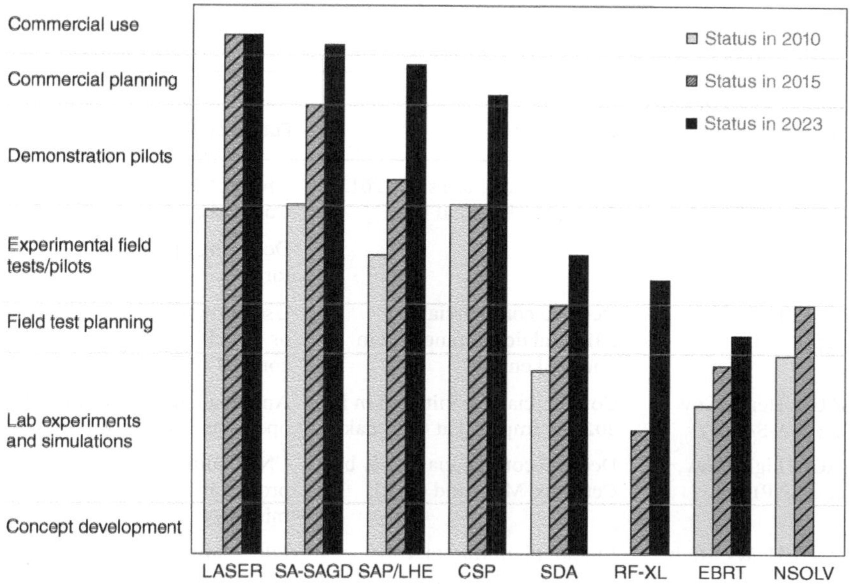

Figure 4.23 Progression of technology development for new recovery technologies since 2010. No 2023 status is shown for the Nsolv process, since field activity was terminated, with abandonment of the wells in 2018 and no further reports of future plans.

4.3.6.2 Current Status and Future Outlook

So now, after two decades of development, solvent-based processes appear to be on the cusp of commerciality, with commercial operations underway in 2024 at Cold Lake (SA-SAGD) and at Kirby North (LHE). Successful outcomes for these projects could be the final hurdle to be cleared, leading to a wider expansion of applications in existing and future projects. Descriptions of the current status and future outlook for the individual technologies are summarized in Table 4.13. Of particular note is the planned high-solvent concentration pilot of the Enhanced Bitumen Recovery Technology (EBRT) process planned by Imperial and Suncor. This process uses a high concentration solvent that has been specifically tailored for operation at lower temperatures. The process has the potential to reduce GHG emission intensity over SAGD by 60%, as well as reducing capital and operating costs by about 50% by 2027 (Emissions Reduction Alberta, 2025). If the industry views this type of process as the true long-term strategy, there may be a continued lull in field activity until the results of the current commercial activities and the higher-concentration pilot are seen.

Table 4.13 2023 Status and future outlooks for new recovery processes: "Heavy" = diluent, condensate, naphtha; "Light" = propane, butane.

Configuration/Solvent/ Concentration	Current status	Future outlook
Cyclic/Heavy/Low (e.g., LASER)	In commercial use since 2012 by Imperial at Cold Lake	Only potential use at Cold Lake
		Declining opportunities for use
Cyclic/Light/100% (e.g., CSP)	Deemed commercial in 2018 and development plan undertaken	Use in thick, low-saturation or thin, high-saturation zones (Cold Lake)
SAGD/Heavy/Low (e.g., SA-SAGD)	Commercial use initiated in late 2023 by Imperial at Cold Lake	Anticipate use by additional operators if successful
SAGD/Light/Low (e.g., SAP)	Deemed commercial-ready by Cenovus, MEG and CNRL	CNRL initiating commercial project at Kirby North in mid-2024
SAGD/Heavy/High (e.g., EBRT)	In pilot planning stage by Imperial and Suncor for Aspen	Pilot planned for 2027, with potential commercialization in mid-2030s
SAGD/Light/High (e.g., SDP)	Ongoing pilot by Cenovus at Foster Creek	Cenovus plans solvent use in future projects

As noted earlier, future progress in solvent development will remain challenged by

- applicability to existing projects with mature pads,
- cost to retrofit projects, especially for light hydrocarbon solvents,
- little impact on emissions due to maximizing use of available steam capacity,
- competition from other initiatives such as CCS, EM technology, and SMRs.

Despite these challenges, the promise remains that solvents can have a substantial impact on GHG emission reduction for new in situ oil sands projects that are brought forward in the next decade. With the aging of the current facilities and their resources and the economic uncertainty of alternate approaches, the next few years of commercial application of solvents (albeit limited) may demonstrate that solvent processes can indeed have a long-term impact on reducing GHG emissions while sustaining heavy oil industry production. At a minimum, they will provide a path for increasing the recovery efficiency of thermal processes and shrink the limitations on their current applicability.

4.4 Other Recovery and Associated GHG Intensity Reduction Technologies

Solvent-based processes will not be a panacea or the only technology available to assist in reducing GHG intensities and emissions. Other technologies have been tested, with some suspended, some already implemented, and others still in the early stages of planning and development. Many of these technologies are more focused on improving operational efficiencies in existing operations, while others examine completely different approaches to reducing GHGs. Several of these technologies are the subject of this section.

It should be noted that some of the developing technologies discussed in this section include information on their assigned Technology Readiness Level (TRL). This number (1–9) is a numerical yardstick of how far along a technology has progressed, as measured against specific types of achievements. For the technologies discussed, the TRL is generally in the range of 6–9, which is defined by the Government of Canada 2024) as summarized in Table 4.14.

4.4.1 Other Technologies – Historical

In addition to the methods discussed earlier, in which injectants such as steam and solvents are heated at surface prior to injection, there are additional processes that have been field-tested in which the heat is introduced into the reservoir by either

- air injection to initiate in situ combustion (ISC), or
- electrical resistance heating of the reservoir around a wellbore.

For reasons such as performance, operational problems, or competition from newer methods, these processes either currently have low levels of production

Table 4.14 Descriptions of more advanced technology readiness levels (TRL).

TRL	Description
6	A model or prototype (laboratory or simulation) has been shown to represent a near-desired operational configuration.
7	Prototype has been constructed at planned operational level and is ready for demonstration.
8	Technology has been proven to work in its final form and under expected operational conditions.
9	Actual application of the technology demonstrated in its final form and under real-life conditions.

and little industry interest (ISC) or did not progress beyond the small-pilot stage (electrical resistance). These processes are presented herein for completeness.

4.4.1.1 In Situ Combustion

This process uses injected oxygen or oxygen-enriched air to burn some of the oil in situ, generating a burn front in the reservoir. Ahead of the burn front, the generated heat lowers the viscosity of the bitumen, which is pushed to a producing well.

ISC has been undertaken at over 230 pilot and commercial-scale projects in the United States. Notable examples were at the Midway Sunset, and Belridge fields in California and at Bellevue in Louisiana, which is the only currently active project in the United States. Several testing initiatives have also been undertaken in Canada, including those at the Battrum, Eyehill, Tangleflags, Lindberg, and Countess projects (Sharma et al., 2021). Outside of North America, field projects were initiated in Europe in the 1980s, particularly in Romania, Hungary, and the former Soviet Union. There are currently a limited number of commercial-scale projects worldwide (Antolinez, 2023). Operational problems, such as control of the burn front, ignition and sustainment of high-temperature oxidation, and production of tight emulsions, have resulted in limited interest by industry.

Publications that have noted the use of ISC to achieve reduced GHG intensities have specifically assumed that Carbon Capture and Sequestration would be incorporated into the process design to achieve such reduction. Two examples are provided by Paurola et al. (2016) and Rodriguez et al. (2023).

4.4.1.2 Electrical Heating

These processes use subsurface electrodes (e.g.,) to directly heat the reservoir and reduce the bitumen viscosity for subsequent production. From a GHG perspective, these processes potentially offer to reduce GHG emissions by using electricity from the power grid rather than burning natural gas on-site to generate steam. However, as noted in a prior section, further scrutiny of the Scope 2 emissions associated with the electricity generation process (coal, natural gas, hydroelectric, or wind/solar) is required to confirm this potential. In general, the field pilots of these processes have been disappointing because they have either

- exhibited performance metrics that would not be economically competitive, or
- experienced significant operational problems that have resulted in termination of the pilot.

Pilots illustrating different approaches to electrical heating are reviewed in the following sections.

4.4.1.2.1 Electrical Preheating

The pilot operated by the PCEJ consortium of companies ran from April 1981 to February 1982 in the Athabasca Lower McMurray sand (Ghannadi, 2016). The objective was to evaluate the performance of a steam flood using 4 wells that were heated by electrodes against unheated performance. The pilot was terminated following major operational challenges of flashing and gas liberation due to heating the reservoir and an inability to balance output from the electrodes.

4.4.1.2.2 Thermal-assisted Gravity Drainage

Athabasca Oil Corporation's thermal-assisted gravity drainage (TAGD) process heated the reservoir using heaters in both the top (heater well) and bottom (producer) (Ghannadi, 2016). Heated oil around the top well drains to the bottom well. The pilot was conducted in the (carbonate) Leduc formation from 2011 to 2105 and achieved inter-well temperatures of 150 °C. The company reported that by the completion of the pilot, 1500 m^3 of bitumen had been recovered using about 55 Terra-Joules of energy. Using only the latent heat of water vaporization for calculation, this is equivalent to an SOR of 16.2. No further work was conducted.

4.4.1.2.3 Electro-thermal Dynamic Stripping Process

This process was developed by ET-Energy (McMillan-McGee) to extract bitumen from mid-range depth reservoirs (with depths of between 75 and 150 m) in which neither mining nor SAGD processes can be economic or allowed under regulatory regimes if there are caprock concerns. (Ghannadi, 2016). This process uses the electric conductivity of connate water to propagate an alternating current between electrodes. The first field test started at Poplar Creek in 2006, in which 14 vertical production wells were placed between 23 heated ("Extraction") wells in patterns. The company reported that 622 m^3 of bitumen was produced using a total of 19,470 GJ over 425 days of operation. Using only the latent heat of water vaporization for calculation, this is equivalent to an SOR of 15. No further work was conducted.

4.4.1.2.4 Low-pressure Electro-thermally Assisted Drive

This field test conducted by Perpetual Energy used electrical resistance heating in a horizontal well to reduce the 50,000-mPa·s bitumen viscosity (Perpetual Energy, 2019). The well was originally planned to operate in a cyclic mode (heat-produce), but the operator determined that it could operate in a continuous mode. The pilot operated from October 2015 to June 2017 and used 11,271 GJ of energy to produce 1,650 m^3 of bitumen. Using only the latent heat of water vaporization for calculation, this is equivalent to an SOR of 3. The company noted that a primary problem encountered was heater failures, and the pilot was terminated after a failure in 2017. No further work has been conducted to date.

4.4.1.2.5 Summary

The pilot data presented on the electrical heating processes can be used with the emission data for various power sources (Table 4.5) to provide estimates of the GHG intensities associated with each power source, as shown in Table 4.15. For comparison, the GHG intensity of an average current thermal process having an SOR = 3 is 0.472 tonnes-CO_2e/m^3-bitumen. The low-pressure electro-thermally assisted drive (LEAD) pilot had the lowest estimated GHG intensity of the three processes and a substantially lower GHG intensity than the current average thermal process if a "green" power source is used to supply electricity. However,

Table 4.15 Estimated GHG intensities for historical electrical heating pilots.

Pilot	Thermal-Assisted Gravity Drainage (TAGD)	Electro-Thermal Dynamic Stripping Process (ET-DSP™)	Low-Pressure Electro-Thermally Assisted Drive (LEAD)
Pilot data			
Energy used (GJ)	55,000	19,470	11,271
Oil produced (m^3)	1,500	622	1,650
Equivalent SOR	16.5	15	3
Electrical power source	**GHG intensity (tonnes-CO_2e/m^3-bitumen)**		
Coal	9.02	7.70	1.68
Natural gas	5.39	4.60	1.00
Solar (photovoltaic)	0.528	0.451	0.098
Solar (mirrors)	0.297	0.254	0.055
Hydroelectric	0.264	0.225	0.049
Nuclear	0.132	0.113	0.025
Wind	0.128	0.110	0.024
Electrical power source	**Reduction in GHG intensity from steam (0.472 tonne-CO_2e/m^3)**		
Coal	0%	0%	0%
Natural gas	0%	0%	0%
Solar (photovoltaic)	0%	4%	79%
Solar (mirrors)	37%	46%	88%
Hydroelectric	44%	52%	90%
Nuclear	72%	76%	95%
Wind	73%	77%	95%

if the electricity for these projects was supplied by power plants that use coal or natural gas, all three processes would show substantially higher GHG intensities than the average current commercial thermal project. If the performance metrics estimated for these pilots reflect their commercial performance, then some form of green power is required for them to be able to reduce GHG emissions from the current average of steam-based thermal processes.

4.4.2 Other Technologies – Current

This section discusses emission-reducing technologies that are currently in commercial use that are either used directly – or as support in thermal recovery projects – to recover heavy oil. In some cases, the impact on CO_2e emissions has been quantified. In other cases, there have not been publications to date to quantify the impact, although the impact would be expected to be favorable and useful in improving the GHG performance of the recovery processes.

4.4.2.1 Polymer Flooding

Polymer flooding is an enhanced oil recovery method for recovering viscous oils that has been practiced worldwide for decades. It consists of adding low concentrations of high MW polymers – usually polyacrylamides – to increase the viscosity of injected water. The higher viscosity provides better sweep efficiently in the reservoir and allows more oil to be recovered with less fluid injection. Historically, the process has been applied in reservoirs with only moderately viscous crude oils having viscosities <1,000 mPa·s. Beginning in the early 2000s, polymer flooding was extended to reservoirs with up to 50,000 mPa·s in Alberta, with ultimate recovery factors of over 20% expected (Canadian Natural Resources, 2020). On a worldwide basis, the application of polymer flooding in such viscous crudes still remains more of an exception to routine practice.

Although the increase in recovery efficiency for heavy oils has been the primary impetus for this technology, recent studies have also highlighted its benefits from a GHG-intensity perspective, noting the following reductions compared to waterflooding: 35% (Skauge et al., 2024), 2–6 times (Dupuis et al., 2021), and 3–6 times (SNF, 2021, 2023). It is important to recognize that the GHG intensity level of waterflooding is a fraction of the GHG intensity for thermal processes.

Table 4.16 shows comparisons between the reduction in GHGs for

- an SA-SAGD project relative to a typical SAGD project (SOR = 3),
- the SNF analysis for a polymer flood compared to a waterflood.

As shown in the table, although the SNF analysis (SNF, 2023) showed more than twice the relative reduction (%) in GHG intensity by polymer flooding (full-life) compared to a solvent-steam process, the solvent-steam process still reduced

Table 4.16 Estimated reduction in GHG intensity by enhanced processes: polymer flooding vs. SA-SAGD.

Base process	**SAGD**	**Waterflooding**
Enhanced process	**SA-SAGD**	**Polymer flooding**
GHG intensity of base process (tonne-CO_2e/m^3)	0.472	0.164
GHG intensity of enhanced process (tonne-CO_2e/m^3)	0.330	0.044
GHG intensity reduction (%)	30%	73%
GHG intensity reduction (tonne-CO_2e/m^3)	0.142	0.120

the intensity by a higher value. This difference is a direct result of the thermal process having almost three times the initial GHG intensity as the waterflooding process.

4.4.2.2 NCG Injection

This practice has evolved into a standard operating practice in the in situ thermal industry in the past decade. Operators have determined that as thermal projects (both steam flood and primarily SAGD) enter their final declines and monthly SORs increase to high levels, the injected steam can be replaced by injection of NCG – usually natural gas – without any material impact on production rates, because so much heat is already stored in the formation and over/underlying strata. The steam that is made available by being replaced by NCG injection into the old, high-SOR pads can thus be used in new, low-SOR pads. This results in a lower, or more slowly increasing, SOR for the overall project, with a corresponding impact on the project GHG intensity as discussed earlier.

On a project-by-project basis, the quantitative level of impact will vary depending on the specific distribution of the steam to the project pads and their SORs. There has been some limited information published for estimates of the specific impact on either SOR or intensity.

- Strathcona Resources (2023) has indicated that the practice reduced monthly steam usage – and thus GHG intensity – by 20% on pads where it had been used, and SORs by up to 0.5-m^3-steam/m^3-bitumen.
- Athabasca Oil Corporation (2020) reported that injection of NCG with steam resulted in a 25% reduction in SOR (from 4.5 to 3.25) in a four well pair pilot at its Hangingstone SAGD project.
- ConocoPhillips Canada (2019) noted that its steam-NCG injection pilot at Surmont was able to reduce the SOR by 35% by mitigating steam losses to a thief zone at the top of the formation.

4.4.2.3 Flow Control Devices

These devices are installed in wellbores to automatically control the inflow of produced fluids or outflow of injected fluids. Historically, they have been used in the conventional oil and gas industry to mitigate coning. Over the past decade, they have found substantial use in the heavy oil industry. In the thermal sector, FCDs are used to more evenly distribute steam along injection wells or to reduce excessive production at points along producers that could result in the production of live steam. By more uniformly contacting the bitumen with steam and producing the heated bitumen (termed "improved conformance"), the thermal efficiency (SOR) of the process is improved, with a corresponding reduction in GHG emissions. Despite their growing level of use, there has been no quantification published to date by the industry on the specific level of impact on SORs or GHGs.

In primary heavy oil production, FCDs can be used to reduce water cuts, leading to less energy usage for treating and transportation. A recent paper (Moradi, 2022) examined the impact of FCDs on mitigation of water production in a "cold flow" primary production project. The result indicated that FCDs were found to reduce GHG emissions and energy consumption in production and surface processing by 67%.

4.4.2.4 Cogeneration of Steam and Electricity

Cogeneration refers to the simultaneous production of steam and electrical energy. Thermal projects that use cogeneration to generate steam can receive GHG emission credits for any power that is sent to the electric grid, because it is displacing some demand (and associated GHG production) from the power grid. The formula currently used to compute the offset is (Alberta Government, 2024):

$$\text{Co-gen adjustedtotal emissions} = (\text{Total emissions} - \text{Co-gen emissions}) + \text{Co-gen heat} \times 0.06299 \text{ tonnes per GJ heat} \tag{4.4}$$

For 2022, which is the latest year for which emission data are publicly available, 9 of the 23 thermal projects in Alberta used cogeneration to produce all or part of the project steam. The contributions of these projects are shown in Table 4.17 (Alberta Government, 2024; geoSCOUT, 2024). Key points from this analysis are as follows:

- Use of cogeneration was credited with reducing total GHG emissions and intensity by 9% on an industry basis, and by an average 19% for those projects that used cogeneration.
- Reductions for the individual projects that used cogeneration ranged from 8% to 44%, depending on the contribution of cogeneration to total steam production.

Table 4.17 Contributions of cogeneration to reducing emissions in thermal industry in 2022.

Data	Co-gen Projects	Non-co-gen Projects	All Projects
Emissions (tonnes CO_2e)			
Total	23,500,000	25,400,000	48,900,000
From Co-gen	9,600,000	0	9,600,000
Heat from Co-gen Heat (GJ)	81,700,000	0	81,700,000
Emissions – Co-gen adjusted (tonnes CO_2e)			
Total	19,000,000	25,400,000	44,400,000
Reduction due to Co-gen	4,500,000	N/A	4,500,000
% Reduction in total due to Co-gen	19%	N/A	9%
Bitumen production (m^3)	54,000,000	34,300,000	88,300,000
GHG intensity (tonnes CO_2e/m^3)			
Before Co-gen adjustment	0.435	0.741	0.554
After Co-gen credit adjustment	0.352	0.741	0.503
GHG intensity reduction (tonnes CO_2e/m^3)	0.083	N/A	0.051

- In general, cogeneration was used in projects (8 of 9) that already had SORs and intensities below the industry average and contributed over 60% of the total production in 2022.

4.4.2.5 Geophysical Monitoring

One recent paper (Passalacqua and Strack, 2020) discussed the potential for geophysical monitoring technologies to assist the thermal industry in acquiring subsurface information on the 3-D development of the steam chamber and conformance within the reservoir. This information can help to optimize the thermal process and thus reduce SORs and GHG intensities and emissions. These geophysical techniques include 4D borehole and surface gravimetry, 4D seismic, and 4D EM. The authors note that "In a daily production of 3 million barrels of heavy oil by thermal injection methods with an SOR of 3.5, this [1% reduction] is equivalent to a reduction of 300 thousand kgs of CO_2 daily."

4.4.2.6 Solar Power

The use of solar power technology to generate energy for steam generation has an obvious environmental benefit over the use of natural gas, in that it would essentially eliminate Scope 1 emissions associated with the thermal recovery process.

The key limiting concerns have been both based on economics (high capital costs for the required solar power) and additional requirements for energy storage (or backup) when solar power is not locally available (night).

An initial economics study in 2012 (Sandler et al., 2012) examined the economics for implementation of solar-driven steam generation for a thermal project in California's Kern River field. The study concluded that solar power could be a viable economic alternative for steam generation, especially in areas that received a high degree of sunlight. A subsequent paper (Yegane et al., 2015) examined the power and CO_2e savings for potential use in Venezuela thermal projects, and came to the same conclusion – that solar power could be an economically viable alternative, even if used only during the day.

In 2015, Petroleum Development Oman announced (Renewables Now, 2017) that the first block of its solar park at Miraah had successfully delivered its first steam to the Amal West oilfield using a "concentrated" (mirror-based) solar collection design (vs. photovoltaic). Upon completion of the remaining solar blocks, the complex would have the capacity to deliver 1,021 MW of thermal energy to produce 6,000 tonnes of steam per day. In 2018, Aera Energy announced the proposed construction of an 850 MW solar facility to generate 1.9 million m^3 of steam annually for its 12,700 m^3/d thermal project at Belridge, California. The same "trough" mirror technology as used at Amal West was proposed. A comparison of the estimated GHG reduction for both projects is presented in Table 4.18.

4.4.3 Other Technologies – Evolving

This section discusses emerging heavy oil recovery processes and supporting technologies that either

- have received some field testing with results that were somewhat encouraging but require further development before they can be considered for commercial use, or
- are in the pre-field-demonstration phase, with the potential for substantial impacts in the future.

4.4.3.1 Electromagnetic Heating

4.4.3.1.1 Electromagnetic-SAGD

In its 2010 regulatory application for a SAGD project at its Clearwater West lease, Alberta Oilsands Inc. included a request to potentially pilot a 2.5 MW electromagnetic (EM)-SAGD process that had been developed by Siemens AG. The technology uses an electromagnetic resonance induction coil around a

Table 4.18 Estimated GHG reductions for Amal West (Prabhu, 2022) and Belridge (Power Technology, 2018) Solar Projects.

	Amal West	Belridge
Commercial thermal project (Existing)		
Daily oil rate (m^3/d)	3,023	12,719
Daily steam rate (m^3/day)	13,920	5,227
Steam-oil ratio (SOR)	4.60	0.41
Solar project		
Plant power (MW)	1,021	850
Daily energy (GJ)[a]	40,840	34,000
Daily steam rate (m^3/day)	6,000	5,227
SOR (assumed from commercial project)	4.6	0.41
Estimated daily oil rate (m^3/day)	1,304	12,749
Typical concentrated solar GHG intensity (tonne-CO_2e/GJ)	0.0081	0.0081
Estimated daily GHG emissions (tonne-CO_2e)	330.8	275.4
Estimated solar GHG intensity (tonne-CO_2e/m^3)	0.2536	0.0216
Estimated steam GHG intensity (tonne-CO_2e/m^3)	0.7234	0.0645
Estimated GHG intensity reduction (tonne-CO_2e/m^3)	0.4697	0.04287
Estimated GHG intensity reduction (%)	65%	66%

a. Assumes 12 hours per day of solar energy.

SAGD well pair. Electrical power to the coil generates an electromagnetic field to heat the surrounding bitumen. Simulations conducted by Siemens indicated a 38% improvement by using EM-SAGD over SAGD in a low-pressure application (Koolman et al., 2008). The pilot could not be conducted because the location of the project was not approved by the regulator. No further testing of this process was pursued.

4.4.3.1.2 ESEIEH Radio Frequency Heating Process

The ESEIEH ("Effective Solvent Extraction Incorporating Electromagnetic Heating") process uses radiofrequency heating around an injection well in which solvent is being injected. In this SAGD configuration, the heated and diluted bitumen drains to the production well. The initial field test of radio frequency heating at Suncor's Steepbank Mine confirmed that electromagnetic energy could penetrate into the oil sands and behaved as predicted by numerical simulations.

Initial estimates suggested that the ESEIEH process could reduce GHG emissions by up to 80% (Emissions Reduction Alberta, 2023).

A small-scale (100-m wells) pilot was next conducted at Suncor's Dover site. Although the initial start-up of the process in July 2015 proceeded successfully with reservoir heating following expectations, a downhole event subsequently occurred that resulted in temperature being lost. After multiple attempts at high-power restarts and shutdowns, which included one resulting from a potential arcing event in the transmitter well, operations were suspended in December 2015, and the decision was made to not pursue the project further Suncor Energy Inc. (2022).

4.4.3.1.3 Radio Frequency-XL

Like the ESEIEH process, Acceleware's radio frequency-XL (RF-XL) process uses radio frequency waves from an antenna in a horizontal well to heat the surrounding formation, allowing the heated bitumen to gravity drain to an underlying well (Acceleware Ltd., 2017). A distinguishing feature is that no solvent is used in the RF-XL process. The company indicates that its patent-pending Clean Tech Inverter (CTI) can deliver highly efficient radio frequency energy to the reservoir. Field pilot operations were begun at Broadview Energy's Marwayne field in March 2022, with expectations that the process could reduce GHG emissions by 50–100% (Acceleware, 2017). Acceleware reported initial success with the CTI delivering sustained heating around the 900-m antenna well. Since July 2022, the operation has experienced downhole problems and is currently suspended pending redeployment of subsurface components. The substantial prize offered by the radio frequency technologies in reducing GHGs continues to keep it within industry's interest.

4.4.3.2 Direct Contact Steam Generation

Direct Contact Steam Generation (DCSG) technology co-injects steam and combustion exhaust gases downhole, simultaneously improving the efficiency of oil recovery (by adding both heat and pressure) and mitigating the atmospheric release of GHGs, permanently storing much of the CO_2 in the reservoir. The developer (General Energy Recovery Inc., 2024) estimates that the process emits up to 67% less emissions than the conventional Once-Through Steam Generators that are currently used by the industry. This process is now at a TRL 9, indicating that the process has been proven through successful deployment in an operational setting.

4.4.3.3 In Situ Upgrading Technology

This process injects nano-catalysts into a SAGD injector, placing the catalysts around the well. The catalyst generates an exothermic reaction in the reservoir, promoting better conductive heat transfer in the near-wellbore region and generating lighter components that provide a solvent effect, diluting and helping to further mobilize bitumen. Simultaneously, gases migrate to the upper parts of the steam chamber, creating an insulating blanket, reducing heat loss to the overburden. The developer (NanosTech, 2024) estimates that the process has several benefits:

- Lowering GHG emissions by up to 35%.
- Increasing oil recovery to up to 90% of the original oil in place.
- Upgrading of the oil quality (and thus requiring less diluent for pipelining).

The company is currently pursuing funding options for a field test.

4.4.3.4 Downhole Steam Generation

The concept of using downhole steam generation (DSG) has been around for decades, and there has been some limited use of the technology. The process has historically used natural gas to be combusted downhole to vaporize co-injected water to produce steam. The lack of process control and the need for high-quality steam, especially in SAGD, have limited its use to small-scale field trials.

A new process developed by eSteam (eSteam, 2024) is claimed to allow extraction of heavy oil and oilsands with significantly less costs than a conventional OTSG, with a much lower SOR from existing wells through the use of its proprietary "Zero-Emission Hydrogen Boiler." Hot heat-transfer fluid descends downhole and transfers heat to the feedwater to create high-quality 80% saturated steam or superheated steam in the wellbore to be injected into the reservoir. Among other attributes, the process claims to achieve

- ~50% less fuel per barrel of steam,
- ~25% lower operating expenses,
- ~45% less water converted to steam,
- a lower SOR.

The technology is currently at TRL 7, indicating that it is at the prototype level in an operational environment.

4.4.3.5 Dimethyl Ether

Dimethyl ether (DME) is a highly soluble mutual (oil and water) solvent that can mix with bitumen in the reservoir much faster than other hydrocarbon solvents. The DME bitumen recovery process injects moisturized vapor DME into the reservoir at 80 °C, making it much more energy efficient than SAGD because of the lower temperature. Experimental studies conducted at InnoTech (Alberta Innovates, 2023) demonstrated that the DME bitumen recovery process could deliver the same performance as SAGD with an 86% reduction in direct GHG emissions while using 79% less water. The major concerns expressed about the process have historically been (1) the unit cost of the DME relative to non-synthetic, hydrocarbon solvents, and (2) the retention of the solvent in the reservoir because of its mutual solubility in both oil and water. The technology is currently at TRL 7, indicating that it is at the prototype level in an operational environment.

4.4.3.6 Small Modular Reactors

SMRs are nuclear reactors that have a power capacity of up to 300 MW per unit, which is about one-third of the generating capacity of traditional nuclear power reactors. SMRs can produce a large amount of low-carbon electricity and are physically a fraction of the size of a conventional nuclear power reactor. In comparison to existing reactors, proposed SMR designs are generally simpler, and the safety concept for SMRs often relies more on passive systems and the inherent safety characteristics of the reactor, such as low power and operating pressure. The International Atomic Energy Agency reports that there are currently four SMRs in either advanced stages of construction or operation in Argentina, China, and Russia, with more than 80 SMRs in the design stage globally (IAEA, 2024).

In 2023, Hatch issued a feasibility assessment of using SMRs to provide energy for steam generation at a SAGD facility. The study was sponsored by Cenovus, TC Energy, and Alberta Innovates. It concluded that "SMRs are a feasible option for the provision of electricity and steam in the oil sands to support net-zero energy production at in situ recovery facilities." (Hatch, 2023) The size of the SMR facility used in this assessment was 400 MW (thermal). For the generic economic assessment, it was assumed that the capital cost for the SMR facility would be $3 Billion, with annual operating costs between $32.5 Million and $97.5 Million. The basis of the design assumed a facility capable of generating 15,757 tonnes per day of steam and producing 5,246 m^3/day of bitumen. It was estimated that it would take 10 years from initial approval until first steam.

The feasibility assessment did not provide a quantitative value for the level of GHG reduction. However, the design information provided above (400 MW to make 5,246 m^3-bitumen per day) can be used with the information in Table 4.4 on the typical life-cycle emissions from nuclear power (0.0033 tonnes-CO_2e/GJ) to provide an estimate of 0.020 tonnes-CO_2e/m^3-bitumen. This represents a 96% reduction over the Scope 1 emissions (0.472 tonnes-CO_2e/m^3) estimated for a steam-based thermal project having an industry-average SOR of 3.

4.4.3.7 Carbon Capture and Storage

While the SMR approach would effectively eliminates GHG emissions at the front end of a thermal project by replacing natural gas-fired steam generators with a small modular nuclear reactor, this approach eliminates the emissions to the atmosphere at the back end of the steam generation facility by capturing the emissions and storing them underground in deep formations in the earth. This approach has been used at sites in the United States and Canada, in particular at:

- the Shell Quest project in Alberta, in which CO_2 from a refinery that upgrades produced bitumen is pipelined to an abandoned oil reservoir and stored.
- the Whitecap Resources project in Weyburn, Saskatchewan, receives CO_2 from the Dakota Gasification Company coal gasification facility in North Dakota and from SaskPower's Boundary Dam CCS Facility. The injected CO_2 is also used to recover additional oil before being finally stored in the oil reservoir itself. This type of application is termed Carbon Capture, Use, and Storage (CCUS).

For the oil sands industry in Alberta, a consortium, named The Pathways Alliance, has been formed by the major oil sand producers to investigate the feasibility of storage of the CO_2 from their operations. The application for regulatory approval was submitted in Q1 2024. With regulatory approval, the Alliance estimated that operations to store the CO_2 could begin by late 2026. (Pathways Alliance, 2023).

There are two important points to recognize regarding CCS projects (Shell Canada, 2023 and Pathways Alliance):

- CCS projects are expensive to build and operate, as shown in Tables 4.19 and 4.20.

Table 4.19 Estimated capital costs and capital efficiencies for CO_2 sequestration.

Project	Capital (Estimated)	MT-CO_2 Stored per year	Tonnes-CO_2 per day	Capital efficiency \$/tonne-CO_2 per day
Quest	C$1,350,000,000	1	2,740	C$492,750
Pathways	C$16,500,000,000	10	27,397	C$602,250

Table 4.20 Operating and total costs for the Quest CCS Project in 2022.

Basis	Operating costs	Total costs
Cost per tonne of CO_2 captured	C$58.83	C$139.52
Cost per tonne of CO_2 avoided[a]	C$75.66	C$167.86
Incremental cost due to associated emissions from CCS	29%	20%

a. "CO_2 Avoided" is calculated as the mass of "CO_2 Captured" net of the mass of CO_2 emissions associated with the capture and sequestration processes.

Table 4.21 Operating and total costs for Shell Quest CCS project in 2022.

Year	MT-CO_2 stored	MT-CO_2 emissions for capture process	Emissions from capture as % of stored emissions
2018	1.066	0.195	18%
2019	1.128	0.182	16%
2020	0.941	0.158	17%
2021	1.055	0.180	17%
2022	0.971	0.167	17%

- There are significant levels of GHG emissions associated with the CCS projects, as shown in Table 4.21.

A rough estimate of the efficiency of CCS in capturing and sequestering CO_2 emissions would thus fall in the range of 70% (based on Table 4.20) to 80% (based on Table 4.20). Note: the capital efficiencies of the projects appear even closer when the capital estimate for the Quest project is inflated by 32% (World Data, 2024) from its estimate in 2009 to 2023 (C$650,430/tonne-$CO_2$ per day).

4.5 Summary of GHG Reduction Estimates

Table 4.22 summarizes all of the estimates for GHG reductions for the technologies presented. These reductions are compared against the average intensity of 0.472 tonne-CO_2e/m^3 (SOR = 3) for current thermal processes.

4.6 Key Takeaways and Concluding Comments

This chapter has attempted to present a broad picture of technologies that have been and are being field tested or in commercial use that can contribute to reducing GHG emissions from the in situ heavy oil sector. Focus has been placed on solvent-based processes because of the amount of time and field testing that have been spent in their development, along with the potential high level of impact that they can have on reducing emissions and intensities in heavy oil production, both in Canada and worldwide. Some key takeaways from this analysis are as follows:

- Historically, significant reductions in the annual GHG intensity of the thermal industry have resulted from the contributions of newer, lower SOR

Table 4.22 Summary of expected contributions of technologies to reduce emissions.

Technology	Reduction	Notes/comments
Solvent-based technologies (field-tested)		
Solvent-steam processes: SAGD configuration, Low-concentration	25–45%	Range based on solvent type used (light vs. heavy)
Solvent-steam processes: SAGD configuration, high-concentration	60–90%	Range based on solvent type used (light vs. heavy)
Solvent-steam process: CSS configuration	25%	LASER process in comparison to CSS process
Solvent-only processes: SAGD configuration	>80%	SDP and EBRT processes
Solvent-only process: CSS configuration	>80%	CSP process
Other technologies – historical, field-tested, but no impact without CCS or "green" power source		
In situ combustion (ISC)	—	Requires/assumes associated CCS project to handle produced emissions
Thermal-assisted gravity Drainage (TAGD)	<73%	No reduction unless green power source used
Electro-thermal dynamic stripping process (ET-DSP)	<77%	No reduction unless green power source used
Low-pressure electro-thermally assisted drive (LEAD)	<95%	No reduction unless green power source used
Other technologies – currently in commercial use		
Polymer flooding	73%	In comparison to waterflooding
Non-condensable gas (NCG) injection	20%	Operator estimate using reduced steam on pad basis
Flow control devices (FCDs)	67%	Cold flow production; no estimate for thermal
Cogeneration of steam and electricity	8–44%	Variation in level of co-gen contributions in projects
Geophysical monitoring	1%	Estimate used only as basis for impact
Solar power	<78%	Solar projects at Amal West and Belridge show substantial GHG reductions, but high costs

(Continued)

Table 4.22 (Continued)

Technology	Reduction	Notes/comments
Other technologies – evolving, with field experience/data		
Electro-magnetic (ESEIEH)	80%	Initial estimate for full-cycle vs. SAGD
Radio frequency XL heating (RF-XL)	50–100%	Developer simulation estimate; field demo underway
Direct contact steam generation (DCSG)	Up to 67%	TRL 9; developer estimate vs. commercial OTSGs
Small modular reactors (SMRs)	96%	Estimated direct emissions; 4 units commercialized
Carbon capture and storage (CCS)	70–80%	Commercial; estimate based on Quest data for 2022
Other technologies – evolving, pre-field demonstration		
Electro-magnetic (EM)-SAGD	38%	Developer simulation estimate; no field testing
In situ upgrading Technology (ISUT^{TM})	Up to 35%	Developer estimate
Downhole steam generation (eSteam)	~50%	TRL 7; developer estimate (reduced fuel use)
Dimethyl ether (DME) solvent	86%	TRL 7; estimated reduction in direct emissions

projects in the high-quality oil sands resources. The trends in decreasing SORs and corresponding GHG intensities appear to have reached plateaus or minimum values.

- After two decades of development, solvent-based processes have reached the point of being commercialized to reduce the emission intensities in two new commercial projects. Past concerns with solvent recovery and performance have been overcome.
- The ability to economically retrofit existing projects to use solvents still needs to be demonstrated, especially for the application of light solvents such as propane and butane. Where implemented, solvent-based projects should be able to deliver emission intensity reductions of 25% to over 50%.
- The economic motivation to maximize utilization of existing steam capacity will always present a challenge to reducing project emissions in existing projects. GHG intensity can be decreased at the same time that total emissions increase.

- Many of the other technologies to reduce GHG intensities in thermal heavy oil production to date have been more operationally focused, such as:
  - NCG injection to free up steam for use in newer pads having lower SORs. Some available data suggest that NCG injection has been able to reduce SORs in more mature projects by 0.5 SOR units, or about 15% of the average project SOR.
  - FCDs to provide more efficient distribution of steam injection and lower SORs. These types of technologies have had a positive directional impact that has maintained downward pressure on the project SORs and GHG intensities, but quantitative assessments of their impact have not generally been published by the thermal industry.
- Based on field performance to date, recovery technologies that use or require electrical energy rather than steam will require green energy sources to be competitive in reducing GHG intensities when viewed from the perspective of Scope 1 and Scope 2 emissions. If green sources are available, these processes can achieve reductions in GHG intensities of over 50%.
- SMRs and CCS initiatives present the highest technical potential for reducing emissions and are thus receiving a high level of interest. They are burdened, however, by social concerns as well as very high costs and long lead times for implementation.

Despite challenges to the development of technologies to reduce GHGs, there are social and political forces that may ultimately provide even stronger impetus for industry to implement these technologies in the near future. Implementation of legislation requiring reductions, such as carbon taxation, may dramatically accelerate implementation of the various technologies. In such a scenario, having a broad arsenal of potential technologies available would thus not only be beneficial but also crucial for the industry to maintain a robust, economic energy supply.

Acknowledgments

The author would like to express his appreciation to the following individuals for their reviews of this manuscript and their input into it.

Dr. Catherine Laureshen
Mr. Robert Dumeah
Dr. Harvey Yarranton

Appendix – Historical Timeline for Solvent-based Pilots

Industry field pilots of solvent-based bitumen recovery technologies.

Start	Operator	Field	Solvent (process)	Reference
1996	Imperial	Cold Lake	Xylene (CSP)	Imperial Oil Resources (2010)
1998	Imperial	Cold Lake	Ethane (CSP)	Imperial Oil Resources (2010)
2001	Encana	Senlac	Butane (SAP)	Gupta (2005)
2001	Suncor	Firebag	Propane, butane, diesel (ETS)	Suncor Energy (2004)
2002	Encana	Foster Creek	Propane (VAPEX)	Gupta (2005)
2003	Imperial	Cold Lake	Diluent (LASER)	Imperial Oil (2008)
2004	CNRL	Primrose	Butane (Cyclic)	Canadian Natural resources (2007)
2005	CNRL	Primrose	Condensate (Low-P Cyclic)	Canadian Natural Resources (2009)
2005	Devon	Dover	Propane/Methane (VAPEX)	geoSCOUT (2024)
2005	Suncor	Firebag	Diluent (ES-SAGD)	Suncor Energy (2008)
2006	Cenovus	Christina Lake	Butane (SAP)	Gupta and Gittins (2006)
2006	CNRL	Primrose	Butane (VAPEX)	Canadian Natural Resources (2008)
2006	Nexen	Long Lake	C7–C12 "Jet B" (Solvent-Steam)	Orr (2009)
2009	Cenovus	Christina Lake	Butane (SAP)	Cenovus FCCL Ltd. (2018)
2009	Imperial	Cold Lake	Propane (CSP)	Imperial Oil Resources (2010)
2010	Imperial	Cold Lake	Condensate (SA-SAGD)	Imperial Oil (2014)
2011	Connacher	Great Divide	Condensate (SAGD+)	Lau et al. (2014)
2012	Cenovus	Christina Lake	Condensate (CondenSAP)	Cenovus Energy (2015)

(Continued)

(Continued)

Start	Operator	Field	Solvent (process)	Reference
2013	Statoil	Leismer	Condensate (Solvent Co-Injection)	Athabasca Oil Corporation (2017)
2013	ConocoPhillips	Surmont	NGL (E-SAGD)	ConocoPhillips Canada (2015)
2013	Devon	Jackfish	Hexane (Solvent SAGD)	Alberta Innovates (2017)
2014	Laricina	Germain	Condensate (SC-SAGD)	Laricina Energy Ltd. (2015)
2014	Shell	Peace River	Diluent (SESD)	Shell (2017)
2014	Suncor	Dover	Butane (BEST)	Suncor
2015	Suncor	Dover	Propane (ESEIEH)	Suncor Energy Inc. (2022)
2016	MEG	Christina Lake	Propane (eMVAPEX)	MEG Energy (2023)
2016	Imperial	Cold Lake	Propane (CSP)	Imperial Oil Resources (2021)
2017	Cenovus	Foster Creek	Propane (SDP)	geoSCOUT (2024)
2018	Cenovus	Foster Creek	Propane (SAP)	geoSCOUT (2024)
2019	Suncor	Firebag	Diluent (ES-SAGD)	Suncor
2020	CNRL	Kirby South	Butane (LHE)	Canadian Natural Resources
2022	ConocoPhillips	Surmont	Butane (WASP)	ConocoPhillips Canada
2022	Suncor	Firebag	Diluent (ES-SAGD)	Suncor
Commercial projects				
2023	Imperial	Cold Lake	Diluent (SA-SAGD)	Imperial Oil (2024)
2024	CNRL	Kirby North	Butane (LHE)	Canadian Natural Resources

References

Acceleware Ltd. (2017). "Alberta-based Acceleware Awarded Funding to Test RF XL", Acceleware News Release November 3, 2017, https://www.acceleware.com/post/alberta-based-acceleware-awarded-funding-to-test-rf-xl

Alberta Energy Regulator (2022). ST3 Statistical Report, https://www.aer.ca/data-and-performance-reports/statistical-reports/st3

Alberta Energy Regulator (2023). ST53 Statistical Report. https://www.aer.ca/data-and-performance-reports/statistical-reports/st53

Alberta Government (2024). Alberta oil sands greenhouse gas emission intensity analysis. https://open.alberta.ca/opendata/alberta-oil-sands-greenhouse-gas-emission-intensity-analysis (accessed 2024).

Alberta Innovates (2017). In situ solvent assisted and solvent-based recovery: workshop 1, solvent leadership series. https://albertainnovates.ca/wp-content/uploads/2019/09/Solvent-Leadership-Series-Workshop-1.pdf

Alberta Innovates (2023). 2022-23 Annual Report, https://albertainnovates.ca/wp-content/uploads/2023/09/AI-2022-23-ANNUAL-REPORT.pdf (accessed 2024).

Antolinez, J.D., Miri, R., and Nouri, A. (2023). In situ combustion: a comprehensive review of the current state of knowledge. *Energies* 16: 6306. https://doi.org/10.3390/en16176306.

Ardala, M., Mamora, D.D., and Barrufet, M. (2010). A comparative simulation study of addition of solvents to steam in SAGD process. CSUG/SPE 138170, presented at the Canadian Unconventional Resources & International Petroleum Conference, Calgary (19–21 October 2010).

Athabasca Oil Corporation (2017). AER Leismer Update, May 2017. https://www.aer.ca/data-and-performance-reports/activity-and-data/in-situ-performance-presentations

Athabasca Oil Corporation (2020). Energy intensity reduction through FCDs and NCG Final project report (Revision 1). https://www.eralberta.ca/wp-content/uploads/2019/02/E0160359_Athabasca-Oil_Final-Report_PUBLIC-1.pdf (accessed 2024).

Bayestehparvin, B., Farouq Ali, S.M., and Abedi, J. (2019). Solvent-based and solvent-assisted recovery processes: state of the art. *SPE Reservoir Evaluation & Engineering* 22 (1).

BP (2020). Statistical review of world energy. https://www.bp.com/content/dam/bp/business-sites/en/global/corporate/pdfs/energy-economics/statistical-review/bp-stats-review-2020-full-report.pdf (accessed 2024).

Business Development Bank of Canada (2024). Scope 1, 2 and 3 carbon emissions, https://www.bdc.ca/en/articles-tools/entrepreneur-toolkit/templates-business-guides/glossary/scope-1-2-and-3-carbon-emissions#:~:text=Scope%201%20represents%20what%20the%20company%20emits%20itself%3B%20Scope%202,are%20emitting%20into%20the%20atmosphere (accessed 2024).

Butler, R.M. (1997). *Thermal Recovery of Oil and Bitumen*. Calgary, Alberta, Canada: GravDrain Inc.

Butler, R.M. and Mokrys, I.J. (1991). A new process (VAPEX) for recovering heavy oils using hot water and hydrocarbon vapour. *Journal of Canadian Petroleum Technology* 30 (01): PETSOC-91-01-09.

California Energy Commission (2020). Petroleum watch. https://www.energy.ca.gov/sites/default/files/2020-12/2020-11%20Petroleum%20Watch_ADA.pdf (accessed 2024).

Canadian Natural Resources (2008). Primrose, wolf lake and burnt lake annual presentation to the EUB.

Canadian Natural Resources (2009). Primrose, wolf lake and burnt lake annual presentation to the EUB.

Canadian Natural (2022). Corporate presentation, Chart 41. https://www.cnrl.com/upload/media_element/1387/2dce002761b5/p_corp_pres_september.pdf

Canadian Natural Resources (2023). Primrose, wolf lake and burnt lake annual presentation to the EUB.

Canadian Natural Resources. (2024). Case study: solvent enhanced oil recovery. CNRL. https://www.cnrl.com/case-studies/case-study-solvent-enhanced-oil-recovery/

Canadian Natural Resources (2020). Annual Performance Presentation In Situ Oil Sands Schemes 9673 / 10147 / 10423 / 10787 / 9404 March 2020

Canadian Natural Resources. Q4 2023 earnings call, https://seekingalpha.com/article/4674949-canadian-natural-resources-limited-cnq-q4-2023-earnings-call-transcript.

Canadian Natural Resources Limited (2020). Application for an experimental oil sands scheme – Primrose South light hydrocarbon-enhanced steamflood pilot project, Application # 1929135.

Cenovus Energy (2015). Christina Lake in-situ progress report scheme 8591 – 2015 update – 2015 D54 annual presentation to the Alberta energy regulator.

Cenovus Energy Inc. (2021). Foster creek in-situ progress report scheme 8623, 2020 update.

Cenovus Energy Inc. (2024). Q4 2023 earnings conference call.

Cenovus FCCL Ltd. (2018). Christina Lake in-situ progress report scheme 8591 – 2017 update – 2017 D54 annual presentation to the Alberta energy regulator.

ConocoPhillips Canada (2015). 2014 e-SAGD Year-End Summary. Presentation to the Alberta energy regulator.

ConocoPhillips Canada (2019). Non-condensable gas co-injection for thief zone mitigation non-confidential final report. https://www.eralberta.ca/projects/details/non-condensable-gas-co-injection-for-thief-zone/

ConocoPhillips Canada. (2022) Surmont In Situ Oil Sands Project, Directive 054 Submission & Annual OSCA Report Commercial Scheme Approval No. 9426, Reporting Period: January 1, 2021 to December 31, 2021

Dupuis, G., AL-Khoury, O., Nieuwirf, J. et al. (2021). Using polymer EOR to reduce carbon intensity while increasing oil recovery. IOR 2021 21st European improved oil recovery symposium.

Edmunds, N. and Chhina, H. (2001). Economic optimum operating pressure for SAGD projects in Alberta. *Journal of Canadian Petroleum Technology* 40 (12), pp. 13–16.

Edmunds, N. and Peterson, J. (2007). A unified model for prediction of CSOR in steam-based bitumen recovery. Petroleum society paper 2007-027. Presented at the 8th CIPC Conference, Calgary.

Emissions Reduction Alberta (2023). https://www.eralberta.ca/projects/details/effective-solvent-extraction-incorporating-electromagnetic-heating-eseieh/

Emissions Reduction Alberta (2024). EMVAPEX Pilot, Phase 3. https://www.eralberta.ca/projects/details/emvapex-pilot-phase-3/ (accessed 2024).

Emissions Reduction Alberta (2025). Enhanced Bitumen Recovery Technology Pilot. https://www.eralberta.ca/projects/details/enhanced-bitumen-recovery-technology-pilot/ (accessed 2025).

eSteam (2024). https://esteamoil.com/introduction

General Energy Recovery Inc. (2024). https://www.geri.com/technology/

geoSCOUT. geoLOGIC Systems Ltd.

Ghannadi, S. (2016). Thermal pressurization in EM-SAGD projects. https://era.library.ualberta.ca/items/092da92e-6333-4736-9b95-1f0a3de2afb/view/21080c79-0127-4b9a-ae77-06d31e12cde2/Ghannadi_Sahar_2016_9_23_v2.pdf

Government of Canada (2024). Greenhouse gas reporting program (GHGRP) – facility greenhouse gas (GHG) data. https://open.canada.ca/data/en/dataset/a8ba14b7-7f23-462a-bdbb-83b0ef629823

Government of Canada (2024). Technology readiness levels. https://ised-isde.canada.ca/site/innovation-canada/en/technology-readiness-levels (accessed 2024).

Gupta, S. (2005). Unlocking the planet's heavy oil and bitumen resources – a look at SAGD. SPE Distinguished Lecturer Series.

Gupta, S.C. and Gittins, S.D. (2006). Christina Lake solvent aided process pilot. *Journal of Canadian Petroleum Technology* 45 (9): 15–18

Harding, T.G. (2024). Methods to reduce environmental impact of in-situ recovery of heavy oil and bitumen. *Canadian Heavy Oil Association Journal.*

Hatch (2023). Small modular reactors - feasibility study for oil sands applications (SAGD facility). https://albertainnovates.ca/wp-content/uploads/2023/09/Cenovus-Non-confidential-SMR-study.pdf

Hein, F. (2016). Geology of bitumen and heavy oil: an overview. *Journal of Petroleum Science and Engineering* 154 (2): 551–563.

IAEA (2024). Small Modular Reactors, https://www.iaea.org/topics/small-modular-reactors (accessed 2024).

Imperial Oil (2008). Cold lake approvals 8558 and 4510, Annual Performance Review.

Imperial Oil (2014). IETP Application No. 03-047 imperial oil resources - Cold Lake solvent assisted - SAGD pilot, Final Project Technical Report.

Imperial Oil (2024). Q1 2024 earnings call.

Imperial Oil Resources (2010). Application for an experimental scheme approval for CSP horizontal well pilot, Application 1658664.

Imperial Oil Resources (2018). Cyclic solvent process pilot non-confidential final report, Prepared for emissions reduction Alberta.

Imperial Oil Resources (2021). Cyclic solvent process pilot 2018 final project technical report.

Imperial Oil (2024), Innovation brochure. https://www.imperialoil.ca/-/media/imperial/files/publications-and-reports/imperial_innovation_brochure.pdf, p. 16 (accessed 2024).

Jeffries-Harris, M.J. and Coppel, C.P. (1969). Solvent stimulation in low gravity oil reservoirs. *Journal of Petroleum Technology* 21 (2): 167–175. SPE-2158-PA.

Koolman, M., Huber, N., Diehl, D. et al. (2008). Electromagnetic heating method to improve steam assisted gravity drainage. Paper 1177481 presented in Proceedings of the International Thermal Operations and Heavy Oil Symposium, Calgary, Alberta, Canada (20—23 October 2008), SPE117481-MS.

Laricina Energy Ltd. (2015). AER 2015 annual presentation.

Lau, E.C., Johnson, M.D., Livesey, D.B. et al. (2014). Optimization of the SAGD+^{R} process technology. World Heavy Oil Congress, WHOC14-153.

Leaute', R.P. and Carey, B.S. (2007). Liquid addition to steam for enhancing recovery (Laser) of bitumen with CSS: results from the first pilot cycle. *Journal of Canadian Petroleum Technology* 46 (9): 22–30. PETSOC-07-09-01.

Lee, J. and Babadagli, T. (2021). Mitigating greenhouse gas intensity through new generation techniques during heavy oil recovery. *Journal of Cleaner Production* 286: 124980. https://doi.org/10.1016/j.jclepro.2020.124980.

Martinez, A.R. (1987). The Orinoco Belt, Venezuela. *Journal of Petroleum Geology* 10 (2): 125–134.

MEG Energy (2023). Christina Lake regional project, 2022 directive 54 performance report. experimental scheme approval no. 12528.

MEG Energy Corp. (2023). Q3 2023 earnings conference call.

Moradi, M. (2022). The Important Role of Flow Control Device Technologies In Delivering Net Zero Production, 83rd EAGE Annual Conference and Exhibition, Madrid, Spain, Jun 2022, Volume 2022, pp. 1–5, DOI: https://doi.org/10.3997/2214-4609.202210855

NanosTech (2024). ISUT® In Situ Upgrading Technology. https://www.nanos.tech/our-solutions/isut (accessed 2024).

OnePetro. https://onepetro.org/ (accessed 2024).

Orr, B.W. (2009). SPE-129518, ES-SAGD; past, present and future. Presented at the SPE International Student Paper Contest at the SPE Annual Technical Conference and Exhibition, New Orleans, Louisiana, USA.

Passalacqua, H. and Strack, K. (2020). Reducing carbon footprint by geophysical monitoring of EOR processes. Paper Number: SEG-2020-3424907 presented at the SEG International Exposition and Annual Meeting, October 11-16, 2020, 3384–3388. https://doi.org/10.1190/segam2020-3424907.1.

Pathways Alliance (2023). Pathways Alliance advances key oil sands CO_2 emissions reduction activities, Pathways Alliance news release, Nov 27, 2023, https://www.newswire.ca/news-releases/pathways-alliance-advances-key-oil-sands-co2-emissions-reduction-activities-860332032.html (accessed 2025).

Paurola, P., Vindspoll, H., Grande, K.V. et al. (2016). In situ combustion process with reduced Co, Emissions. US Patent Number US009470077B2.

Perpetual Energy (2019). Panny LEAD Pilot Project (Low-Pressure Electro-Thermally Assisted Drive). IETP Project Approval No. 06-095, Final Report.

Power Technology (2018). Belridge solar thermal power plant, California. https://www.power-technology.com/projects/belridge-solar-thermal-power-plant-california/?cf-view (accessed 2024).

Prabhu, C. (2022). Share of enhanced oil recovery in oil production set to triple by 2030, Oman Observer. https://www.omanobserver.om/article/1125665/ business/energy/share-of-enhanced-oil-recovery-in-oil-production-set-to-triple-by-2030-pdo (accessed 2024).

Renewables Now (2017). Miraah solar thermal plant in Oman delivers 1st steam for EOR. https://renewablesnow.com/news/miraah-solar-thermal-plant-in-oman-delivers-1st-steam-for-eor-589407/ (accessed 2024).

Rodriguez, F., Llamedo, M., Belhaj, H. et al. (2023). Workflow of the in situ combustion EOR method in Venezuela: challenges and opportunities. *ACS Omega* 8: 28060–28079.

Sandler, J., Fowler, G., and Cheng, K. (2012). Solar-generated steam for oil recovery: reservoir simulation, economic analysis, and life cycle assessment. SPE Western Regional Meeting, Bakersfield, California, 21–23 March. SPE-153806- MS. https://doi.org/10.2118/153806-MS.

Schlömer, S., Bruckner, T., Fulton, L. et al. (2014). Annex III: technology-specific cost and performance parameters. In: *Climate Change 2014: Mitigation of Climate Change. Contribution of Working Group III to the Fifth Assessment Report of the Intergovernmental Panel on Climate Change* (ed. O. Edenhofer, R. Pichs-Madruga, Y. Sokona, et al.). Cambridge, United Kingdom, New York, NY, USA: Cambridge University Press. https://www.ipcc.ch/site/assets/uploads/2018/02/ipcc_wg3_ar5_annex-iii.pdf

Sharma, J., Dean, J., Aljaberi, F. et al. (2021). In-situ combustion in Bellevue field in Louisiana – history, current state and future strategies. *Fuel* 284: 118992, https://doi.org/10.1016/j.fuel.2020.118992

Shell (2017). Peace River in Situ Oil Sands Progress Report. Commercial Scheme Approval.

Shell Canada (2023). Quest Carbon Capture and Storage Project. Annual Summary Report – Alberta Department of Energy.

Skauge, T., Skauge, A., Lugo, N. et al. (2024). How polymer flooding reduces CO_2 emissions and energy consumption -an exergy return on exergy investment case study. SPE-218231.

SNF (2001). SNF – Reducing the Carbon Footprint of Enhanced Oil Recovery., https://storymaps.arcgis.com/stories/f08e47f950b843258f3d47563745df71 (accessed 2024).

SNF (2023). Polymer enhanced oil recovery: the best of both worlds. Winter 2023 OG Innovation Magazine, February 8, 2023. https://www.snf.com/media/global-communication/oginnovation-winter2023/ (accessed 2024).

Strathcona Resources (2023). Lindbergh Project - 2022 Annual Performance Report. https://static.aer.ca/prd/documents/oilsands/insitu-presentations/2023-strathcona-lindbergh-6410.pdf

Suncor (2024). https://www.suncor.com/en-ca/news-and-stories/our-stories/advancing-technologies-for-in-situ-extraction (accesssed 2024).

Suncor Energy (2004). Firebag Project, Enhanced Thermal Solvent Pilot Project (Approval No. 8663), Annual EUB Review Presentation.

Suncor Energy (2008). 2008 ERCB Annual Review. Suncor Firebag.

Suncor Energy Inc. (2022). Effective Solvent Extraction Incorporating Electromagnetic Heating (ESEIEH®). Emissions Reduction Alberta Final Project Report: Project ID: F101150.

Suncor (2015). BEST (Nsolv) Pilot Project 2015 AER Performance Presentation Experimental Scheme Approval No. 11825 January 18, 2016 Reporting Period September 1, 2014–August 31, 2015.

Suncor, (2017) Performance Presentation for In-Situ Oilsands Experimental Scheme Approval No. 11825, Reporting Period September 1, 2016–August 31, 2017.

Suncor (2018). Performance Presentation for In-Situ Oilsands Experimental Scheme Approval No. 11825, Reporting Period September 1, 2017–August 31, 2018.

Svrcek, W.Y. and Mehrotra, A.K. (1988). One parameter correlation for bitumen viscosity. *Chemical Engineering Research and Design* 66 (4): 323–327.

World Data (2024). https://www.worlddata.info/america/canada/inflation-rates.php (accessed 2024).

Yegane, M.M., Ayatolllahi, S., Bashtani, F. et al. (2015). Solar generated steam injection in Hamaca, Venezuelan extra heavy oil reservoir; simulation study for oil recovery performance, economical and environmental feasibilities. SPE-174305-MS. presented at the 2015 EUROPEC Conference.

5

Energy Intensity in Mineable Oil Sands

Danuta M. Sztukowski and Elco D. Hollander

Coanda Research & Development Corporation, Tetra Tech Company, Calgary, AB, Canada

5.1 Introduction to Mineable Oil Sands

This chapter presents advances in the reduction of energy intensity (EI) and Greenhouse Gas (GHG) emissions in mineable oil sands processing. Oil sands ores are mixtures of heavy hydrocarbons, silica sand, clay, heavy minerals, and water. The hydrocarbons are classified as "bitumen" based on their relatively high density and viscosity and require significant processing before becoming a marketable product. Oil sands ores can be found across the globe, but large-scale commercial exploitation of these deposits is primarily found in the Athabasca oil sands region in Northern Alberta, Canada. A key feature of the Athabasca deposit is that the ore is "water-wet," meaning that the bitumen is separated from the sand by a thin layer of water. As a result, the initial bulk separation of sand and bitumen can be achieved relatively easily by exposing the ore to hot water and mechanical energy. In contrast, the initial separation of oil-wet ores generally requires the use of solvents. The Athabasca oil sands have been studied widely and have gone through six decades of process evolution and were therefore chosen as the focus for this chapter.

Alberta's oil sands deposits contain 165 billion barrels of bitumen and are some of the world's largest crude oil deposits (Alberta Energy Regulator, 2024) with the fourth largest proven oil reserves in the world, after Venezuela, Saudi Arabia, and Iran. The production facilities in Northern Alberta are some of the most commercially developed and technologically advanced, with numerous in situ and mineable schemes in the region producing 3.3 million barrels of crude bitumen per day as of 2022 (Government of Alberta, 2024a).

Cleaner Petroleum Production and Refining Technologies, First Edition, Edited by M. R. Riazi and Harvey W. Yarranton.

Exploiting bitumen as an energy source was not contemplated until the early 20th century, with the first research and development work to separate bitumen from the oil sand matrix conducted in the mid-20th century by Dr. Karl Clark at the Research Council of Alberta, the University of Alberta, and the large-scale experimental Bitumount facility (Government of Alberta, 2024b; Sheppard, 1989). This research was used to demonstrate that bitumen production was possible at a commercial scale. The process step that achieves primary separation of the bitumen is based on intimate contact between ore, hot water, and air bubbles. The hot water reduces the bitumen viscosity, making it easier for the bitumen to be liberated from the sand fraction and the air bubbles to be incorporated into a bitumen "froth." The froth has a reduced density compared to the water/sand slurry, making it amenable to gravity separation. The bitumen froth is considered an intermediate product stream that requires further purification, which is in practice achieved by a solvent-based froth-treatment process step. All current commercial oil sands operations are based on the Clark Hot Water Process (CHWE), often referred to as primary extraction.

Alberta contains three major oil sands deposits:

- Athabasca, the largest deposit, located to the north, west, and south of the Municipality of Fort McMurray
- Cold Lake, located approximately 100 km north of the town of Lloydminster and on the border of Alberta and Saskatchewan
- Peace River, located in the north-central region of Alberta near the town of Peace River

Only the Athabasca deposit is surface-mineable, that is, within 75 m of the surface (Kasperski, 2001; Government of Alberta, 2024c). Surface mining accounts for 20% of Alberta's recoverable bitumen reserves. In situ recovery methods are used elsewhere where the resource is found at depths that prohibit surface mining (typically >200 m). In situ processes are discussed in detail in Chapter 4.

At present, there are six commercial surface-mining oil sands operations in the Athabasca region. The first commercial facility, Suncor Base Plant, has been operating since 1967. Syncrude Base Plant and the Aurora North Expansion have been in operation since 1978 and 2001, respectively. Both the Syncrude and Suncor facilities use Naphthenic Froth Treatment (NFT) to purify the intermediate bitumen froth stream described earlier. Here, a naphtha cut is used to dilute the froth and reduce the bitumen viscosity, enabling separation of the remaining water and solids from the froth stream. As will be discussed in more detail in Section 5.2.4, NFT-based flowsheets produce a diluted bitumen (dilbit) product that requires on-site upgrading because the dilbit is not of sufficient quality in terms of, e.g., Base Sediment and Water (BS&W) content to be transported by pipeline for further processing.

The Muskeg River and Jackpine Mines, collectively referred to as the Albian mines, have been in operation since 2003. These mines are owned by the Athabasca

Oil Sands Project (AOSP), which has been majority owned and operated by Canadian Natural Upgrading Ltd. (CNUL) since 2017. These were the first operations to deploy Paraffinic Froth Treatment (PFT). Here, the bitumen froth is contacted with a linear alkane mixture rather than a naphtha cut. The use of a paraffinic solvent results in partial deasphalting of the bitumen froth, which further ensures more efficient separation of the remaining water and solids. Dilbit quality from this flowsheet allows for long-distance transport, which enables upgrading of this material off-site. In the case of AOSP, upgrading occurs at the Scotford Upgrader in Fort Saskatchewan, Alberta. The selection of NFT versus PFT in commercial flowsheets depends on numerous technical, logistical, and commercial factors. It is for this reason that there is no consistent implementation of newer facilities. For example, Horizon Oil Sands, owned and operated by Canadian Natural Resources Ltd. (CNRL), is an NFT-based facility that has been in operation since 2009. Kearl Oil Sands, in operation since 2011 and operated by Imperial Oil Limited (IOL), and Suncor's Fort Hills, in operation since 2018, are both PFT-based.

The EI of mineable oil sands operations is generally reported on an aggregate basis, meaning that the EI and emissions associated with specific unit operations are generally not delineated. This delineation is reflected in life-cycle models that compare EI performance for oil sands surface mining operations that either do or do not employ on-site cogeneration (Bergerson et al., 2012; Nimana et al., 2015). Comparisons between models are difficult in practice because the flowsheet boundaries may differ (Xing et al., 2021). However, it can be said that the largest emissions are associated with steam and electricity generation or, in the absence of cogeneration with power, withdrawal from the grid. For example, Bergerson et al. (2012) gives a range of 2.9–8.9 g CO_2e/MJ synthetic crude oil (SCO) for a surface mining operation without cogeneration and a range of 2.5–8.1 g CO_2e/MJ SCO for an operation with cogeneration. In addition, it is noted that emissions associated with dilbit transport vary between 0.4 and 5.6 g CO_2e/MJ SCO depending on the pipeline transfer distance. Further, the GHG intensity of bitumen recovery and extraction from surface mining is on average lower than that from in situ bitumen recovery. In contrast, the land disturbance of surface mining is significantly larger than that of in situ operation.

This chapter reviews current process flowsheets, their evolution since first commercial deployment almost six decades ago, and potential next-generation implementations to improve EI. Where possible, EI improvements and GHG emission reductions are quantified.

5.1.1 Mineable Oil Sands Characteristics

There are numerous excellent references that describe physical and (biogeo-) chemical properties of oil sands, bitumen, and intermediate and waste streams present in the oil sands flowsheets as they relate to bitumen extraction and upgrading (Masliyah et al., 2011; Czarnecki et al., 2013; Foght et al., 2017; Cossey et al., 2021). This section provides a general description of the oil sands ore and the bitumen product in order to explain the need for the CHWE process and subsequent purification in froth treatment.

Oil sands ore consists of a matrix of sand, silt, and clay, residual quantities of water, and bitumen. The ore grade (bitumen content of ore) varies between less than 4 and in excess of 16 wt% (Kasperski, 2001; Masliyah et al., 2011). The connate water content is typically around 5–10 wt% and contains ions such as sodium, calcium, magnesium, potassium, chloride, (bi-)carbonate, and sulfate, as evidenced by what accumulates in process-affected water (Cossey et al., 2021). The majority of oil sand (70–90 wt%) is composed of mineral solids, which are primarily silica quartz and various clays (kaolinite, illite, smectite, montmorillonite, etc.), as well as lesser quantities of heavy mineral solids (ilmenite, rutile, pyrite, zircon) and other minerals (Kaminsky, 2008). It is found in practice that bitumen recovery in water-based extraction is sensitive to factors like salinity, clay content, bitumen grade, extraction temperature, and pH (Masliyah et al., 2011). For a surface mining bitumen extraction operation to be commercially viable, the long-term average ore grade typically exceeds 10 wt%, while the fines content (i.e., the mass fraction of particles less than 44 μm on a dry-solids basis) is generally less than 25 wt%. Ores are commonly defined according to the particle size and sand-to-fines ratio (SFR), the latter of which is given by the fraction of material greater than 44 μm over the fraction less than 44 μm.

Oil sand solids also contain monazite and lesser quantities of xenotime, which host rare earth elements (REE) as documented by various authors (Ciu et al., 2003; Kaminsky, 2008; Roth et al., 2017; Lindsay et al., 2019). Neither heavy minerals such as titanium and zircon sands nor REE have been recovered commercially from Alberta's oil sands, although technology to recover both from oil sands waste streams has progressed significantly in the last 20 years (Oxenford and Coward, 2001; Ciu et al., 2003; Marshall et al., 2014; Titanium Corporation, 2019; CVW CleanTech, 2023).

The bitumen found in the ore body is a heavy (~1000 kg/m^3) and viscous (10^5 mPas at STP) petroleum hydrocarbon and is typically viewed as the lowest grade of crude oil (Masliyah et al., 2011). Its density, viscosity, and composition result from biodegradation of conventional crude oils over geological time scales. It contains a wide range of high-molecular-weight aliphatic, aromatic, and naphthenic compounds, as well as nitrogen, oxygen, and sulfur heteroatoms. Bitumen can be classified in various ways depending on the application, e.g., through chemical composition, boiling range, or solubility behavior. The solubility classification defines bitumen into two broad groups (maltenes and asphaltenes). This classification is relevant to the difference in bitumen behavior during NFT and PFT as described later.

In oil sands processing, proportionally little product (around 10 wt% bitumen on an ore basis) is separated from a large matrix. At current bitumen prices, the operation can only be made economically viable at large mass flow scales (typically 10^4 tonnes per hour of ore). This scale implies a large mining footprint and, as a consequence, a large tailings inventory. The tailings inventory will require remediation in a timeline acceptable to stakeholders. Tailings handling will be the focus of Chapter 6. The scale of operation, the waste management aspect of bitumen processing, and the climatic conditions in which the facilities are operated lead to several design constraints (e.g., mine haul distances, hot water requirements, water containment, and recycle requirements) that drive EI of the overall process. These design considerations are the subject of the next sections.

5.1.2 General Process Overview

A surface mining oil sand operation is composed of several processing steps that separate and concentrate the bitumen from the oil sand matrix to an upgradable product (Figure 5.1). The process is "once-through" for the ore, but the processing units are interconnected via heat integration and water recycling. Since the facilities currently operate under a zero water-discharge requirement, it should be recognized that any water (e.g., precipitation or river water import) brought into contact with the process needs to be stored on-site in tailings structures. It is therefore beneficial to recycle as much water as practicable. Additionally, freshwater

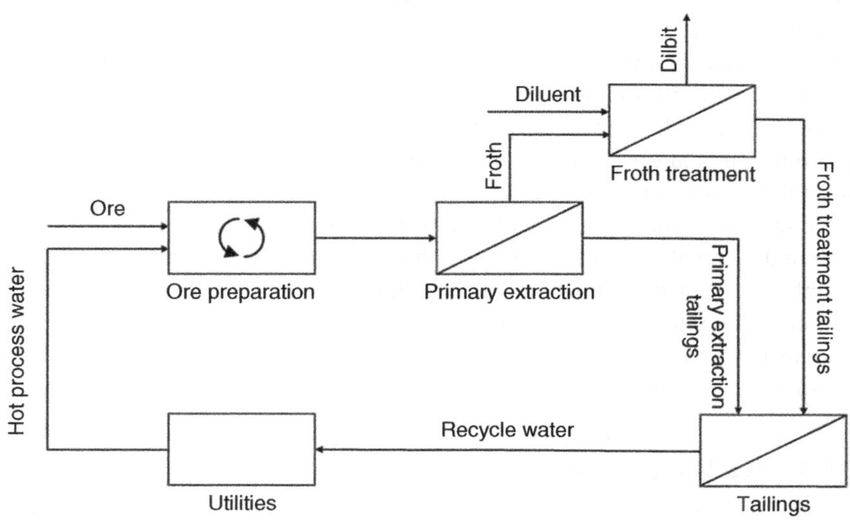

Figure 5.1 Conventional oil sands process overview showing major unit operations and key process streams. Streams signifying minor losses are omitted for simplicity.

withdrawal from the Athabasca River is regulated with limits on individual and combined operator withdrawal rates, further emphasizing the benefit of water recycling.

The recovery of bitumen from oil sands ore is achieved through a series of mixing, transportation, and separation steps in which the bitumen content progressively increases. In the mining step, bitumen ore is separated by shovel from waste material and transported by truck to the processing plant. The waste material is used in various mine and tailings containment construction or otherwise stockpiled for ultimate reclamation activities. In the ore preparation unit, the virgin ore is crushed and mixed with hot water, chemical process aids, and entrained air. In some flowsheets, oversize material is rejected and hauled to a stockpile. The ore-water slurry is transported to the primary extraction unit. In primary extraction, the bitumen froth is separated by gravity from the sand/water fraction in various separation vessels. Bitumen froth is de-aerated and transferred to the froth treatment unit, while the waste streams are transferred to tailings treatment (and ultimately tailings storage).

De-aerated froth is contacted in froth treatment with either naphthenic or paraffinic solvent to reduce the bitumen viscosity and allow for the removal of residual solids and water from the hydrocarbon phase. The dilbit is then prepared for upgrading, while froth treatment tailings (FTT) are subject to solvent recovery prior to being transferred to tailings storage.

In these process steps, bitumen content increases from a nominal 10% bitumen in ore to approximately 98% bitumen for NFT and >99% bitumen in PFT. Overall bitumen recovery is nominally 90%. Some nuances regarding ore grade, ore layering, and accounting for asphaltene rejection in PFT are discussed later in this chapter. The utilities required to operate the separation processes are electricity, water, steam, and natural gas (primarily for power generation, as well as safeguarding). In addition, make-up solvent and freshwater are required to compensate for permanent losses to tailings and evaporative losses.

Oil sands operations are typically multi-train lineups that run 24/7, 365 days per year under climatic conditions that vary from −40 to + 35 °C ambient. Production is only interrupted for maintenance activities required to counter erosion, corrosion, and fouling. Larger maintenance turnarounds typically occur once every four to five years and are coordinated with similar maintenance activities in downstream processing.

5.1.3 Regulatory Considerations

An aspect that is rarely discussed in a process engineering context is the regulatory considerations that apply to oil sands facilities. However, regulations pose real constraints on mine development, day-to-day operations, and reclamation activities. Some context is provided here to show the complexity of acquiring initial approval for developing a site, as well as modifying or supplementing existing infrastructure to improve both EI and broader environmental performance.

The development and operation of a successful oil sands mine incorporates short, intermediate, and long-term plans for mining, bitumen extraction, and closure activities. These are required by legislation under the Mines and Minerals Act (Province of Alberta, 2023) and defined by regulatory regimes and approvals under the Oil Sands Conservation Act (OSCA) (Province of Alberta, 2022a). The OSCA outlines the requirements needed to approve a mining scheme, the processing plant, and any amendments to an approval.

OSCA requirements are supported by numerous other regulations, including the Oil Sands Conservation Rules (Province of Alberta, 2022b), which define applications and approvals as related to the mine site, storage and disposal within that site, and the operations of the processing plant. The OSCA is administered by the Alberta Energy Regulator (AER, formerly the Energy Resources Conservation Board, ERCB). Alberta's Environmental Protection and Enhancement Act (EPEA), Alberta's Water Act, and Canada's Water Act also govern oil sand mining facilities. As part of EPEA requirements, an Environmental Impact Assessment (EIA) is conducted prior to approval for resource development. The EIA is done in consultation with regional stakeholders and First Nations. In addition to providing project context as related to proposed mining operations, the resource base, its associated geology, and proposed industrial activities, EIAs describe the environmental impacts to air, noise, environmental and human health, and aquatic and terrestrial resources expected as a result of the operation. Major flowsheet changes can trigger a requirement for updates to an EIA, which can take years to develop and acquire approval, since potential regional impacts in addition to local site-specific aspects need to be considered.

Operators are required to provide periodic updates to AER-approved plans that guide the development and progressive reclamation of the mine. These include the OSCA and EPEA site-specific requirements, Tailings Management Plan (TMP), and Life of Mine Closure Plan (LMCP), amongst others. These documents are reviewed by the AER and other government bodies. Regional and local stakeholders are included in the review process, and updated plans are accepted when the provided information is deemed to adequately demonstrate that the planned production and progressive site reclamation profile will be met.

Two AER guiding documents in particular have guided resource development and progressive closure and reclamation: Directive 082 and the Lower Athabasca Region Tailings Management Framework (TMF) for the Mineable Athabasca Oil Sands. Section 2 of Directive 082 (Alberta Energy Regulator, 2016a) outlines the four criteria that have historically been employed to define the part of the ore body that is considered recoverable in a water-based extraction process:

- The in situ cutoff grade.
- The minimum mining thickness separating ore and waste.
- The cutoff for total volume to bitumen in place (TV/BIP) to determine pit crest limits.
- The processing plant recovery, describing minimum recovery based on as-mined ore grade.

The intent of Directive 082 is to ensure that maximum value is extracted from the resource, and that the selective mining of high-grade ore is prevented. It is in the interest of the province to ensure that the total recovery from the resource is maximized within practical and environmental constraints. It is also noted, however, that lower-quality ore will require more energy to meet recovery requirements. Here, lower-quality ore refers to materials that are generally lower in bitumen content (ore grade) and/or contain higher clay content, making the material more difficult to process using the CHWE process. Section 3 of the directive states that sterilization of ore (i.e., opting to leave ore behind that technically should be processed under Directive 082) should be minimized, except in the case of valid environmental, technical, economic, or operational reasons. Recognizing that valid reasons exist, some mining operations have recently secured waivers from Section 2 of Directive 082, as they have demonstrated that problematic ores that technically qualify as recoverable result in disproportionately high EI and fluid tailings volume increases. An example that demonstrates how environmental performance (e.g., in terms of EI) can be improved without physical changes to the processing facilities is discussed in Section 5.3.3.

The TMF (Government of Alberta, 2015) and accompanying Directive 085 (Alberta Energy Regulator, 2016b) were developed to ensure that reclamation and closure activities are being addressed during the operating phase of the mine. Directive 085 succeeded Tailings Directive 074 (Energy Resources Conservation Board, 2009) in 2016. Where Directive 074 strictly addressed "fines management" in an attempt to reduce Mature Fine Tailings (MFT) inventory (process-focused), Directive 085 puts more emphasis on the ultimate goal of the development of a self-sustaining Boreal Forest ecosystem (outcome-based). The impact of this change is addressed in more detail in Chapter 6.

5.2 Evolution of Process Flowsheets

In this section, current operational flowsheets are presented in more detail to show the natural evolution of oil sands processing and illustrate the significant process improvements that have been made in the past decades.

5.2.1 Mining

A typical oil sands ore deposit consists of up to several tens of meters of topsoil and waste material that must be removed before the bitumen ore is reached. Mine operations expose the ore and isolate it from waste, after which the ore is transported to a crushing facility for size reduction and processing. The mining step also creates tailings footprint and manages waste and reclamation soil storage for haul road and dyke construction, as well as future pit closure and reclamation activities.

Mine operations are typically based on solids-handling equipment (i.e., truck and shovel), which differs from tailings operations that generally handle materials in slurry form. In order to close the site material balance, the mine and tailings operations are heavily integrated. The need for integration is clear when considering an idealized mining sequence:

- At the start of operation, the first ore pit is excavated. The tailings generated from this initial ore need to be stored in an external tailings facility (tailings pond) due to unavailable in-pit storage.
- For the majority of the operation, tailings are placed in mined-out ore pits for containment.
- At the end of the operation, the final ore pit will become a pit lake as there is no further tailings requiring storage.

Waste management poses significant logistical challenges, as the ratio of ore to waste is typically around 1:1. Rehandling of material must be minimized as much as possible, and direct placement of materials in the final location is critical. Sequencing requires careful planning and design, as the final contents placed in an excavated pit must result in a closure landform that becomes a self-sustaining boreal forest ecosystem of equal pre-disturbance capability. However, in practice, there often are material balance conflicts due to, e.g., mine sequencing requirements and geotechnical constraints for mine pit and tailings pond construction, as well as changes in regulatory and stakeholder expectations. These constraints make the mine/tailings planning efforts a challenging aspect of oil sands operations.

Historically, mining was based on drag lines and bucket-wheel reclaimers and excavators (with associated conveyors) that originated in the coal mining industry. This equipment proved to show poor ore/waste selectivity as well as low reliability under the climatic conditions in the Athabasca region and was made obsolete in the 1990s. All current oil sand surface mining operations in Alberta are "truck and shovel" based, signifying a substantial improvement in operational, environmental, and economic performance through improved stability of the operation. Development of the mine and its benches, dykes, and storage structures also relies on other equipment such as dozers, graders, backhoes, and loaders. Ore is separated from waste using hydraulic shovels when increased selectivity is required to separate the oil sand and electric shovels otherwise. Shovels excavate ore simultaneously from several locations within the mine to generate an ore blend suitable for stable plant operation and adequate bitumen recovery. Additionally, production rates from the shovels are optimized for ore and waste haul distances and mine pit development. Optimization in the mine can lead to conflicting requirements, as the development of the mine often requires excavation of lower-grade materials. A typical oil sands mine employs 240–400 metric tonne haul trucks that supply 5,000–10,000 tph of ore per production train to the processing plant.

Diesel consumption for surface mineable oil sands operations varies between 5 and 15 L/m^3, as reported in various life-cycle models such as the GHOST and FUNNEL models (Bergerson et al., 2012; Nimana et al., 2015). Much of the emissions originate from the haul fleet, where the CO_2 emissions are estimated as 3,700 tpa per truck (Council of Canadian Academies, 2015). Overall, it is estimated that diesel for trucks and shovels accounts for 12 to 20% of the total emissions associated with a mineable oil sands plant (whether electricity is imported or produced via cogeneration), or between 0.5 and 1.1 g CO_2e/MJ of bitumen (Bergerson et al., 2012; Nimana et al., 2015).

The change from bucket wheels and drag lines to truck and shovel operation in the 1990s marked the only major change in the mining aspect of the flowsheet since commercial operation commenced. Currently, a technology under evaluation by Suncor Energy is the implementation of Autonomous Haul Trucks. This technology is already in use in other mining applications and promises fuel efficiency gains by up to 15% (Council of Canadian Academies, 2015). Reductions in GHG emissions are also anticipated; for example, a reduction of 30 to 50% in NO_x emissions is expected. Secondary benefits include an improvement in the stability of feed supply to the plant operation. Other technologies under consideration are related to changes in fuel sources for trucks to hydrogen, natural gas, electric, and even small modular nuclear reactors. These options are all primarily aimed at reducing CO_2 emissions from the truck fleet.

5.2.2 Ore Preparation and Slurry Handling

The ore preparation plant receives ore from the mine and converts the dry material into a pumpable slurry (Figure 5.2). Size reduction is required to avoid plugging downstream pumping equipment and generate surface area for bitumen liberation and air contact. The process starts with a dry crushing step where large clumps of (often frozen) ore are dumped from the truck bed into a double-roll crusher or sizer, where it is crushed into a loose mixture. This mixture is transferred to a silo that provides several hours of surge capacity between dry crushing and wet ore preparation steps. From there, the ore is transported by conveyor to the wet processing steps. Oil sand operators use mix boxes (Syncrude's Aurora mine, CNRL's Horizon mine, and

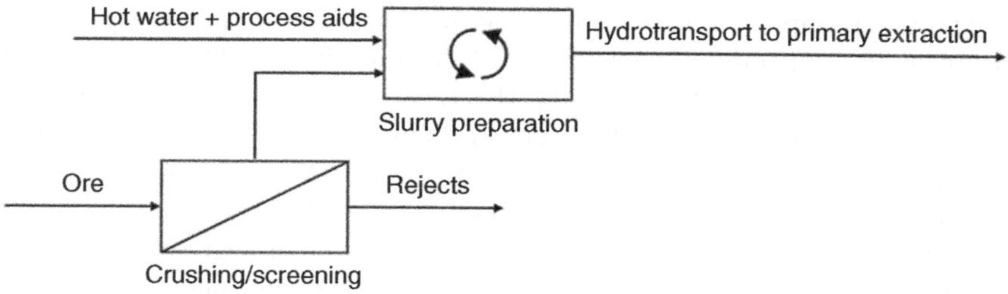

Figure 5.2 Ore preparation processing steps.

Table 5.1 Illustrative mass balance for a single train, 7,500 tph ore, ore preparation unit flowsheet.

		Stream			
Parameter	**Units**	**Ore**	**Rejects**	**Hot water**	**Slurry**
Mass flow	tph	7,500	75	3,519	10,944
Bitumen content	kg/kg	0.110	0.110	—	0.075
Water content	kg/kg	0.050	0.050	1.000	0.355
Solids content	kg/kg	0.840	0.840	—	0.570
Sand/fines ratio (SFR)	kg/kg	5.7	5.7	—	5.7
Density	kg/m^3	2,097	2,097	1,000	1,550
Volumetric flow	m^3/h	3,577	36	3,519	7,061

IOL's original Kearl facility), rotary breakers (Suncor's Steepbank and Millenium mines and CNUL's Albian mines), or wet crushers (Syncrude's Mildred Lake mine replacement and IOL's Kearl expansion) to further reduce the ore size and contact the ore with hot process water and in some instances process aids (e.g., caustic soda, sodium citrate). Each technology uses mechanical and thermal energy to create a slurry that allows for bitumen liberation, coalescence, and air attachment, which is required to separate a bitumen froth from the oil sands matrix.

An illustrative mass balance for a single train ore preparation unit is presented in Table 5.1. A bitumen recovery of greater than 99% is usually realized for the crushing/screening step. Ore temperature varies seasonally, and the ore can include frozen lumps. Slurry flow conditions are typically 50 °C, and slurry densities of 1,550–1,600 kg/m^3 are preferred. Hot water temperature is adjusted to achieve these slurry conditions and generally exceeds 75 °C.

Factors that improve bitumen liberation and coalescence include high shear and adequate temperature (in excess of 40 °C), while air attachment is promoted through control of the water chemistry. This is particularly important for systems with high clay content. A dispersive chemistry (slurry pH 8–9) is required to ensure that bitumen does not adhere to solids. (Masliyah et al., 2011). The ability to maximize the benefit of shear, heat, and process aids is vastly improved if the slurry residence time is increased. In order to form air-bitumen aggregates, successful collisions between the two phases must occur. Generally, increasing specific power input increases the collision rate, and higher retention time increases the number of collisions, which can be achieved naturally with increased transportation distances and high slurry line velocities. Hydrotransport technology was first developed by Syncrude in the 1990s and is now universal in oil sands processing plants. These 2–3 km pipelines provide an additional 15–20 minutes of slurry residence time compared to the typical 3–5 minute retention time in mix boxes, breakers, or wet crushing operations. The incorporation of hydrotransport into the bitumen extraction flowsheet marked one of the most significant improvements in EI reduction since the initial commercialization, with reported reductions of EI of the bitumen extraction process of 40% (IHS Energy, 2011).

It cannot be understated how important steady-state operation is for minimizing EI in ore preparation. The surge bin in the process lineup is critical to ensuring steady feed rates and operation at design capacity. Variability in both feed rate and composition can result in reduced recovery and throughput. What is often not appreciated is that the maximum turndown for coarse slurry pipelines is limited since the sand fraction has a tendency to settle out (increasing the risk of plugging the process) under low flow conditions. When the ore feed rate is reduced, the process water feed rate must be increased to guarantee minimum line velocities. Since ore preparation has only hot water available, an increased amount of hot water is pumped directly to tailings without producing bitumen. If throughput drops below a critical value, the system will require complete flushing for process safety reasons. Current operational efforts to improve energy efficiency are therefore aimed specifically at stabilizing the operation, e.g., by focusing on eliminating throughput reduction during shift change. A single hour of downtime per day amounts to approximately 4% energy inefficiency loss for the primary extraction circuit. Autonomous Haul Trucks are an avenue to stabilize ore feed, as they are unaffected by operator shift changes.

5.2.3 Primary Extraction

The Primary Extraction step separates the oil sand slurry into a concentrated froth containing the majority of the bitumen and a tailings stream with most of the mineral solids (Figure 5.3). Gravity separation occurs in several flotation stages.

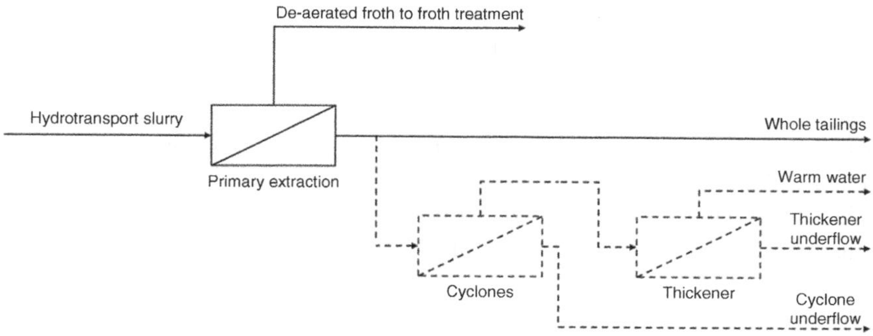

Figure 5.3 Primary extraction processing steps. Depending on tailings requirements, operators produce either whole tailings or a coarse- and fines-dominated tailings product.

Primary, secondary, and tertiary steps are deployed to further increase bitumen recovery and also enable warm water recovery for reuse in the extraction circuit.

The majority of the bitumen is recovered from the hydrotransport slurry in large (~30 m diameter), cone-shaped vessels called primary separation vessels (PSVs) or cells (PSCs). The feed slurry is diluted with process water and recycle streams and fed into the vessel through a feedwell with either a radial or downcomer design. The feedwell creates a quiescent zone, which allows for the separation of the aerated bitumen "froth" from the slurry. The froth is collected in a launder and pumped to a deaerator, where it is, in some flowsheets, contacted with steam before it is sent to the froth treatment facility. Air removal is necessary to improve froth pumpability and remove oxygen before the material enters the light hydrocarbon processing portion of the flowsheet. The de-aerated froth is pumped to storage tanks, which provide surge capacity between the aqueous portion of the bitumen extraction plant and the froth treatment facility. High-quality froth (i.e., in excess of 55% bitumen content) is typically formed from high-quality ore (bitumen content in excess of 11% and fines content less than 20%). Over the past decades, several flowsheet additions have been made to try and improve bitumen recovery and froth quality. These operations mimic scavenging circuits found in conventional mineral processing and are typically based on air flotation circuits.

The waste streams from the Primary Extraction process contain mostly coarse sand, fines, and water, as well as unrecovered bitumen. Depending on the operator, the stream can further be split into coarse-dominated and fines-dominated streams. The fines-dominated material is then usually processed further in a thickener circuit for warm water recovery. It is noted that thickeners were implemented for warm water recovery at PFT operations, since these facilities lack waste heat from local upgrading. The tailings streams are transferred to tailings impoundment for storage and ultimate process water recovery.

Some operators further process the tailings streams to improve the tailings properties. CNRL Horizon, for instance, recombines the coarse and thickened tailings streams to form Non-Segregating-Tailings (NST), while IOL's Kearl post-treats the thickener underflow stream. Flocculants and coagulants are added at various stages (including shortly before discharge) in an attempt to increase the capture of fine solids within the coarse sand matrix. Increased fines capture helps reduce the formation of fluid tailings inventory. The challenges associated with the fluid tailings behavior will be discussed in more detail in Chapter 6.

An illustrative mass balance for a single train, 7,500 tph ore, primary extraction unit is presented in Table 5.2. Certain dilution water streams have been omitted (e.g., dilution water for cyclone underflow pumping) to show the relative material distribution on an ore basis. In addition, froth deaeration and associated steam addition have been excluded for simplicity, and it is further assumed that no sand reports to the froth. A bitumen recovery of 92 to 95% (assumed 94% here) is usually realized for the separation step. Preferred froth quality exceeds 55 wt% bitumen content. Coarse and whole tailings slurry densities of 1,550–1,600 kg/m^3 are targeted, where thickener underflow varies between 35 and 50 wt% solids content. For cyclone performance, 95% sand recovery to the underflow is assumed, at an underflow solids content of 72 wt%. Identical splits for water, bitumen, and fines are assumed.

In the past decades, an increase in the amount of bitumen recovery equipment in addition to the primary separation cells has occurred. While the equipment increases the bitumen recovery from the ore, the equipment itself is energy-intensive and typically creates a lower quality froth than the primary froth, which negatively impacts downstream processing. It is therefore not clear-cut whether the EI per produced barrel of bitumen increases or decreases with the addition of more flotation equipment.

5.2.4 Froth Treatment and Solvent Recovery

Froth treatment uses a light hydrocarbon diluent to isolate the bitumen from the remaining mineral solids and water entrained in froth and further concentrates the dilbit to recover the light diluent for reuse. Dilbit is separated from water and mineral solids by gravity in a counter-current operation. Solvent is recovered from both the product and tailings by evaporative techniques (Figure 5.4).

Froth treatment is the last step of the bitumen extraction process. It generates a product that is nearly free of mineral solids and water, and a tailings stream that contains effectively all the mineral solids and water contained in froth, trace unrecovered bitumen, and residual quantities of diluent. The diluent losses must meet regulatory requirements for tailings disposal. The disposal limit for all oil

Table 5.2 Illustrative mass balance for a single train, 7,500 tph ore, primary extraction unit flowsheet.

		Stream					
Parameter	**Units**	**Slurry**	**De-aerated froth**	**Whole tailings**	**Cyclone underflow**	**Thickener underflow**	**Warm water**
Mass flow	tph	10,944	1,396	9,548	7,605	1,201	742
Bitumen content	kg/kg	0.075	0.550	0.005	0.004	0.008	0.014
Water content	kg/kg	0.355	0.300	0.364	0.276	0.542	0.971
Solids content	kg/kg	0.570	0.150	0.631	0.720	0.450	0.015
SFR	kg/kg	5.7	0.0	7.3	11.5	1.0	0.0
Density	kg/m^3	1,550	1,103	1,648	1,813	1,389	1,010
Volumetric flow	m^3/h	7,061	1,266	5,795	4,196	865	735

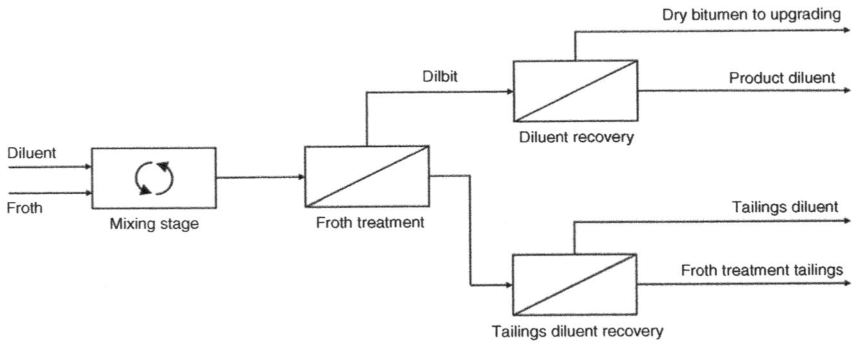

Figure 5.4 Froth treatment processing steps.

sand operations is a condition of each site's EPEA and is set at 4 volumes of light hydrocarbon for every 1,000 volumes of produced bitumen.

An illustrative mass balance covering both the gravity separation and solvent recovery steps for a single train, 7,500 tph ore, NFT unit is presented in Table 5.3. Certain simplifying assumptions have been made to show the relative material distribution on an ore basis, including the exclusion of direct steam injection for heating purposes in the tailings diluent recovery process. A bitumen recovery of 98% or greater is usually realized for an NFT application. Differences in bitumen recovery and product quality between NFT and PFT operations are discussed later. A separate mass balance for PFT is omitted for brevity. Tailings slurry densities of 1250 kg/m^3 are shown here. In practice, this density is closer to 1,100 kg/m^3 as a result of steam condensate from heating, as well as the addition of various non-NFT-related water streams.

5.2.4.1 Froth Treatment (Gravity Separation)

The goal of froth treatment is to produce a bitumen product suitable for upgrading and subsequent refining. It was previously noted that virgin bitumen has a density similar to water. Therefore, gravity separation can only be employed if the density difference between bitumen and water is increased while the viscosity of the bitumen is substantially reduced. Two different hydrocarbon diluents are used in commercial froth treatment: a light paraffinic diluent (referred to as "solvent" in PFT), composed of pentane or a pentane/hexane mixture, or naphtha (referred to as "diluent" in NFT). The term "solvent" is somewhat of a misnomer in PFT: the solvent is added to bitumen in a ratio that exceeds the solubility limit for asphaltenes and therefore behaves more like an anti-solvent for the heaviest fraction. Partial deasphalting is a primary benefit of the PFT process, as the deasphalting step effectively increases the quality of the bitumen product. PFT is employed at CNUL's Albian mines, IOL's Kearl mine, and Suncor's Fort Hills

Table 5.3 Illustrative mass balance for a single train, 7,500 tph ore, froth treatment unit flowsheet.

		Stream					
Parameter	**Units**	**De-aerated froth**	**Diluent**	**Dry bitumen**	**Product diluent**	**Tailings diluent**	**Froth treatment tailings**
Mass flow	tph	1,396	539	776	527	11	622
Bitumen content	kg/kg	0.550	—	0.970	—	—	0.025
Diluent content	kg/kg	—	1.000	—	1.000	1.000	0.003
Water content	kg/kg	0.300	—	0.020	—	—	0.648
Solids content	kg/kg	0.150	—	0.010	—	—	0.324
D:B	kg/kg	—	0.700	—	—	—	0.110
Density	kg/m^3	1,103	750	1,006	750	750	1,251
Volumetric flow	m^3/h	1,266	719	771	702	14	497

operation, while NFT is used at the Suncor base plant, the Syncrude operations, and CNRL's Horizon mine.

Both PFT and NFT rely on increasing the settling velocity for the mineral solids and water fraction. Recall that Stokes' law shows that the settling rate scales with $\Delta\rho g d^2/\mu$, where $\Delta\rho$ denotes the density difference between continuous and dispersed phase, d is the characteristic size of the dispersed phase, g is the gravitational acceleration constant, and μ is the dynamic viscosity of the continuous phase. The practical NFT operating condition is at a dilution ratio of approximately 0.7 kg diluent/kg bitumen. At this ratio, the density of the hydrocarbon phase decreases from that of pure bitumen to 850–900 kg/m^3, while the hydrocarbon phase viscosity drops by several orders of magnitude. It is found, however, that the bitumen product specification cannot be met via gravity settling alone. Both the water and solids content in the product would be higher than corrosion and fouling limits in downstream processing can accept. NFT operators increase separation efficiency by employing equipment that increases acceleration forces (e.g., by means of scroll or disc centrifuges) or reduces the settling distance for the underflow (i.e., inclined plate settlers, IPS). Even with this additional equipment, the bitumen product will contain 2–3% water and 1% solids. This quality is not sufficient to meet pipeline specifications, and therefore all NFT operators have on-site upgrading facilities that employ coking technology.

The PFT philosophy is similar to NFT with respect to the role of solvent in the reduction of hydrocarbon density and viscosity. However, PFT further relies on the precipitation of asphaltenes and the resulting creation of large-diameter water/solids/asphaltene aggregates (Zawala et al., 2012; Long et al., 2002). The increase in diameter increases the settling velocity such that simple gravity settlers can be used. The solvent-to-bitumen (S/B) mass ratio is elevated to 1.5–2.5 kg solvent/kg bitumen, at which approximately half the asphaltenes precipitate (Xu, 2018). The bitumen product contains less than 0.1% water and solids, which is sufficient to meet pipeline specifications for transport to off-site upgrading using techniques other than coking. It is noted that Shell's decision to commercialize PFT, for instance, was linked to a desire to locate the upgrading facility next to its refinery in Fort Saskatchewan, Alberta.

At their core, all froth treatment process blocks are based on multi-stage counter-current mixer-settler arrangements, referred to as counter-current decantation (CCD, Figure 5.5). The steps in this block flowsheet are as follows:

- Fresh light hydrocarbon travels from the last to previous stages via recycle overflow streams.
- Water and solids travel from the first stage to the last stage via underflow streams.
- At every stage, the bitumen is contacted with light hydrocarbon. The overflow of the first vessel is the dilbit product, while the tailings of the final vessel is the FTT.

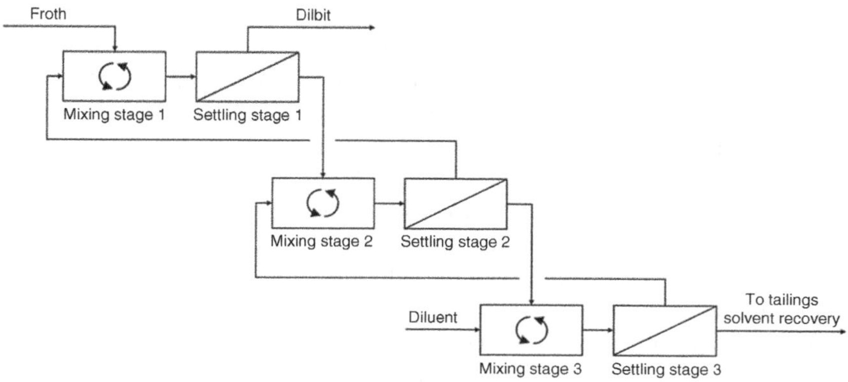

Figure 5.5 Three-stage counter-current mixer-settler arrangement.

Bitumen froth can show both water- and oil-continuous behavior, and bringing the froth in effective contact with the light hydrocarbon therefore requires significant mixing intensity (Zawala et al., 2012). However, over-shearing can lead to emulsion formation, which in turn leads to processing issues. While a detailed description of the mixing implementation is beyond the scope of this chapter, it is worth noting that commercial flowsheets use static inline mixers for contacting.

Both the dilbit product and the tailings streams are sent to dedicated solvent recovery units (SRUs) to ensure that the valuable light hydrocarbon is recovered and recycled. These units are detailed later. Major unit operations used in commercial NFT and PFT counter-current arrangements at each site are presented in the following text for awareness.

Suncor froth treatment flowsheets deploy IPS, centrifuges, and cyclones. The process evolved from a two-stage centrifuge operation to one that incorporated IPS as part of the Millenium expansion. The Syncrude froth treatment flowsheet is similar to Suncor's in that it employs IPS and centrifuges. Note that froth produced at the Aurora mine is pipelined to Syncrude Mildred Lake, and no Froth Treatment process exists at the Aurora facility itself. Suncor's second phase of froth treatment deployed hydrocarbon cyclones for processing intermediate froth treatment streams that were later deployed as an alternative to centrifuges. The Horizon NFT facility is the most recent naphthenic operation and, like Suncor's second phase, employs IPS and cyclones. Facilities that exclude energy (and maintenance)-intensive centrifuges are considered an improvement in NFT technology from the perspective of energy input. Cyclones were found to be more effective than other separation equipment at breaking undesired emulsions (Czarnecki et al., 2013).

Although originally patented by Syncrude, the first commercial PFT process was deployed at the MRM mine by Shell, currently majority owned and operated by CNRL. Two PFT processes exist at MRM and its expansion facility: a Low Temperature Froth Treatment (LTFT) in operation since 2003 and a High Temperature Froth Treatment (HTFT) process in operation since 2011. Froths from the Muskeg River Mine (MRM) and Jackpine Mine (JPM) bitumen extraction plants are distributed amongst two LTFT trains and one HTFT train. Both froth treatment plants are physically located at MRM. Similar to the Aurora mine, JPM froth is pumped to the MRM facility, and Tailings Solvent Recovery Unit (TSRU) tailings originating from JPM froth are disposed at the MRM site.

In its relatively short commercial life, PFT has already seen significant process evolution. The original LTFT process relied on the use of conventional thickener designs for its settlers that were equipped with a non-pressure-rated roof to avoid solvent evaporation. This design was not rated for elevated pressures and therefore required the operating temperature to remain significantly below the boiling point of the lowest boiling fraction in the solvent (~33 °C). Since the bitumen froth is nominally at 45–55 °C, the solvent must be *cooled* to achieve the needed mixture temperature. This cooling step leads to an intrinsic heat integration inefficiency, specifically because both the dilbit and the tailings streams need to be heated for solvent recovery and reuse. It was found in later research that settling rates at elevated temperatures (75–80 °C) under pressurized conditions were one to two orders of magnitude faster than settling rates at LTFT conditions (Kosior et al., 2018). Since throughput scales with the product of settling rate and available cross-sectional area of the separation vessel, the equipment cost penalty for using a pressurized vessel was offset by the fact that the diameter could be reduced significantly at equal production rates. Further benefits included:

- Optimized heat integration, due to the elimination of the cooling step.
- Operation at reduced S/B ratios, which ultimately enabled a change from a three-stage to a two-stage counter-current operation, further reducing capital expenditure.
- Elimination of settler internals, enabled by reduced vessel diameter.

Three HTFT commercial facilities are currently in operation: the HTFT facility at Albian, the Kearl facility, and Fort Hills. The HTFT facility at Albian is a one-train, two-stage counter-current operation that operates at approximately 75–80 °C, and at an S/B ratio slightly lower than the S/B of 2.0–2.5 w/w for LTFT. The S/B ratio is selected to achieve a bitumen product stream such that approximately half of the asphaltenes are rejected to TSRU tailings. In practice, the S/B can be optimized to achieve higher product quality (towards a higher S/B ratio) while minimizing hydrocarbon rejection to tailings (towards lower S/B ratios).

Further flowsheet evolutions are evident in the implementations at both the Kearl and Fort Hills facilities. Both operations use a mixture of normal and iso-pentane, which leads to S/B ratios even lower than the Albian HTFT facility. However, all three facilities operate at a similar temperature of 75–90 °C. This apparent sweet spot suggests that the main benefit of aggregate size and morphology (~1 mm dense particles) is achieved at this temperature, while a further increase in temperature will not lead to further reduction in overall capital expenditure or improvements in energy efficiency.

A key difference between NFT and PFT (besides the process equipment noted earlier) is the difference in product quality. Where NFT recovers the bitumen phase without selectivity, PFT rejects about half of the asphaltenes while recovering essentially all maltenes. It can therefore be considered a partially upgraded bitumen. This apparent waste of the resource in the PFT operation is offset by the avoidance of coking losses in NFT upgrading. Because the loss of these bottom-of-the-barrel hydrocarbons occurs in froth treatment for PFT, but in upgrading for NFT, it is difficult to compare bitumen recovery for the two process lineups. To facilitate a fair comparison, it is worth mentioning that for regulatory purposes an *asphaltene credit* is applied to PFT production accounting. Through this method, PFT is not penalized for the precipitated asphaltenes lost to tailings. These credited asphaltenes are added to the bitumen recovery in order to evaluate overall bitumen recovery.

5.2.4.2 Solvent and Diluent Recovery

In Section 5.2.4.1, the heart of the froth treatment operation was described, and the two process streams leaving the unit were defined. In both the product and waste stream, the light hydrocarbon concentration must be reduced in subsequent process steps. Light hydrocarbon recovery is achieved with evaporative separation to ensure that the bulk of the diluent used in froth treatment is recycled. Since light hydrocarbon is more valuable than the bitumen itself, it is critical to remove the material prior to upgrading. For the tailings stream, the light hydrocarbon needs to be removed for both economic and environmental reasons. Finally, vapor streams are captured from process vessels and recycled, diverted to Cogeneration fuel, or flaring. Makeup diluent is added to the circuit to overcome any losses.

While the process functions are essentially identical, the naming convention for light hydrocarbon recovery from the product is different for NFT and PFT. NFT operations use the term Diluent Recovery Unit (DRU), while PFT operations use the term SRU. Note that for Albian's PFT, the bitumen product is pipelined with some residual solvent to off-site upgrading, which is needed to meet pumping requirements. The final solvent recovery occurs at the Scotford upgrader, and the solvent is pipelined back to the mine site. The Kearl operation is similar to the Albian operation, but the product is shipped to market differently. To avoid pentane loss, Kearl removes all solvent in the SRU and subsequently dilutes the bitumen product with condensate to reduce the viscosity for pipeline transport. Fort Hills uses yet another method by transferring Hot Bitumen (Hot Bit) to the East Tank Farm terminal for blending and shipment to market. NFT operators use upgrading local to the mine site, as the bitumen is not directly suitable for market.

The DRU and SRU implementation can be considered as conventional flash technologies in relatively clean service. Opportunities for heat integration, such as waste heat from on-site NFT upgrading and the use of hot condensate return to preheat process water for primary extraction, have already been incorporated in existing flowsheets. It was previously noted that HTFT is more energy efficient than LTFT due to the avoidance of solvent cooling upstream of the counter-current extraction circuit. Here, an additional benefit is observed in that the streams produced from the CCD circuit (operated at around 80 °C) only require minimal additional heating for solvent recovery.

NFT operators denote diluent recovery from tailings as the Naphtha Recovery Unit (NRU), while PFT operators refer to this step as the TSRU. Light hydrocarbon removal from the tailings stream is more complex than product solvent recovery, due to the highly erosive nature of the slurry and the challenge of processing clay-stabilized emulsions in the froth treatment underflow. For all recovery technologies that process FTT, the process fluid must be exposed to a combination of elevated temperatures and/or reduced pressures to flash off the light hydrocarbons. Conventional heating techniques (e.g., heat exchangers) and column internals (e.g., trays) are not suitable, however, due to bitumen and clay fouling, solids erosion, and plugging potential. The only practical method of slurry heating is direct steam injection, which is intrinsically energy inefficient, as the steam condensate is lost to tailings.

In practice, FTT streams exit the process at greater than 80 °C, making them a target for potential heat and/or hot water recovery initiatives (see Section 5.3.1). Another concept for heat recovery is the use of slurry heat exchangers, e.g., for heating the TSRU *feed* with the TSRU tailings stream. Key operational challenges with slurry heat exchangers are the intrinsic erosive nature of the slurry and the potential for plugging small clearances with coal and wood debris known to be present in FTT streams. These factors have thus far prevented implementation.

5.2.5 Tailings Treatment

The bitumen extraction process as a whole generates two major waste streams. The primary extraction tailings account for more than 95 wt% of the solid waste. This stream typically contains less than 1 wt% residual bitumen and no light hydrocarbons. FTT are smaller in particle size compared to extraction tailings and do contain residual diluent. In addition, FTT shows elevated concentrations of oleophilic minerals like metal sulfides and Naturally Occurring Radioactive Materials (NORMs) (Kuznetsov et al., 2015; Lindsay et al., 2019). These compounds may pose additional risk at closure. The biogeochemical risks posed by FTT, as well as the challenges around operational management of tailings in general, are discussed separately in Chapter 6.

5.2.6 Utilities

Bitumen extraction is known to require large amounts of energy in the form of hot water and steam, as well as electrical power. For context, hydrotransport and tailings pumps are of the order of 2–3 MW per pump, with typically four to six pumps in series on a single tailings train. In addition, electric shovels have a typical power demand of several tens of MW. The utilities unit within oil sands operations is among the largest found in the process industry when considering the scale of power generation.

While a detailed description of the utilities system is beyond the scope of this chapter, two opportunities for environmental performance improvement are mentioned for completeness. First, the requirement for both power and steam generation provides a natural fit for cogeneration. All mineable oil sands operators either have or are in the process of deploying cogeneration power plants. For example, Suncor's coke-fired boiler replacement is expected to provide up to 800 MW of power and reduce GHG emissions at the bitumen extraction and upgrading plants by 25% (Government of Alberta, 2020). Life cycle analysis models such as the GHOST and FUNNEL models indicate that the EI and GHG emissions are reduced when on-site cogeneration can replace electricity drawn from the grid. For example, Bergerson et al. (2012) report that surface mining without cogeneration emits between 2.9 and 8.9 g CO_2e/MJ, whereas surface mining with cogeneration emits 2.5–8.1 g CO_2e/MJ. The direct site emissions increase with on-site cogeneration, but the increase is more than offset by the reduction in indirect emissions associated with grid withdrawal. Further, cogeneration using gas turbines reduces emissions more than steam turbines, because the former exports more electricity to the grid (Nimana et al., 2015).

Less-known opportunities are the utilization of petroleum coke (generated as a waste stream at NFT operators) as a reclamation substrate and in the treatment of oil sands process water (Zubot et al., 2012; Small et al., 2012; Pourrezaei et al., 2014). Petcoke has lost its value as a boiler feed fuel due to the high carbon-to-hydrogen ratio (and therefore proportionally high CO_2 emissions during combustion). Since the material is abundant and requires on-site storage, the waste material provides a unique opportunity for creative reuse applications. Research in this area is ongoing.

5.3 Continuous Improvement Opportunities

Up to this point, a description of conventional flowsheets as implemented in the Athabasca region has been described, which included discussion of the evolutions leading to EI and GHG reduction efforts *within* specific unit operations. This section focuses on similar efforts that transcend single process blocks or are aimed at additional infrastructure required for either mine expansions or significantly improved environmental performance at existing operations. Operators continue to look for opportunities to improve energy efficiency, and it is therefore impossible to provide a complete list. Rather, a selection of three broadly different approaches to performance improvement are described. The reader is encouraged to review other literature, including initiatives documented in, e.g., the Canadian Oil Sands Innovation Alliance's (COSIA) annual reports with respect to air, land, water, and tailings Environmental Priority Areas (EPAs).

5.3.1 Diluent and Bitumen Recovery from Tailings

Hydrocarbon losses from Primary Extraction and Froth Treatment represent both a lost economic opportunity and an environmental impact that may require potentially energy intensive mitigation during reclamation activities. While the risks associated with residual hydrocarbons in post-mining landscapes are detailed in the next chapter, it is noted here that hydrocarbon migration and biodegradation from tailings can compromise bio-performance of both aquatic and terrestrial closure structures. The liability associated with these potential outcomes is not well understood and is a topic of ongoing research.

In order to reduce this liability, operators continue to contemplate options for further removal of hydrocarbons in "bolt-on" processing units. Both "live" tailings (i.e., material directly from the processing plant) and fluid tailings inventory (as stored in tailings ponds) are being considered as feedstock for such processes. Although bolt-on hydrocarbon removal technologies have not been deployed commercially at this time, concepts have been progressed significantly by operators and third-party collaborators. An overview is presented in the following text.

TSRU and NRU tailings (often referred to as FTT) are obvious candidates for bolt-on hydrocarbon removal, as these contain valuable light hydrocarbon, are concentrated in bitumen relative to Primary Extraction tailings, and leave the process at elevated temperature (~80–90 °C). Treatment aimed at recovering hydrocarbons and hot process water may result in several near- and long-term EI benefits, such as:

- Hot water recycling into base operations, offsetting hot water make-up requirements.
- Reduction in tailings footprint: dense tailings are created in parallel with hot water recovery.

- Avoidance of GHG (particularly methane) emissions due to biodegradation of diluent.
- Avoidance of potentially energy-intensive (and lengthy) remediation activities associated with hydrocarbon migration in the closure landscape.

It is notable that FTT is enriched in heavy minerals and REE (Roth et al., 2017). Historically, the pursuit of the recovery of these components was developed in parallel with hydrocarbon recovery (Ityokumbul et al., 1988; Oxenford and Coward, 2001; Ciu et al., 2003; Liu et al., 2006; Marshall et al., 2014; Duyvesteyn et al., 2006). A bolt-on that targets all benefits described above is the "Creating Value from Waste" (CVW) technology developed by CVW Cleantech (CVW CleanTech, 2023). CVW technology was developed and refined in collaboration with oil sand operators, including a front-end engineering and design (FEED)-level study with CNRL (Titanium Corporation, 2019). It uses mineral processing and solvent recovery technologies to separate organics, hot water, and heavy minerals from FTT. An overview of the process is shown in Figure 5.6.

The CVW process intercepts the FTT stream and splits the material into fines- and coarse-dominated streams using cyclone separation. The fines-dominated overflow is processed in a Concentrator Plant (CP), where hydrocarbons are separated from water and fine mineral solids using conventional flotation and solvent extraction. The counter-current extraction operation uses light hydrocarbons supplied by the base operation. Hydrocarbons recovered in the CP are routed back to the host operation for further processing. Tailings are sent to a Tailings Distillation Unit (TDU), recovering residual light hydrocarbon using evaporative techniques. This light hydrocarbon is recycled into the CP process. TDU and flotation tailings are sent to a thickener, described

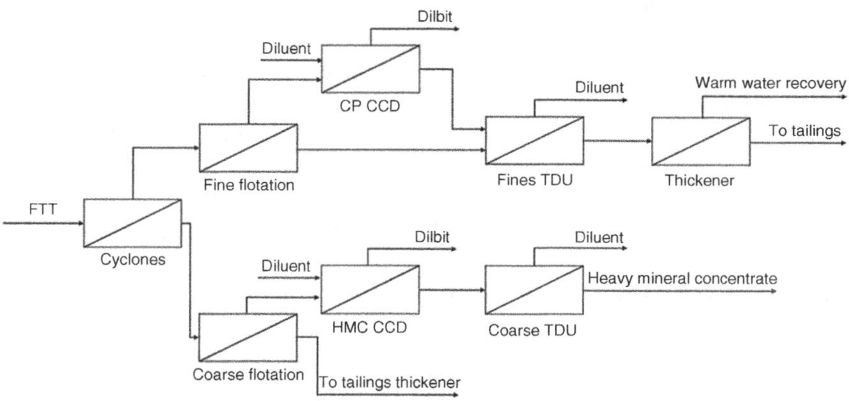

Figure 5.6 CVW flowsheet as described in CVW CleanTech (2023). *Source*: Adapted from CVW CleanTech (2023).

later. The TDU offers superior performance compared to conventional distillation in base operations as a result of bitumen removal in the CP. TDU tailings are presumed to contain lower light hydrocarbon concentrations than current commercial operations.

The coarse cyclone underflow is processed in a Mineral Separation Plant (MSP). Here, valuable heavy minerals (VHM) like titanium oxides and zircon sand, as well as REE, are separated from the slurry matrix. First, flotation and solvent extraction produce a Heavy Mineral Concentrate (HMC). The flotation step is selective and removes gangue such as silica, while multi-stage countercurrent solvent extraction removes bitumen. As in the CP, the cleaning steps are performed with light hydrocarbons supplied by the base operation. Coker-feed quality bitumen is routed back to the base operation, while the slurry is processed in a dedicated distillation column to recover light hydrocarbons (for reuse) and produce an HMC suitable for additional treatment. Second, the HMC is processed in a variety of conventional mineral processing technologies, e.g., gravity spirals, shaker tables, wet magnets, high tension rolls, and magnetic separators, creating VHM and REE concentrates. The concentrates can be transported and sold to the market or processed further at the host site. Flotation tailings from the CP and MSP and TDU tailings from the CP are sent to a thickener where hot water is recovered for reuse in the CVW process or integration into the base operation.

It is worth noting that CVW at a PFT site may be more challenging than NFT implementation. For a host using PFT, treatment of the tailings (which contain asphaltenes precipitated in the base PFT operation) with paraffinic solvent would only recover residual maltenes (and associated soluble asphaltenes). CVW has proposed an option where naphtha is utilized even for PFT operators in the CVW solvent extraction steps. Such a "solvent switch" may require additional technical and regulatory evaluation. These challenges are not present for an NFT implementation. Naphtha will dissolve all hydrocarbons in NRU tailings, and the diluent recovered in CVW distillation units is easily returned to base operation.

CVW Cleantech estimates the benefit of a model implementation for a 250,000 bbl/d SCO operation as follows:

- GHG reduction by 5–10%, mainly due to avoidance of biodegradation of diluent in tailings.
- reduction of volatile organic compounds (VOCs) emitted at tailings discharge as a result of deeper naphtha recovery.
- a net raw water use reduction of 2.8 million m^3 per year, based on efficiencies in the utilities plant. Further, around 15 million m^3 recycled water per year is made available at plant operating temperatures, as opposed to tailings temperatures for base implementation.

Potential further benefits include reduced tailings treatment intensity due to improved thickening, as well as the ability to isolate pyrite in the MSP for dedicated treatment. Pyrite is an Acid Rock Drainage (ARD) precursor that concentrates in FTT. The cost associated with geochemical risks in the closure landscape is not fully understood at this point.

The actual benefits would be dependent on the extent of implementation and integration at a site. For example, hot water recovery will only be beneficial if it can be used to offset hot water production in the base operation. Further, the full benefit will not be realized if operators choose to forego mineral separation. An example where the requirement is to recover hydrocarbons and hot water only is CNRL's Naphtha Recovery Unit Treated Tailings (NRUTT) process (Canadian Natural Resource Ltd., 2024). This technology has recently been disclosed in an application to the AER for the Horizon mine. The process has an intent similar to the CVW CP. In NRUTT, NRU tails are intercepted and processed through dilbit recovery and cleaning, and water separation steps. The process can recover greater than 50% bitumen and naphtha and is anticipated to reduce emissions through the avoidance of biogenic gas generation from residual diluent in tailings.

Bolt-on technologies can provide both EI reductions and environmental benefits, as illustrated by the abovementioned examples. They are most beneficial if adopted from the start of operation on "live" streams. Benefits are reduced when the technology is applied later in the life of mine (LOM), when the remaining FTT make is small relative to tailings already in inventory and rehandling of the inventory volume is prohibitive. Opportunities for implementation may exist near the end of mine life, however. As an operation transitions into reclamation activities, localized management of Constituents of Potential Concern (CoPC) may become critical for certain deposits, e.g., for material in contact with the reclamation cover. Removal of CoPCs like residual hydrocarbons or sulfide-bearing minerals from these tailings may be necessary to meet closure outcomes.

It is noted that heavy mineral recovery from tailings deposits has been contemplated but has not progressed commercially at this point. However, hydrocarbon removal options from fluid tailings during rehandling have recently been described in a series of Suncor patents (Hollander et al., 2019; Brown et al., 2019; Hollander et al., 2020). One option uses flotation and sparging of a rehandled fluid tailings stream, while another option uses air introduction into fluid tailings for in situ aerobic biodegradation of residual naphtha. The benefits are again related to the avoidance of GHG emissions and hydrocarbon liberation due to biogas generation in water-capped tailings deposits. Further details on tailings treatment options focused on reclamation outcomes are described in Chapter 6.

5.3.2 Modular and Mobile Design

Any new mining operation is typically developed with the following considerations regarding the ore body:

- The processing facility will be in close proximity to high-grade ore to ensure high production during the initial mining phase, to allow for high return on capital invested.
- Since no in-pit tailings storage is available at the start of the operation (no ore has been mined yet), tailings will need to be placed in an externally constructed tailings facility. This structure needs to be placed on low-grade ore, as the material underneath the tailings pond will become permanently inaccessible.

This combination has the undesired effect that, as the mine progresses, the haul distance to the plant increases while the average ore grade decreases. In addition, tailings placement becomes more energy intensive as the tailings pond increases in elevation and tailings placement distances generally increase. To counter this effect, various operators have considered moving the processing facility closer to the mine face as the mine advances. If the processing plant is designed as a modular/mobile concept, then relocation can occur each time a new mine pit is opened. One example is CNRL's In-Pit Extraction Process (IPEP). While not commercialized at this point, the concept not only enables plant relocation but also deploys tailings technologies aimed at producing dry, stackable tailings.

IPEP was developed as a way to eliminate fluid tailings impoundments. Besides the tailings benefit, a lower environmental process footprint is anticipated due to reduced energy consumption related to ore and tailings transportation distances. As per non-confidential information (Canadian Natural, 2021), CNRL documented findings collected during the operation of a 500 tph pilot that operated during a two-year period at its Horizon facility. For a realized throughput of 300–325 tph, a primary extraction bitumen recovery of 90.7% was achieved, and a tailings waste stream with a solids content between 75 and 80 wt% was created. While these outcomes were slightly below target, opportunities to optimize the flowsheet and improve reliability were identified. IPEP is based on commercially available equipment already used in mining, which implies that any risks associated with equipment scale-up are largely mitigated. A brief summary of the process lineup is provided in Figure 5.7.

Ore is crushed and mixed with hot process water to form a slurry that is fed to washer screws. These screws separate bitumen and water from the coarse mineral fraction. The bitumen-rich stream is processed in a three-stage counter-current flotation circuit and transported to a centrally located froth treatment plant. Underflows from the washer screws and the flotation circuit are processed in a

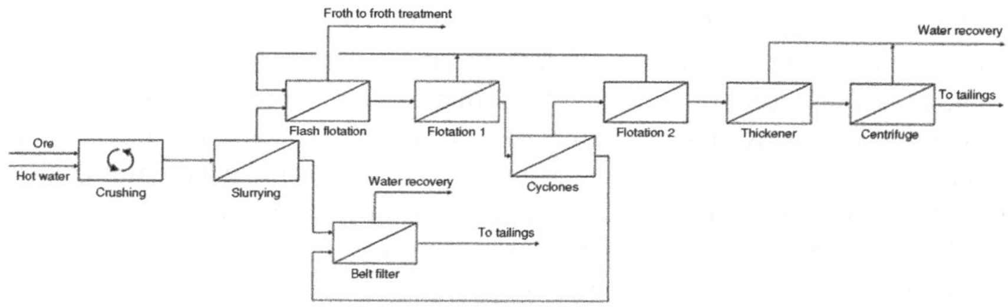

Figure 5.7 IPEP flowsheet as per Phase 3b pilot configuration. *Source*: Adapted from Canadian Natural (2021).

series of dewatering steps in order to recover hot water for reuse and separate dry, stackable tailings. Dewatering occurs through cycloning, centrifugation, thickening, and belt filtration. The combined solids product is then transported by conveyor or truck to the mined-out area.

The GHG reduction offered by IPEP was evaluated by comparing the power requirements of Horizon's current operation to that of a commercial-scale IPEP design for the South Mine. It was claimed that IPEP showed a net reduction of 18%, with an EI of 4.9 kWh/tonne processed compared to a base plant EI of 6.0 kWh/tonne. This comparison excluded froth treatment. CNRL projects that the IPEP could reduce GHG emissions by up to 40% (Canadian Natural, 2022). As of 2022, the FEED of the demonstration plant had been completed, with the intent of proceeding with a 750 tph commercial demonstration unit (Government of Alberta, 2021).

While the concept of localized ore and tailings processing is elegant, there are deployment challenges for commercial implementation. As was noted in pilot plant operation, an average froth quality of approximately 42 wt% bitumen was realized. CNRL noted that the IPEP froth would need further processing in a froth cleaning step to raise the quality to a level compatible with NFT (~55 wt% bitumen). Further, another challenge related to long-distance transport of high air content froth was identified.

With respect to dewatering performance, CNRL noted that the components functioned well individually, but off-spec material negatively impacted conveyor operation. Further, belt filter operation for tailings with elevated bitumen and clay content could suffer from plugging of the filter cloth. CNRL identified potential improvements such as the inclusion of belt-wetting systems and the application of more loose cloth weave in the belt filter. While reliability and availability were lowest in the crushing and conveying portions of the pilot, the equipment used is well-established in the oil sands industry, suggesting that reliability can be improved.

A final challenge noted by CNRL was the build-up of clays in the reclaim water collected from the dewatering circuit. Without a periodic dump of the reclaim water, fines were accumulating to a level that the water became unsuitable for reuse. Build-up of clays and other finely divided mineral solids would not be compatible with commercial operation. Mitigations proposed included optimization of the thickener operation. In a commercial IPEP operation, this water would be recycled to the main plant for heating, and hot process water would be pumped back to the IPEP unit for bitumen extraction. The fines build-up further suggests that not all solids were combined in the stackable tailings product. Fines capture efficiency is of critical importance in this process flowsheet, as the implementation implies that tailings impoundments are to be eliminated.

While IPEP is innovative in its goal to eliminate tailings ponds, the concept of a mobile ore preparation and bitumen extraction plant at the mine face is not new. For example, Suncor's deferred Voyageur South project was based on a mobile ore preparation unit called Slurry at Face (SAF) (Suncor Energy Inc., 2007). The goal of SAF was primarily to reduce diesel fuel consumption by reducing the haul fleet (shorter haul distances require fewer trucks). Additional benefits such as reduced capital, operating and maintenance costs, as well as improved safety due to decreased large-vehicle traffic were also anticipated. Similarly to IPEP, SAF would be located at the mine face and would produce a slurry comparable to stationary ore preparation plants. Unlike IPEP, SAF did not contain modular and mobile primary extraction components. Rather, a conventional stationary primary extraction plant was planned. Syncrude also explored the concept of a mobile ore preparation plant. Their In-Pit Bitumen Production process, or "In-bit," also intended to move the ore preparation process closer to the mine site in order to reduce GHGs through the reduction in the number of haul trucks. Despite a number of development initiatives, none of these technologies have been commercialized at this point.

5.3.3 Resource Optimization: Definition of "Bitumen Ore"

It is often assumed that improvements in energy efficiency require entirely different technology implementations or major improvements to hardware within the unit operations. Historical examples of such changes were previously described: the switch from bucketwheel and drag line mining to truck and shovel operation and the evolutions within froth treatment. However, due to the capital intensity of changes to the existing equipment, operators are generally reluctant to implement such changes. Below, an entirely different approach to EI reduction is described: rather than changing the equipment, the definition of the ore body is addressed. Since poorly processing ore requires higher processing intensity and contributes disproportionally to the tailings challenge, there is an economic and environmental case to be made for eliminating the worst materials from the feed blend. While the sterilization of low-quality ore may at a glance appear to resemble high grading, this framework is supported by *both* the operator and the regulator.

In late 2016, Shell Canada submitted a request for a waiver to the mining criteria as outlined in Section 2 of AER's Directive 082 (Alberta Energy Regulator, 2016a). This request followed a pilot program at its Albian mines, during which problematic ores were segregated from the ore blend and improvements in energy consumption and environmental outcomes across the operation were monitored. Shell's Resource Optimization Pilot (ROP) was executed in 2015–2016 (Shell Canada Energy, 2016). It was demonstrated that problematic ores were detrimental to the extraction process and resulted in increased GHG emissions per barrel of bitumen produced, as well as an increase in the volumes of fines and clay suspensions requiring tailings treatment.

Based on these results, Shell proposed to utilize criteria listed in Section 3 of the directive and define ore as materials that should be *selected* for processing due to "environmental, technical, economic, or operational" reasons. This approach is in contrast with defining ore according to established criteria defined in Section 2 of the directive, which gives guidance on when materials should be *rejected* for processing. This change in emphasis allows for a more balanced and flexible development of the ore body and should lead to reduced reclamation liability related to fluid tailings inventory.

To identify problematic ores, Shell developed the Ore Processability Index (OPI), linking extraction recovery to ore features such as fines/clay content and process water chemistry. Facies associated with reduced primary extraction recovery were also identified. Aspects such as geological complexity and separation practicality were also used to identify and segregate problematic materials. Collectively, these practices enabled a refined definition of the ore body that accounts for a range of criteria broader than the Directive 082 definition. Delineation between ore and waste in this manner led to significant improvements in mining and plant performance.

Mining improvements included higher equipment productivity and advancement in mining construction activities. While the average feed blend during the ROP resembled the average feed blend prior, significant improvements in plant performance were observed during the ROP. It was concluded that during the ROP, plant feed blends consistently contained a lower proportion of materials with excessive (>30 wt%) fines content. As a result, primary extraction recovery at the MRM mine increased during the ROP and remained stable above 93%. Operation where very good extraction recovery was achieved (set at 96% or higher) doubled, while very poor recovery operation (<90%, leading to non-compliance under Directive 82) decreased five-fold. The interrelation of bitumen recovery and production to the characteristics of the ore blend was evaluated using a metric called the Plant Performance Factor (PPF). The PPF was shown to increase during ROP operations relative to pre-ROP operations, but only limited benefits were observed during partial ROP operation.

ROP operations not only increased bitumen recovery and production but also resulted in further performance improvements. According to Shell's waiver application, the following improvements were noted:

- Reduced hot process water temperature requirements at both MRM and JPM.
- Increased froth quality at MRM and sustained quality at JPM.
- Reduced flushing requirements at both MRM and JPM extraction plants.
- The subsequent ability to operate the LTFT unit at a lower S/B ratio, which in turn resulted in less asphaltene rejection and lower solvent losses to tailings.
- Lower steam consumption in the TSRU.
- Improved plant stability with reduced downtime and maintenance activities.

Environmental benefits were examined in terms of EI and fluid tailings volume make. According to Shell, the overall EI dropped by 9% in summer (from 436 GJ/kbbl to 398 GJ/kbbl) and dropped by 6% during winter months (from 535 to 501 GJ/kbbl when operating under "partial" ROP with some problematic ores reintroduced into the blend).

ROP further suggested a significantly lower volume of fluid fine tailings due to the avoidance of high clay-content feed. To quantify this, poor OPI materials from facies known to result in poor extraction performance were processed in a laboratory-scale Batch Extraction Unit (BEU), which mimics the ore preparation and primary extraction process. Settling performance of the BEU tailings was measured. The fluid tailings production correlated strongly to the feed fines content, and Shell was able to show that for every percent increase in fines in the feed blend, the fluid tailings make increased by ~1.3 m^3/tonne. Historical cross-referencing of the fines content and ore grade at the MRM mine revealed that fluid tailings generation increased sharply once the ore grade fell below 11% bitumen, which is consistent with the laboratory-scale findings.

Shell developed two LOM scenarios comparing GHG intensity and fluid tailings reduction relative to a scenario based on Directive 082 requirements. An "additional heat" scenario covered anticipated increased energy needed to process problematic ores. The "ore flexibility" scenario described sterilization of problematic ores in a manner consistent with the principles of ROP. Shell concluded that under the "ore flexibility" scenario the total produced bitumen would decrease by approximately 5% while the mine life would decrease by 3.5 years. These impacts were offset by a reduction in GHG intensity of 2.7 kg CO_2e/bbl, which is 8% relative to the assumed baseline of 35.6 kg CO_2e/bbl and amounts to a total GHG reduction of 8.5 Mt CO_2e over the LOM. In contrast, the "additional heat" scenario resulted in an *increase* in GHG intensity of 1.4 kg CO_2e/bbl, 4% higher than the status quo and higher than the "ore flexibility" case. Materials segregated as waste under the ore flexibility scenario had a GHG intensity of 88 kg CO_2e/bbl, 2.5 times larger than the baseline.

The "ore flexibility" scenario further showed fluid tailings benefits: it was 18% less intense at 79 kg dry solids/bbl, compared to 91 kg dry solids/bbl under the additional heat scenario. The fluid tailings intensity of the sterilized material was 330 kg dry solids/bbl for comparison.

The Albian mines currently define ore according to ROP principles, incorporating adaptive management strategies to ensure that plant feed consists of materials that do not adversely impact GHG intensity and fluid tailings profile in a disproportionate manner. The Horizon mine uses a similar strategy and incorporates criteria to define ore other than those defined by Section 2 of Directive 082. The Alternative Ore Criteria (AOC) pilot application focuses on the reduction of fluid tailings generation (Canadian Natural Resources Ltd., 2023). As outlined by CNRL, ore that does not meet bitumen cut-off criteria of greater than or equal to 6–7 wt% and fines less than or equal to 30–35 wt% (on a dry solids basis) are classified as waste. In a previous iteration of this concept, CNRL estimated that this strategy would result in a fines placement reduction of 52 Mt (141 Mm^3 of fluid tailings equivalent) in its tailings impoundments. It is estimated that as a result 121 Mm^3 less water will be taken in from the environment (Canadian Natural Resources Ltd., 2016). Finally, in an integrated application submitted to the AER (Imperial Oil Resources Ltd., 2022), Imperial Oil describes its AER Directive 082 Pilot program, which commenced in 2019 with the goal to apply a more holistic method by which to classify ore and waste. Similarly to the ROP, the basis by which the resource was defined considered operating constraints, costs, and environmental outcomes. Production matrices were used to describe the processability of the oil sand material based on attributes in addition to ore grade. The pilot was anticipated to result in a GHG intensity reduction of 2% and an FFT generation (on a m^3/bbl basis) reduction of 17%. Realized outcomes were similar to those predicted as part of a simulation exercise.

By revisiting the definition of bitumen ore, operators have been able to improve both economic and environmental performance of the assets without the need for additional process equipment. This form of optimization is easier and quicker to implement and does not require further capital investment. It is expected that these operational opportunities will be the focus in the near term.

5.4 Step-out Technologies

So far, all technologies described have been based on relatively small improvements on the base oil sands processing flowsheet, using CHWE as the heart of the process. In this section, concepts are discussed that are based on a fundamentally different approach to bitumen extraction and utilization.

5.4.1 Solvent Extraction

Non-Aqueous Extraction (NAE) is a significant step-out from CHWE-based oil sand operations in that it replaces water with an organic solvent as the extraction medium. The intent of eliminating the aqueous process step is that it, at least in concept, eliminates the need for water recovery from tailings and thereby tailings ponds. Solvent is used to dissolve the bitumen from the oil sands matrix. Since the dilbit is significantly less dense than the solids fraction, gravity separation can be deployed without the need for air attachment to the bitumen phase, as is required in the CHWE process.

NAE has not been commercially deployed currently, but laboratory and pilot-scale studies have been active since the 1960s (Lin et al., 2017; Council of Canadian Academies, 2015). Several process variables have been investigated experimentally, including the impact of ore quality (grade, fines content), solvent selection, dilution ratio, ore contacting and mixing, as well as process temperature (Yuan and Elektorowicz, 2021; Hooshiar et al., 2012a; Hooshiar et al., 2012b; Geramian et al., 2019; Wu and Dabros, 2011; Nikakhatari et al., 2013; Wang et al., 2014; Li et al., 2012; Tamayo et al., 2024). It is noteworthy that NAE research originated in the United States for oil-wet tar sands, which are not processable using CHWE. Most oil sands operators have explored this technology at some point, as is evident by numerous patents assigned to Shell, Syncrude, IOL (Lin et al., 2017), and more recently Suncor (Huq et al., 2022). The NAE concept is attractive to new mine sites or expansions, when the benefits of mine development and tailings management can be integrated seamlessly and no commitment to CHWE infrastructure has yet been made.

For NAE to be selected, it must offer significantly better economic and environmental value than the incumbent technology, including:

- Improved bitumen recovery performance over a broader range of ore types, eliminating blending requirements and expanding the resource base. This approach can be contrasted to the previously discussed ROP philosophy.
- Eliminate water recovery from tailings ponds and subsequent accelerated progressive reclamation, reducing the overall land disturbance of the mine.
- Reduced EI, since hot water production for primary extraction is eliminated.
- Lower capital intensity.

A conceptual NAE flowsheet is shown in Figure 5.8. It is noted that the process shows similarities in mining, ore preparation, and solvent recovery but combines the function of primary extraction and froth treatment in a single unit operation.

Mined ore is delivered to a crushing facility for size reduction and increase in contact area. The crushed ore is contacted with solvent in a digestion step for bitumen dissolution. Depending on the volatility of the solvent, this process step

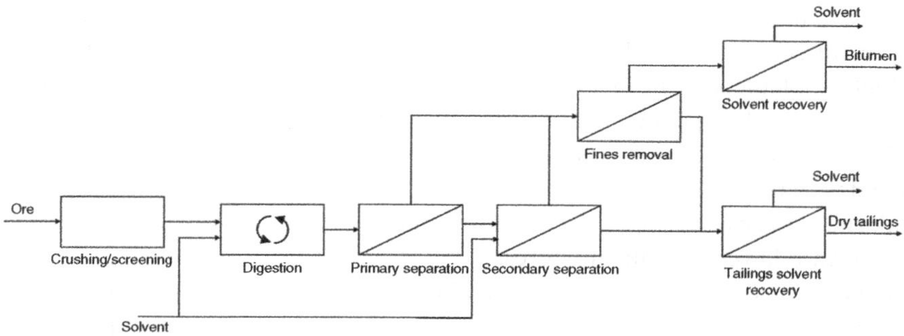

Figure 5.8 NAE flowsheet as described in Lin et al. (2017). *Source:* Adapted from Lin et al. (2017).

may require an inert atmosphere. Proposed processing equipment includes dissolution tanks, tumblers, or troughs with rotating elements. The slurry is fed to separation and washing stages where coarse mineral solids are isolated from dilbit. Depending on the number of contact stages, equipment proposed includes gravity settlers, IPS, centrifuges, classifiers, and in some permutations, thickeners and belt filters.

The dilbit is subsequently cleaned in a fines removal step. Fines removal is critical for pipeline transport and cannot be accomplished with conventional gravity settling. Aggregation techniques for fine solids particle size increase need to be employed. These include agglomeration with small quantities of water, aggregation through the addition of controlled amount of antisolvent, or addition of polymeric additives (Lin et al., 2017). Solvent is then removed from the dilbit in a solvent recovery step that uses evaporative techniques. The solvent is recycled into the process, while bitumen proceeds to upgrading.

The fines treatment underflow is combined with the coarse underflow from the separation and washing steps and is subsequently sent to solvent recovery from the waste stream. Solvent is removed using evaporative techniques and may include steam or gas stripping, washing with a surfactant, the assistance of a second solvent, and drying in a fluidized bed (Lin et al., 2017). Here, the goal is to recover sufficient solvent so that dry tailings can be transported to the mined-out pit. Transportation could be accomplished with either conveyors or by truck.

NAE performance will depend strongly on the selected solvent within the process. Solvents considered range from alkanes, cycloalkanes, aromatics, or combinations of these solvents (Wang et al., 2014; Nikakhatari et al., 2013; Pal et al., 2015; Wu and Dabros, 2011; Yuan and Elektorowicz, 2021). The solvent must be partially or fully miscible with the bitumen. However, the volatility of the solvent needs to be optimized for two competing requirements. Low volatility would enable a reduction in safeguarding requirements related to ensuring the process atmosphere remains below the Lower Explosion Limit (LEL) but would make it more difficult to recover the solvent from the dilbit and waste stream. Conversely, a highly volatile solvent increases safeguarding requirements (e.g., by requiring the process to be operated above the Higher Explosion Limit (HEL)) but makes solvent recovery from the waste stream easier (Nikakhtari et al., 2014). It should be noted that bitumen ore is highly abrasive, and pipeline erosion and equipment leaks need to be anticipated. Process safety requirements may drive solvent selection toward lower volatility mixtures.

For completeness, it is mentioned that Figure 5.8 presents a simplified "solvent-alone" flowsheet. Various permutations of this flowsheet have been proposed and include Solvent Extraction Spherical Agglomeration (SESA, fines agglomeration by water addition), Ionic Liquid Assisted Solvent Extraction (ILASE), Switchable Hydrophilicity Solvent Extraction (SHSE), and hybrid options that combine the concept of NAE and CHWE (which would still require tailings containment areas) (Lin et al., 2017).

While NAE has clear process benefits, there are several implementation challenges that remain. Two examples are provided here. First, it is recognized that, in this process, the full waste stream has been exposed to light hydrocarbons, which is in contrast with the conventional flowsheet where only 2–5% of the solids are contacted in froth treatment. Residual solvent in tailings poses a significant environmental risk that requires mitigation. Solvent recovery from FTT is known to be energy-intensive, and it is therefore likely that the intensity for solvent recovery from the NAE waste stream would also be significant. It is not immediately clear what the energy requirements will need to be to create environmentally acceptable dry tailings from this process. There is little public information on the reduction in EI and GHG emissions that NAE may provide. For example, Alberta Innovates (2020) suggests that NAE may reduce the total annual GHG emissions for an oil sands mining operation by 25%. Furthermore, it is noted that IOL's NAE pilot indicated a reduction in water consumption of 80%. However, it has also been noted that little information is available on the costs and environmental impacts of the NAE process (Council of Canadian Academies, 2015).

Second, off-normal operation will be significantly more challenging than what is experienced in current commercial operation. Any solvent slurry will show segregating behavior, which implies that the process equipment will be prone to plugging under upset conditions. In aqueous extraction, such upsets are known but can be mitigated by water flushing strategies. It is unlikely that such a strategy can be adapted with solvent, and it is therefore plausible that water-based flushes and steam treatment will be required. This material will need to be diverted to fluid containment areas, as it will be too wet to convey. The infrastructure required for these activities will add to the cost and complexity of the NAE flowsheet. After almost six decades of operation, the CHWE process still represents the most robust bitumen extraction method. While the obvious drawback of the aqueous process is the tailings footprint, it is unclear if this liability outweighs the uncertainty of a solvent-based process.

5.4.2 Carbon Capture and Storage

An alternative way to reduce site emissions from oil sands operations is by deploying Carbon Capture and Storage (CCS) of CO_2 generated in the process. CCS provides emission reduction opportunities for industrial sectors that are difficult to decarbonize through process modifications. Such a philosophy focuses on the mitigation rather than the prevention of GHG emissions. This option is of particular interest for interception of emissions from power generation activities, as these are localized and show relatively high CO_2 concentrations. CCS is applicable to processes such as thermal generation stations, refineries, and cement manufacturing. CCS is a viable option for certain aspects of oil sands processing due to the availability of nearby geologically stable storage capacity (Council of Canadian Academies, 2015).

Recently, a CCS network, including a 400 km pipeline from the mining operators to the Cold Lake area, has been proposed by the Pathways Alliance for this purpose (Government of Alberta, 2024d). A final investment decision is anticipated in 2025. An example of CCS is found at the Scotford Upgrader complex in Fort Saskatchewan, AB. The Quest project commenced operation in 2015 and has captured over 7.7 Mt of CO_2 as of 2022 (Shell Canada Energy, 2023). It is based on CO_2 recovery from the hydrogen production process required to upgrade the PFT bitumen product received at the Scotford Upgrader. Amine-based scrubbing technology is used to capture the CO_2, and the liquified stream is subsequently pipelined and injected into 2 km deep wells (Shell Canada Energy, 2024).

5.5 Concluding Remarks

This chapter presented an overview of the main process flowsheets in operation within mineable oil sands, the key evolutions seen since initial deployment, and the major bolt-on and step-out opportunities that are currently being explored. It is striking to see that after almost 60 years of operation, the CHWE process essentially remains unchanged. The process has proven to be robust, and energy integration within the operating units has led to a system that is optimized within what can be achieved in a harsh process environment.

Recent emissions from oil sands mining operations alone have been estimated at approximately 15 Mt CO_2e/yr, which is 15–20% of the total emissions attributed to oil sands (in situ, mining, and upgrading operations) (Canadian Association of

Petroleum Producers, 2024). In the period 2012–2021, the production from mineable oil sands increased by 71% to approximately 1.5 MB/day, while the emissions increased by 68%, roughly proportional to the production increase. Over the same period, the overall oil sands EI reduced by 20%, driven primarily by reductions in upgrading (−19%) and in situ operation (−9%). With no appreciable increase in mining operations capacity anticipated in the next 25 years (Canada Energy Regulator, 2021), no major shift in GHG emissions associated with mineable bitumen extraction is to be expected.

Given the capital intensity of these mining projects, it seems unlikely that operators will be abandoning current assets, which means that process improvements are most likely to occur at the periphery of the process (e.g., autonomous haul trucks, opportunities associated with FTT). Such changes would require limited modifications within the heart of the brown-field operation and are therefore easier to implement. Similarly, improvements in operating philosophy may be preferred over bolt-on technologies. Further optimization opportunities of heat integration beyond what is already in place are limited and are difficult to implement in a running production facility. An added drawback is that tighter integration often leads to a system more sensitive to upset conditions. The gains in efficiency would therefore be offset by a reduction in plant stability.

The anticipated energy savings from bolt-on process technologies are likely to be offset by the increased energy requirements for reclamation. In fact, in the foreseeable future, the overall EI for oil sands operations is likely to increase, e.g., as a result of the energy requirements associated with reclamation activities. These activities become more significant as the end of mine life approaches, which applies particularly to Suncor Base Plant and Syncrude Mildred Lake. Further, management of water inventory and potential treatment may also increase energy needs. The operations need to balance not only the environmental performance requirements for emissions but also the requirements for land reclamation and water quality. The challenges in the trade-offs between performance criteria are the subject of Chapter 6.

References

Alberta Energy Regulator (2016a). *Directive 082. Operating Criteria: Resource Recovery Requirements for Oil Sands Mine and Processing Plant Operations*. Alberta: Alberta Energy Regulator.

Alberta Energy Regulator (2016b). *Fluid Tailings Management for Oil Sands Mining Projects*. Alberta: Alberta Energy Regulator.

Alberta Energy Regulator (2024). Oil sands. https://www.aer.ca/providing-information/by-topic/oil-sands (accessed September 2024).

Alberta Innovates (2020). *Clean Resources - Advanced Hydrocarbons.* Edmonton: Alberta Innovates.

Bergerson, J.A., Kofoworola, O., Charpentier, A.D. et al. (2012). Life cycle greenhouse gas emissions of current oil sands technologies: surface mining and in situ applications. *Environmental Science & Technology* 46: 7865–7874.

Brown, W., Hollander, E.D., Derakhshandeh, B. et al. (2019). Pretreatment of froth treatment affected tailings with floatation and stripping prior to tailings dewatering and containment. Canada Patent No. CA 3037959.

Canada Energy Regulator (2021). Canada's energy future 2021. https://www.cer-rec.gc.ca/en/data-analysis/canada-energy-future/2021oilsands/2021-oilsands.pdf (accesssed September 2024).

Canadian Association of Petroleum Producers (2024). GHG emissions. https://www.capp.ca/wp-content/uploads/2024/03/GHG-Emissions-v032724.pdf (accessed September 2024).

Canadian Natural (2021). *IPEP 500 Non-Confidential Report.* Calgary: Canadian Natural.

Canadian Natural (2022). *Creating Value through Technology and Innovation 2022 Case Studies.* Calgary: Canadian Natural.

Canadian Natural Resource Ltd. (2024). *Naphtha Recovery Unit Tailings Treatment Facility.* Calgary: Canadian Natural Resource Ltd.

Canadian Natural Resources Ltd. (2016). *Application for the Approval of Canadian Natural's Tailings Management Plan (TMP).* Calgary: Canadian Natural Resources Ltd.

Canadian Natural Resources Ltd. (2023). *Alternative Ore Criterial Pilot Application.* Calgary: Canadian Natural Resources Ltd.

Ciu, Z., Liu, Q., Etsell, T.H. et al. (2003). Heavy minerals In the Athabasca oil sands tailings - potential and recovery processes. *Canadian Metallurgical Quarterly* 42 (4): 383.

Cossey, H., Batycky, A., Kaminsky, H. et al. (2021). Geochemical stability of oil sands tailings in mine closure landforms. *Minerals* 11: 830.

Council of Canadian Academies (2015). *Technological Prospects for Reducing the Environmental Footprint of Canadian Oil Sands.* Ottawa: Council of Canadian Academies.

CVW CleanTech (2023). *CVW CleanTech Process and Technology Overview.* CVW CleanTech.

Czarnecki, J., Masliyah, J., Xu, Z. et al. (2013). *Handbook on Theory and Practice of Bitumen recovery from Athabasca Oil Sands. Vol II: Industrial Practice.* Calgary: Kingsley Publishing Services.

Duyvesteyn, W., Kift, J., and Morley, R. (2006). Processing asphaltene-containing tailings. Canada Patent No. CA 2645450 A1.

Energy Resources Conservation Board (2009). *Directive 074: Tailings Performance Criteria and Requirements for Oil Sands Mining Schemes.* Calgary: Energy Resources Conservation Board.

Foght, J., Gieg, L., and Siddique, T. (2017). The microbiology of oil sands tailings: past, present, future. *FEMS Microbiology Ecology* 93 (5): fix034.

Geramian, M., Liu, Q., Ivey, D. et al. (2019). Influence of oil sands composition on bitumen quality during non-aqueous bitumen extraction from the Athabasca deposit. *The Canadian Journal of Chemical Engineering* 97: 269–280.

Government of Alberta (2015). *Lower Athabasca Region Tailings Management Framework for the Mineable Athabasca Oil Sands*. Alberta: Government of Alberta.

Government of Alberta (2020). Suncor power cogeneration units. https://majorprojects.alberta.ca/details/Suncor-Power-Cogeneration-Units/3918 (accessed September 2024).

Government of Alberta (2021). CNRL In-pit extraction process (IPEP) demonstration. https://www.majorprojects.alberta.ca/details/CNRL-In-Pit-Extraction-Process-Demonstration/6521 (accessed September 2024).

Government of Alberta (2024a). Oil sands facts and statistics. https://www.alberta.ca/oil-sands-facts-and-statistics#jumplinks-0 (accessed 1 August 2024).

Government of Alberta (2024b). Bitumount provincial historic resource. http://www.history.alberta.ca/energyheritage/bitumount/default.aspx (accessed September 2024).

Government of Alberta (2024c). Oil sands 101. https://www.alberta.ca/oil-sands-101#:~:text=Surface%20mining%20is%20used%20to,to%20400%20tonnes%20of%20ore (accessed 1 August 2024).

Government of Alberta (2024d). Pathways Alliance carbon capture storage hub (Phase 1). https://majorprojects.alberta.ca/details/Pathways-Alliance-Carbon-Capture-Storage-Hub-Phase-1/10695

Hollander, E.D, Derakhshandeh, B., Brown, W. et al. (2020). Processing of a bitumen froth treatment overflow stream from froth treatment tailings for enhanced recovery of bitumen. Canada Patent No. CA 3094533.

Hollander, E., Derakhshandeh, B., Brown, W. et al. (2019). Active biodegradation of light hydrocarbons from fluid tailings. Canada Patent No. CA 3059673.

Hooshiar, A., Uhlik, P., Ivey, D. et al. (2012b). Clay minerals in nonaqueous extraction of bitumen from Alberta oil sands part 2. Characterization of clay minerals. *Fuel Processing Technology* 96: 183–194.

Hooshiar, A., Uhlik, P., Liu, Q. et al. (2012a). Clay minerals in nonaqueous extraction of bitumen from Alberta oil sands part 1. Nonaqueous extraction procedure. *Fuel Processing Technology* 94: 80–85.

Huq, I., Abbaspour, A., Doucette, B. et al. (2022). Non-aqueous extraction of bitumen from oil sands. US Patent No. US 2023/0159833 A1.

IHS Energy (2011). *Special Report: Oil Sands Technology, Past, Present and Future*. Cambridge: IHS Energy.

Imperial Oil Resources Ltd. (2022). *Mine Plan Amendment and Tailings Management Plan Integrated Application*. Calgary: Imperial Oil Resources Ltd.

Ityokumbul, M., Bulani, W., and Kosaric, N. (1988). Flotation kinetics for titanium, zirconium and bitumen recovery from oilsand tailings. *Canadian Journal of Chemical Engineering* 66: 382–385.

Kaminsky, H. (2008). *Characterization of an Athabasca Oil Sand Ore and Process Streams*. Edmonton: University of Alberta.

Kasperski, K. (2001). *Review of Research on Aqueous Extraction of bitumen from Mined Oil Sands*. Devon: Canadian Centre for Mineral and Energy Technology (CANMET) Western Research Center.

Kosior, D., Ngo, E., and Xu, Y. (2018). Aggregates in paraffinic froth treatment: settling properties and structure. *Energy Fuels* 32: 8268–8276.

Kuznetsov, P., Kuznetsova, A., Foght, J. et al. (2015). Oil sands thickened froth treatment tailings exhibit acid rock drainage potential during evaporative drying. *Science of The Total Environment* 505: 1–10.

Li, X., He, L., Wu, G. et al. (2012). Operational parameters, evaluation methods, and fundamental mechanisms: aspects of nonaqueous extraction of bitumen from oil sands. *Energy Fuels* 26: 3553–3563.

Lin, F., Stoyanov, S., and Xu, Y. (2017). Recent advances in nonaqueous extraction of bitumen from mineable oil sands: a review. *Organic Process Research & Development* 21: 492–510.

Lindsay, M., Vessey, C., and Robertson, J. (2019). Mineralogy and geochemistry of oil sands froth treatment tailings: implications for acid generation and metal(oid) release. *Journal of Applied Geochemistry* 102: 186.

Liu, Q., Cui, Z., and Etsell, T. (2006). Characterization of Athabasca oil sands froth treatment tailings for heavy mineral recovery. *Fuel* 85: 807–814.

Long, Y., Dabros, T., and Hamza, H. (2002). Stability and settling characteristics of solvent-diluted bitumen emulsions. *Fuel* 81: 1945–1952.

Marshall, G., Patarachao, B., Moran, K. et al. (2014). Zircon mineral solids concentrated from Athabasca oil sands froth treatment tailings: surface chemistry and flotation properties. *Minerals Engineering* 65: 79.

Masliyah, J., Czarnecki, J., and Xu, Z. (2011). *Handbook on Theory and Practice of Bitumen recovery from Athabasca Oil Sands. Vol I: Theoretical Basis*. Calgary: Kingsley Publishing Services.

Nikakhatari, H., Vagi, L., Choi, P. et al. (2013). Solvent screening for non-aqueous extraction of alberta oil sands. *Canadian Journal of Chemical Engineering* 91: 1153–1160.

Nikakhtari, H., Wolf, S., Choi, P. et al. (2014). Migration of fine solids into product bitumen from solvent extraction of Alberta oilsands. *Energy Fuels* 28: 2925–2932.

Nimana, B., Canter, C., and Kumar, A. (2015). Energy consumption and greenhouse gas emissions in the recovery and extraction of crude bitumen from Canada's oil sands. *Applied Energy* 143: 189–199.

Oxenford, J. and Coward, J. (2001). *Heavy Minerals from Alberta's Oil Sands*. SGS Mineral Services.

Pal, K., Nogueira Branco, L.D.P., Heintz, A. et al. (2015). Peformance of solvent mixtures for non-aqueous extraction of Alberta oil sands. *Energy Fuels* 29: 2261–2267.

Pourrezaei, P., Alpatova, A., Chelme-Ayala, P. et al. (2014). Impact of petroleum coke characteristics on the adsorption of the organic fractions from oil sands process-affected water. *International Journal of Environmental Science and Technology* 11: 2037–2050.

Province of Alberta (2022a). *Oil Sands Conservation Act*. Edmonton: Alberta King's Printer.

Province of Alberta (2022b). *Oil Sands Conservation Rules*. Edmonton: Alberta King's Printer.

Province of Alberta (2023). *Mines and Minerals Act*. Edmonton: Alberta King's Printer.

Roth, E., Bank, T., Howard, B. et al. (2017). Rare earth elements in Alberta oil sand process streams. *Energy Fuels* 31 (5): 4714–4720.

Shell Canada Energy (2016). *Resource Optimization Pilot – One Year Progress and Project Projections*. Calgary: Shell Canada Energy.

Shell Canada Energy (2023). *Quest Carbon Capture and Storage Project Annual Summary Report - Alberta Department of Energy: 2022*. Fort Saskatchewan: Shell Canada Energy.

Shell Canada Energy (2024). Quest carbon capture and storage. https://www.shell.ca/en_ca/about-us/projects-and-sites/quest-carbon-capture-and-storage-project.html (accessed September 2024).

Sheppard, M. (1989). *Oil Sands Scientist: The Letters of Karl A. Clark 1920-1949*. Edmonton: The University of Alberta Press.

Small, C., Hashisho, Z., and Ulrich, A. (2012). Preparation and characterization of activated carbon from oil sands coke. *Fuel* 92: 69–76.

Suncor Energy Inc. (2007). *Voyageur South Project Volume 1 Project Application*. Fort McMurray: Suncor Energy Inc.

Tamayo, C., Baydak, E., and Yarranton, H. (2024). Effect of ore quality on non-aqueous oil sand extraction performance. *Canadian Journal of Chemical Engineering* 102: 1750–1763.

Titanium Corporation (2019). *Reducing Methane Emissions and other Environmental Impacts from Oil Sands Tailings and Ponds - Deployment of Sustainable Technology*. Titanium Corporation.

Wang, T., Zhang, C., Zhao, R. et al. (2014). Solvent extraction of bitumen from oil sands. *Energy Fuels* 28: 2297–2304.

Wu, J. and Dabros, T. (2011). Process for solvent extraction of bitumen from oil sand. *Energy Fuels* 26: 1002–1008.

Xing, R., Chiappori, D.V., Arbuckle, E.J. et al. (2021). Canadian oil sands extraction and upgrading: a synthesis of the data on energy consumption, CO_2 emissions, and supply costs. *Energies* 14 (19): 6374.

Xu, Y. (2018). Asphaltene precipitation in paraffinic froth treatment: effects of solvent and temperature. *Energy Fuels* 32: 2801–2810.

Yuan, J. and Elektorowicz, M. (2021). Factors influencing non-aqueous extraction process of bitumen to mitigate the environmental impacts. *Process Safety and Environmental Protection* 153: 75–83.

Zawala, J., Dabros, T., and Hamza, H. (2012). Settling properties of aggregates in paraffinic froth treatment. *Energy Fuels* 26: 5775–5781.

Zubot, W., MacKinnon, M.D., Chelme-Ayala, P. et al. (2012). Petroleum coke adsorption as a water management option for oil sands process-affected water. *Science of The Total Environment* 427-428: 364–372.

6

Tailings and Waste Management in Mineable Oil Sands

Elco D. Hollander and Danuta M. Sztukowski

Coanda Research & Development Corporation, Tetra Tech Company, Calgary, AB, Canada

6.1 Introduction to Tailings and Waste Management

Technical and operational aspects of the management of tailings and other waste streams for mineable oil sands in the Athabasca region are discussed in this chapter. Tailings ponds contain a significant portion of both the fluid tailings and recycled process water inventory of the mine sites and are often some of the most visible aspects of oil sands mining operations. As of 2016, tailings ponds covered 257 km^2 of surface area, roughly comparable to half the size of the city of Edmonton (Cossey et al., 2021). Tailings storage facilities are frequently in the news for their surface footprint, impact on wildlife, negative impact on local air quality, and potential impact on off-site water quality. Operators are mandated to ultimately return the leases to a locally common boreal forest ecosystem of "equal or better capability" relative to pre-disturbance (Government of Alberta, 2015). Reducing tailings liability through progressive reclamation is in the interest of both the operators and the public. It can therefore be difficult to understand why progress in this area appears to be slow as public concerns continue to increase. Here, an attempt is made to describe the conflicting requirements that need to be managed to ensure safe and efficient operation on a day-to-day basis, while protecting closure and reclamation outcomes over timescales of decades and centuries. Environmental net effects for air (including Greenhouse Gas [GHG] emissions), water, and land must be prioritized and in practice are often found to be in tension. GHG emissions typically fall into two categories for tailings management: emissions associated with transport and treatment of materials and

Cleaner Petroleum Production and Refining Technologies, First Edition, Edited by M. R. Riazi and Harvey W. Yarranton.

emissions associated with biodegradation of residual diluent (and bitumen to a much lower extent) associated with Froth Treatment Tailings (FTT). Both will be addressed in more detail in the following text.

Mineable oil sands differ from in situ operations in that the latter are based largely on molecular transport phenomena over the scale of the full reservoir, whereas the former act on turbulent mass and heat transfer effects on the length scales of the screening size in the ore preparation units (refer to Chapter 5 for details). As a rule of thumb, turbulent transfer is much faster than molecular transfer. This gives mining operations the benefit of fast production ramp-up and turndown. For instance, the lag in production due to the required reservoir warmup time is avoided in mining. The main drawback is that the entire ore body must be processed, with only approximately 10% of the material being of recoverable ore value. This limitation makes the operation logistically challenging: the run-of-mine ore fed to the bitumen extraction facility and the waste stream leaving the plant are essentially equal in size. Further, the waste streams are highly erosive and prone to heat loss and therefore challenging for energy recovery. Both aspects contribute to the energy intensity of oil sands operations.

Contrary to the bitumen product, almost all waste generated at an oil sands mine remains on-site. As will be detailed further below, the characteristic tailings flow rate is in the order of 10,000–15,000 m^3/h (roughly one Olympic-sized swimming pool every four to six minutes). Any form of buffer capacity inside the processing facility is impractical at this scale, and the waste stream must therefore be handled in real time. Without direct tailings storage, bitumen production of this closed system would come to an immediate halt. It is not guaranteed that material can be placed in locations compatible with reclamation requirements, and material rehandling may be necessary.

Successful waste management is dependent on several operational disciplines within and outside the bitumen production process itself.

- **Mine operations** are responsible for ore delivery to the plant but must also ensure that footprint is cleared for waste storage. Therefore, the mine plan cannot consider ore quality and bitumen production only. Mine operations must also ensure that the pit is completely excavated before advancing to new mining areas.
- **Plant operations** must ensure that the tailings waste is produced to a specification that is compatible with required tailings performance. Ideally, the extraction circuit should be operated as a facility that generates two product streams: the bitumen stream and a tailings product that is compatible with future reclamation. In practice, however, the revenue-generating bitumen production requirements are usually prioritized.
- **Tailings operations** must ensure that the waste material is placed in a location consistent with short- and long-term operational needs. Mine

sequencing and the timescale for developing new storage locations (tailings ponds external or within the mine open pit) are not always compatible with these requirements.

Oil sands tailings management occurs over timescales varying from hours in plant operation to decades during site closure activities. Integrated mine-tailings-reclamation plans are required to ensure that both short- and long-term constraints are met. The interfaces between the disciplines and integrated activity planning can be difficult to manage, however, since their primary accountabilities are different. Therefore, the waste management challenge is often approached as either a strict "process engineering" problem (focusing on e.g., thickeners, centrifuges, and flocculant/coagulant dosage) or a "reclamation" problem (focusing on e.g., relative surface area of uplands, wetlands, and lakes). This chapter tries to take a more integrated approach and to present the criticality of the link between mining, the processing plant, and successful reclamation outcomes.

To illustrate the conflicting requirements that need to be resolved, two examples are given to set the stage. First, for bitumen recovery to be successful, ore must be converted into a clay-dispersive state during the extraction process (see Chapter 5). Because of the dispersed state, the fines fraction holds a significant amount of process water upon discharge to the tailings containment areas. This trapped water results in the generation of large volumes of fluid tailings (historically referred to as Mature Fine Tailings or MFT) that require long-term storage. The ore body has undergone densification over geological timescales before reaching a soil state, whereas it is dispersed in minutes in the ore preparation step. There is a desire, and even an expectation, that the tailings will be reconstituted to the original state over a period of years to decades during the reclamation process. However, there is no guarantee that this is technically, much less economically, feasible because consolidation timescales for clay-rich materials can be in the order of centuries to millennia (McKenna et al., 2016). As will be shown later, several tailings treatments aimed at increasing soil strength further rely heavily on the addition of coagulants such as gypsum or alum. However, water released from these tailings, which is reused in the process, tends to have negative impact on extraction performance *because of* this coagulating behavior. This dilemma has led to the adage in oil sands that "everything good for extraction is bad for tailings management and vice versa."

Second, operators are required to store water on-site during the life of mine (LoM) under the current zero-discharge requirements for process-affected water. During the operational phase, the mine is constructed such that the flow of seepage and precipitation is "inward," and water is intercepted and stored on-lease within tailings containment areas. To meet reclamation requirements, however, the land must be integrated into the regional watershed, which implies that water must flow *through* the lease freely. For this, surface-water flow must be reversed to allow water to flow "outward," or off-lease. This change requires significant modifications to the landscape and soil characteristics. As a minimum, rehandling of material will re-disturb the landscape and will lead to significant energy input.

The two examples described above illustrate technical challenges associated with tailings management. The authors acknowledge that significant socio-political and economic sensitivities exist and recognize that this venue cannot do justice to these aspects. By describing the conflicting technical requirements of tailings and reclamation management, this chapter attempts to show that even if issues were strictly technical in nature, the challenges would still be significant.

6.2 Tailings Mass Balance and Waste Material Types

The reader is referred to Chapter 5 for a description of oil sands mining operations and individual process unit mass balances. This chapter focuses specifically on the waste streams generated during the bitumen extraction operation and the fate of the material within the operating site.

Various waste streams can be discerned, originating from mining, plant operation, upgrading, and process water inventory management. In open-pit mining, overburden and reclamation material (e.g., peat, soil) is removed to access the ore. These materials are partly stored in stockpiles for future reclamation activities and partly used for dike construction and haul road infrastructure. The ore itself typically consists of 4–16% bitumen and 5% connate water by mass, with the balance made up of sand, silt, and clay (Kasperski, 2001). In the conversion of raw bitumen to an upgraded hydrocarbon product stream, waste products such as petroleum coke and sulfur are generated. These streams can represent a modest value as a fuel source (Environment and Climate Change Canada, 2020) or precursors to, e.g., fertilizers (Natural Resources Canada, 2008), and require on-site storage. The most obvious waste streams, however, are the primary extraction and FTT and their associated process water inventory. These streams are contained in tailings ponds, which form highly visible aspects of the mining operation. This chapter focuses primarily on the tailings streams, Oil Sands Process Water (OSPW), and to some extent the fugitive emissions associated with the FTT.

6.2.1 Tailings Material Balance

To gain a deeper understanding of the practical constraints under which tailings management operates, it is insightful to develop a simplified tailings mass balance from basic assumptions about tailings behavior:

- The ore consists of bitumen, sand (size fraction larger than 44 μm), fines (size fraction smaller than 44 μm), and water. For simplicity, the solids are assumed to have a density of 2.65 t/m^3, and water and bitumen a density of 1.0 t/m^3. Fines can be thought of as particles that act as fluid tracers during process operation, whereas sand is a settling component. It is convenient to define the sand-to-fines ratio (SFR) and the ratio of fines/(fines + water), FoFW. The latter ratio can be thought of as the slurry density of the non-settling fraction, often referred to as the carrier fluid.

- As discussed in Chapter 5, ore preparation converts the ore into a slurry with a density of 1.55 t/m^3. This process step sets the FoFW throughout the bitumen extraction process and tailings placement. For simplicity, it is assumed that bitumen recovery is complete.
- Upon tailings discharge, the sandy fraction of the tailings will settle out quickly. The sand fraction compacts until a sand porosity $(1 - m^3_{\text{sand}}/m^3_{\text{total}})$ equal to the porosity in the original ore is reached. The coarse fraction is saturated with carrier fluid.
- Excess carrier fluid will report to fluid tailings containment, where it will settle over time to a typical solids content of 30–40% by mass (Canadian Oil Sands Innovation Alliance, 2022).

At this point, no distinction is made between primary extraction and FTT. The mass balance for the tailings is illustrated in Figure 6.1. For a 100 Mt/a operation (7500 tph, two-train operation at 75–80% utilization), the resulting mass balance is summarized in Table 6.1. Note that for tailings management, it is often more useful to express the flows in volumetric equivalents.

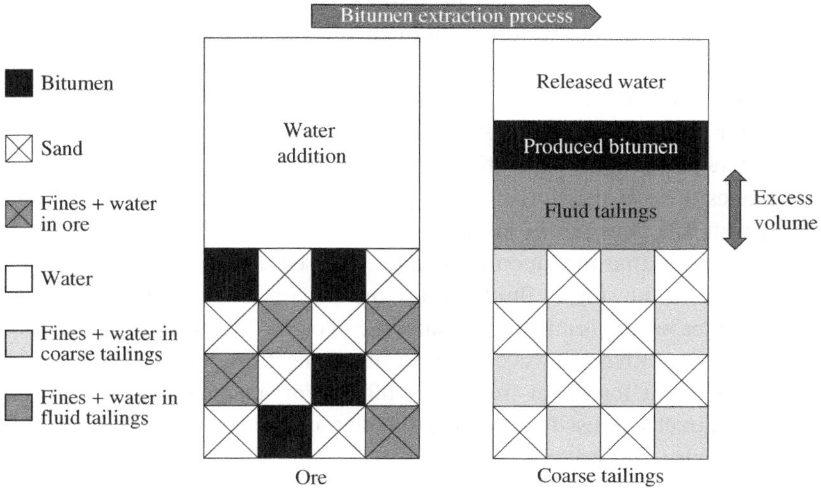

Figure 6.1 Visual representation of tailings mass balance presented in Table 6.1.

Table 6.1 Conceptual mass balance for the tailings stream from a 100 Mt/a process.

Parameter	Units	Ore	Water addition	Produced bitumen	Coarse tailings	Fluid tailings	Released water
Bitumen	kg/kg (m^3/m^3)	0.110 (0.231)	—	1.000 (1.000)	—	—	—
Water	kg/kg (m^3/m^3)	0.050 (0.105)	1.000 (1.000)	—	0.200 (0.399)	0.650 (0.831)	1.000 (1.000)
Sand	kg/kg (m^3/m^3)	0.714 (0.565)	—	—	0.752 (0.565)	—	—
Fines	kg/kg (m^3/m^3)	0.126 (0.100)	—	—	0.048 (0.036)	0.350 (0.169)	—
SFR	kg/kg	5.7	—	—	15.6	—	—
FoFW	kg/kg	0.716	—	—	0.194	0.350	—
Mass flow	Mt/a	100	47.4	11	95.0	22.9	18.5
Volume flow	Mm^3/a	47.7	47.4	11	47.7	17.9	18.5

Some key observations can be made from the conceptual mass balance. First, the bitumen fraction makes up approximately 23% of the ore by volume. The 47.7 Mm^3/a ore produces 11 Mm^3/a of bitumen, or around 190 kbpd of crude bitumen. Despite the removal of this bitumen volume, the total tailings stream shows a volumetric *excess* of 17.9 Mm^3/a, in the form of fluid tailings. The root cause of this excess is the water addition in the ore preparation step, which dilutes the FoFW from 0.716 in the ore to 0.194 in the bitumen extraction process (in practice the FoFW is often even less as a result of additional dilution, not accounted for in this simplified mass balance). Consequently, the tailings volume generated does not fit in the excavation from where the ore originated. This excess volume makes the construction of above-grade storage inevitable. Further, fluid tailings dewatering does not in itself reduce the overall fluid inventory. Denser fluid tailings imply that the amount of water released during initial dewatering has increased, but the sum of fluid tailings and release water remains constant. Additional water release is only beneficial if it can offset the intake of fresh water from the Athabasca River. However, in most situations, the quality of the released water is insufficient for direct freshwater offset. For instance, tailings recycle water cannot directly be used as a source of boiler feed water in utilities, which is one of the primary freshwater users.

Second, it is noted that of the 47.4 Mm^3/a water added during ore preparation, only 18.5 Mm^3/a is released from the fluid tailings. Around 15 Mm^3/a remains trapped within the fluid tailings, while the balance (19.0 Mm^3/a) remains in the coarse tailings as pore water. Somewhat counterintuitively, it is the coarse material (e.g., the dike structures) that acts as the largest process water sink. Since most of the water in both the coarse fraction and the fluid tailings is unavailable during the operational phase, continuous water import is unavoidable.

The excess volume of tailings originates from the dilute state of the carrier fluid during tailings placement: it is the low FoFW in the coarse tailings that accounts for most of the excess. Any form of increased fines capture in the coarse tailings will therefore directly reduce the fluid tailings inventory. Operationally, fines capture can be improved by maximizing the FoFW in the carrier fluid. In practice, this means operating the ore preparation circuit at a higher density. However, this strategy is unpopular in bitumen extraction, as it increases the risk of sanding off (i.e., clogging) the process units and can compromise bitumen recovery due to the increased carrier fluid viscosity.

Alternatively, fines capture can be increased in the tailings containment area itself by attempting to encapsulate fluid tailings inventory by placing coarse material overtop. Such "beaching" operations must be sufficiently gentle to avoid erosion and displacement of the inherently weak fluid tailings layer. Importantly, densification of the coarse tailings beach, e.g., by means of track packing operation (Martens et al., 2008), will have the opposite effect on fines capture. Recall that the fines fraction acts as a fluid tracer. If the sand fraction is compacted by mechanical means, the water released from the substrate will see a proportional release of fines from the deposit. If beach densification is necessary to stabilize the tailings dike structure, fines loss must be accepted because geotechnical performance of the structure takes priority.

A final method to increase fines capture is based on the elimination of differential settlement between the fines and coarse fraction. Stokes' law for particle settling states that the settling rate for laminar flow conditions scales with $\Delta\rho g d^2/\mu$ ($\Delta\rho$ denotes the density difference between particle and carrier fluid, d is the particle size, g is the gravitational acceleration constant, and μ is the dynamic viscosity of the carrier fluid). Particles of different sizes will show different settling rates, *unless* the carrier fluid viscosity is "infinitely large." Practically this means that the carrier fluid must have a yield stress. Such a condition would have to be created after the bitumen extraction process is complete, but before tailings are placed. This step would require dedicated tailings treatment equipment. Such technology alternatives will be described further in Section 6.5.1.

6.2.2 Process-affected Water

The tailings mass balance in Table 6.1 summarizes that a significant amount of process water is required during bitumen extraction. OSPW is essential during the operational phase of the mine, and therefore, strictly speaking is not a waste stream. Most compounds in process water are either not harmful to the bitumen extraction process or manageable through the addition of extraction process aids. However, significant treatment may be required to remove these constituents prior to release during the reclamation phase. These treatment options are addressed further in Section 6.5.5.

Any water that has been in contact with the mining and bitumen extraction processes is considered process-affected and must therefore be stored on-site. Because of the zero-discharge requirement, the reuse of process water is essential for oil sands operations. From a fluid containment perspective, it would be ideal to operate under maximum process water recycle. In practice, however, this is not possible due to water quality constraints. Since the site operates as a closed system for water, recycle inevitably results in an upcycling of soluble ions originating from the ore body, sodium chloride in particular. Chloride ions are implicated in corrosion-related reliability concerns, specifically in upgrading services. Increases in sodium levels eventually lead to conditions where the clay fraction in the oil sands slurry is no longer dispersed. Clay coagulation can subsequently negatively impact bitumen recovery (Wallace et al., 2004).

Accumulation of Constituents of Potential Concern (CoPC) has the potential to hinder future reclamation activities. Concerns regarding the environmental fate of, e.g., naphthenic acids, metals and metalloids, and Polycyclic Aromatic Hydrocarbons (PAH) are well-documented (Cossey et al., 2021; Hussain and Stafford, 2023; Lindsay et al., 2019). It is worth mentioning that mitigation for sodium chloride is expected to be most complicated, as there is no natural sink for this CoPC. Prolonged salt release from coarse tailings structures is anticipated as the pore volume is flushed through natural precipitation or water infiltration events. Treatment for sodium chloride generally requires membrane or evaporative techniques, which are energy-intensive, and will be discussed further in Section 6.5.5.

6.2.3 Other Material Types

Up to this point, the main waste streams generated in the bitumen extraction process have been discussed: coarse and fine tailings, as well as OSPW. Several additional wastes are also stored on-site. Reclamation stockpiles are in a sense considered "waste" during mine operation but form an integral component of the reclamation mass balance. Stockpiles are usually stored on sections of the mine that do not contain economic ore or represent ore that can be mined after the material has been used for progressive reclamation.

Operators with on-site upgrading facilities generate two additional process streams: petroleum coke and sulfur. Coke has traditionally been used as a source of power generation but has fallen out of favor due to the high GHG penalty during combustion. However, from a regulatory perspective, this material is still considered a hydrocarbon resource. It therefore needs to be stored in a manner that enables future utilization. Currently, coke is being considered as an alternative reclamation material for capping of soft tailings deposits, as described in Section 6.5.4. A portion of the sulfur is sold to market and is primarily used for the production of fertilizers. Any remaining inventory will need to be addressed during closure activities.

For completeness, it is noted that various forms of point source and fugitive emissions exist within the oil sands operation. These include stacks, flares, volatile emissions from tailings discharge points, as well as tailings ponds and the truck fleet. More detail will be provided on the emissions associated with tailings ponds in Section 6.5.3. A comparison of the emissions associated with material handling, i.e., hauling versus pumping, is provided in the following text.

6.2.4 Energy Intensity Related to Material Transport

One aspect of the energy intensity of tailings placement is the fact that the streams are transferred to the tailings containment areas at process temperatures. For primary extraction tailings, this temperature is around 45 °C, while FTT are discharged at 70–80 °C. At a typical single-train flow rate of 6,000 m^3/h, the heat loss is well over 100 MW per train. This loss is in part accepted as a way to prevent the tailings lines from freezing during winter operation (tailings lines are not insulated/heat traced for ease of maintenance). Further, there is no practical use for such low-grade heat, and heat recovery from highly erosive slurries is technically challenging. The only tailings stream under consideration for heat recovery is the FTT, due to its elevated temperature and relatively low abrasiveness, as previously discussed in Chapter 5. Strictly, this energy loss should be allocated to the bitumen extraction process, as the energy input is not the result of tailings management.

The energy intensity that *is* directly attributable to tailings is the energy input related to material transport. Transportation of ore and waste (e.g., overburden, interburden, coke) is accomplished with the haul fleet, while tailings are transported by pipeline. To gain insight into where optimization can occur, it is useful to quantify the relative energy intensity of these means of transport. It is important to note that detailed optimization is strongly site-specific, and the example provided is for illustrative purposes only.

First, we consider the transportation of dry waste using the largest available haul truck size of 400 tons (363 tonnes) since higher haul tonnages lead to increased energy efficiency. Assuming an average haul truck velocity of 35 km/h, a moderate diesel consumption of 215 l/h (Caterpillar, 2022), and a diesel energy content of 38.6 MJ/l, an energy intensity of 0.65 MJ/(tonne·km) can be derived; that is, it takes 0.65 MJ of energy to transport 1 tonne of material over a distance of 1 km. This value can be compared to typical values for diesel consumption in oil sands surface mining, as discussed in, e.g., Nimana et al. (2015) and Bergerson et al. (2012). These authors use the diesel consumption as inputs to GHG emission models such as FUNNEL-GHG-OS and GHOST for oil sands mining. Values of 5–15 l diesel/m^3 (8 l diesel/m^3 reported average) of produced bitumen are quoted. These values include all aspects of the mining operation, including ore and waste haul, shovel operation, and support fleet. For an average fuel consumption of 8 l diesel/m^3 of produced bitumen and the average diesel usage of 6.1 l/km, an equivalent ore haul round-trip distance of 57 km is calculated. Since this distance includes the diesel consumed by non-ore hauling equipment, this number is inflated if applied to ore haulage only. If it is assumed that roughly a third of the consumption is associated with ore transport, a haul distance of 19 km (round-trip) is derived. This number is comparable to ore haul distances found in practice and shows that the order of magnitude of the energy intensity of 0.65 MJ/(tonne·km) is reasonable.

Similar to the energy intensity of ore transportation, we can derive the energy intensity for slurry transport. The mechanical energy of transport is given by $Q\Delta P$, with Q is the volumetric flow rate [m^3/s] and ΔP is the pipeline pressure drop [Pa]. Slurry pipeline pressure drops are in the range of 200–400 Pa/m for typical operating conditions. Various pipeline transport models exist for the estimation of pressure drops. Here, we use the SRC pipeflow model (Saskatchewan Research Council, 2011). This model predicts a pressure drop of 365 Pa/m for the whole tailings flow conditions described in Table 5.2 in Chapter 5 and an assumed pipeline diameter of 24.5″ ID. A similar energy intensity can be derived as $(\Delta P/L)/(\rho_{sl}\chi_s)$, where $\Delta P/L$ is the hydraulic gradient [Pa/m], ρ_{sl} is the slurry density [kg/m^3], and χ_s is the solids mass fraction [kg/kg]. A numerical value of 0.35 MJ/(tonne·km) is found for this example. In practice, the actual energy intensity is slightly higher due to, e.g., efficiency losses in the pumping system. Regardless, the intensity numbers for haul transport and slurry transport are surprisingly close, indicating that there is no obvious mode of transportation that is preferred from an energy intensity perspective. What is clear, however, is that *rehandling* of material will increase GHG emissions. It will therefore always be preferable to place the materials in a reclamation-compatible location from the start. In practice, immediate placement of tailings in their final location is rarely achieved, however, due to fundamental limitations of the available tailings treatment technologies. These challenges are detailed in Section 6.3.

6.3 The Tailings Management Challenge

The main challenge in tailings management is the need to manage the production and site reclamation requirements that play out on significantly different time-scales. On the one hand, the bitumen production profile sets the requirements as to the amount and quality of ore that must be processed in the near term. The large variability of the ore body dictates operational parameters such as the mine advance, tailings production rates, and clearing of tailings footprint. On the other hand, tailings treatment rates and tailings product performance (e.g., in terms of consolidation behavior and geochemical stability) define the reclamation schedule.

The requirements are further complicated by the climatic context of the oil sands region and the nature of the materials themselves. For instance, water transfer is complicated during the deep winter period where temperatures can reach below −40 °C. Further, the region is net evaporative, receiving less water through precipitation (around 450–500 mm/year) than it loses through evaporation (575–600 mm/year) (Government of Alberta, 2013). The net evaporation is significant for reclamation landscape design, in particular the design of aquatic features such as lakes and wetlands (Cumulative Environmental Management Association, 2014). Under net-evaporative conditions, any stagnant areas will be prone to drying. Under these conditions, dissolved salts will concentrate, which will stress aquatic life. Land features therefore require sufficient flow-through of water to flush out accumulated salt to remain viable, meaning that aquatic features must remain connected to the watershed as the landscape evolves as the tailings substrates settle. Post-placement differential settlement of the substrates can lead to areas losing their connectivity, thereby becoming ecologically unviable over time. It is therefore critical to develop landforms that are temporally robust. In principle, one can account for post-placement settlement by "over-building" land features, i.e., placing sufficient reclamation material to ensure connectivity even when the material settles. Overbuilding increases the requirements for geotechnical competency of the tailings substrate and becomes impractical if the substrate is too soft.

To avoid the development of stagnant areas during consolidation, it may be tempting to consider using pit lakes as a primary alternative for storing fluid tailings. Similar salinization effects will occur here, however, if insufficient recharge is available from the surrounding landscape. It is recommended to design catchment areas of 8–20 times a lake's surface area for it to be viable in the closure landscape (Cumulative Environmental Management Association, 2012). This requirement shows that there is only a limited amount of lake area that can be sustained in the region. It should further be noted that under the climatic conditions in the region, ice cover reaches a maximum thickness of around 1.5 m annually. Hence, water bodies deeper than about 2 m will remain at least partially fluid year-round. As a rule of thumb, if an aquatic feature becomes deeper than 2 m over time due to differential settlement, it will therefore show lake characteristics (McKenna et al., 2016). This estimate provides an upper bound for the differential settling that can be accommodated before a feature changes its nature from terrestrial to aquatic.

With the abovementioned constraints, one can develop a conceptual outline of minimum tailings substrate performance requirements that enable viable reclamation:

- Given that the ratio of catchment-to-lake area should be as an order of magnitude 10:1, it follows that 90% of the reclaimed area should be terrestrial in nature.
- For terrestrial reclamation to be sustained, differential settlement should be limited to no more than 2 m as a rule of thumb. This can easily be achieved with sand-dominated materials but is challenging for fines-dominated deposits.
- To construct terrestrial closure landscapes, the mechanical strength of the substrate must be compatible with heavy haul truck traffic for practical placement of the reclamation soil. Several meters of sand and topsoil must be placed on the tailings substrate to protect the root zone from saline water breakthrough. For a surface area in the order of 100 km^2, 10^8–10^9 m^3 of material is required, which can only be placed practically by means of heavy haulers.

Ensuring the fines-containing deposits are compatible with these requirements poses significant technical, economic, and regulatory challenges. Further, the challenges are not limited to geotechnical performance but also include the need to manage the long-term biogeochemical outcomes within the deposits. These aspects are discussed in the next sections.

6.3.1 Operational Requirements

Chapter 5 presented a description of the bitumen extraction process, detailing requirements for ore quality and throughput to ensure production targets can be met. To assure sustainability on a day-to-day basis, the focus is primarily on the revenue-generating part of the business. In contrast, the focus of tailings operations is mainly to prevent impact to bitumen production. To ensure that the production goals and longer-term tailings storage and reclamation needs remain synchronized, it is critical to develop an integrated mine and tailings plan that satisfies the following requirements:

- **Ore availability:** Ore faces must be opened to sustain production, allowing for sufficient quality blends to meet primary and secondary extraction needs. This constrains maximum blended fines content, minimum blended ore content, and blended connate water salinity.
- **Tailings containment footprint:** Since one can only mine from the exposed mine face (i.e., what is "in front of the shovel"), it is the cleared ore that creates the space for future in-pit tailings storage. As mentioned in Chapter 5, operators have historically been regulated under minimum bitumen recovery requirements through Alberta Energy Regulator (AER) directive 082 (Alberta Energy Regulator, 2016a). Leaving lower quality ore behind could therefore violate D082 requirements, but mining material *in future* would delay the availability of tailings and process water storage. The sequencing of ore processing and tailings containment planning is therefore critical.
- **A closed and achievable mass balance:** While obvious on paper, material scheduling is not trivial in practice. The lack of buffer capacity within the plant makes immediate tailings placement necessary. However, mine and tailings infrastructure development (e.g., haul roads, tailings dikes, pumping/pipeline infrastructure) requires months, or even years to construct. Delays in infrastructure availability, and even short-term off-specification tailings streams, can lead to sub-optimal placement and rehandling requirements in future. Note that material adequacy refers to both geotechnical competency and geochemical stability, as described later.

These short-duration requirements are sensitive to operational and climatic upset conditions, such as equipment breakdown in both the mine and downstream operating units, rain events, etc. The mine and tailings plan therefore needs to be sufficiently robust and flexible to accommodate short-term variability. It is noted that operational requirements for the development of the reclamation landscape are less directly linked to the plant operation and are therefore in some ways more flexible. However, since the haul fleet for reclamation handles disturbed soil (i.e., less competent material), operational performance will be less predictable and will rely more on the success of tailings treatment technologies.

6.3.2 Reclamation Considerations

Considerations for the planning and execution of site reclamation activities are provided in this section. The discussion focuses on aspects such as the practicality of execution of the design and consistency with the ultimate closure goals. Since the reclamation of fluid tailings is considered the most challenging from a geotechnical perspective, one emphasis will be on this material type. In addition, details will be provided on the reclamation of FTT, as it is one of the more difficult materials from a geochemical perspective.

A simplified but useful schematic illustrating the importance of these aspects and the link to the overall geotechnical and geochemical sustainability of the closure landscape is given in Figure 6.2. The diagram describes the common features of reclamation landforms, by assessing the long-term stability of the underlying material in terms of physical and chemical characteristics. The horizontal axis gives a measure of the geotechnical performance of the tailings substrate, which ranges from a fluid to a soil state. The vertical axis describes the intrinsic geochemical mobility of the CoPCs in the landscape, which ranges from highly mobile (requiring continuous monitoring and intervention) to immobilized. The diagram identifies four quadrants. The bottom-left quadrant denotes fluid tailings with high geochemical mobility, stored inside a conventional tailings pond. This quadrant describes the operational state of the tailings. From here, one can consider the application of dewatering (moving right), immobilization of CoPCs (moving up), or a combination of both (moving diagonally up). Historically, tailings technology development has focused on fluid tailings dewatering techniques, e.g., through centrifugation, thickening, or atmospheric drying. Dewatering techniques are of value when the tailings material can be converted to an in situ material strength that is compatible with a terrestrial landform. To achieve a terrestrial landform,

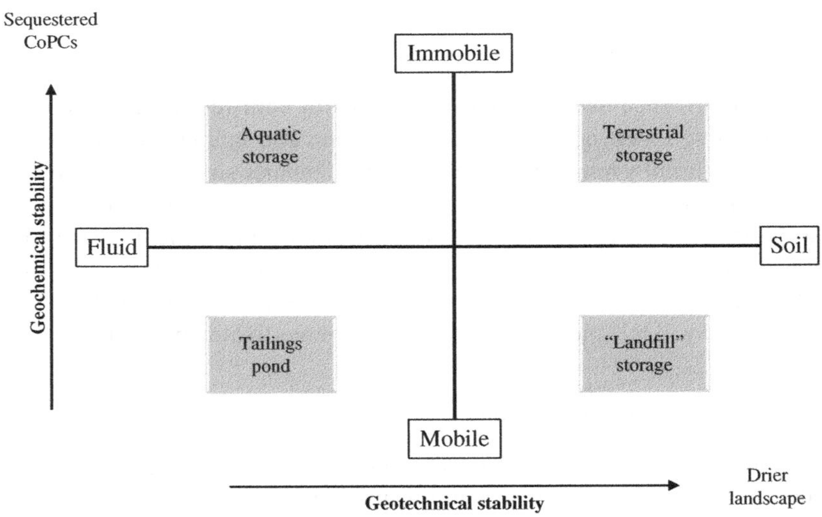

Figure 6.2 Common characteristics of sustainable landscape, and the link to tailings management. *Source*: Omotoso (2019)/Canadian Institute of Mining, Metallurgy and Petroleum.

differential settling must be limited, and the surface strength of the material must be sufficient for heavy equipment access.

A critical aspect of reclamation certification is the requirement to delicense dike structures. Any material stored behind dikes will need to be brought to a non-liquefiable state to classify as a terrestrial landform and meet Dam Safety Guidelines, i.e., typically 75–80% solids content by mass (Hyndman and Sobkowicz, 2010). *Partial* dewatering has limited value in this context, as the tailings would still be considered a fluid from a geotechnical perspective. Tailings treatment technologies that bring MFT from a nominal 35–50% solids to 65% solids therefore show some benefit from a water reuse perspective but are usually insufficient for terrestrial reclamation purposes.

To appreciate the importance of the previously mentioned post-placement settlement and consolidation behavior, one can evaluate the shrinkage of fluid tailings with decreasing water content. Here we consider a unit solid mass of fluid tailings from the initial discharge state to the terminal density in the landscape. Following from the mass balance in Table 6.1, the FoFW of the fluid tailings fraction upon discharge is 0.194 kg/kg, which corresponds to a solids-based specific volume of 4.53 m^3/tonne of solids. Initially, the material dewaters relatively quickly (i.e., in the order of years) to form typical fluid tailings at a FoFW of 0.35 kg/kg, equivalent to 2.23 m^3/tonne of solids. Fluid tailings dewatering techniques to be described in Section 6.5.2 can typically reach FoFW levels of 0.60 kg/kg, yielding specific volumes of around 1.04 m^3/tonne of solids. Finally, if the material is to reach its original density in the ore body, the FoFW must increase to 0.72 kg/kg, or a specific volume of 0.77 m^3/tonne of solids. The volume reduction between the fluid tailings inventory at 35% solids content and the consolidated material equivalent to its original state at 72% is almost a factor of 3. Hence, over time, a fluid tailings column will settle to about one-third of its original height. For a threshold settlement of 2 m, the placed fluid column therefore cannot exceed approximately 3 m in thickness to guarantee that the landscape remains compatible with terrestrial closure. For comparison, sand-dominated materials undergo much lower post-placement shrinkage (typically less than 1% (Cumulative Environmental Management Association, 2014)), making terrestrial reclamation of sand-dominated materials significantly less challenging than fluid tailings.

The relatively recent introduction of the Tailings Management Framework (TMF) and AER Directive 085 (refer to Chapter 5 and Section 6.3.4) has placed emphasis on the geochemical behavior of the tailings material and its implications on closure design. Geochemical risks are captured by the vertical axis in Figure 6.2. A walk-away solution (i.e., one that does not require long-term intervention) can only be found in the top-half of the diagram, suggesting that geochemical stability is a necessary component of successful reclamation. CoPCs must be permanently locked into the reclamation structures such that biogeochemical processes cannot remobilize these constituents over time. In principle, immobilization can be achieved in both aquatic and terrestrial settings if the following conditions can be met:

- Soluble ions have been sufficiently flushed out of the tailings structures such that residual salt loading is compatible with the ecosystem.
- Redox conditions have stabilized to the point that residual change can be accommodated by the landscape.
- Residual biogas generation does not negatively impact the ecosystem. Note that biodegradation of residual light hydrocarbons is the dominant (post-placement) GHG emission process in tailings.

These requirements place constraints on the composition and deposition strategy of the tailings streams. When considering construction of sandy structures from coarse tailings, for instance, the pore space between grains will be filled initially with OSPW. This relatively high salinity water will be displaced over time by water from precipitation and drain into the site's water circuit. Due to the slow migration of water through the sand structures, displacement may occur over decades. The landscape is therefore exposed to prolonged salt loading even after bitumen extraction has ceased. Saline drainage must be managed during reclamation through either active and/or passive water treatment. Displacement can in principle be accelerated by active flushing through the import of fresh water. Under the zero-discharge requirement, however, this water must be stored on-site, which further strains fluid containment. The use of natural precipitation is preferable, as this stream already requires storage and therefore does not increase containment requirements.

In the tailings mass balance discussed in Section 6.2.1, no distinction was made between primary extraction and FTT. The FTT stream represents a small fraction of the total fines budget and therefore does not require specific consideration from a volumetric perspective. However, FTT is geochemically dissimilar to extraction tailings and hence poses unique challenges for site operation and closure. Due to their oleophilic nature, heavy minerals such as titanium, iron, zircon, and sulfide-bearing minerals tend to partition into the froth phase in primary extraction (Ciu et al., 2003; Cossey et al., 2021; Lindsay et al., 2019; Kaminsky, 2008). In froth treatment, these minerals are rejected to tailings. Naturally Occurring Radioactive Materials (NORMs) associate with the oleophilic minerals and are therefore also enriched in the FTT stream. Finally, due to thermodynamic limits on the maximum practical light hydrocarbon recovery in froth treatment (refer to Chapter 5), FTT also contains residual naphtha or paraffinic diluent, depending on the process flowsheet. These constituents pose risks to reclamation outcomes that have conflicting preferred mitigation strategies.

Sulfide minerals such as pyrite are known Acid Rock Drainage (ARD) and Metal Leaching (ML) precursors. When exposed to atmospheric conditions, sulfide oxidation is initiated by microorganisms and subsequently sustained by oxidation from atmospheric oxygen. This process generally occurs at circum-neutral conditions and releases sulfate, ferrous iron, and some acidity. Ferrous iron is oxidized to ferric iron, creating a regenerative reaction where ferric iron oxidizes pyrite. This releases further sulfate and significantly more acidity. The reaction generally continues until all sulfide is converted and is difficult to stop once initiated (International Network Acid Prevention, 2009). If insufficient neutralization capacity is available in the matrix, continuous acid generation has the potential to dissolve other mineral phases and consequently release metals and metalloids into the pore water. These mechanisms will negatively affect the water quality and may increase the treatment requirements during reclamation. An obvious method to prevent ARD is to limit contact between oxygen and the mineral phase. Therefore, FTT placement below the phreatic surface is a preferred mitigation. Note that this placement strategy also provides mitigation for NORMs, as wet storage prevents airborne mobilization of radioactive dust.

A property that makes FTT unique in the mining industry is the presence of trace light hydrocarbons (i.e., naphtha, paraffinic diluent) that associate with bitumen, resulting from incomplete diluent recovery in the froth treatment process. These hydrocarbons are a bioavailable source of carbon for both aerobic and anaerobic microbial activity. Aerobic degradation is the fastest and thermodynamically most favorable process, in which the residual diluent is converted into CO_2. In the absence of oxygen, iron and sulfate reduction are the next preferential degradation pathways and result in the generation of CO_2 and (in the case of sulfate reduction) H_2S. It is found in practice that a limited amount of ferric iron is available under anaerobic conditions, and sulfate levels are limited for operators that do not use coagulants such as gypsum or alum as tailings process aids.

The least thermodynamically favorable degradation pathway, but dominant mechanism when all other terminal electron acceptors (TEAs) are depleted, is methanogenesis (Van Dongen et al., 2021; Foght et al., 2017; Holowenko et al., 2000). Here, fermentation of residual diluent shows a net production of methane and CO_2. The acetoclastic pathway produces 3–4 moles of methane per mole of CO_2, whereas ratios above 3–4 are observed when carbonate reduction (hydrogenotrophic pathway) occurs. While there is evidence of methanotrophy (conversion of methane to CO_2) in the aerobic portion of the tailings sediment (Burkus et al., 2014; Saidi-Mehrabad et al., 2012), FTT-containing deposits are still methane-producing (Small et al., 2015). Because methane is a more potent GHG gas with a global warming potential (GWP) of 25, methane emissions from tailings ponds have recently received attention (Baray et al., 2018).

The low solubility of methane in water gives rise to the additional risk of bubble formation from FTT-containing deposits. Biogas release from aquatic tailings storage has been implicated in the remobilization of hydrocarbons and the formation of surface oil sheens (Suncor Energy (Syncrude) Operating Inc. (SESOI), 2023; Francis et al., 2022). From a reclamation management perspective, it is critical to determine the duration of biogas release. Limited data is available for the expected in situ degradation timescale, and laboratory-scale kinetic studies may not be representative of field conditions (Siddique et al., 2008; Kong et al., 2019). Analogue systems suggest that the degradation rate constant is surprisingly well-correlated with the age of the sediment layer only (Jørgensen et al., 2022). Based on this observation, it is suggested that the reactivity (i.e., biogas generation) will gradually decrease over time, but that degradation timelines in the range of 100 years are conceivable.

If methane generation is to be avoided, two potential strategies can be deployed. First, FTT-containing deposits can be exposed to electron acceptors that lead to the formation of CO_2 rather than methane (Van Dongen et al., 2021; Ramsay et al., 2021). These compounds would need to be added at least stoichiometrically to be effective, which in practice can be challenging. The simplest implementation is to expose the FTT deposit to oxygen. However, oxygen introduces the potential for ARD as described previously. It is apparent that the preferred mitigation for ARD-ML and NORMs on one hand and residual diluent on the other are in conflict. As an alternative, the second option is to remove the majority of the labile carbon prior to placement of the material. Diluent removal strategies are discussed further in Section 6.5.3.

6.3.3 Economic Context

One reality that is often under-appreciated in technical discussions is the importance of the economic context of oil sands mining operations. These projects have been under scrutiny for their capital intensity, with project costs in the order of C$10 billion for 100 kbpd production. Any form of (what could be considered) upfront discretionary spending is generally avoided to improve project economics. At the start of mine life, exposure to reclamation cost will be negligible due to (1) the initially small disturbed footprint, and (2) the fact that the Net Present Value (NPV) of a "do-nothing" scenario (where rehandling of the tailings deposits near the end of mine life is accepted) will appear favorable. However, as one progresses further into the mine life, any flexibility to rehandle the material is reduced due to constraints imposed by existing infrastructure (e.g., tailings ponds, plant and pipeline infrastructure), existing tailings inventory, and the value of the remaining ore inventory. At that point, implementation of rehandling solutions may no longer be workable, and insufficient value is left to implement high-cost solution suites. Finally, at the end of mine life when no further revenue is generated, little capital is available for the implementation of any treatment technology platform. This gives rise to at least one important aspect of the Mine Financial Security Program (Alberta Energy Regulator, 2024) to assist with asset retirement obligations.

The economic context presented above demonstrates how challenging it is to secure funding for non-revenue-generating scope at any point in the operation. Operating costs for mineable oil sands are typically in the range of \$25–40/bbl at the time of writing. If an increase in cost for tailings management of \$1/bbl (2.5–4%) is accepted, the all-inclusive treatment and reclamation costs would be limited to approximately \$4/m^3 of fluid tailings. Actual operating costs for fluid tailings treatment are not publicly available, but approximate costs in the range of \$15/m^3 MFT can be inferred from financial statements from operators. Further, the Pembina Institute quoted a fluid tailings treatment cost of \$12.3–18.2/m^3 (Pembina Institute, 2010). These numbers do not include capital expenditure nor the costs of rehandling and placement of reclamation materials, and only consider fluid tailings.

To increase the likelihood of implementation, any tailings solution should ideally be economically neutral compared to a "do-nothing" scenario and create at least some form of revenue in terms of, e.g., additional bitumen recovery or reduction in taxation of GHG emissions. It should also be realized that implementation of "partial solutions" (e.g., incremental improvement on dewatering performance without guaranteed terrestrial closure outcomes) would be difficult to justify, as they would be competing for the limited financial resources.

6.3.4 Regulatory Context

As discussed in Section 6.3.3, the regulatory framework plays a critical role in the way oil sands operators approach tailings management. Recent changes in regulatory requirements have shifted the focus from "fines treatment" to a more holistic "outcomes-based" approach. Until 2015, oil sands operators were regulated under Tailings Directive D074 (Energy Resources Conservation Board, 2009). This directive was based on two main criteria: the need to process fines by an "approved technology" and the storage of treated fines in Dedicated Disposal Areas (DDAs). Here, an approved technology refers to one that has been accepted by the regulator through a specific approval process. Several unintended consequences of how the directive was formulated made adherence difficult. One illustrative example is the definition of *fines* that was used in the directive. It assumed that 50% of the fines in the ore body were naturally captured in the tailings beaches, and the operator would therefore need to prove capture of the remaining fines. However, it proved impossible in practice to close the fines balance over the sites. First, ore bodies are not typically analyzed in sufficient detail for fines content during exploration and development coring programs. Therefore, the fines mass flow into the site is poorly defined. Representative sampling of the ore feed in the dry and wet parts of the ore preparation unit is not practical due to the hazards associated with the mass flow rates. Further, fines measurement techniques on the tailings and product streams were not standardized across industry. Therefore, operators used different techniques to measure fines (laser diffraction, sieve hydrometer, etc.). Finally, measurement of fines content in the beaches was inconsistent, making it difficult to confirm the assumed 50% beach fines capture. These issues led to significant discrepancies in the overall fines balance.

More problematic was that Directive 074 was not "outcome-based": it did not necessarily support desired reclamation outcomes. For instance, if an operator could capture more than 50% fines in the beaches (which would be a beneficial outcome overall), the operator would be *penalized*, as it would appear the fines treated by the approved technology were insufficient. Conversely, the directive stipulated minimum strength criteria for the treated fines deposits. The strength targets were on the one hand not practically achievable and on the other hand insufficient to meet geotechnical performance requirements. Finally, the directive was in tension with Directive 082 (Alberta Energy Regulator, 2016a) for ore definition requirements, as processing higher fines ores resulted in increased demand to capture fines in tailings treatment. Regardless of the definitions and assumptions built into Directive 074, the directive led to material investments in various tailings technologies by industry. For instance, Syncrude invested C$1.9 billion in a fluid tailings centrifuge plant at their Mildred Lake site, while Canadian Natural Resources Ltd. (CNRL) invested around C$2 billion in their Non-Segregating Tailings (NST) plant at the Horizon site. These technologies were deployed to specifically address the fines capture requirements.

In 2015, the TMF (Government of Alberta, 2015) was released as a policy for the environmental management of the Lower Athabasca Region. This framework encompasses management of air quality and surface- and groundwater quality. In addition, it provides "*...direction to manage fluid tailings volumes during and after mine operation in order to manage and decrease liability and environmental risk resulting from the accumulation of fluid tailings on the landscape....*" In 2016, Directive 074 was replaced by Tailings Directive 085 (D085) (Alberta Energy Regulator, 2016b).

The intent of the current regulatory framework is to steward toward desired reclamation outcomes for tailings and manage legacy tailings inventory, with the tailings "ready-to-reclaim" (RTR) within an acceptable timeframe. One aspect of the framework is that a cap on total fluid inventory is established for each operating site, with the intent to reduce overall tailings liability. For material to be considered RTR, fluid tailings need to "... *have been processed with an accepted technology, placed in their final landscape position and meet performance criteria....*" Only then can the fluid tailings be removed from inventory. The RTR requirements are encoded in two sub-objectives. Sub-objective 1 addresses the deposit's physical properties and is linked to the geotechnical outcomes of the reclamation landscape. Sub-objective 2 focuses on minimizing the effect the deposit has on the surrounding environment, which is linked to the biogeochemical outcomes and the "... *ability to reclaim to a locally common, diverse, and self-sustaining ecosystem....*" In practice, the focus has been on water chemistry.

The previously discussed concept in Figure 6.2 is consistent with the TMF objectives and highlights the importance of both geotechnical and geochemical stability of the reclamation landscape. The recognition of the criticality of biogeochemical behavior of the tailings substrates has been the primary driver for subsequent research on FTT performance in the closure landscape.

6.4 Tailings Technology Trade-offs

Section 6.3 described constraints that must be considered in a successful tailings management strategy. Optimal implementation is usually site-specific, and multiple options can often be considered. The vision for the ultimate closure landscape design depends heavily on the performance of the tailings treatment that is chosen, as well as the amount of legacy inventory that must be reprocessed. Further, the site-wide mass balance must be capable of supporting the construction of the envisioned closure landscape. This section will show how these constraints lead to several technical and operational decisions that an operation must resolve in order to successfully implement such a strategy.

6.4.1 Above-grade Versus Below-grade Storage

Any open-pit mine commences with the construction of an out-of-pit tailings storage structure. This out-of-pit structure is needed to contain initial tailings production, as no space in a mined-out pit is available yet. This step is followed by progressive backfilling of the mined-out sections with tailings materials for the bulk of the mine life. One final mined-out section remains unfilled at the end of mine life, essentially accounting for the initial tailings that were stored out-of-pit. This section will typically become a pit lake.

The out-of-pit tailings structure qualifies as a dike that eventually must be delicensed. Any liquefiable material within the dike structure must either be transferred or treated to a non-liquefiable state. The major advantage of in-pit, below-grade storage is that the pit usually does *not* qualify as a dike and therefore does not require delicensing. Hence, it removes the strength requirement from the stored tailings material. This fact explains the desire to place tailings below-grade as quickly as practicable. The disadvantage of accelerated in-pit placement is that it reduces mining flexibility, as it is not possible to reenter the mining area once tailings placement commences.

It is noted that there may be geotechnical performance requirements for the material that is stored in-pit for reasons other than the avoidance of dike delicensing. If, for example, the closure design requires the area to be a tall terrestrial structure, the tailings material must be able to support the placement of the cap by means of heavy earthworks equipment and remain geotechnically stable under the load of the reclamation material itself.

6.4.2 Combined Versus Segregated Storage

Referring to the tailings mass balance presented in Table 6.1, 15% of the solids fraction on average is classified as fines. It is the fines fraction that generates fluid tailings and leads to inventory build-up in tailings ponds and results in slow dewatering performance and poor soil strength. The remaining 85% of the solids (i.e., the sand fraction) is a valuable construction material, e.g., for dike construction. There are two opposing philosophies as to how the sandy fraction should be used. The first option is to generate a combined tailings product that mimics the original ore body (minus the bitumen). Here, the geotechnical strength of the final tailings product is generated through grain-to-grain contact between the sand particles, while the pore space is used to store the fines fraction. Such options are referred to as NST or composite tailings (CT). If successful, these options have a high probability of approaching the original ore volume. The success is reliant on the ability to capture all fines in the pore spaces and avoid segregation during tailings placement. Since the combined tailings product will be fluid during placement, the stream still needs to be stored in a tailings impoundment, for which alternative construction materials (e.g., overburden waste) must then be found: the majority of the sand will be consumed as part of the combined tailings mixture. Examples of the combined tailings technologies are described in more detail in Section 6.5.1.

Alternatively, the sand stream can be allocated as construction material while the fines stream is handled separately. Such options include the conversion of fluid tailings to geotechnically competent products through (atmospheric) drying techniques, or alternatively the material is accepted as geotechnically *incompetent*. In the latter option, the material must be permanently stored below grade in a geotechnically competent containment area such as a mined-out pit.

The choice between the two fines-dominated options is heavily dependent on site-specific conditions and utilization of the available lease area. To employ drying techniques, it is critical that sufficient drying surface area can be made available within the lease. Conversely, sufficient deep deposit storage volume is required if the material is stored in a geotechnically incompetent state. For either option to be sustainable, long-range mine-tailings plans must accommodate these storage requirements. Examples of fines-dominated technologies are detailed further in Section 6.5.2.

6.4.3 Wet Versus Dry Storage

Fluid tailings will require different forms of treatment and placement strategies for wet and dry storage. The main benefit of dry storage is that it is easier to develop geotechnically competent materials in this setting. Dry storage implies that the phreatic surface is located below the tailings material and oxygen ingress into the deposit is plausible. The resulting redox-positive conditions will make oxidation of any sulfide-bearing minerals and resulting ARD-ML effects possible. Oxidation is particularly notable in tailings deposits formed from FTT but may also occur in other tailings streams with low net neutralization potential.

In wet tailings storage, the water table is located above the tailings material, and oxygen ingress is generally hindered. As noted in Section 6.3.2, an anoxic environment results in chemically reducing conditions that are less prone to ML effects. Reducing conditions, however, do support anaerobic biodegradation of residual light hydrocarbons associated with FTT, which leads to the formation of hydrogen sulfide, methane, and carbon dioxide. In contrast, aerobic degradation in dry storage leads to the formation of carbon dioxide and water. It should be noted that gas formation in general has impacts on the viability of the reclamation landform. Further, H_2S release can negatively impact air quality in the region during reclamation activities due to its odorous nature. Finally, biogas bubble release in aquatic tailings storage is implicated in resuspension of solids and bitumen, as described previously.

6.4.4 Integrated Processing Versus Rehandling

From a process development perspective, a key design choice is whether tailings treatment will be operated as an integrated part of the bitumen extraction flowsheet, or if it will be designed as a separate process unit that harvests previously placed tailings. Both options have advantages and drawbacks, and the design philosophy depends largely on the trade-off between operational efficiency and robustness of an integrated flowsheet.

Process integration has the advantage that the tailings stream is available year-round at operating pressures and temperatures. Power, water, and process control infrastructure are easier to secure within the processing plant. A major disadvantage, however, is that the tailings processing facility needs to be robust against (highly common) process variability. Since the revenue-generating process will always have priority during day-to-day operation, the bitumen extraction process cannot be impacted by off-normal tailings operation. As a result, any integrated tailings process would need either greater redundancy and operational robustness compared to bitumen processing, or the tailings deposit would need to accept some level of off-specification material as part of standard operation. Oil sands is already considered highly capital-intensive. It is therefore unlikely that waste treatment would be designed to a higher degree of redundancy than the bitumen extraction process. Therefore, off-spec material will need to be accepted in an integrated flowsheet. Significant rehandle and/or remediation of off-spec material may be required to meet reclamation needs.

Disconnecting the tailings process from the revenue stream eliminates concerns related to the propagation of process upsets to the bitumen extraction process, thereby reducing the need for process redundancy. However, a decoupled flowsheet comes at the price of a lack of available process infrastructure. Stand-alone tailings treatments are generally operated closer to the tailings deposits and therefore require investment in power and control infrastructure, etc. Further, decoupling adds the complexity of harvesting materials from previously deposited tailings. The fluid tailings that are desirable for tailings treatment (i.e., FoFW of 35% or more) exhibit yield stresses in the order of 100 Pa and are therefore difficult to mobilize without dilution (Hajieghrary and Frigaard, 2024). To date, the only practical method to harvest the material at the treatment flow rates required (on the order of 1,000 m^3/h) is by dredging operations. High yield-stress fluids tend to show coning behavior during dredging, when the recharge rates of fluid tailings toward the dredge location are lower than the harvesting rate. Coning in turn leads to breakthrough of the water cap and resulting dilution of the fluid tailings. Coning can be mitigated in principle by relocating the dredge system frequently, but this method is impractical during winter when an ice cover is present. If one opts to run tailings treatment as a seasonal operation, then the instantaneous treatment capacity must be doubled to compensate for the six-month operating window. On the other hand, if year-round operation is chosen, then the process must be compatible with feed temperatures as low as 0.5 °C in the winter months due to the lack of process heat. This requirement is particularly problematic if the tailings treatment is based on flocculation processes, as flocculants become exceedingly inefficient below 10 °C (Wang et al., 2020a).

6.4.5 Reclamation Timeline Versus Energy Intensity

A complex and sensitive reclamation metric is the maximum acceptable timeline for oil sands site reclamation and the associated energy intensity of the treatment technologies. In one extreme, it is noted that long-term consolidation of fluid tailings *will* eventually lead to geotechnically competent reclamation substrates. However, depending on the depth of the deposit, the timelines can extend from years to millennia (McKenna et al., 2016). Since natural consolidation occurs under gravity (and perhaps freeze-thaw and evaporative actions), the required external energy input for this extreme is negligible. From a GHG emissions perspective, natural consolidation can therefore be considered the most favorable alternative. If one desires accelerated reclamation timelines, then an increase in energy intensity for tailings treatment, with the associated increase in GHG emissions, is inevitable.

The contradicting requirements discussed above lead to the question of how to weigh the competing environmental priorities of GHG emission reduction and accelerated land reclamation. The authors acknowledge that such trade-offs extend beyond technical considerations and carry significant impact to local and regional communities.

6.5 Technologies Deployed

Given the constraints described in previous sections, industry has deployed several tailings treatment technologies over the past three decades. These technologies broadly fall in the categories of "combined tailings" and "fines-dominated" tailings options. More recently, technologies specific to FTT have also been developed to manage their unique biogeochemical challenges. Finally, dedicated technologies for the treatment of OSPW have gained attention. This section provides a description of deployed technologies and technologies with high technology readiness levels only in the context of mineable oil sands.

6.5.1 Combined Tailings Technologies

The combined tailings technology concept aims to reproduce a tailings product stream with a composition similar to that in the original ore. The fluid tailings inventory is minimized by maximizing the fines capture in the pore spaces of the coarse stream. If successful, this concept allows for terrestrial storage of fines, as the geotechnical strength of the tailings product is generated by the sand skeleton. Two permutations of this concept exist, the distinction being the source of the fines fraction in the mixture (Alberta Energy Regulator, 2023a):

- NST use the fines fraction as it is received from the run-of-plant. Since the fines content in the ore is variable, the SFR of the NST product will vary as well. This implementation is currently deployed at CNRL's Horizon mine.
- Combined, or CT, recombine fluid tailings harvested from tailings ponds with the primary extraction tailings stream. The process aims at keeping the product SFR constant. This technology platform has been deployed at both Syncrude and Suncor and is currently used at Syncrude's Aurora North and Mildred Lake mines. It has also been used at Canadian Natural Upgrading Ltd.'s (CNUL) Muskeg River Mine (MRM).

The main challenge with combined tailings technologies is that the benefit can only be realized if the SFR of the deposit post placement matches the SFR of the process mixture. Within the process equipment and tailings pipelines, a homogeneous mixture can usually be guaranteed due to the highly turbulent process conditions. Strictly speaking, turbulence is not sufficient to guarantee homogeneous flow (as opposed to stratified flow). Pipeline stratification is governed by the densimetric Froude number, expressing the ratio of fluid inertia forces to gravity forces. Homogeneous flow can only be ensured if the Froude number is sufficiently high. Upon discharge, however, the main resuspension mechanism ceases, and particle settling of the coarser fraction ensues. This "segregation" behavior leads to deposits with higher SFR levels on the discharge (upstream) side and fluid runoff with low SFR levels in the ponds. This condition is very similar to conventional tailings placement and will *not* lead to the desired fines capture behavior.

Segregation can only be avoided under two conditions: either by generating a high enough yield stress in the carrier fluid such that sand settling is avoided, or by operating at mixture sand volume fractions that show grain-to-grain contact. Under the latter condition, no settling occurs as a sand skeleton is already formed: no more carrier fluid is released from this material. While no fluid release may sound beneficial, it should be noted that such mixtures exhibit dilation under deformation and are therefore not pumpable, leaving transport by heavy haulers as the only alternative. Hauling the material is not practical either, however, e.g., due to the challenges posed by winter operation, management of off-specification materials, and accessibility of the tailings placement areas. This limitation leaves the generation of high yield-stress carrier fluids as the preferred method for creating non-segregating mixtures. In principle, this can be achieved by deep dewatering and the addition of coagulants such as gypsum or alum. The determination of the segregation boundary (i.e., the mixture composition and shear flow conditions) that guarantees non-segregating behavior has been subject to extended research (Azan and Scott, 2005; Van Der Merwe et al., 2007; Moussavi Nik, 2013).

In principle, non-segregating mixtures can be generated through conventional coarse and fines dewatering techniques. For coarse material, cyclones and dewatering screens have been deployed, while thickeners, centrifuges, and filter presses can be used for the fines fraction. For fluid tailings harvesting, operation is based on dredging of sufficiently dewatered fluid inventory while avoiding dilution of the material with the water cap.

While the combined tailings mixtures will show characteristics similar to the ore body *after* consolidation is complete, the material will act as a fluid during discharge. As a result, the material still must be placed in a tailings containment structure. Only after dewatering can it be treated as a soil. Since this operation requires both sand for the combined tailings product and construction material for the containment structure, it must be ensured that the site material balance can support both requirements. Equally important, it must be recognized that segregation from off-specification material will manifest itself through the generation of fluid tailings inventory in the deposit. This material will ultimately need to be reprocessed.

Some sites deploy operational methods to create inter-layered tailings deposits that may exhibit similar characteristics to process-based combined tailings streams. For example, CNUL has co-deposited coarse tailings and thickener underflow to form a layered tailings deposit in their external tailings facility, as well as in-pit DDAs. In some instances, FTT is also co-deposited (Canadian Natural Upgrading Ltd., 2020). These operating strategies lack the process control one can expect within a dedicated processing plant but have the advantage of being less costly and are quicker to deploy. A key question is how tightly the streams need to be inter-layered before the material acts as a combined product as opposed to the individual constituents. This question is the subject of current research (Moussavi Nik and Sakuhuni, 2022; Talmon et al., 2018; Zambrano-Narváez et al., 2018).

6.5.2 Fines-dominated Technologies

Contrary to combined tailings, fines-dominated technologies aim at developing the required reclamation performance in the materials without the use of sand. Fines-dominated tailings storage techniques can be classified into two categories (Alberta Energy Regulator, 2023a):

- Surface drying techniques, such as Suncor's Tailings Reduction Operation (TRO) and CNUL's Atmospheric Fines Drying (AFD). These techniques are intended for terrestrial closure of the fluid tailings source.
- Deep deposit storage, such as Syncrude's Base Mine Lake (BML) and Suncor's Permanent Aquatic Storage Structure (PASS). These techniques envision aquatic closure of the fluid tailings. All operators propose some version of water-capped tailings storage. However, the AER has not approved this form of closure at the time of writing.

Interestingly, the techniques show opposing optima for land utilization: drying techniques require significant surface area at shallow depths, while deep deposit storage minimizes surface area but requires as much depth as practical.

Between 2005 and 2020, significant progress had been made for surface-drying techniques for fluid tailings (Salma Rima and Beier, 2021; Matthews et al., 2011; Wells et al., 2011; Kabwe et al., 2017). The concept is based on the observation that clay-rich tailings can be dried comparatively easily by exposing the material to regional climatic conditions for 12–18 months. Recall that the region is net-evaporative, which implies that during summer evaporative drying reduces moisture content. This mechanism stalls, however, when a surface crust develops. The key benefit occurs during the winter months when frost penetration leads to the formation of cracks. These cracks create channels that allow for efficient dewatering during the following spring. In principle, further benefit could be created from underdrainage of the deposit, but this is difficult to operationalize at commercial scale. The dewatering performance can be accelerated by flocculating the fluid tailings prior to placement in the drying cells. This flocculation step adds process complexity, as well as significant operating cost. Most flocculants are anionic polyacrylamide based (PAM), and some concerns about the ultimate environmental fate of the acrylamide monomers remain (Intrinsik Corp., 2022; Li et al., 2023).

This technique was considered a promising solution for fluid tailings management but encountered operational challenges on commercial scale. The key finding was that reliable dewatering required placement thickness (or “lifts”) of a maximum of 1 m per season: frost penetrates around 1 m during winter, meaning that lower portions of thicker lifts would not see the benefit of freeze-thaw. Referring to the mass balance in Table 6.1, it is observed that the annual fluid tailings production is in the order of 18 Mm^3. Thin-lift drying therefore requires 18 km^2 of active drying area per year, which is comparable to the footprint of multiple tailings ponds. Further, the area would need to be reasonably well graded/sloped to ensure that the water can run off freely during thaw and inevitable rain events. An 18 km^2 square surface with a 1% slope would require an elevation difference of approximately 40 m. These constraints can typically not be met for processing the full fluid tailings production.

Operational pressures to meet treatment targets resulted in increased unit area loading (i.e., placement of more material per m^2 drying area). This increase, however, resulted in significant reduction in dewatering performance: only the top meter reached desired solids content, while the deeper layers essentially remained unchanged. Hence, the main benefit of terrestrial storage of treated fluid tailings was lost. Attempts were made to improve performance through optimized flocculation, but these attempts did not result in sufficient performance improvement, e.g., due to increased process control requirements to manage feed variability.

Since the partially dewatered material did not meet geotechnical requirements, the material could not be left in place, and rehandling and storage in dedicated waste dumps was required. Since the product was still partially fluid, it could only be moved by haul truck when frozen. As discussed in Section 6.2.4, this outcome is an example of waste rehandling leading to significant increases in energy intensity. The high operating cost of drying operations (material rehandling + flocculation cost) has made this option fall out of favor for treatment of the full fluid tailings stream, even though niche applications on smaller scales (without rehandling and flocculation) may still be feasible.

The Achilles' heel of fines drying techniques is the need to place thin lifts to ensure geotechnical performance of the product after drying. If this threshold cannot be met operationally, then the diametrically opposite solution becomes more appealing. Rather than focusing on strength gain for the fluid tailings, one can consider reducing the storage footprint as much as practical while accepting that the tailings will remain in a fluid state. Geotechnical strength must then be provided by a containment structure (e.g., mined-out pit or dike structure). Such deep deposit storage has the advantage of avoiding rehandling, while geotechnical performance is not necessary for the treated tailings. It must be accepted that material will settle for extended periods of time and will, by definition, become a low spot in the landscape: inevitably such a feature will become a lake. If the release water rate and chemistry are compatible with the aquatic closure outcome, lake formation does not have to pose a problem, however (Omotoso et al., 2017).

For both fines-dominated scenarios, there is an opportunity for footprint reduction by dewatering the fluid tailings prior to placement in the landscape. Conventional process units such as thickeners (CNUL Jackpine, MRM, IOL Kearl) and centrifuges (Jackpine, Syncrude Mildred Lake) can be deployed for initial dewatering (Cossey et al., 2021). These process units add capital and operating expenses and increase the complexity of operation and GHG intensity, and these additions need to be justified based on site-specific conditions. Note that both thickeners and centrifuges can in principle be deployed for both fluid tailings treatment and dewatering of virgin fine tailings from the extraction process.

For completeness, two additional technologies under consideration for fluid tailings dewatering are mentioned. First, the use of filter presses has been explored by multiple operators as an extension of initial dewatering. This method was most notably piloted by CNUL (Canadian Natural Upgrading Ltd., 2020). It was found that residual bitumen in the tailings feed could cause blinding of the filter cloth, and specific process chemistry was needed to mitigate this issue. Practical throughput limitations make it less suitable for full-scale deployment. A similar technique is based on the use of geotextile tubes. Fluid tailings are treated and transferred to geobags and left to dewater locally. Primary advantage is that the bag provides initial strength and the fluid tailings can be stored above-grade.

Since the geotextile acts similarly to the filter cloth in pressure filtration, similar operating issues with respect to bitumen blinding can be expected (Canadian Natural Upgrading Ltd., 2020).

6.5.3 Technologies Specific to FTT

In previous sections, the complexity of reclaiming deposits containing FTT has been mentioned. For both the ARD and NORMs mitigation, the preferential reclamation strategy is to store the tailings below the phreatic surface to limit oxygen contact with the material. It was noted, however, that anoxic conditions are favorable for methanogenesis of the residual light hydrocarbons associated with FTT. Avoiding methane gas generation is most simply accomplished by ensuring aerobic environments in which hydrocarbons are degraded to CO_2 and water. It is apparent that tailings cannot be stored both wet and dry at the same time, and that, therefore, the reclamation of FTT is particularly challenging.

Operators are required to report annually on fugitive emissions from tailings ponds. Summary information is presented in Government of Alberta (2023). A complication with this dataset is that it provides aggregated data for all ponds per operator, not distinguishing between tailings material. Emissions from ponds broken down by material type are cited in Small et al. (2015). It is shown there that methane emissions are strongly dependent on age and tailings type, with highest methane emissions observed at the two oldest FTT-containing ponds (9–26 tonne CH_4/ha/a, compared to around 0.2 tonne CH_4/ha/a for non-FTT-containing ponds). Further, the methane to CO_2 molar ratio is highest in FTT-containing ponds. More recent evaluations have shown that interpretation of the data needs to account for the measurement methodology (flux chambers, inverse dispersion modeling, etc.) (Baray et al., 2018; Staebler et al., 2024; You et al., 2021). A detailed discussion of emission measurement methodologies is beyond the scope of this chapter.

Technology concepts that aim to address the root cause of the challenges typically try to remove either the sulfide minerals to prevent ARD or the diluent fraction to limit methanogenesis. The CVW Cleantech process (refer to Chapter 5) is an example of a technology platform that aims to accomplish both (CVW CleanTech, 2023). The relative complexity of the flowsheet must be justified against the predicted 5–10% GHG reduction through the avoidance of methanogenic bioconversion of diluent. An alternative mitigation strategy is described in a series of Suncor patents. Technology concepts for the in situ aerobic biodegradation of residual diluent, as well as recovery techniques for diluent containing bitumen from tailings deposits (Brown et al., 2019; Hollander et al., 2019, 2020), are disclosed.

An upper-bound estimate for methane emissions from FTT-affected deposits can be derived from the regulatory limits on diluent loss of 4 volumes of diluent per

1,000 volumes of produced bitumen (Small et al., 2015; Burkus et al., 2014; Burkus, 2014). Assuming that diluent can be approximated as octane (703 kg/m^3), the stoichiometric ratio of methane to CO_2 for methanogenesis is 6.25:1.75, and the GWP of methane is 25, the maximum emission is 10.1 kg CO_2e per barrel. This represents a scenario where all lost diluent is converted through methanogenesis. In practice, it is found that some diluent is volatilized during discharge and that several other biodegradation pathways (aerobic, sulfate reduction) take place. These factors reduce the anticipated emissions to approximately half of this theoretical maximum (Burkus, 2014). Clearly, lowering the diluent loss, as well as steering the biodegradation toward pathways other than methanogenesis, provides direct methods for reducing the overall emission potential.

6.5.4 Capping Strategies

For completeness, the final step toward reclamation of the landscape is briefly discussed in this section. After the tailings stream is treated and transferred to its intended final storage location, the material is capped and isolated. Two means of capping tailings deposits have been deployed by industry: water caps and soil covers. Water-capped tailings have been discussed implicitly in Section 6.5.2. The intent is to flush the water layer covering the tailings for a period of time to remove CoPCs. Over time, the residual CoPC loading from the sediment layer (e.g., due to water release resulting from sediment consolidation) will decrease. Once the natural flow-through from the watershed is sufficient to accommodate this residual loading, the lake can be considered self-sustaining.

Syncrude's BML and Suncor's PASS are two commercial implementations of this strategy. BML started in 2013 as the first commercial demonstration of water-capped tailings in oil sands. Several publications exist describing the performance of this lake system (Suncor Energy (Syncrude) Operating Inc. (SESOI), 2023; Francis et al., 2022; Slater et al., 2021). PASS is a more recent development, with operation starting in 2018. The primary difference between the two implementations is that all fluid tailings transferred to PASS were treated with coagulant/flocculant prior to placement. PASS was intended to decrease the likelihood of Total Suspended Solids (TSS) resuspension and the remobilization of bitumen and other CoPCs (Suncor Energy Inc., 2023).

As an alternative to water-capped tailings, several methods for capping soft deposits with solid materials have been suggested. A technical challenge for these methods is to develop techniques that can place load on the fluid tailings without displacing the substrate. Note that these fluid tailings are of very low strength and generally are unable to support foot traffic, let alone heavy machinery. If placement is too aggressive, the cover may simply displace the fluid tailings rather than encapsulate them. The displaced material would ultimately have to be rehandled in place or removed to an alternative location. It is therefore key to place the cover in small enough increments such that the substrate can support it.

An example of a terrestrial cover can be found at Suncor's Pond 5, where a petroleum-coke-based cover was deployed (Abusaid et al., 2011; Zhang et al., 2024). First, the tailings deposit was naturally frozen during winter to gain access for placement of a geosynthetic layer. Subsequently, layers of petroleum coke were placed on top of the geosynthetics. Due to the relatively low density of the coke, the submerged load on the underlying tailings layer was small. Subsequent installation of wick drains ensured expedited dewatering by reducing dewatering distance. Dewatering led to strength gain of the deposit, which in turn enabled the placement of additional material. Suncor is currently in the process of testing an alternative placement method that eliminates the need for freezing the deposit. By raining in the coke from a floating barge, the minimum requirements for initial surface strength are further reduced (Suncor Energy Inc., 2023).

The use of petroleum coke as a cover material is elegant, as the material is local and abundant at sites such as Suncor that have on-site upgrading. Other operators are considering similar strategies but using sand as an alternative placement material (Fredlund et al., 2018; Langseth et al., 2015). However, the specific gravity of sand is much higher than that of petroleum coke, putting further constraints on the maximum lift that can be placed. Cover design is a discipline in itself, and a full description is beyond the scope of this chapter. The preceding serves as an illustration of the long-term activities needed to bring the landscape from initial tailings treatment and transfer to the ultimate self-sustaining ecosystem.

6.5.5 OSPW Treatment

The total fluid inventory (OSPW + fluid tailings) in oil sands operations typically consists of 25% OSPW and 75% fluid tailings, with more than 400 Mm^3 of OSPW in inventory between the sites as of 2022 (Alberta Energy Regulator, 2023a). OSPW is a circum-neutral stream (pH 8–8.5) that shows a conductivity range of 1,700–4,400 μS/cm, TSS levels of 61–1,588 mg/l, and qualifies as moderately brackish water (Glaeser et al., 2021). Water composition is dependent on ore characteristics and connate water quality and the use of process aids in both bitumen extraction (e.g., caustic soda) and tailings treatment (e.g., gypsum, alum). Further, as a result of the zero-discharge requirement, components without natural sinks in the environment will increase over time.

The CoPCs within the material can broadly be classified into four categories: residual organics (e.g., PAH, Napthenic Acids), metal and metalloids, TSS, and dissolved solids (i.e., salinity, TDS). Several references for OSPW composition exist in open literature, and operators report compositional information on recycle water quality in annual tailings management reports (Alberta Energy

Regulator, 2023b). A recent summary of OSPW compositions for tailings deposits with ponded surface water was given by Cossey et al. (2021), and showed the following major ion concentrations: Na^+ (193–947 mg/l), Ca^{2+} (12–54 mg/l), Mg^{2+} (8–27 mg/l), K^+ (11–28 mg/l), Cl^- (<50–767 mg/l), SO_4^{2-} (215–850 mg/l), and HCO_3^- (300–1,326 mg/l). These results are limited to tailings ponds and water-capped structures.

OSPW is not directly compatible with water quality parameters as described (Canadian Council of Ministers of the Environment, 2024). Several strategies are being considered to mitigate this incompatibility, one of which is the implementation of process water treatment. However, questions exist regarding the viability for treatment. For instance, most conventional water treatment technologies may not be compatible with the relatively high oil/grease and TSS levels in OSPW, which may result in additional requirements for pretreatment of the feed. Further, it must be recognized that no mechanism exists to segregate "treated" process water from the OSPW inventory, as this would require separate fluid storage and pumping infrastructure. This has the somewhat counterintuitive consequence that process water treatment during the operational phase of a mine has limited value, unless a method is found to dispose of the treated water off-site (e.g., transfer to in situ operators, deep-well injection, zero-liquid discharge with solid waste disposal).

Depending on the CoPC that is targeted with the water treatment option, different waste management strategies may be necessary. In the management of residual organics, for instance, chemical oxidative techniques (e.g., hydrogen peroxide) could be used in principle to simply convert the CoPCs into CO_2 and water. Biological oxidative processes could equally treat residual organics but would generate bio-sludge that would require disposal. Metal(loid) treatment could be considered through, e.g., Activated Carbon (AC), ion exchange, or precipitation techniques. In the case of AC, spent AC would require storage such that metal(loids) would not remobilize over time.

A surprisingly challenging CoPC is chloride. If CCME guidelines are applied to OSPW, long-term chloride concentrations need to be limited to less than 120 ppm (Canadian Council of Ministers for the Environment, 2011). Since there is no natural sink for NaCl, two main categories of treatment are feasible: membrane-based treatment and dilution with freshwater.

- Membrane technologies include Reverse Osmosis (RO), Nanofiltration (NF), and Electrodialysis (ED). Significant pretreatment is likely required due to incompatibility of organics and TSS in OSPW with membrane techniques. Furthermore, the brine waste is soluble in water and therefore prone to remobilization. If the waste remobilizes, then it would ultimately become part of the OSPW inventory again, effectively undoing the treatment. In this case, some form of on-site isolation or off-site storage would likely be required.

- Dilution with a freshwater source effectively reduces the TDS concentration. On-site dilution is not feasible in practice due to storage constraints. For a 5- to 10-fold reduction in Cl^- concentration, a proportional dilution of a typical 100 Mm^3 OSPW inventory would require additional storage of 0.5–1 Bm^3 NaCl-free dilution water. Continuous concentration of NaCl from the ore during operation will exacerbate this problem with time.

It is possible to provide a lower-bound estimate for the energy requirements for NaCl removal. The treatment can be expressed in terms of the minimum specific energy consumption (SEC_{min}) as follows (Wang et al., 2020b):

$$SEC_{min} = 2RT\left[\frac{c_f}{R_w}\ln\left(\frac{c_b}{c_f}\right) - c_p\ln\left(\frac{c_b}{c_p}\right)\right]$$

Assuming an OSPW feed $[Cl^-]$ of 750 ppm and a product $[Cl^-]$ of 120 ppm, a water recovery R_w of 84% is required. These conditions result in a brine $[Cl^-]$ of 15,625 ppm and an SEC_{min} of 326 kJ/m^3. For a treatment flow rate of 10 Mm^3/a, the minimum power consumption would be around 100 kW. This power consumption would go up with increased feed concentration. Note that this estimate is for treatment only, and does not include energy intensity related to fluid transfer, pretreatment, etc.

OSPW treatment has the potential to result in an energy intensity increase for the mining operation, specifically during the reclamation phase. Quantifying this effect is difficult; however, as long as the extent of treatment and the ultimate end user of the treated water stream have not been identified. It is noted that passive treatment methodologies (e.g., engineered wetland treatment, photocatalytic treatment, etc.) are currently being investigated. However, since these treatments do not address dissolved salt, these technologies can only partially offset the energy intensity increase.

6.5.6 Energy Intensity Related to Tailings Treatment

The previous sections described a variety of deployed tailings technologies. In terms of process complexity, the technologies varied from as simple as a fluid tailings transfer step with coagulant/flocculant addition to as complex as a full fluid tailings centrifugation process. From an energy intensity perspective, technologies varied from relying fully on atmospheric input (solar + freeze-thaw) to high levels of mechanical energy input (centrifuges, thickeners). Clearly, the overall energy intensity of these technologies will vary significantly (Oil Sands Tailings Consortium, 2012). For illustration purposes, rough lower and upper bound intensities can be estimated.

For in-line flocculation, the tailings treatment intensity is limited to preparation and injection of the flocculant solution into the fluid tailings transfer system. Often, the flocculant suppliers propose proprietary technologies for dry and wet preparation of the flocculant solutions. Generally, the power input for these systems is in the range of 1–10 kW only, which is negligible compared to the energy intensity of the fluid transfer itself. In this estimate, the energy footprint of the production of the flocculant itself has not been included.

On the upper end of process energy intensity, fluid tailings centrifugation can be assessed. Power consumption for the Lynx-1000 decanter centrifuge (as deployed by both Syncrude and CNUL) can be found in Alfa Laval (2024). Operating conditions for throughput and power demand will vary depending on the specific composition of the fluid tailings. As a point of reference, the maximum hydraulic flow rate of 540 m^3/h and total motor power of 440 kW can be used, giving an energy intensity of 2.9 MJ/m^3, comparable to the fluid tailings transport intensity.

From the above, it is clear that the selected tailings treatment will impact the site's energy intensity, particularly near the end of mine life. Meeting the intended treated tailings performance is critical, however, as any rehandling of the material will further increase the energy intensity.

6.6 Concluding Remarks

In this chapter, an overview was presented of the main aspects of oil sands tailings management. Emphasis was placed on the competing requirements between bitumen production and site reclamation. Conflicts are not limited to tensions between, e.g., short-term production goals and long-term environmental outcomes, but include the trade-offs between different environmental performance metrics, such as accelerated land reclamation and overall energy intensity.

Three areas for current and future GHG intensity were explored. The first (and largest) category is the transportation of waste material. Mineable oil sands operate on mass flow scales of tens of thousands of tonnes per hour, as well as length scales of tens of kilometers. Material handling in solid form requires heavy haul equipment, while slurry-based transport implies the need for extensive pumping power. Direct placement of the tailings is desirable, but often impossible when the geochemical and geotechnical performance of the tailings material is not compatible with the ultimate land use of available placement space. The inability of direct placements leads to *interim* storage and future retreatment and rehandling of the material. Rehandling will in principle repeat for all materials that do not meet reclamation intent. Rehandling will be biased toward the end of mine life, leading to increased energy intensity.

The second source of energy intensity is driven by the tailings treatment process itself. Compared to transportation, this contribution is relatively modest if treatment is limited to the addition of flocculants and coagulants only. If deeper treatment is chosen, e.g., to reduce bitumen and/or light hydrocarbon content in the tailings, to recover valuable minerals, or to mechanically dewater fluid tailings, this intensity will increase. Given the large tailings inventory, there is a compelling driver to minimize treatment as much as possible while using natural energy sources (solar effects, freeze-thaw effects) as much as practical. This choice in turn leads to tension with respect to reclamation timelines.

The last source of energy intensity is related to the potential need for future process water treatment. For successful reclamation, sites must ultimately be tied back into the regional watershed, meaning water will flow through the landscape freely. The degree of treatment will further increase energy intensity, specifically closer to the end of mine life.

Technology development for tailings treatment has been ongoing for decades and is expected to continue in the foreseeable future. In the near term, it is expected that the emphasis for technology development will be on the need to balance the required reclamation outcomes and the inevitable increase in energy intensity that will result. A key enabler to achieve this balance is to progressively reclaim the mine site as the tailings areas become available. However, as illustrated in this chapter, several technical and non-technical barriers must be resolved before such reclamation strategies can be implemented successfully.

References

Abusaid, A., Fear, C., McRoberts, E. et al. (2011). *An Update to the Construction of the Suncor Oil Sands Tailings Pond 5 Cover*. Vancouver: Tailings and Mine Waste.

Alberta Energy Regulator (2016a). *Directive 082. Operating Criteria: Resource Recovery Requirements for Oil Sands Mine and Processing Plant Operations*. Alberta: Alberta Energy Regulator.

Alberta Energy Regulator (2016b). *Fluid Tailings Management for Oil Sands Mining Projects*. Alberta: Alberta Energy Regulator.

Alberta Energy Regulator (2023a). *State of Fluid Tailings Management for Mineable Oil Sands, 2022*. Calgary: AER.

Alberta Energy Regulator (2023b). Tailings management. https://www.aer.ca/providing-information/by-topic/tailings/tailings-management (accessed September 2024).

Alberta Energy Regulator (2024). Mine financial security program. https://www.aer.ca/regulations-and-compliance-enforcement/liability-management-programs/mine-financial-security-program (accessed September 2024).

Alfa Laval (2024). High performance decanter centrifuge. https://www.alfalaval.ca/products/separation/centrifugal-separators/decanters/lynx/ (accessed September 2024).

Azan, S. and Scott, D. (2005). Revisiting the ternary diagram for tailings characterization and management. *Geotech News* 23 (4): 43–46.

Baray, S., Darlington, A., Gordon, M. et al. (2018). Quantification of methane sources in the Athabasca oil sands region of Alberta by aircraft mass balance. *Atmospheric Chemistry and Physics* 18 (10): 7361–7378.

Bergerson, J.A., Kofoworola, O., Charpentier, A.D. et al. (2012). Life cycle greenhouse gas emissions of current oil sands technologies: surface mining and in-situ applications. *Environmental Science & Technology* 46: 7865–7874.

Brown, W., Hollander, E.D., Derakhshandeh, B. et al. (2019). Pretreatment of froth treatment affected tailings with floatation and stripping prior to tailings dewatering and containment. Canada Patent No. CA 3037959.

Burkus, Z. (2014). *GHG Emission from Oil Sands Tailings Ponds: Overview and Modelling Based on Fermentable Substrates Part II: Modelling of GHG Emissions from Tailings ponds based on fermentable substrates.* Edmonton: Alberta Environment and Sustainable Resource Development (AESRD).

Burkus, Z., Wheler, J., and Pletcher, S. (2014). *GHG Emission from Oil Sands Tailings Ponds: Overview and Modelling Based on Fermentable Substrates Part I: Review of the Tailings Ponds Facts and Practices.* Edmonton: Alberta Environment and Sustainable Resource Development.

Canadian Council of Ministers for the Environment (2011). *Canadian Water Quality Guidelines for the Protection of Aquatic Life: Chloride.* CCME.

Canadian Council of Ministers of the Environment (2024). Water - aquatic life. https://ccme.ca/en/resources# (accessed September 2024).

Canadian Natural Upgrading Ltd. (2020). *Muskeg River Mine Fluid Tailings Management Report.* Calgary: CNUL.

Canadian Oil Sands Innovation Alliance (2022). *Deep Deposit Design Guide for Oil Sands Tailings.* Calgary: COSIA.

Caterpillar (2022). *Caterpillar performance handbook.* Irving: Caterpillar Inc.

Ciu, Z., Liu, Q., Etsell, T.H. et al. (2003). Heavy minerals in the Athabasca oil sands tailings - potential and recovery processes. *Canadian Metallurgical Quarterly* 42 (4): 383.

Cossey, H., Batycky, A., Kaminsky, H. et al. (2021). Geochemical stability of oil sands tailings in mine closure landforms. *Minerals* 11: 830.

Cumulative Environmental Management Association (2012). *End pit lakes guidance document.* CEMA.

Cumulative Environmental Management Association (2014). *Guidelines for wetland establishment on reclaimed oil sands leases.* Fort McMurray: CEMA.

CVW CleanTech (2023). *CVW CleanTech Process and Technology Overview.* CVW CleanTech.

Energy Resources Conservation Board (2009). *Directive 074: Tailings Performance Criteria and Requirements for Oil Sands Mining Schemes*. Calgary: Energy Resources Conservation Board.

Environment and Climate Change Canada (2020). *Draft Screening Assessment Petroleum Coke Group, 2020*. Government of Canada.

Foght, J., Gieg, L., and Siddique, T. (2017). The Microbiology of oil sands tailings: past, present, future. *FEMS Microbiology Ecology* 93 (5): fix034.

Francis, D.J., Lee Barbour, S., and Lindsay, M.B.J. (2022). Ebullition enhances chemical mass transport across the tailings-water interface of oil sands pit lakes. *Journal of Contaminant Hydrology* 245: 103938.

Fredlund, M., Pham, N., Donaldson, M. et al. (2018). *Preliminary Study for Sand Capping of Treated Tailings*. Banff: IOSTC.

Glaeser, L.C., House, M., and Vitt, D.H. (2021). Reclaiming to brackish wetlands in the Alberta oil sands: comparison of responses to sodium concentrations by carex atherodes and carex aquatilis. *Plants* 10 (8): 1511.

Government of Alberta (2013). *Evaporation and Evapotranspiration in Alberta*. Edmonton: Government of Alberta.

Government of Alberta (2015). *Lower Athabasca Region Tailings Management Framework for the Mineable Athabasca Oil Sands*. Government of Alberta.

Government of Alberta (2023). Area fugitive emissions from oil sands mines. https://open.alberta.ca/opendata/area-fugitive-emissions-from-oil-sands-mines (accessed September 2024).

Hajieghrary, O. and Frigaard, I. (2024). Trapping of bubbles in oil sands tailing ponds. *Energies* 17: 4943.

Hollander, E., Derakhshandeh, B., Brown, W. et al. (2019). Active biodegradation of light hydrocarbons from fluid tailings. Canada Patent No. CA 3059673.

Hollander, E., Derakhshandeh, B., Brown, W. et al. (2020). Processing of a bitumen froth treatment overflow stream from froth treatment tailings for enhanced recovery of bitumen. Canada Patent No. CA 3094533.

Holowenko, F.M., MacKinnon, M.D., and Fedorak, P.M. (2000). Methanogens and sulfate-reducing bacteria in oil sands fine tailings. *Canadian Journal of Microbiology* 46 (10): 927–937.

Hussain, N.A.S. and Stafford, J.L. (2023). Abiotic and biotic constituents of oil sands process affected waters. *Journal of Environmental Sciences* 127: 169–186.

Hyndman, A. and Sobkowicz, J. (2010). *Oil Sands Tailings: Reclamation Goals & the State of Technology*. Calgary: Geo2010.

International Network Acid Prevention (2009). Global acid rock drainage (GARD) guide. https://www.gardguide.com/

Intrinsik Corp. (2022). *Environmental Fate and Behaviour of Anionic Polyacrylamides in Oil Sands Tailings Management: A State of Knowledge Synthesis*. Mississauga, ON.

Jørgensen, B.B., Wenzhöfer, F., Egger, M. et al. (2022). Sediment oxygen consumption: role in the global marine carbon cycle. *Earth-Science Reviews* 228: 103987.

Kabwe, L., Wilson, W., Beier, N. et al. (2017). *Effects of Atmospheric Drying and Consolidation of Flocculated Thickened Tailings and Centrifuged Cake on Near-Surface Shear Strength*. Dallas, TX: PanAm Unsaturated Soilds 2017.

Kaminsky, H. (2008). *Characterization of an Athabasca Oil Sand Ore and Process Streams*. Edmonton: University of Alberta.

Kasperski, K. (2001). *Review of Research on Aqueous Extraction of Bitumen from Mined Oil Sands*. Devon: Canadian Centre for Mineral and Energy Technology (CANMET) Western Research Center.

Kong, J.D., Wang, H., Siddique, T. et al. (2019). Second-generation stoichiometric mathematical model to predict methane emissions from oil sands tailings. *Science of The Total Environment* 694: 133645.

Langseth, J., Luger, D., Jacobs, W. et al. (2015). *Placement of Caps on Soft and Fluid Tailings*. Vancouver: Tailings and Mine Waste.

Li, J., How, Z., and Gamal El-Din, M. (2023). Aerobic degradation of anionic polyacrylamide in oil sands tailings: impact factor, degradation effect, and mechanism. *Science of The Total Environment* 856: 159079.

Lindsay, M., Vessey, C., and Robertson, J. (2019). Mineralogy and geochemistry of oil sands froth treatment tailings: implications for acid generation and metal(oid) release. *Journal of Applied Geochemistry* 102: 186.

Martens, S., Lappin, T., and Godwaldt, R. (2008). *Compaction of Upstream Construction Tailings Dam Beaches using Dozers*. Edmonton: GeoEdmonton'08.

Matthews, J.G., Dhadli, N., House, P. et al. (2011). *Field Trials of Thin-Lift Deposition of Amended Mature Fine Tailings at the Muskeg River Mine in Northern Alberta*. Perth: Australian Center for Geomechanics.

McKenna, G., Mooder, B., and Jamieson, A. (2016). *Shear Strength And Density of Oil Sands Fine Tailings for Reclamation to a Boreal Forest Landscape*. Lake Louise: IOSTC.

Moussavi Nik, R. (2013). *Application of Dewatering Technologies in Production of Robust Non-Segregating Tailings*. Edmonton: University of Alberta.

Moussavi Nik, R. and Sakuhuni, G. (2022). *Enhanced in-line flocculation of FFT and its Co-Deposition with CST*. Edmonton: IOSTC.

Natural Resources Canada (2008). *Canadian Minerals Yearbook*. Ottawa: Minerals and Metals Sector.

Nimana, B., Canter, C., and Kumar, A. (2015). Energy consumption and greenhouse gas emissions in the recovery and extraction of crude bitumen from Canada's oil sands. *Applied Energy* 143: 189–199.

Oil Sands Tailings Consortium (2012). *Technical Guide for Fluid Fine Tailings Management*. OSTC.

Omotoso, D. (2019). *PASS – A New Technology Solution for Tailings Management.* Montreal: CIM.

Omotoso, O., Revington, A., Goranson, M. et al. (2017). Treatment of fine tailings including chemical immobilization, polymer flocculation and dewatering. Canada Patent No. CA3036315A1.

Pembina Institute (2010). *Toxic Liability, How Albertans Could End Up Paying for Oil Sands Mine Reclamation*. Pembina Institute.

Ramsay, J.A., de Lima e Silva, M.R., Tawadrous, A.R. et al. (2021). Does addition of phosphate and ammonium nutrients affect microbial activity in froth treatment affected tailings? *Microorganisms* 9 (11): 2224.

Saidi-Mehrabad, A., He, Z., Tamas, I. et al. (2012). Methanotrophic bacteria in oil sands tailings ponds of northern Alberta. *The ISME Journal* 7: 908–921.

Salma Rima, U. and Beier, N.A. (2021). Effects of seasonal weathering on dewatering and strength of anoil sands tailings deposit. *Canadian Geotechnical Journal* 59: 447–457.

Saskatchewan Research Council (2011). *Pipeflow 10.2*. Saskatoon: SRC.

Siddique, T., Gupta, R., Fedorak, P.M. et al. (2008). A first approximation kinetic model to predict methane gas generation from an oil sands tailings settling basin. *Chemosphere* 72: 1573–1580.

Slater, G. F., Goad, C.A., Lindsay, M.B. et al. (2021). Isotopic and chemical assessment of the dynamics of methane sources and microbial cycling during early development of an oil sands pit lake. *Microorganisms* 9: 2509.

Small, C.C., Cho, S., Hashisho, Z. et al. (2015). Emissions from oil sands tailings ponds: review of tailings ponds parameters and emission estimates. *Journal of Petroleum Science and Engineering* 127: 490–501.

Staebler, R.M., Liggio, J., Hayden, K.L. et al. (2024). Aircraft-derived CH4 emissions from surface and in-situ mining activities in the Alberta oil sands region. *Atmospheric Environment: X* 23: 100280.

Suncor Energy (Syncrude) Operating Inc (SESOI) (2023). *2023 Pit Lake Monitoring and Research Report (Base Mine Lake Demonstration Summary: 2012-2022)*. Syncrude.

Suncor Energy Inc. (2023). *2023 Base Plant Fluid Tailings Management Report*. Calgary: Suncor Energy Inc.

Talmon, A.M., Hanssen, J.L.J., Van Maren, D.S. et al. (2018). *Numerical Modeling of Tailings Flow, Sand Segregation and Sand Co-Deposition: Latest Developments and Applications*. Edmonton: IOSTC.

Van Der Merwe, S., Hollander, E.D., and Penner, B.J. (2007). *A Method for Producing a Non-Segregating Waste Stream*. Canada, Patent No. 2585059.

Van Dongen, A., Samad, A., Heshka, N.E. et al. (2021). A deep look into the microbiology and chemistry of froth treatment tailings: a review. *Microorganisms* 9 (5): 1091.

Wallace, D., Tipman, R., Komishke, B. et al. (2004). Fines/Water interactions and consequences of the presence of degraded illite on oil sands extractability. *The Canadian Journal of Chemical Engineering* 82: 667–677.

Wang, D., Zhang, Q., Chen, Q. et al. (2020a). Temperature variation characteristics in flocculation settlement of tailings and its mechanism. *International Journal of Minerals, Metallurgy and Materials* 27 (11): 1438.

Wang, L., Violet, C., DuChanois, R.M. et al. (2020b). Derivation of the theoretical minimum energy of separation of desalination processes. *Journal of Chemical Education* 97: 4361–4369.

Wells, P.S., Revington, A., and Omotoso, O. (2011). *Mature Fine Tailings Drying — Technology Update*. Perth: Australian Center for Geomechanics.

You, Y., Staebler, R.M., Moussa, S.G. et al. (2021). Methane emissions from an oil sands tailings pond: a quantitative comparison of fluxes derived by different methods. *Atmospheric Measurement Techniques* 14: 1879–1892.

Zambrano-Narváez, G., Wang, Y., Chalaturnyk, R.J. et al. (2018). *Assessment of Self-Weight Consolidation of Co-Deposition Deposits in the Geotechnical Beam Centrifuge*. Sixth International Oil Sands Tailings Conference, Edmonton, Alberta, December 9-12. Edmonton: IOSTC.

Zhang, Y., Aboudiab, A., Abusaid, A.H. et al. (2024). *Integration of Geosynthetics in Reclaiming an Oil Sands Tailings Pond*. Pan-American Conference on Geosynthetics.

Part II

Cleaner Petroleum Refining

7

Refining of Unconventional and Heavy Oil

M. R. Riazi[1] *and Dicho Stratiev*[2]

[1] *Department of Chemical and Natural Gas Engineering, Texas A&M University-Kingsville, Kingsville, TX, USA*
[2] *Institute of Biophysics and Biomedical Engineering-Bulgarian Academy of Sciences, Sofia, Bulgaria*

7.1 Introduction

In this chapter, the fundamental and practical aspects of technologies used to process and refine unconventional crude oils are presented. In particular, we are focusing on the heavy and unconventional oil refining technologies such as fluid catalytic cracking (FCC), hydrocracking, coking, hydrotreating, and residue processing. The focus would be to maximize light and midstream products to increase refining margins while minimizing the environmental pollutants. Due to the value of aromatics in production of petrochemicals, a greater emphasis is given to the cracking units. This chapter is divided into two sections – Section 7.1: Refining of heavy and extra heavy oils by M. R. Riazi and Section 7.2: Refining of shale oil by Dicho Stratiev.

7.2 Heavy and Extra Heavy Oil Refining Technologies

According to a recent data, oil, gas, and coal combined represent approximately 81% of the world's energy production in 2023, with oil accounting for roughly 32%, coal at 26%, and natural gas at 23% of the total energy mix (Harvey, 2024). The market value of crude oil is at $2.1 trillion, and it is bigger than the 10 largest metal markets combined (Bhutada, 2023).

Cleaner Petroleum Production and Refining Technologies, First Edition, Edited by M. R. Riazi and Harvey W. Yarranton.

In this section, nature of heavy oil and extra heavy oils and their importance in providing clean fuels and chemicals in the coming decades are presented. Recent developments in the refining technology from conventional to present and future refining configurations are reviewed. Important units that play key roles in shaping refineries to process unconventional and heavy oils, such as cracking, hydrotreating, and hydrogen production units, are briefly reviewed.

Processing and uses of heavy oils produce large quantity of CO_2. In reducing carbon emission, with the use of energy CO_2 can be converted to many chemicals. Converting CO_2 to methane and methanol is technically available. Methanol can be converted to olefins through known processes. From ethylene and propylene alone, more than half of all chemicals and polymers can be produced (LeClerc et al., 2024). We also review configuration of units with respect to reducing CO_2 and GHG emissions. Refining is a mature but a competitive industry in which, with changing market demand and environmental regulations, it is important to properly configurate the processes to make it sustainable and profitable. Generally, refineries that convert heaviest oils into lightest products are more sustainable.

7.2.1 Definition of Unconventional Oils, Resources, and Importance of Crude Oil Market

Distribution of various types of oil is shown in Figure 7.1. As shown in this figure, unconventional oil contributes to about 70% of global proven oil reserves. This indicates that heavy oil, bitumen, and extra heavy oil will play an important role in oil production when conventional type of oil is not sufficient to meet world oil demand. Definition of these types of oil based on their API gravity is given in Table 7.1. Oils with an API gravity between 22 and 10 will be considered as heavy crude. Extra heavy crudes are defined as oils with an API gravity of <10°API and a viscosity of <10,000 cSt that are mobile at initial reservoir conditions (ASTM MNL58).

According to Energy Institute (EI, 2024), the world energy consumption in 2023 was equivalent to 620 EJ and the oil is contributing 196 EJ or 31.6% of this

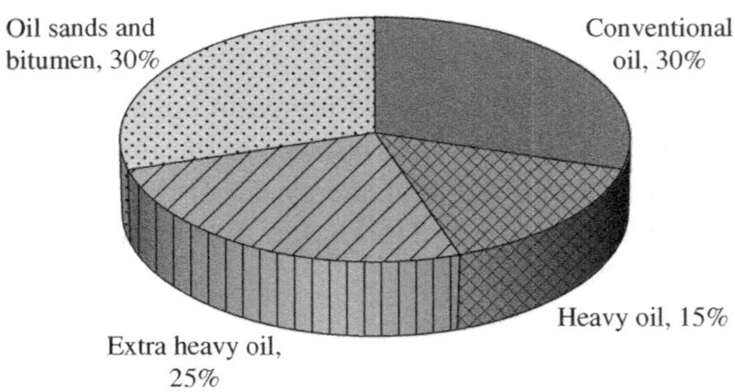

Figure 7.1 Distribution of various types of oil resources. *Source*: Gomaa (2023)/with permission of Elsevier/CC BY-NC-ND 4.0.

Table 7.1 Classification of oils and definition of heavy oil, extra heavy oil, and bitumen.

Type of oil	API gravity of oil	Viscosity at the reservoir conditions	Definition of oil
Conventional oil	>45°		Condensate
	22°–45°		Medium – light oil
Unconventional oil	10°–22°	<10,000 cSt Mobile at reservoir conditions	Heavy oil
	<10°	<10,000 cSt Mobile at reservoir conditions	Extra heavy oil
	<10°	>10,000 cSt Immobile at reservoir conditions	Bitumen

Source: Riazi and Eser, ASTM MNL 58 (2013)/ASTM International.

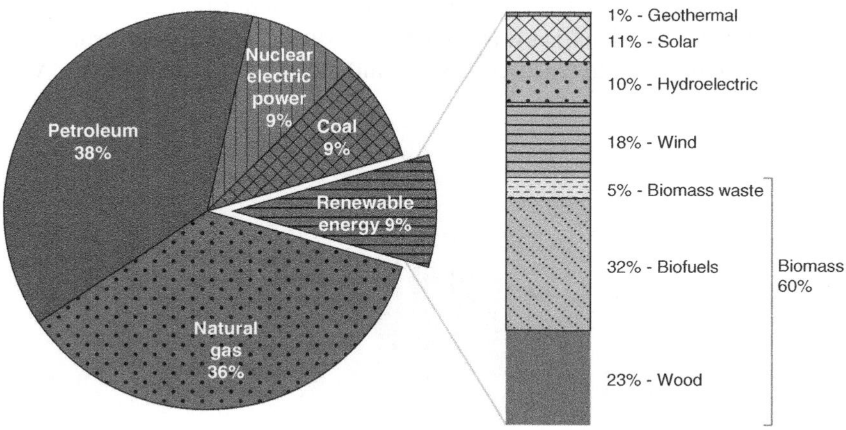

Figure 7.2 Various forms of energy consumption in the United States. *Source*: Adapted from EIA (2024a).

demand, followed by coal (164 EJ) and natural gas (144 EJ). In the United States, petroleum contributes to 38% of energy demand followed by natural gas with 36%, as shown in Figure 7.2 (EIA, 2024). Production of coal contributes to about 9% of total U.S. energy consumption, although in recent years and after pandemic, its production has decreased by about 3%. Coal liquids are rich in aromatics and heteroatoms, very similar to extra heavy oils (Riazi and Gupta, 2015).

Crude oil has been one of the main resources of energy during the past and this century. Despite many projections in the past that oil production could reach to a peak over the turn of the century, now with increase in population and demand, oil production reached all time high of 102.9 mb/d (million barrels per day) in April 2024. It is expected that in 2025, the production will further increase by 1.6 md/d (IEA, 2024a). According to the U.S. Department of Energy (EIA) data, the United States produced 13.4 million barrels per day, a record high in the U.S. oil industry, during August 2024. Figure 7.3 shows the demand on crude oil from 2005 to 2024. Except for the year 2020 that the demand decreased due to the pandemic, there has been a steady increase in demand and production. Projected crude oil consumption in the United States through 2050 is shown in Figure 7.4.

Ferrari, in a recent article (JPT, June 2024), correctly argues that oil remains an indispensable and irreplaceable part of the global energy mix. As convenient, efficient, and highly energy-dense resources, natural gas and oil are simply unmatched by any currently available alternatives. In addition to energy supply, oil and natural gas are sources of many petrochemicals and added-value chemicals produced from oil and gas that keep demand on oil and gas to grow in the coming decades. It is anticipated that between 2022 and 2030, the market for petrochemicals will increase at a compound annual growth rate (CAGR) of 5.4%.

The global Heavy Oil market was valued at US$1030.5 million (>$1 trillion) in 2023 and is anticipated to reach US$1580.6 million by 2030, witnessing a CAGR

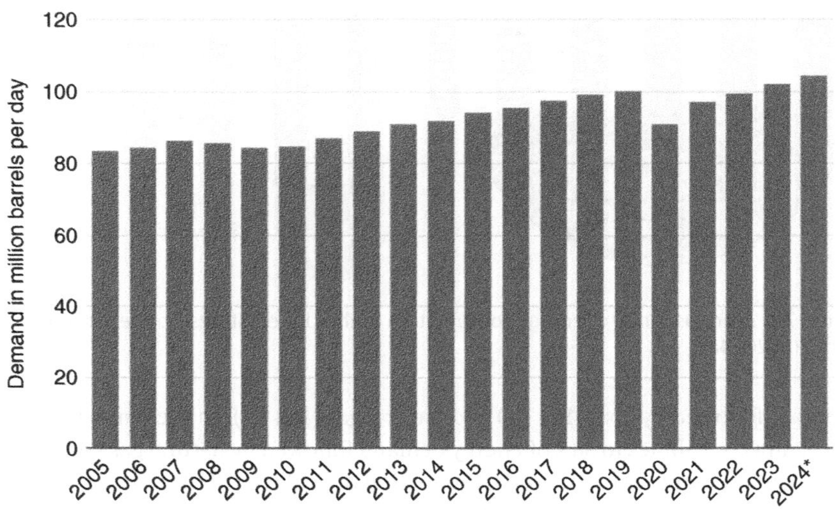

Figure 7.3 Demand for crude oil worldwide 2005–2024. *Source*: Adapted from IEA, Statista (2024b).

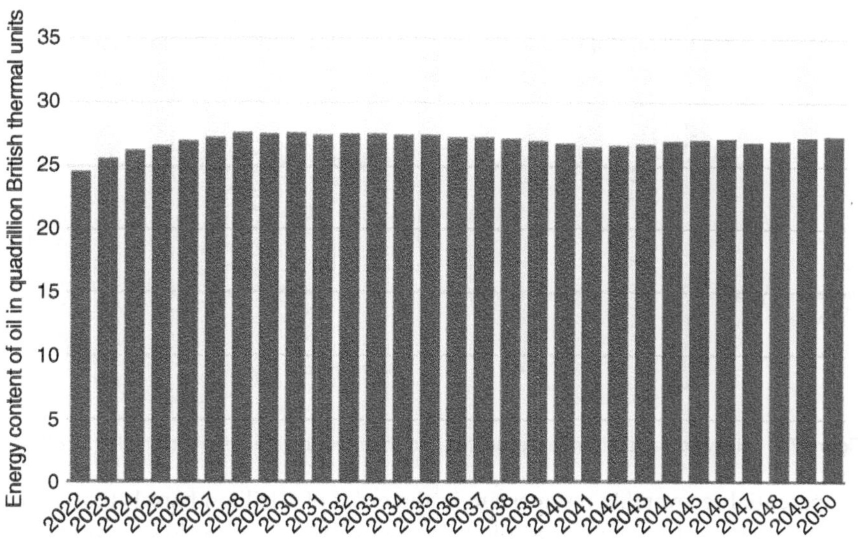

Figure 7.4 Projected crude oil production in the United States from 2022 to 2050 (IEA, Statista, 2024c).

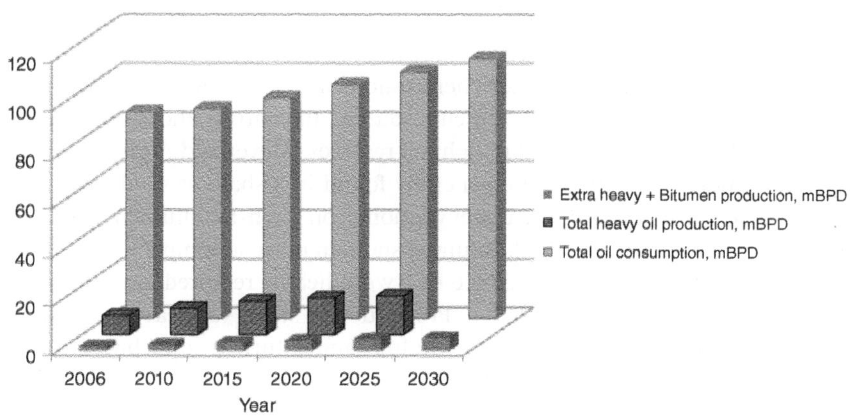

Figure 7.5 Projections of total oil consumption and total production of heavy oil and extra heavy oil/bitumen production (Outlook, 2010; Rana et al., 2013).

of 6.3% during the forecast period 2024–2030 (IEA and RRW, 2024). Projection of production of unconventional oil until 2030 is shown in Figure 7.5.

Heavy oil and bitumen are widely distributed worldwide, as shown in Figure 7.6. The total estimated resources of heavy crude and bitumen are 3.3 trillion barrels and 2.6 trillion barrels, respectively. According to U.S. DOE EIA (2023 and 2024),

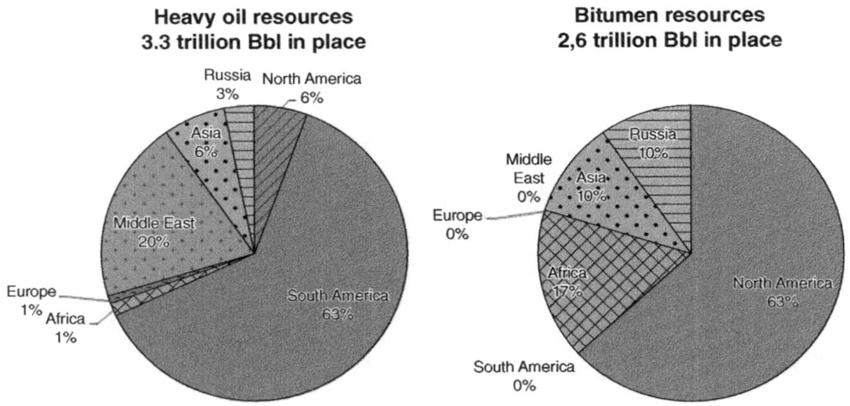

Figure 7.6 Heavy oil and bitumen resources and their worldwide distribution EIA (2024b).

the potential recoverable reserves are in the order of 1 trillion barrels that are in the same order of magnitude as the proven reserves of the world's conventional (medium and light) crude oils.

As shown in Figure 7.6, most of the heavy crude and bitumen produced today come mainly from North and South America: Canada, Venezuela, Mexico, USA, Brazil, and Ecuador. The Canadian oil sands (bitumen) and the Venezuela Orinoco Oil Belt (heavy crude) are the largest accumulations of non-conventional oil reserves in the world. The western hemisphere has 69% of the world's heavy oil reserves and 63% of the natural bitumen resources. In contrast, the eastern hemisphere has about 85% of the world's light oil reserves (Meyer and Attanasi, 2003).

Unconventional oil deposits are generally found in a shale or sandstone, and under the reservoir conditions, unconventional oil such as bitumen does not flow similar to a conventional oil. Bitumen starts to move at temperatures above 200 °C, and for this reason, to produce heavy oil, heat is required, and this heat is usually provided by steam injection. Production of steam for heavy oil production requires combustion of a hydrocarbon fuel, and in turn, it contributes to CO_2 emission (Madugula et al., 2021). The United States and Canada together produced about 17.5 million bbl/d of crude oil in 2023, and about two-thirds of this came from unconventional sources (Oil Sands Magazine, 2024).

Most of heavy oil reserves have not been massively extracted until recent times because of higher production, transport, and processing costs of this type of crude. Declining availability of conventional crude reserves has sparked recent interest in processing and converting the heavier crudes and bitumen. Major oil companies have considered the heavier crudes as replacement for light and medium crudes.

According to EIA (2023), the total oil consumption will increase from 85 mBPD in 2006 to 107 mBPD in 2030, while the total heavy oil production will increase from 8 mBPD in 2006 to 16 mBPD, and the extra heavy oil + bitumen production

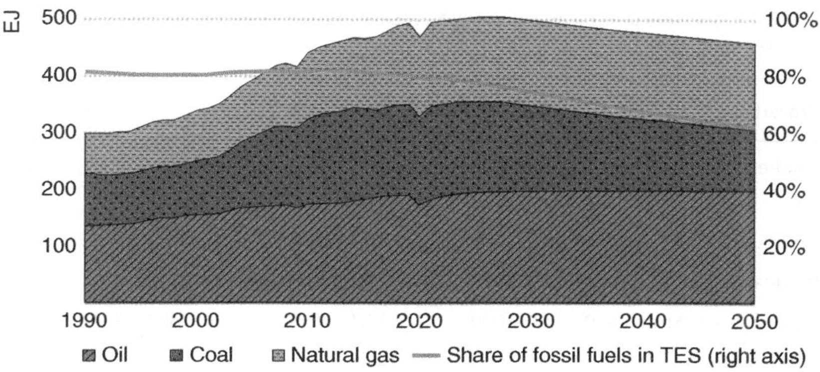

Figure 7.7 Oil, gas, and coal as main source of energy for coming decades EIA (2024c).

will increase from 2 mBPD in 2006 to 6 mBPD by 2030. Currently, a large part of the heavy oil produced in Venezuela and Canada is upgraded to produce a light synthetic crude oil of high quality or diluents to mix with the untreated heavy crude for transportation to the markets because of the lack of light crudes or naphtha for dilution. In the future, the scarcity of the diluents will grow much; most of the heavy crude will need to be commercialized as synthetic crude oil.

Hydrocarbon-type fuels, mainly oil, gas, and coal, have been the main sources of energy during the past century and will remain important contributors to the world energy demand at least for the next several decades. Although various organizations project different scenarios, oil and gas at minimum will provide about 40–50% of world energy demand by 2050 (DNV-GL, 2017).

Estimates of proven global oil and gas reserves increase almost each year. According to OGJ annual assessments, proven oil reserves in 2022 were 1,757 billion barrels, up from 1,735 billion barrels a year earlier and much higher than a decade ago (Xu and Bell-Hammer, 2023). According to 2021, BP statistical report, this estimate for the year 2000 was 1,300 billion barrels. According to Energy Information Administration (US EIA), the role of oil as a source of energy will remain strong at least through the middle of this century, as shown in Figure 7.7.

7.2.2 Characteristics and Properties of Unconventional and Heavy Oil

Heavy oils are generally characterized by their API gravity between 10 and 20, as given in Table 7.1. They are rich in aromatics and have high asphaltene contents in comparison with conventional oils. Characteristics of asphaltene, resins, and crude oil are given in Table 7.2. In this table, M is molecular weight in g/mol, H% is the hydrogen content in weight %, H/C is the hydrogen to carbon atomic ratio, V is the liquid molar volume at 25 °C, *d* is the molecular diameter calculated from

Table 7.2 Properties of typical asphaltenes, resins, aromatics, saturates, oils, and whole vacuum residual oils.

Hydrocarbons	M	H%	H/C	V	d, °A	D
Asphaltene	1000–5000	6.0–11.8	0.8–1.6	900	14.2	4–8
Resin	800–1600	8.9–12.5	1.4–1.7	700	13	2–3
Aromatics	500–1000	10.1–12.4	1.4–1.7			
Saturates	400–900	13.2–15.2	1.8–2.1			
Oil (Saturates + Aromatics)	400–1000	12.5–13.1	1.7–1.8	200–500	8–12	0–0.7
Whole vacuum residue	600–1300	9.2–12.7	1.3–1.7			

Note: In addition to the data from Riazi (2007), Table 7.2 also summarizes data after Stratiev et al. (2019), Stratiev et al. (2025), van den Berg et al. (2003), and Goosens (1996).

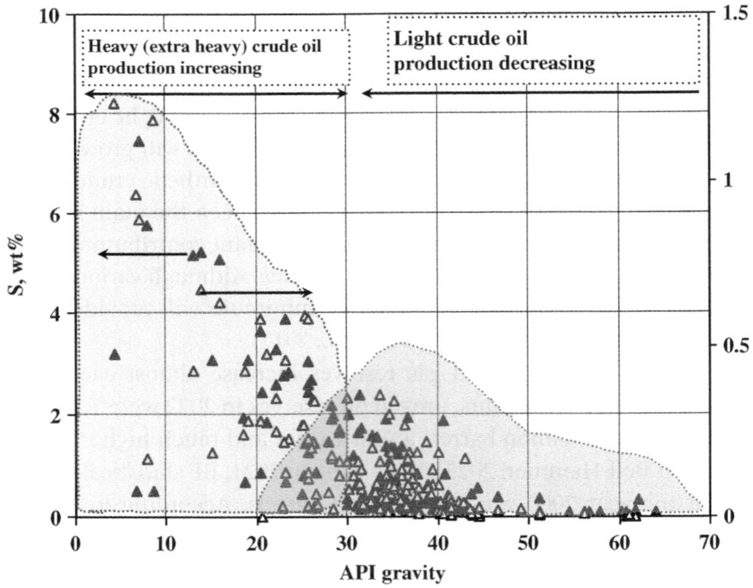

Figure 7.8 Sulfur content of heavy and extra heavy oils. *Source*: ASTM MNL58/with permission of ASTM International.

average molar volume, which for methane molecules is about 4 °A (10^{-10} m), and D is the dipole moment in Debye. These values are approximate and represent properties of typical asphaltenes and oils (Riazi, 2007). Heavy oils are also characterized by high sulfur, viscosity, nitrogen, and metal contents. This is shown in Figures 7.8 and 7.9.

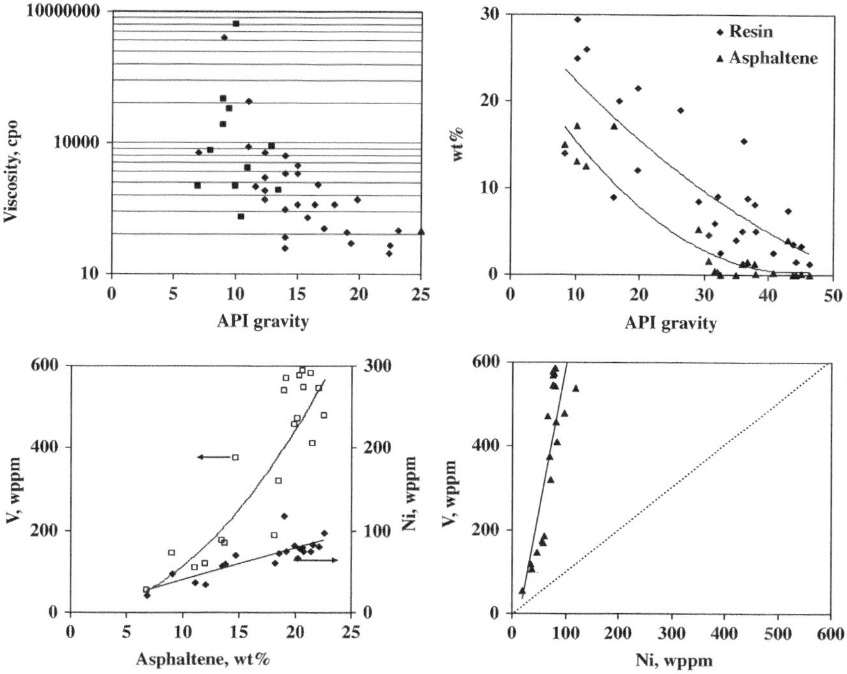

Figure 7.9 Relationship between different properties and composition of various crude oils. *Source*: ASTM MNL58, Riazi et al. (2013)/with permission of ASTM International.

A heavy oil also has higher vacuum residues. For example, light Arabian crude gives vacuum residue of about 20 wt%, while for heavy Arabian oil, the residue is 30 wt%. Similarly, residue of light Arabian oil has nickel content of 26 ppm and vanadium content of 92 ppm, while these numbers for the residue of heavy Arabian oil are 47 and 171 ppm, respectively. A conventional oil such as Alaska crude oil may contain more than 50% atmospheric residue and about 20% vacuum residue, as shown in Figure 7.10.

General specifications of light, heavy, and extra heavy crudes/residue are given in Table 7.3. Crude with less than 0.5 wt% sulfur is considered as sweet crude oil (ASTM MNL50). Heavy and extra heavy oils are characterized by low API degree, high viscosity, as well as high sulfur, nitrogen, metals, and asphaltene contents, as shown in Figure 7.11. These are characteristics of low-quality fuels, and their presence increase cost of crude oil processing and refining. Maximum sulfur content of fuels such as gasoline and diesel in Europe is at 10 ppm, and in the United States should be below 15 ppm. As shown in Figure 7.12, lowering sulfur in a fuel significantly reduces pollutant gases such as unburned hydrocarbon, CO, and NO_x from automobile exhaust tail released to the atmosphere. When sulfur content of a fuel is reduced from 500 to 50 ppm, the HC and CO release

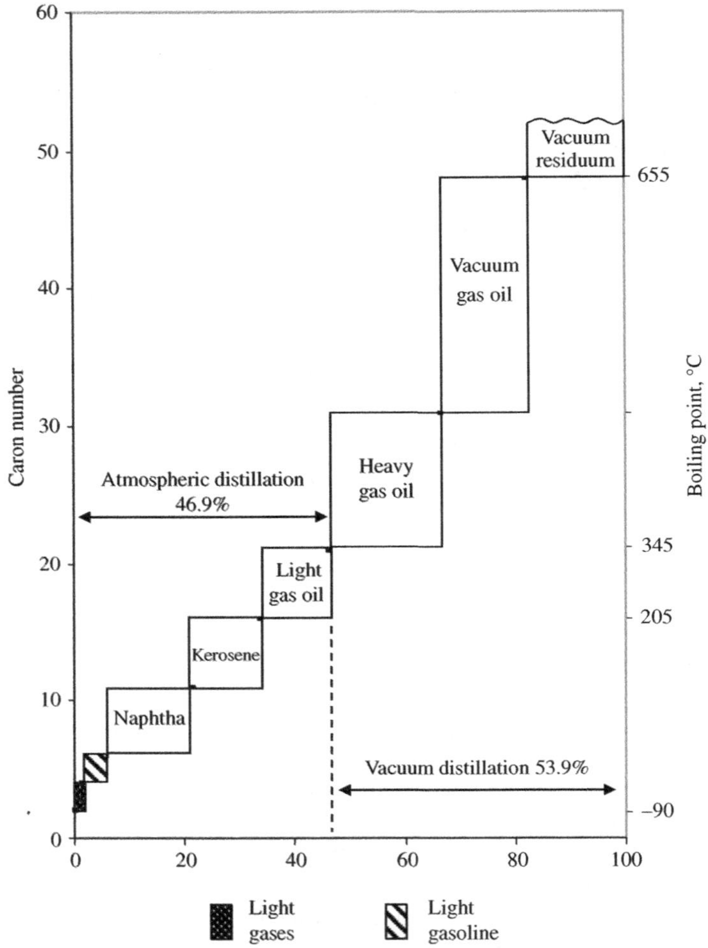

Figure 7.10 Residue production from typical crude oils in distillation columns. *Source*: Riazi (2007)/with permission of ASTM International.

to atmosphere can be reduced by up to 18%. In summary, heavy and extra heavy oils are characterized by high asphaltene, S, N, and metal contents. Typical residues from heavy oil contain about 30 wt% asphaltene and heteroatoms. Therefore, deasphalting, desulfurization, denitrogenation, and demetallation processes are fundamental aspects of heavy oil processes.

Properties of two heavy oil samples from Canada and Colombia and their residues are given in Table 7.4. Distillation data for a bitumen sample from Canada is shown in Table 7.5 (Rana et al., 2013).

Table 7.3 Nature of different types of crude oil (ASTM MNL50).

Classification	Definition
Extra light	High API gravity, low S, N, and negligible asphaltene, and metals
Light crude	Medium range API gravity, low S, N, metals, and moderate asphaltene
Heavy crude	Medium range API gravity, high S, N, high metals, and asphaltene
Extra heavy/ residue	Low API gravity and very high contaminants (S, N, metals, and asphaltenes)

Source: ASTM MNL50/with permission of ASTM International.

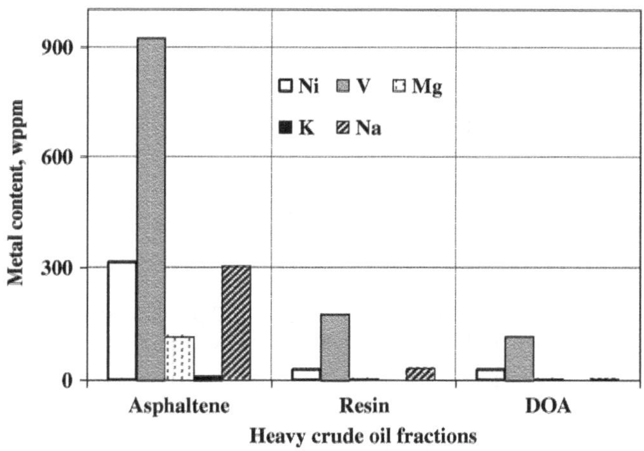

Figure 7.11 Metal content of different fractions of crude oils.

Analysis of a crude oil is essential in optimizing and selecting right configuration of units and processes to refine the oil or its residues. As oil becomes heavier and heavier, its quality characterized by the API gravity decreases, and specific gravity, density, molecular weight, and aromatic and asphaltene contents increase. Distillation data for heavy oils is measured under sub-atmospheric pressures to avoid cracking during distillation. As carbon number of a hydrocarbon increases, the number of its isomers increases exponentially, and this makes the nature of heavy oil different from light and conventional oils, requiring different analytical methods. One of the main techniques to analyze heavy oils is through solvent separation method such as SARA analysis, which gives the concentrations of **S**aturates, **A**romatics, **R**esin, and **A**sphaltene content of the oil. In this method, *n*-heptane is added to the crude to separate asphaltenes as the insoluble part of the crude. The soluble part known as

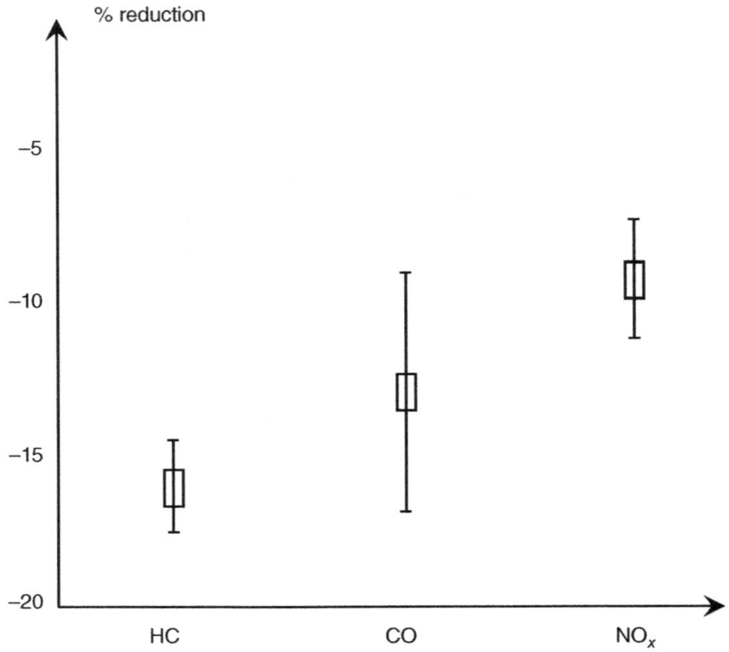

Figure 7.12 Influence of sulfur content in gasoline (from 500 to 50 ppm) in reduction of pollutant gases. *Source*: ASTM MNL50, Riazi (2007)/with permission of ASTM International.

Table 7.4 Properties of two heavy oil samples and their residues.

		Lloydminster, Alberta, Canada		Hamaca, Colombia	
Production properties		**Whole crude**	**Residue**	**Whole crude**	**Residue**
TBP cut point	°C	C5+	585+	C5+	585+
Yield	wt%	100	44.5	100	59.5
Gravity	API	15	5.1	7.9	−0.07
Specific gravity		0.958	1.038	1.016	1.077
Sulfur %	wt%	3.6	5.5	4.1	4.8
CCR	wt%	10	22.1	14.5	24.2
Vanadium	ppmw	120	270	466	783

Table 7.5 Distillation analysis of bitumen (Canadian Tar Sand Bitumen).

Equivalent distillation range at 1 atm temp., °C	Fraction, wt%	Cumulative, wt%	Specific gravity of fraction, wt%	Sulfur wt% in fraction
IBP–200	1.4	1.4	0.816	1.52
200–250	2.2	3.6	0.856	1.02
250–333	9.7	13.3	0.904	1.78
333–418	17.7	31.0	0.955	2.98
418–524	17.5	48.5	0.989	3.8
>524	51.5	100.0	1.073	6.39

deasphalted oil (DAO) is further separated to saturates, aromatics, and resins through liquid chromatography methods. Standard method for SARA analysis is described by ASTM D4124 test method, and some of these methods are discussed by Alboudwarej et al. (2002).

7.2.3 Configuration of Refineries for Processing Heavy Oil and Residues

A general schematic of refineries for processing heavy oil is shown in Figure 7.13. These refineries are characterized by carbon rejection and hydrogen addition. Heavy oils have high C/H ratio, and in order to produce value-added products, carbon must be removed, and hydrogen must be added. Residue-upgrading processes can be divided into two groups: non-catalytic and catalytic processes. Non-catalytic processes include solvent deasphalting (SDA) and thermal processes such as gasification, delayed coking, flexicoking, and visbreaking. Catalytic processes include residue fluid catalytic cracking (RFCC) and hydroprocessing such as fixed bed hydrotreating and hydrocracking. In thermal processes, carbon is rejected, while in hydrotreating and hydrocracking, hydrogen is added (ASTM MNL58, da Silva, 2022).

As shown in Figure 7.13, the main separation units are atmospheric and vacuum distillation columns, followed by dewaxing and deasphalting. In conversion processes (catalytic or thermal), the chemical nature of compounds in the mixture changes due to catalytic reactions or decomposition and cracking due to heating. Catalytic units may include reforming, isomerization, and FCC units. The main thermal processing units are hydrocracking, visbreaking, and coking, which we will discuss later, as they are directly related to heavy oil and residue processing. Hydrocracking or hydrogenation is a thermal decomposition at high temperature and at the presence of hydrogen.

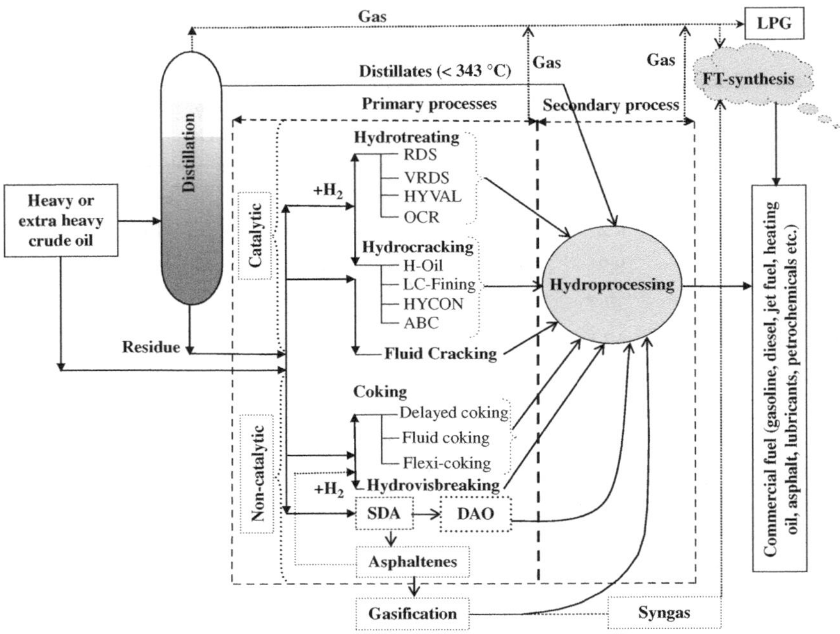

Figure 7.13 Schematic of a refinery to process heavy oil and residue. *Source*: ASTM MNL58 (2013)/with permission of ASTM International.

7.2.3.1 Deasphalting Process

Deasphalting is one of the main non-thermal processes used to remove asphaltene from a heavy oil or vacuum distillation residue (VDR). SDA is commonly used to separate asphaltene from oil by use of a paraffinic solvent. Output of a SDA unit is DAO and byproduct pitch or asphalt. Pitch contains the majority of heavy oil or residue's contaminants such as sulfur, metals, and CCR (Gillis and VanTine, 2003). A SDA process licensed by Foster Wheeler/UOP is shown in Figure 7.14 (Romero, 2020).

Use of lower molecular weight paraffins such as n-pentane or even propane gives higher yields for asphaltenes and better-quality DAO. As sulfur and nitrogen are the main heteroatoms in asphaltenes, the amounts of S and N in produced DAO depend on the type of solvent used, as shown in Figure 7.15. With low paraffinic solvents the concentration of S and N in oil can be significantly reduced. With use of a light paraffin such as propane, high quality (low in sulfur and metals) can be obtained at the expense of lower DAO yields. Integration of deasphalting process in upgrading heavy oil and residues is shown in Figure 7.16.

Asphaltene may also be gasified at 800–1400 °C into syngas, where after going through Fischer–Tropsch (FT) process, it is converted into commercial fuels such

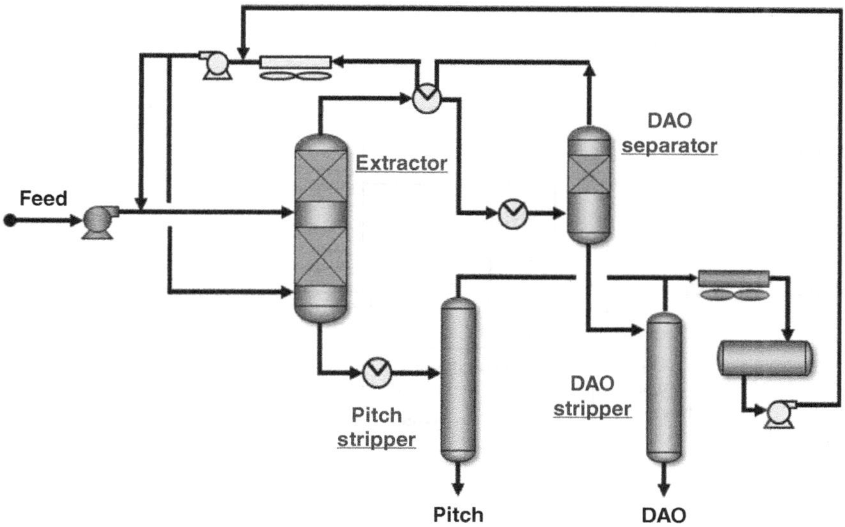

Figure 7.14 Deasphalting process licensed by Foster Wheeler/UOP. *Source*: RefComm, Romero (2023)/RefComm.

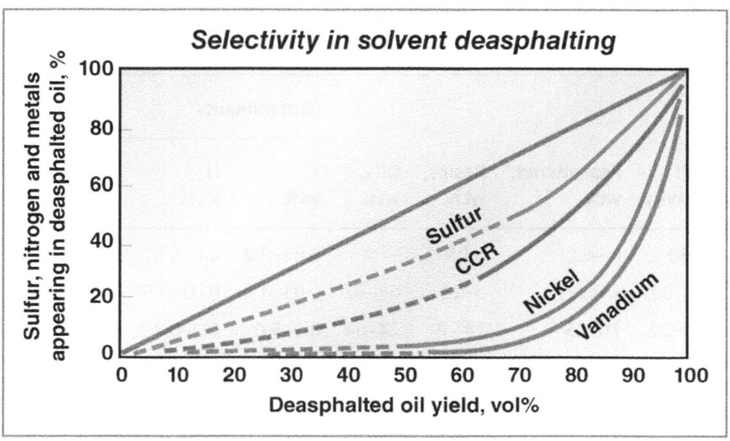

Figure 7.15 Selectivity in SDA process (Riazi, 2013; Romero, 2020).

as petrochemicals, gasoline, and jet fuel (Figure 7.13). In this hybrid process scheme, syngas produced from gasification of asphalt can be converted into liquid fuels through a FT unit. Gas-to-liquid (GTL) processes will be discussed in detail later in Chapter 12.

Distribution of asphaltene in different types of oil is given in Table 7.6.

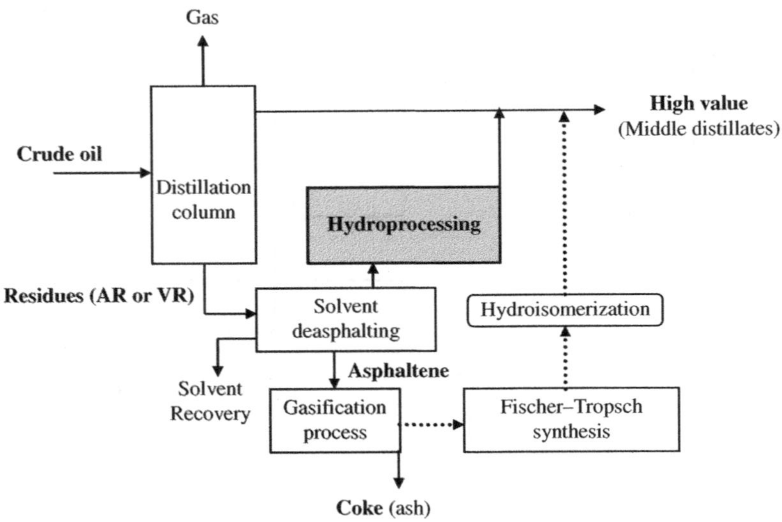

Figure 7.16 Proposed hybrid process scheme for upgrading of heavy petroleum. *Source*: ASTM MNL58 (2013)/with permission of ASTM International.

Table 7.6 Properties and amount of asphaltenes, resins, and oils in light, heavy crudes, and residue.

					Contaminants		
	API gravity	**Asphaltenes, wt%**	**Resins, wt%**	**Oils, wt%**	**S, wt%**	**N, wt%**	**Metals (Ni+V), ppm**
Extra light	>50	0–<2	0.05–3	–	0.02–0.2	0.0–0.01	<10
Light crude	22–32	<0.1–12	3–22	67–97	0.05–4.0	0.02–0.5	10–200
Heavy crude	10–22	10–25	14–39	24–64	0.1–5.0	0.2–0.8	50–500
Extra heavy	<10	15–40	–	–	0.8–6.0	0.1–1.3	200–600
Residue	–	15–30	25–40	<49	–	–	100–>1000

Data taken from various literature sources as discussed in ASTM MNL58.
Source: Eser and Riazi (2013)/with permission of ASTM International.

7.2.3.1.1 FCC, Hydrocracking and Delayed Coking Units

FCC, hydrocracking, and delayed coking units are important units to convert heavy oil and heavy residues into lighter and more valuable fuels. According to GlobalNewsWire (2024), it is expected that the global refinery fluid catalytic cracker units (FCCU) capacity will increase from 14,153 mbd in 2023 to

16,870 mbd in 2028 at an AAGR of 3.5%. The United States, China, India, Russia, and Canada are the major countries that accounted for 66% of the total global FCCU capacity in 2023. Petroleum refiners consider FCC not only for production of a high-octane gasoline but also as a source for production of light olefins and petrochemical feedstock using zeolite or silica alumina-based catalysts. The feed to FCC unit is a gas oil from either atmospheric distillation or vacuum distillation column.

In general terms, FCC process can be divided into three components: reactor, catalyst regenerator, and product fractionator. In the reactor, the cracking reactions take place with the help of catalysts. These reactions are endothermic and require heat. The byproduct of catalytic cracking, which is rich in carbon and is named coke, deposits on the catalyst surface and reduces the activity of catalyst by blocking the pores. Coked catalysts are sent to regenerator to burn off the coke, and regenerated catalysts are returned to the reactor. A schematic of an FCC unit with description of parts is shown in Figure 7.17. Typically, FCC unit feeds worldwide are blends of the following streams: atmospheric and vacuum gas oils as the major components, atmospheric residues, coker and visbreaking gas oils, hydrocracking residues, hydrotreated gas oils and residues, furfural extracts, demetalized oil

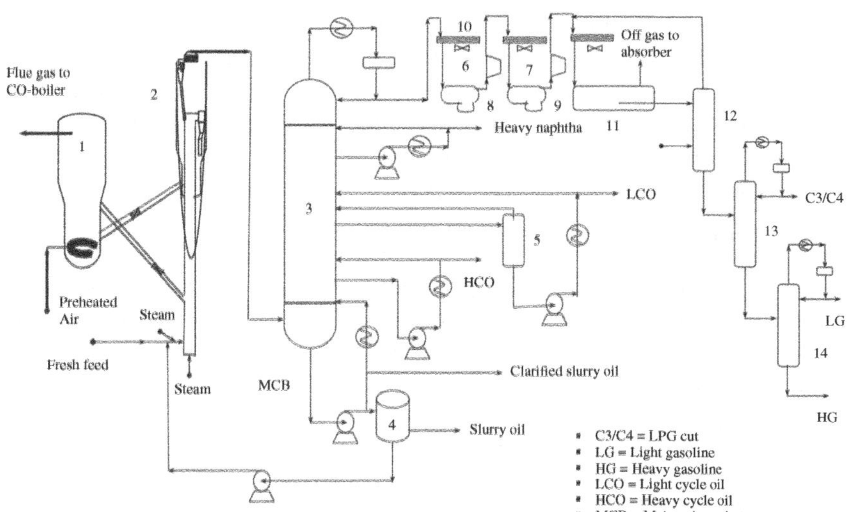

Figure 7.17 FCC complex configuration including product separation. *Source*: ASTM MNL58 (2013)/with permission of ASTM International. 1 = regenerator, 2 = FCC reactor, 3 = fractionator, 4 = slurry settling drum, 5 = LCO stripper, 6 and 7 = compressor stage I and II, 8 = overhead product collection drum, 9 = interstage collection drum, 10 = air fin cool system, 11 = high-pressure separator, 12 = stripper, 13 = debutanizer, and 14 = gasoline splitter.

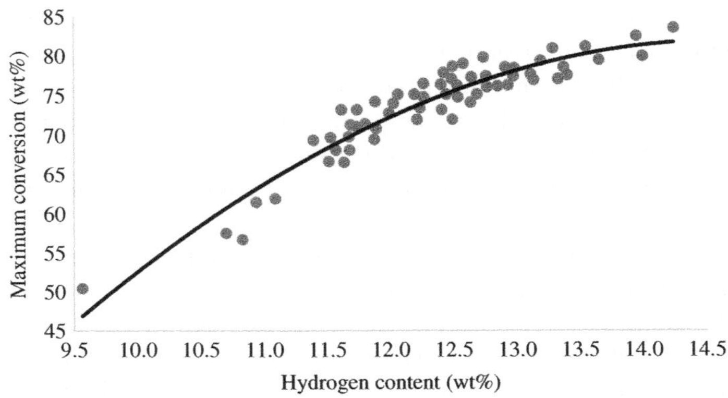

Figure 7.18 Relationship between FCC feed hydrogen content and the maximum conversion. *Source*: Navarro et al. (2015)/Crambeth Allen.

(DMO), etc. Their specific gravity varies between 0.850 and 0.993, and the saturate content oscillates between 30 and 85 wt% (Navarro et al., 2015). The FCC feed conversion fluctuates between 50 and 85 wt% depending on hydrogen content (Figure 7.18).

The degree of crackability and coking tendency of the feed depends on its aromatic carbon content (Bollas et al., 2004; Voolapalli et al., 2013). As a feed's aromatic carbon content increases, its tendency to crack decreases. The FCC conversion of DAOs was found to increase by the hydrogen content (H) enhancement and decrease by the aromatic carbon content (C_A) magnification as displayed by Equation 7.1 (Stratiev et al., 2013).

$$\text{DAO Conversion} = -0.572 \times C_A + 5.595 \times H + 33.96 \tag{7.1}$$

The yield of gasoline is promoted by the contents of saturate (Sat) and aromatic (Aro) compounds (the effect of saturates being twice of that of aromatics) and depressed by the resins (Res) during the DAO catalytic cracking, as shown in Equation 7.2 (Xu et al., 2005).

$$\begin{aligned}\text{DAO Gasoline yield (wt\%)} = &-0.1161 + 0.7070 \times Sat + 0.3304A \\ &\times Aro - 0.0769 \times Res.\end{aligned} \tag{7.2}$$

The yield of diesel is promoted mainly by aromatics, to a much lesser extent by saturates, and is depressed by resins as indicated in Equation 7.3 (Xu et al. 2005).

$$\begin{aligned}\text{DAO Diesel yield (wt\%)} = &-0.6127 + 0.0676 \times Sat \\ &+0.4067 \times Aro - 0.1366 \times Res\end{aligned} \tag{7.3}$$

The main contributor to coke formation during the DAO catalytic cracking is the resins, followed by aromatics (5 times less effective than resins), and the least coke-making saturates (10 times less effective than resins) as exhibited in Equation 7.4 (Xu et al. 2005).

$$\begin{aligned} DAO\ \text{Coke yield (wt\%)} = & -0.0114 + 0.0287 \times Sat \\ & +0.0514 \times Aro + 0.2520 \times Res \end{aligned} \tag{7.4}$$

There are several configurations of FCC process depending on the positions of the reactor and regenerator. In one configuration, feed gas oil is mixed with steam and enters from the bottom of the reactor. UOP (Honeywell) engineers also proposed two-stage regenerators for the FCC plants, which make it easier to operate the unit, and uniform coke burn, which will result in lowering carbon residue, CO, and NO_x emissions in addition to reducing catalysts deactivation. If a third stage is added to the regenerator, it could reduce particulate matter (PM) in flue gases to meet 50 mg/Nm^3 requirement. Addition of catalyst cooler to FCC units could help to control catalysts/oil ratio, increase conversion ability, and improve better catalyst activity and maintenance (Rana et al., 2013).

7.2.3.1.2 Hydrocracking Unit

The second important unit in processing of heavy oil is hydrocracking unit. According to a global research and market company (Research and Market, 2024), the global refinery hydrocracking units capacity increased from 10,926 thousand barrels per day (mbd) in 2017 to 12,498 mbd in 2022 at an Average Annual Growth Rate (AAGR) of 2.7%. China, the United States, India, Russia, and Saudi Arabia are the major countries that accounted for 56% of the total global hydrocracking unit capacity in 2022. In Asia, the increase from 2022 to 2027 is expected to be around 8.8%, where China, India, South Korea, and Japan are the key countries contributing to 90% of total hydrocracking units capacity in Asia in 2022. Europe accounted 16.7% of global hydrocracking units capacity, but it is expected to slightly decrease from 2,086 mbd in 2022 to 2083 mbd in 2027. Schematic of a two-stage hydrocracking unit is shown in Figure 7.19.

The advantages of hydrocracking include its ability to handle a wide range of feedstocks that may be difficult to process by catalytic cracking and its flexibility in selectivity between light and middle distillates. As a hydrogen-addition process, hydrocracking provides high yields of valuable distillates without producing low-grade byproducts such as coke or heavy gas oil, as experienced in carbon-rejection processes such as coking. Usually hydrocracking processes include two reaction stages: the first stage is to remove heteroatoms, and the second stage is hydrocracking to increase H/C ratio. The catalysts for the first stage are cobalt–molybdenum, and for the second stage are nickel–tungsten catalysts (for hydrogenation) supported on alumina/silica (for cracking) as discussed in ASTM MNL58 (Eser and Riazi, 2013).

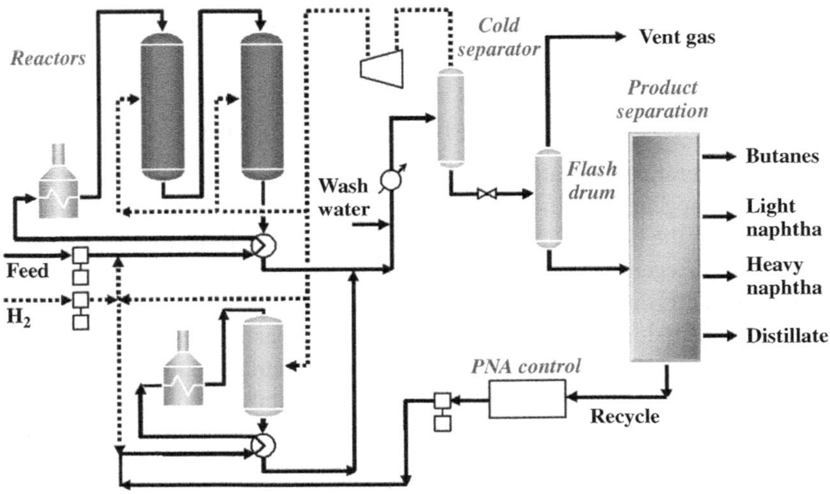

Figure 7.19 A two-stage UOP hydrocracking process (UOP, 2011).

Lababidi et al. (2011) designed a pilot plant for hydrocracking process and conducted series of experiments to develop a model for the prediction of products quality based on the feed specifications and reactor operating conditions, especially temperature and residence time, as shown in Figure 7.20. As temperature and residence time increase, product density decreases while its quality increases. In Figure 7.21, distillation data predicted by their proposed model for three types of products are compared with actual experimental data. Details of the experiments and modeling approach are discussed by Lababidi et al. (2011).

7.2.3.1.3 Delayed Coking Process

Coking, and in particular delayed coking, is a kind of severe thermal cracking to reject carbon from vacuum residues that does not require a catalyst. It produces large quantities of coke as carbon-solid byproduct and light and medium distillates that are processed for fuel pools such as naphtha and gasoline. The price of fuel coke depends on its carbon purity and is traded at a price comparable to that of the coal. The delayed coking is considered as a unique process to maximize production of good-quality coke from very heavy feedstocks. Delayed coking can upgrade a very heavy feed with high sulfur, nitrogen, and metal contents that cannot be upgraded by any other refinery processes. Generally, up to 30 wt% of the sulfur in the feed ends up in the coke. The environmental impacts of delayed coking include emissions from burning fuel for heating the feed and water pollution from the fractionator. The main advantage of delayed coking unit is its ability to process any kind of feed including extra heavy crudes and bitumen to produce distillate products

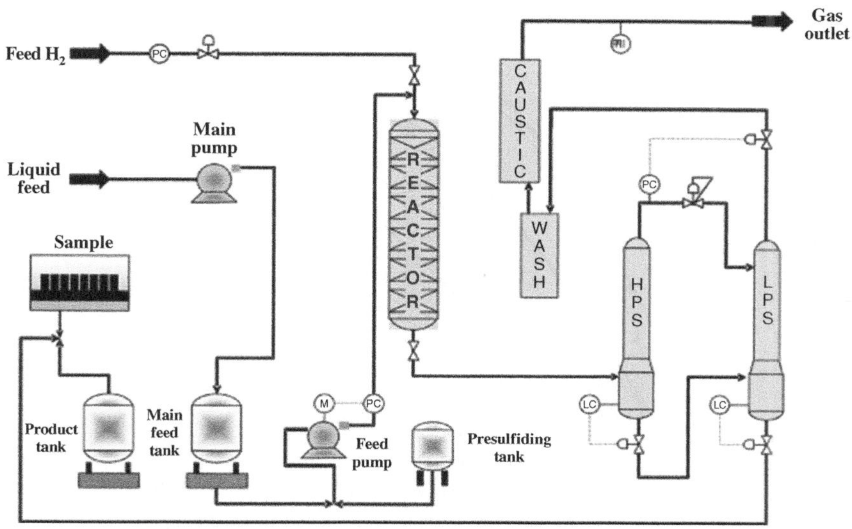

Figure 7.20 Experimental laboratory pilot plant for hydrocracking of a gas oil sample. *Source*: Lababidi and Riazi (2011)/with permission of Elsevier.

with complete rejection of metals and large portions of sulfur and nitrogen. The main characteristic of delayed coking feed affecting product yields is the Conradson carbon content (CCR), or microcarbon residue (MCR) (Muñoz et al., 2013). Comparing different correlations for prediction of delayed coking yields, Muñoz et al. (2013) found that the correlations of Volk et al. (Volk et al., 2002) exhibited the highest accuracy. The delayed coking yield predictions of Volk et al. from feed MCR, reactor temperature (T, in °F), reactor pressure (P, in psia), and liquid space velocity (LSV, in min^{-1}) are presented as Equations 7.5–7.10.

$$\text{Liquid (wt\%)} = -1.1139MCR + 0.0419 \times T - 0.2897 \times P + 1103.09 \times LHSV + 41.59 \tag{7.5}$$

$$\text{Coke (wt\%)} = 0.940 \times MCR - 0.0609 \times T + 0.1529 \times P - 319.759 \times LHSV + 65.075 \tag{7.6}$$

$$\text{Gas (wt\%)} = 0.1729 \times MCR + 0.0191 \times T + 0.13646 \times P - 786.319 \times LHSV - 6.762 \tag{7.7}$$

$$Naphtha\text{ (wt\%)} = -0.3086 \times MCR + 0.0137 \times T + 0.1571 \times P - 819.63 \times LHSV + 16.461 \tag{7.8}$$

$$\text{Diesel (wt\%)} = -0.3339 \times MCR - 0.02635 \times T - 0.0392 \times P + 70.957 \times LHSV + 50.452 \tag{7.9}$$

$$Gas\text{oil (wt\%)} = -0.4714 \times MCR + 0.0546 \times T - 0.4076 \times P + 1851.76 \times LHSV - 25.315 \tag{7.10}$$

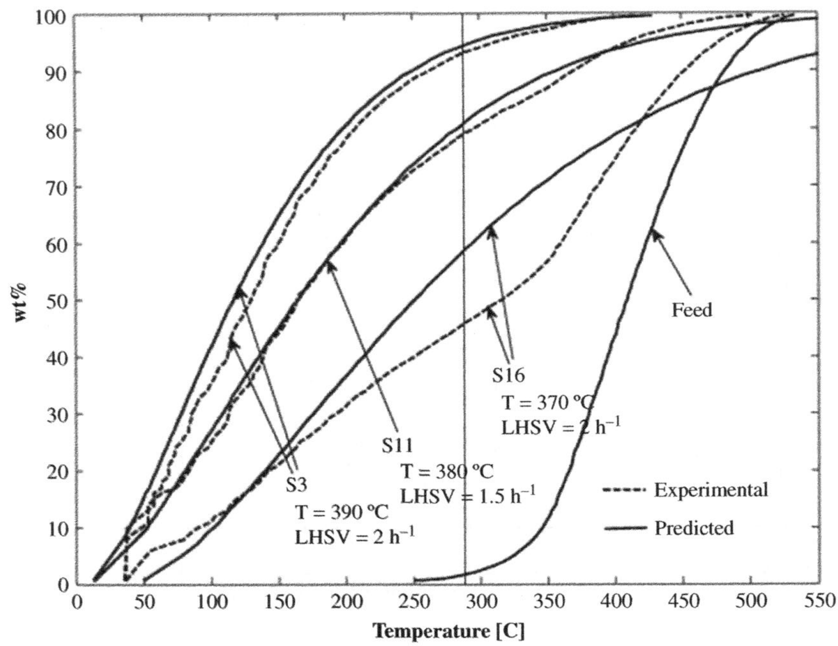

Figure 7.21 Comparing experimental and predicted distillation data for three samples based on the proposed model by Lababidi et al. (2011) and Based on Distribution Model by Riazi (1989, 1997). *Source*: Lababidi and Riazi (2011)/with permission of Elsevier.

Delayed coking unit has a prominent place in a refinery. Delayed coking processes residues from vacuum distillation and FCC decant oil to produce liquids for further processing and more middle distillates, gasoline, and kerosene. Hydrotreatment processes are needed to remove heteroatoms, as it is discussed in Section 7.2.3.1.2. A general schematic of the delayed coking unit is shown in Figure 7.22. The feed is introduced in a distillation column, and the bottom product of the fractionator is mixed with steam and after heating to about 485 °C at a pressure of 2.5 barg, it is introduced to one of the coke drums (drum A), where the coking takes place during filling of the drum for a period of 16–18 hrs. When drum A is being filled, drum B is decoked. Similarly, when drum B is filled, drum A is decoked. The drums produce coke, which is sent back to the fractionator (Kaiser et al., 2019; Eser and Guittian, 2013). The off-gases from FCC and delayed coking units are treated with amine solution to remove H_2S and CO_2. Recycle ratio is about 3–5%, and it is variable to control the endpoint of coker heavy gas oil. Humidity removal helps to avoid freezing in cold processes. There are some other

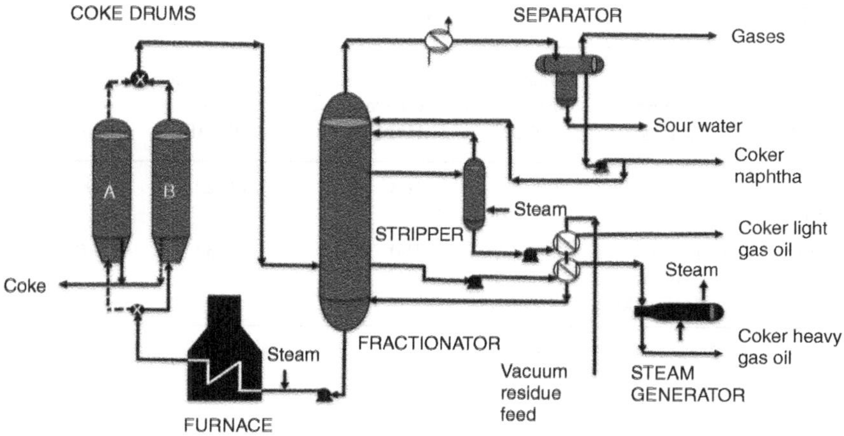

Figure 7.22 A schematic of delayed coking process. *Source*: Eser and Guittian (2013)/ ASTM International.

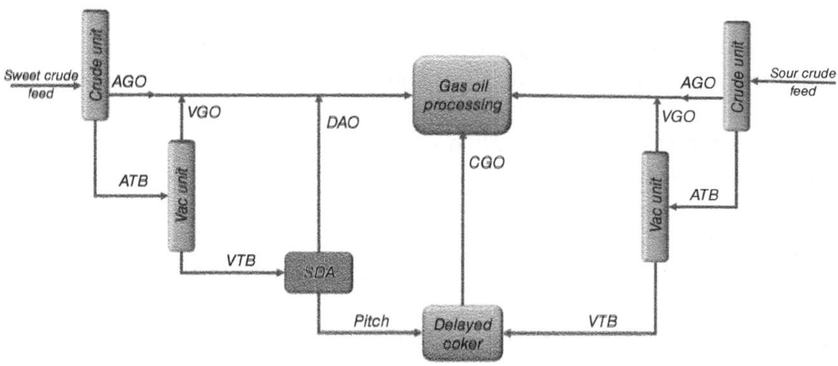

Figure 7.23 Integration of SDA and delayed coking units in upgrading heavy oil. *Source*: Romero (2020)/BECHT.

contaminants from FCC units such as nitriles, mercury, oxygen, ammonia, and nitrogen oxides. These contaminants need to be removed, as they cause catalyst deactivation in subsequent processes. Integration of delayed coking and deasphalting units in the refinery configuration is shown in Figure 7.23. Residue from vacuum tower (VTB) and pitch from SDA unit are the feeds to a delayed coker unit.

Table 7.7 shows the impact of type of heavy oil and the upgrading process on two Venezuelan heavy oil samples. Combination of delayed coking and hydrocracking processes is suitable for upgrading extra heavy oils such as Hamaca oil as shown in this table.

Table 7.7 Upgrading of heavy and extra heavy oils from Venezuela and Coke Yield.

Upgrader	Upgrading processes	Heavy oil production, Mbbl/d	Synthetic crude oil Production, Mbbl/d	Synthetic crude oil, °API	Coke production, Mt/d
Cerro Negro	Delayed coking	120	108	14–22	2,000
Hamaca	Delayed coking and hydrocracking	200	180	27–32	5,000

Source: Adapted from Scheffer et al. (1998).

The petroleum coke produced through the delayed coking process has reached 100 million tons/year in the world. In the United States, average produced petroleum coke from 2014 to 2023 was 46 million tons per year, in which about 41 million tons are exported (EIA, 2024). Nearly 20% of produced cokes are calcined for use in metallurgy for processing alumina titania as well as raw material in the carbon industry for producing graphite electrodes for electric-arc furnaces and nuclear graphite (Eser and Riazi, 2013).

7.2.3.2 Hydroprocessing/Hydrotreating

Seven compounds and metals (nickel and vanadium). Sulfur content of crude oils normally ranges from 0.05 to 9 wt%, nitrogen content varies from 0.05 to 3.0 wt%, oxygen from 0.05 to 2.0 wt%, and metals from 10 to 1500 ppm (Shishkova et al., 2022). The heteroatom compounds are mainly associated with heavy oil and aromatic-rich residues. The heteroatom compounds must be removed or processed as they contribute to the emission of pollutants and damage engines. Heteroatom compounds in refinery streams, such as basic nitrogen compounds, sulfur compounds, and metals, can deactivate catalysts and promote coke formation. Hydrotreating process is one of the major refinery processes that has significant impact on the environment. Figure 7.12 shows how reducing sulfur from 500 to 50 ppm can reduce hydrocarbon, CO, and NO_x pollutions. Similarly, reductions in metal and nitrogen contents significantly reduce emission of these elements in the environment. This is in addition to improvement in catalyst activity in refinery units and engine performance in automobiles caused by presence of heteroatoms.

Hydrotreating processes refer to a set of operations that remove sulfur, nitrogen, metals, and aromatics from fuels through catalytic reactions. Location of hydrotreating units within a typical refinery is shown in Figure 7.24. According to a Global Data report published in 2022, size of global market for hydrotreating units in 2021 was 60,337 mbd with expected growth of more than 2% from 2022 to

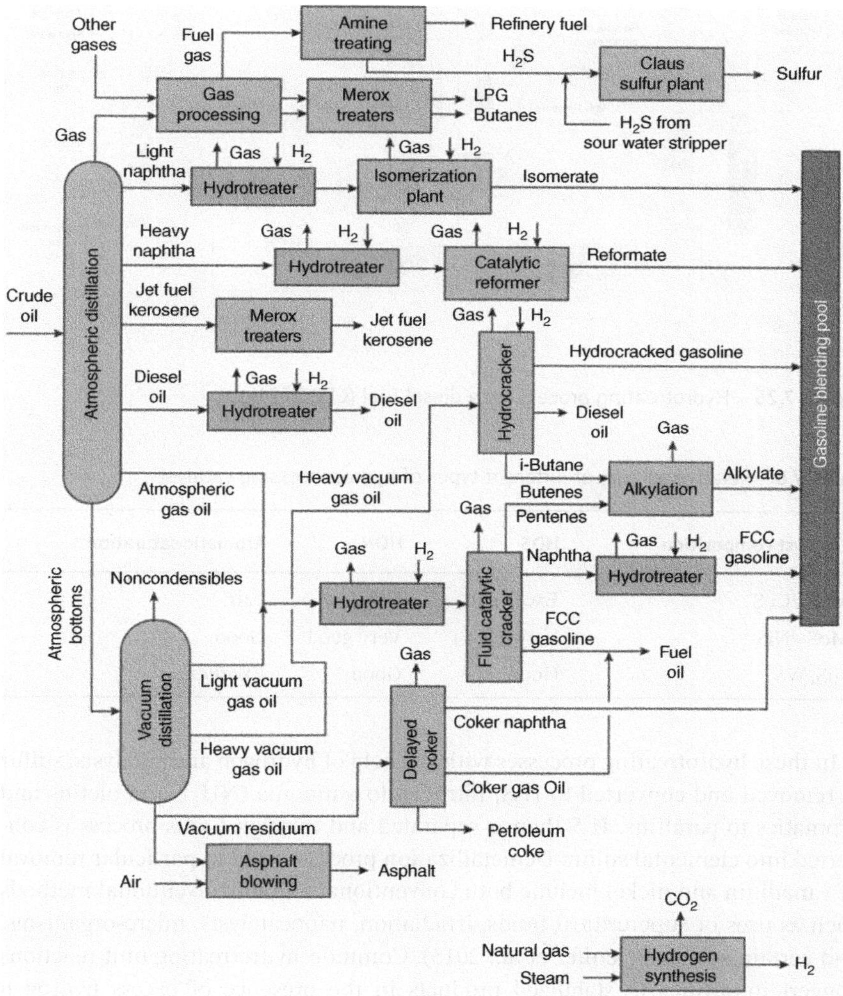

Figure 7.24 An overview of hydrotreating within a refinery. *Source*: Ortega (2021)/ American Institute of Chemical Engineers.

2026. The United States has the highest hydrotreating capacity, followed by China and Japan (GlobalData, 2022).

Middle distillate such as jet fuel or diesel fuel produced from cracking units such as FCC, hydrocracking, and coking units are rich in sulfur and heteroatoms and must go through hydrotreating processes. As an example, a configuration of hydrotreating processes for diesel fuel is shown in Figure 7.25.

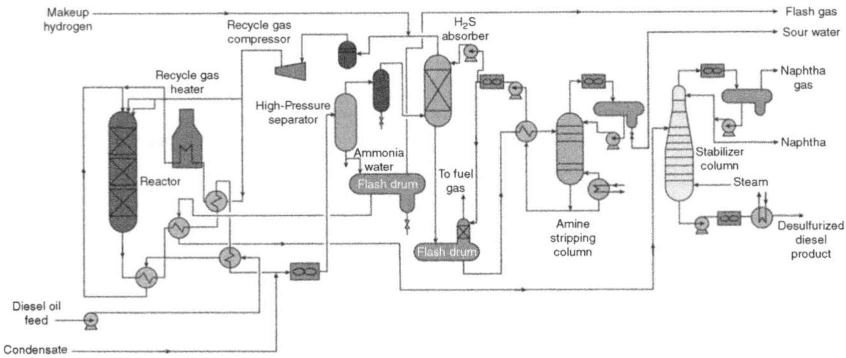

Figure 7.25 Hydrotreating process for a diesel fuel (CEP, 2021).

Table 7.8 Relative activity of different types of hydroprocessing catalysts.

Catalyst composition	HDS	HDN	Aromatic saturation
MoS_2/CoS	Excellent	Good	Fair
MoS_2/NiS	Very good	Very good	Good
NiS/WS	Good	Good	Excellent

In these hydrotreating processes with the help of hydrogen and catalysts, sulfur is removed and converted to H_2S, nitrogen to ammonia (NH_3), and olefins and aromatics to paraffins. H_2S then is separated and through Claus process is converted into elemental sulfur. Demetallization processes and in particular removal of vanadium and nickel include both conventional and unconventional methods such as uses of supercritical fluids, irradiation, nanocatalysts, micro-organisms, and certain solvents (Jenifer et al., 2015). Common hydrotreating unit reactions convert impurities to stabilized products in the presence of excess hydrogen (Ortega, 2021).

For hydrodesulfurization (HDS) processes, usually molybdenum sulfide (MoS_2) promoted with cobalt sulfide (CoS) and supported on alumina is commonly used. To saturate aromatics, WS catalyst is used, while NiS/MoS_2 catalyst is used for hydrodenitrogenation (HDN) processes. A summary of activity of these catalysts for various hydrotreating processes is given in Table 7.8 (data modified from Eser and Riazi, 2013).

A typical fixed bed residue hydrotreating configuration for hydrodemetallization (HDM) and HDS processes is shown in Figure 7.26. Typical operating conditions are pressure: 70–150 bar; temperature: 360–420 °C; LHSV: 0.1–0.5 h^{-1}; nickel

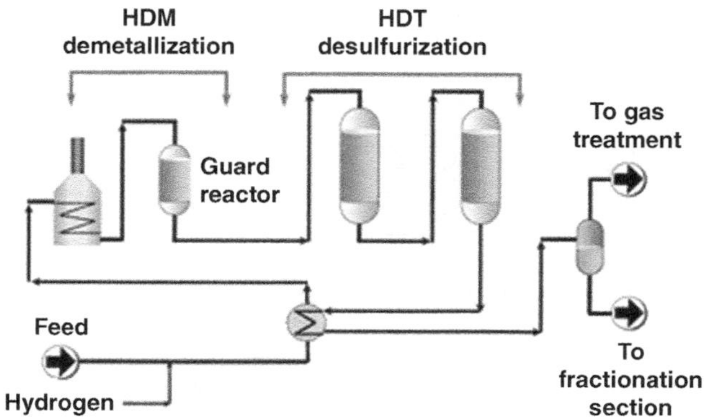

Figure 7.26 Typical fixed bed residue hydrotreating configuration employing guard reactor (Plain, 2004; Rana et al., 2013).

and Va in feed: <200–300 ppm. Conversion is generally low, but residue stability is good (Eser and Guittian, 2013). An alternative to fixed bed reactors is to use guard reactors with moving beds. In this type of reactors, catalyst is circulated by gravity with periodic addition of fresh catalyst from the top of the reactor. The advantage of this type of reactor over fixed bed reactor is its ability to process feeds with high metal content. Further discussion on the role of hydrotreating processes in the petroleum refining is presented in Chapter 8.

7.2.4 Direction of Future Refineries

A high-sulfur and heavy crude costs up to one-third less than lighter or better-quality crude oils. However, high-sulfur crudes require more processing and therefore more fixed expenses and labor. They also require more energy. Energy accounts for roughly half of the operating cost of a refinery. A refinery close to crude oil sources and a high-demand market requires lower transportation cost. A refinery can make a good profit if it is well equipped to run some of the heaviest, cheapest crudes in the world and to convert them into lightest and highest quality products. There are five drivers that have greatest effect on the petroleum industry in the coming decade: (1) environmental concerns, (2) market and demand, (3) competitive forces, (4) process improvements, (5) energy efficiency (Rana et al., 2013).

7.2.4.1 Market Demand, Cost Factors, and Capital Investment

Globally, fuel demand continues to grow slowly at a rate of less than 1% for both gasoline and diesel by the end of this decade. Figure 7.27 clearly shows the demand for gasoline is mainly from outside of OECD countries. Refineries are extending

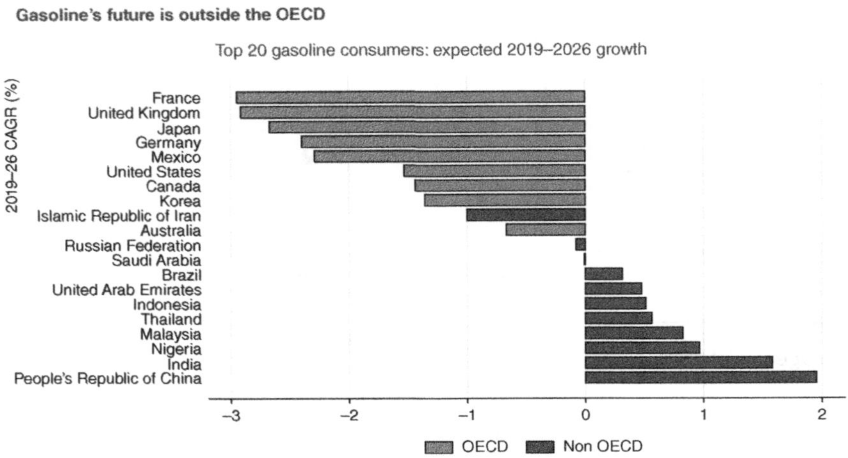

Figure 7.27 Gasoline demand by region (OSHA, 2024).

into petrochemicals to improve their project internal rate of return (IRR), as net cash margins for petrochemicals tend to be higher than those for fuels given the value of the product stream (OSHA, 2024).

Chaudhuri and Singh (2021) prepared an interesting chart on fuels and chemicals derived from crude oil with added values, as shown in Figure 7.28. Therefore, configuration of a refinery is a very important planning parameter for a refinery to survive.

Processing of heavy oil generally requires more energy resulting, in higher fuel consumption and greater carbon emission than a typical refinery processing light and medium oils. Even for desalting of a crude oil, the amount of water wash and temperature of mixing process for oils with API gravity less than 20 is more than twice of what is needed for lighter oils with API of about 40. In addition to energy cost for the processing, hydrogen consumption required for deep conversion, and hydrotreatment processes of heavy oils is an important factor and a major contributor to refinery cost and investment.

Environmental and economical aspect of petroleum refining are two important elements in sustainable operation of a refinery. Projection of price of crude oil based on 2007 prices for three growth cases is shown in Figure 7.29. According to the reference price the projected price in 2022 is about \$115, while the actual average price of crude oil in 2022 was about \$101 per barrel. Considering the impact of pandemic on the energy market this prediction for the period of 15 years is quite reasonable.

Oil refining is a capital-intensive business. Planning, designing, permitting and building a new medium-sized refinery is a five to seven years process with

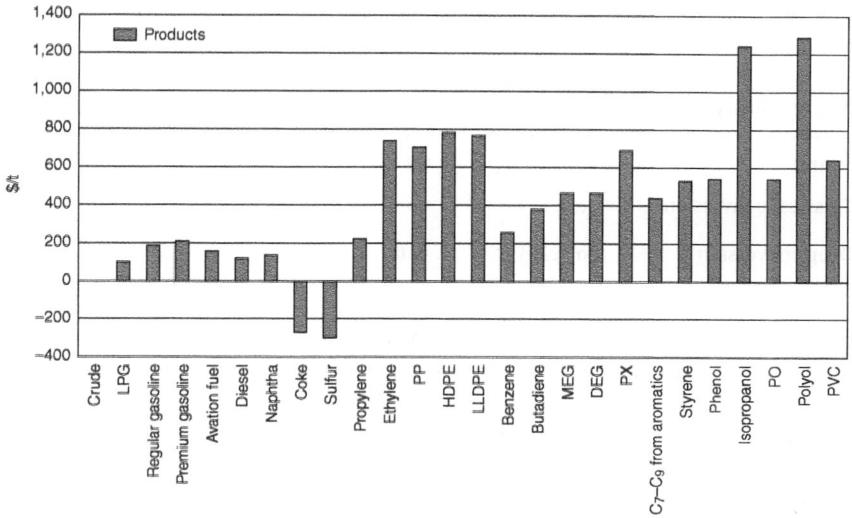

Figure 7.28 Added value on crude oil. *Source*: Chaudhuri and Singh (2021)/Hydrocarbon Processing.

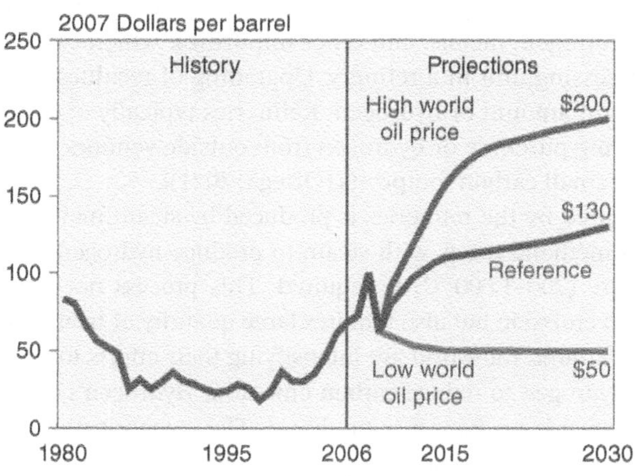

Figure 7.29 World oil prices in three economic growth cases. *Source*: IEA: International Energy Agency in 2007.

a cost of billions of dollars, as given in Table 7.9. Refinery cost is directly related to the degree of conversion of heavy compounds in residues and heavy oil into lighter compounds and fuels. The numbers in this table are not absolute, as the hydrogen and therefore values of ß will depend on operating conditions (nature of crude and sulfur content of products, etc.). Investment costs are given for a typical

Table 7.9 Investment, energy, and hydrogen consumption according to refinery complexity.

Type of refinery	Investment (billion $)	Energy consumption (wt% of crude)	Hydrogen consumption
Simple	$\alpha(\approx 5)$	≈ 4	ß
Classical conversion	$3\alpha(\approx 15)$	≈ 6	2ß
Deep conversion	$5\alpha(\approx 25)$	≈ 10	5ß

refinery of about 200,000 bbl/d capacity. Values of α are based on prices before pandemic with a benchmark of 30,000 $US/bbl of crude oil processed. These are modified values from those given in ASTM MNL58. Further discussion on the refining investment is given in Chapter 14.

7.2.4.2 Hydrogen Production

One main specification of quality of a clean fuel is its high H/C (low C/H) ratio. Fuels with high H/C ratio produce less carbon and have higher heating values. Heavy oil and residues have low H/C (high carbon content), and they are converted into lighter fuels through cracking processes. Heavy oils are also rich in heteroatom such as sulfur, nitrogen, metals, and other impurities, which can be removed through hydroprocessing unit in a refinery. Upgrading of residues and heavy oil requires a significant amount of hydrogen. Refineries typically operate at a hydrogen deficit, requiring purchase of hydrogen from outside vendors or to produce from methods with small carbon footprints (Ortega, 2021).

Majority of hydrogen required by the refineries is produced by steam methane reforming (SMR), in which methane reacts with steam to produce hydrogen and CO_2, and a high temperature (700–1,000 °C) is required. This process not only directly contributes to carbon emission but also requires large quantity of heat. For this reason, research groups around the world are intensifying their efforts to produce more clean and blue hydrogen to reduce carbon emission. Hydrogen's environmental impact mainly depends on how it is produced. The categorization of H_2 into different "colors" – such as grey, blue, green, and turquoise – reflects both the energy source and production methods used (Kollmuss et al., 2008). Most of the world's hydrogen is produced from fossil fuels, primarily through natural gas steam reforming (76%) and coal gasification (23%) (Hermesmann and Müller, 2022; IEA, 2019). This method of production is known as "grey hydrogen." Hydrogen, produced via SMR, where the carbon emissions generated during SMR process are captured and stored, rather than being released into the atmosphere, is called "blue hydrogen" (Oni et al., 2022). Hydrogen manufactured by electrolysis, where water is split into hydrogen and oxygen using electricity from renewable energy sources like wind, solar, or hydropower, is called "green hydrogen" (Newborough and Cooley, 2020). Electrolysis, which produces green hydrogen, is considered the cleanest but more costly method, as it generates no direct emissions when powered by renewable electricity (Newborough and Cooley, 2020). In short, blue hydrogen leverages carbon capture and storage technologies (CCS) to limit the carbon intensity of hydrogen production from natural gas. Green hydrogen is produced through the electrolysis of water using renewable energy (Roy et al., 2025; Tasleem and Alsharaeh, 2025). TotalEnergies and Air Products have signed a 15-year agreement for the annual supply in Europe of 70,000 tons of green hydrogen starting in 2030. This first long-term deal follows TotalEnergies' call for tenders for the supply of 500,000 tons per year of green hydrogen to decarbonize TotalEnergies' European refineries (TotalEnergies, 2024).

Partial Oxidation (POX) is an alternative method that can produce syngas from oxygen and hydrocarbons such as methane, as shown in Figure 7.30. In the POX method, pure oxygen is used (less than required stochiometric for full oxidation), followed by water–gas shift reaction (www.energy.gov).

$$CH_4 + \tfrac{1}{2}O_2 \rightarrow CO + 2H_2(+\text{heat})$$

$$CO + H_2O \rightarrow CO_2 + H_2(+\text{small amount of heat})$$

Use of nickel-based catalysts (such as Ni/CeO_2) for the POX process can reduce required temperature and consequently heat requirement. POX requires use of pure oxygen for efficient oxidation. Reactions involved are: (Amiri, 2024; Li et al., 2024a).

$$\text{Air} \rightarrow \text{Air Separation Unit (ASU)} \rightarrow O_2$$

$$O_2 + \text{Heavy Oil} + \text{Steam} \rightarrow \text{Partial Oxidation (POX)} \rightarrow \text{Syngas}$$

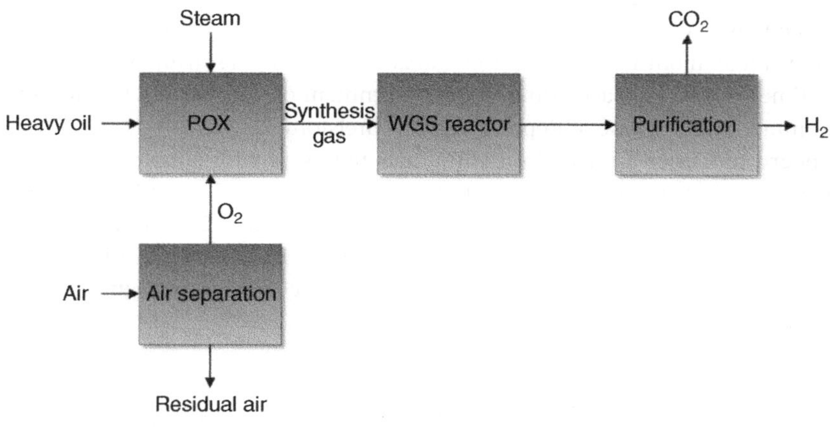

Figure 7.30 Methane production through POX method for heavy oil upgrading. *Source:* Li et al. (2024a)/MDPI/CC BY 4.0.

Most recently Li et al. (2023) suggested to decarbonize fossil fuels by recovering and sequestering solid carbon instead of gaseous CO_2. In this context, methane pyrolysis splits CH_4 directly into its components, i.e., hydrogen and carbon. Unlike other technologies involving fossil fuels, the major benefit of methane pyrolysis is the production of CO_2-free hydrogen (Hermesmann and Müller, 2022). In this method, solid carbon is produced, which has many applications in the industry. This method is attractive but not yet commercialized. Production of hydrogen from renewable sources is discussed in Chapter 9.

7.2.4.3 Next Generation Refineries

Globally, fuels demand continues to grow slowly with CAGR from 2018 to 2028 of <1% for both gasoline and diesel. Large-scale, modern refineries will be built in regions where demand for fuel exists, and they will shift the global supply and demand balance. These changing dynamics will put downward pressure on gasoline and diesel cracks in oversupplied markets and will drive down margins for refiners.

In 2022, Houston Chronicle reported that the $5.6 billion project proposed by El Campo-based Prairie Energy Partners, a subsidiary of Southern Rock Energy Partners, to build a new 250,000-barrel-per-day state-of-the-art refinery near Victoria, Texas (Drane, 2022). The refinery would process light, sweet crude into gasoline, ultra-low sulfur diesel, and bunker fuel and use carbon capture to produce blue hydrogen. In addition, renewable power generated by a 150 megawatt-per-hour plant would be used in conjunction with recycled wastewater to make green hydrogen. In comparison to traditional Gulf Coast refineries, it has reduction in CO_2, SO_x, NO_x, volatile organic compounds, and greenhouse gas emissions by at least 95% (Houston Chronicle, August 8, 2022). Recently, the Texas-based operator selected Cushing, Oklahoma, for the site of this last U.S. refinery that was planned to be built in Victoria, Texas. This decarbonized refinery is planned to be built over a 36-month period beginning in 2024 for targeted startup of operations sometime in late 2027, according to the government of Oklahoma's Department of Commerce and House of Representatives (Brelsford, 2023).

Modern refineries are based on UltraFuels philosophy for simple design that is focusing on upgrading of three liquid products: (1) Ultra Low Sulfur Gasoline (with octane number 87 regular and 93 premium gasoline), (2) Ultra Low Sulfur Diesel Fuel, and (3) Very Low Sulfur Fuel Oil. Hydrogen is required internally, and minimum blending is required. A schematic of such simple design refineries is shown in Figure 7.31.

7.2.4.4 Changing Market

In a 2021 article by Carp & Couth of Honeywell UOP (Carp and Keith, 2021), they emphasize changing market from fuels to petrochemicals. In mature markets like the United States, refineries are revisiting their business models, asset base

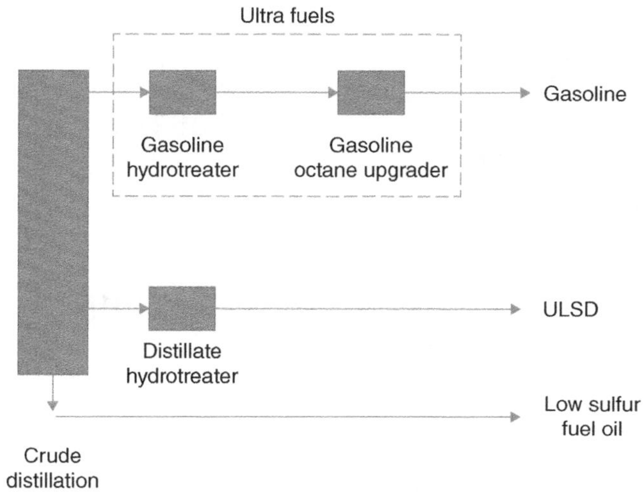

Figure 7.31 Simple design with focus on light liquid products. *Source*: Get UltraFuels/www.getultrafuels.com/accessed on July 24, 2025.

and technology options. This is due to regional and global changes in both fuels specifications and demand. It is anticipated that fuels demand will continue to soften and that petrochemical demand growth will remain strong throughout the world. The Refinery of the Future will have the flexibility to meet rapidly changing market needs and will be integrated into petrochemicals to unlock new value. The strategy utilizes the latest bottom-of-barrel, maximum conversion hydrocracking, and aromatics technologies. Integrating these technologies enables an existing refiner to profitably diversify into petrochemicals and create their own Refinery of the Future.

7.3 Shale Oil Production and Processing

7.3.1 Shale Oil Sources, Characteristics, and Classification

Shale oil is a type of unconventional source of hydrocarbons that exists in free, dissolved, or adsorbed states in any type of sedimentary rock, where some residual oil is retained after hydrocarbon generation and expulsion from the mudstones or shales (Jiang et al., 2016). According to the EIA (EIA, 2015), the estimated technically recoverable reserves of global shale oil are approximately 5.7×10^{10} t, with the United States possessing the richest shale oil resources (1.1×10^{10} t), followed by Russia (1.0×10^{10} t) and China (0.44×10^{10} t). The resource potential

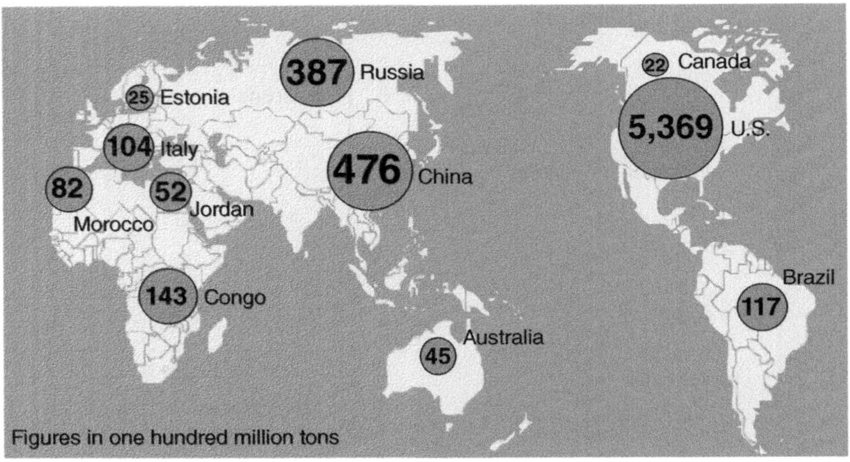

Figure 7.32 Shale oil-equivalent reserves in the world. *Source*: Yang et al. (2020)/with permission of Elsevier.

of immature and low-maturity shale oil is approximately ~3 times that of conventional petroleum resources (Li et al., 2024b), making them important energy substitution fields in the future (Zhou et al., 2018; EIA, 2021). In 2021, the production of shale oil (tight oil) in the United States reached 350 million tons, accounting for 63% of the U.S. crude oil production (Zhou et al., 2018; Wang et al., 2023a). Figure 7.32 presents shale oil-equivalent reserves in the world (Zhou et al., 2018; Yang et al., 2020).

A common characteristic of oil shale fields is the presence of insoluble organic matter called kerogen (Vandenbroucke, 2003), from which an artificial petroleum is obtained by using thermal destruction processes (Saberi and Hosseini-Barzi, 2024; Wang et al., 2023b; Wang et al., 2024; Kang et al., 2020). The kerogen component in oil shale dependent on the region and depth of exploitation was reported to vary between 3 and 81% (Kang et al., 2020; Yan et al., 2022; Bhargava et al., 2005; Solomon and Miknis, 1980). Table 7.10 presents the characteristics of some oil shales obtained from different places all over the world (Solomon and Miknis, 1980). One can see from this data that the oil shale characteristics can fluctuate in a very wide range. The oil yield, determined by using Fischer assay (Dyni, 2006), which is essentially a pyrolysis procedure, strongly correlates with the content of aliphatic carbon and total hydrogen (R = 0.98) as exemplified in the correlation matrix in Table 7.11. The data in Tables 7.10 and 7.11 exhibit that oil shale is a mixture of kerogen with a broad ambit of minerals and that logically the oil yield obtained from the pyrolysis negatively correlates with the total content of minerals.

Table 7.10 Characteristics of oil shale.

Sample	Origin	Org. C (wt %)	'Apparent' Cal (wt %)	Total H (wt %)	Hal (wt %)	Oil yield (wt %)	Kaolin (wt %)	Illite (wt %)	Quartz (wt %)	Carbonate (wt %)	Total mineral	Oil yield gal/ton
1	Colorado	26.23	21	3.53	3.25	22.4	0	10	12	48	70	59.2
2	Australia	27.6	9.4	2.4	1.53	8.4	0	20.4	30	17	67.4	21.2
3	Australia	81.44	66	10.46	10	74.9	0.9	0	0	0	0.9	200
4	Brazil	16.48	12.4	2.79	2.63	11.5	8	57	11	0	75	31.3
5	South Africa	42.72	20.1	3.45	2.35	13	16	22	11	0	49	33.3
6	So Africa	42.2	27.4	4.3	4.08	26.2	17	22	8	0	47	68.8
7	Alaska	57.48	41.96	7.91	8.52	50.5	10	19	0	0	29	136
8	Alaska	42.82	31.69	5.54	6.44	32.4	24	20	0	0	45	88.4
9	New Zealand	45.69	31.98	5.64	5.65	24.8	11	25	1.2	0	37	66.2
10	Nova Scotia	34.26	9.25	2.67	1.13	5.4	13	50	8	0	71	13.2
	min	**16.48**	**9.25**	**2.4**	**1.13**	**5.4**	**0**	**0**	**0**	**0**	**0.9**	**13.2**
	max	**81.44**	**66**	**10.46**	**10**	**74.9**	**24**	**57**	**30**	**48**	**75**	**200**

Source: Solomon and Miknis (1980)/with permission of Elsevier.

Table 7.11 Correlation matrix of characteristics of oil shale from Table 7.10.

	Org. C	Cal	Total H	Hal	Oil yield	Kaolin	Illite	Quartz	Carbonate	Total mineral	Oil yield gal/ton
Org. C	1										
Cal	0.93	1									
Total H	0.93	**0.99**	1								
Hal	0.85	**0.96**	**0.98**	1							
Oil yield	0.89	**0.98**	**0.98**	**0.96**	1						
Kaolin	−0.01	−0.09	−0.09	−0.01	−0.16	1					
Illite	−0.65	−0.68	−0.62	−0.60	−0.67	0.27	1				
Quartz	−0.62	−0.68	−0.70	−0.73	−0.63	−0.44	0.15	1			
Carbonate	−0.39	−0.25	−0.29	−0.27	−0.18	−0.58	−0.32	0.44	1		
Total mineral	**−0.97**	**−0.96**	**−0.95**	−0.91	−0.91	−0.01	0.66	0.64	0.41	1	
Oil yield gal/ton	0.89	**0.98**	**0.98**	**0.96**	1.00	−0.16	−0.67	−0.64	−0.18	−0.91	1

Note: The numbers in bold mean the presence of strong relations ($R \geq 0.95$)

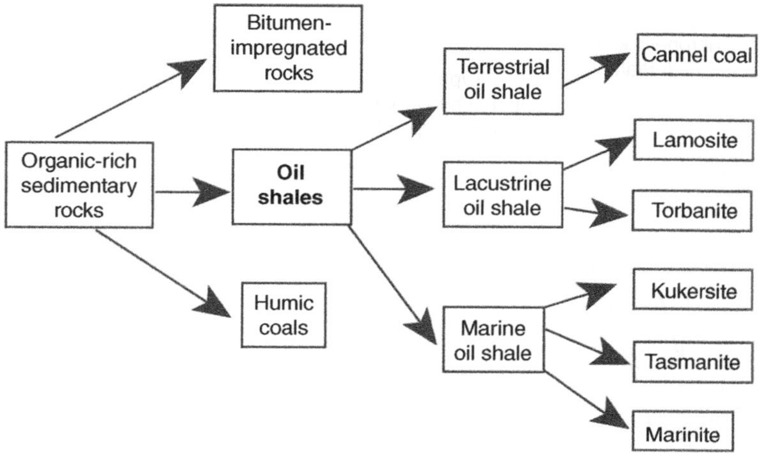

Figure 7.33 Classification of oil shales. *Source*: Adapted from Hutton (1987).

Oil shale has been formed in terrestrial, lacustrine, and marine environments. It can be presented as one of three general classes of organic-rich sedimentary rocks: (1) humic coal and carbonaceous shale, (2) bitumen-impregnated rock, and (3) oil shale and partitioned into three groups based upon their environments of deposition – terrestrial, lacustrine, and marine (Figure 7.33) (Hutton, 1991). The kerogens from lacustrine, marine, and terrestrial oil shales can be classified as types I, II, and III, respectively, based on the kerogen classification established by researchers from the French Petroleum Institute (Vandenbroucke, 2003).

Kerogen type I is highly aliphatic with atomic H/C ratio > 1.5, and O/C ratio between 0.03 and 0.1 (Vandenbroucke, 2003). The pyrolysis oils obtained from kerogen type I are very rich in long chain *n*-alkanes, extending over 40 carbon atoms (Vandenbroucke, 2003). The source rocks containing kerogen type I are not prevalent, and they make only 2.7% of the world's original reserves in petroleum and gas (Klemme and Ulmishek, 1991).

Kerogen type II features with atomic H/C ratio ~1.3 and O/C ratio of about 0.15. It distinguishes with a higher concentration of cyclic aliphatic structures than type I and contains more aromatics, whose structure involves various aliphatic cycles. Kerogen type II pyrolysis oils contain n-alkanes with a carbon number less than 25 (Vandenbroucke, 2003).

Kerogen type III has H/C ratio lower than 0.8 and O/C ratio between 0.05 and 0.30 (see Figure 7.34).

As obvious from Equation 7.11, developed by using the data in Table 7.10, the different oil shales can generate distinct oil yield dependent on their H/C ratio and inorganic content.

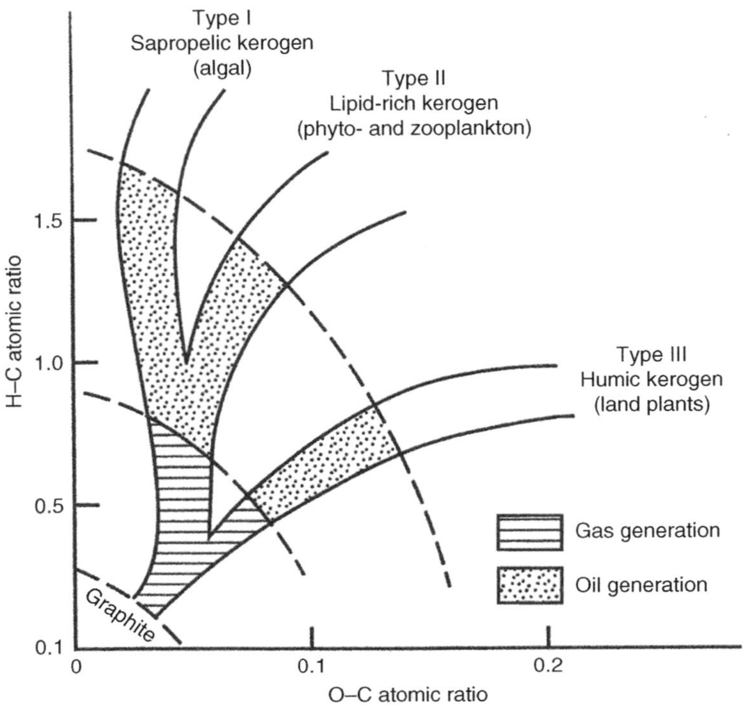

Figure 7.34 Determination of kerogen types and pathways of their development. *Source*: Speight (2012a)/with permission of Elsevier.

$$\text{Oil yield} = 44.4 + 16.19 \times \frac{H}{C} - 0.82 \times \text{Total minerals} \quad R = 0.95 \tag{7.11}$$

where,

Oil yield = Pyrolysis yield of oil from oil shale, wt%;

$\frac{H}{C}$ = Atomic H/C ratio of oil shale, determined by 13C cross polarization n.m.r. technique:

Total minerals = Content of minerals in oil shale, wt%.

7.3.2 Kerogen Type and Production of Shale Oil

Based on Equation 7.11 and the H/C ratio of the different kerogen types one may deduce that at the same mineral content, the oil yield will follow the order: Kerogen I > Kerogen II > Kerogen III. The reported higher yields of aromatic hydrocarbons from the pyrolysis of continental source rocks relative to those

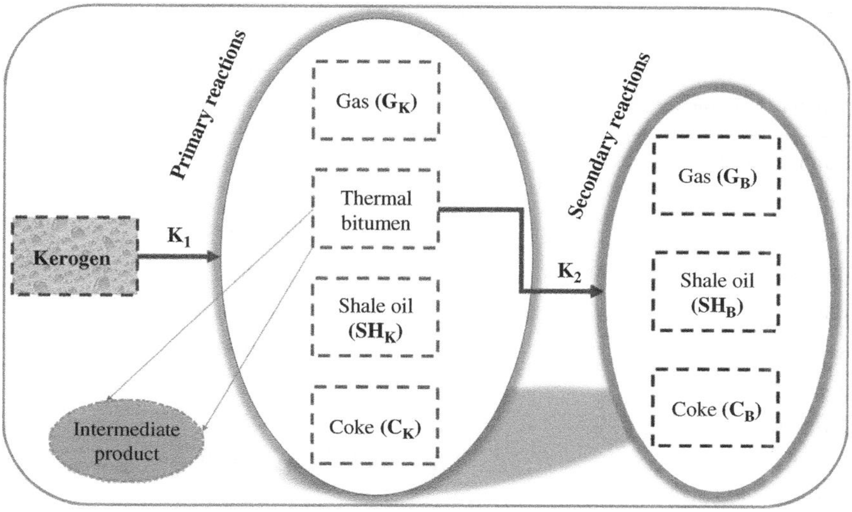

Figure 7.35 Reaction network of oil shale retorting. *Source*: Abdi-Khanghah et al. (2024)/with permission of Elsevier/CC BY 4.0.

of marine source rocks reported in Vandenbroucke, (2003) are in line with the lower H/C ratio of the terrestrial kerogen type III in contrast to the marine kerogen type II. Therefore, the aromaticity of the oils produced from pyrolysis of the diverse kerogen types should be opposite to the order of kerogen oil yield shown above.

The kerogen in oil shale can be transformed into liquid shale oil only by the process of pyrolysis (retorting) (Kang et al., 2020). Figure 7.35 exemplifies the oil shale pyrolysis reaction network with bitumen considered as the intermediate component. As apparent from Figure 7.35, the kerogen is thermally converted to bitumen, gas, shale oil, and coke, and the intermediate bitumen to gas, shale oil, and coke.

Oil shale retorting process can be classified as conventional ex situ (aboveground) and in situ (underground) (Kang et al., 2020; Chian et al., 2016). Details about both technologies can be found in Kang et al. 2020), Chian et al., (2016), Crawford and Killen, 2010; Felix et al., 2024). The properties of derived by retorting process shale oil are dependent on oil shale from which it was obtained and on the retorting method (Shekarifard and Taheri-Shakib, 2016; Järvik et al., 2021; Jin et al., 2012). In contrast to the conventional petroleum, shale oil is heavy, viscous, and possesses a higher amount of organic compounds containing heteroatoms (S, N, O, etc.). The heteroatom-containing compounds typically bring about troubles when shale oil is used, related to fuel instability during transportation and storage, increased catalyst deactivation rate in downstream treatment, and

Table 7.12 Ambit of property fluctuation of shale oils derived from various oil shales and using distinct retorting methods.

Content of oil fractions in shale oil, vol%	Min	Max
Naphtha (IBP-200 °C)	2	50
Kerosene (200–250 °C)	5	18
Diesel (250–350 °C)	26	42
Atmospheric residue (>350 °C)	30	54
Elemental analysis, wt%		
Sulfur	0.41	1.31
Nitrogen	0.3	2.19
Hydrogen	9.2	12.2
Carbon	79.50	85.39
Oxygen	0.7	7.6
H/C atomic ratio	1.3	1.8
Specific gravity	0.879	1.01
Content of olefins, wt%	2	44

fouling and pollutant emission in utilization (Zhang et al., 2017). The relatively high content of unsaturated hydrocarbons is another reason for the instability of shale oils. Table 7.12 summarizes the range of variation of properties of shale oils obtained from diverse oil shales and using different retorting methods (Järvik et al., 2021; Landau et al., 1998; Chen et al., 2012; Guo, 2009; Zhang et al., 2020; Speight, 2020).

Typically, shale oils have obnoxious odor related to their high nitrogen compound content (pyridine, quinoline, acridine, amine, and their alkylsubstituted derivatives, pyrrole, indole, carbazole and their derivatives, and the nitrile and amide homologues). They are the most problematic species in shale oils because they deactivate catalysts during upgradation, inflict appearance of instability in gasoline, jet, and diesel fuels, and produce NO_x emissions in burners. Complaints of nasty odor were the reason to interrupt the production of shale oil in Australia (Rundle oil shale) in November 1999 during commissioning of a 250×10^6 USD R&D demonstration plant (the Stuart stage 1 project) (Pope, 2000). Shale oil must therefore undergo upgrading as a pre-treatment step to reduce impurities and improve its physicochemical properties in order to be used as an acceptable refinery feedstock. Various upgrading processes such as acid neutralization, complexation, solvent extraction, and hydrotreatment have been proposed to remove N, S, and O heteroatoms from shale oil (Li et al., 2020; Chishti and

Williams, 1999; Williams and Chishti, 2001; Yu et al., 2010; Bello et al., 2020; Zhang et al., 2020; Raja et al., 2017a; Wang et al., 2017; Raja et al., 2017b; Landau, 1997; Johannes et al., 2006; Chen et al., 2015). Shale oil can be also used as a feedstock for production of value-added chemicals or specialty chemicals (e.g., pyridine base and rust-preventing oil) (Pope, 2000; Chen et al., 2015). Similar to conventional oil refining, the most mature and researched technology for upgrading shale oil is hydrotreating (Zhang et al., 2017; Landau et al., 1998; Chen et al., 2012; Guo, 2009; Zhang et al., 2020; Chishti and Williams, 1999; Williams and Chishti, 2001; Yu et al., 2010; Bello et al., 2020; Zhang et al., 2020; Raja et al., 2017b; Wang et al., 2017; Johannes et al., 2006; Chen et al., 2015; Gao et al., 2019; Bello et al., 2021; Chishti and Williams, 1990; Benyamna et al., 1991; Zhao et al., 2007; Su et al., 2008; Yoshida et al., 1996; Luik et al., 1999a; Luik et al., 1999b; Luik et al., 1999c). The hydrotreating is a basic refinery process that is utilized to refine various streams such as naphtha, middle distillates, vacuum gas oils, vacuum residues, and more recently crude oils. It is employed to decrease the level of oil impurities such as sulfur, nitrogen, oxygen, metals, and saturation of aromatic rings and olefins. Reactions of HDS, HDN, hydrodeoxygenation (HDO), hydrodearomatization (HDA), and HDM all take place simultaneously in the hydrotreatment process (Gao et al., 2019). It is carried out under many operation variables, such as temperature, pressure, space velocity, and H_2/oil ratio, in the presence of a catalyst that ensures the functionality of hydrorefining. Every variable affects the process in a specific way, but the properties of feed oil and the required quality of the product assign the severity to which they are applied (Bello et al., 2021). The higher level of impurities, especially nitrogen, makes the hydrotreatment of shale oil to run at higher severity (Li et al., 2020; Benyamna et al., 1991; Zhao et al., 2007; Su et al., 2008; Yoshida et al., 1996; Luik et al., 1999a, 1999b, 1999c). While the level of nitrogen varies between 0.1 and 0.9 wt% in petroleum (Bello et al., 2021), that in shale oil, as evident from the data in Table 7.12, can reach 2.2 wt%. In addition to contributing to NO_x emissions during oil combustion, nitrogen compounds in oil inhibit other reactions and poison catalysts during oil refining. Catalyst poisoning could be reversible or irreversible. The irreversibility comes from the low desorption rate of nitrogen compounds on active sites, indicating their stability and rigidity with respect to the reactions. They also present the main obstacle to attain ultralow sulfur level by competing for active catalytic sites with sulfur species (Bello et al., 2021). Ni–Mo catalysts exhibit superior performance over Co–Mo ones when shale oil is hydrotreated (Zhang et al., 2017). By using positive-ion mode APPI FT-ICR MS, it is found that N_1 species decrease after hydrotreatment but are still the dominant N compound. The main N_1 species are indole, carbazole, acridine, and their derivatives, and all other N compounds are very few. In hydrotreated shale oils, the content of S_1 species is very low, and that of S_2 species is close to zero. For aromatic hydrocarbons of mono- and double-ring aromatics, they still have high relative abundances, especially the mono-ring aromatics. Due to the high relative abundances of N_1 and N_1O_1 species in shale oil, the transformation between N_1O_1 and N_1 species allows a better understanding of HDN for shale oil. The relative abundances of aromatic hydrocarbons are increased after hydrotreatment, a consequence of the transformation of N_1, N_1O_1, and S_1 species into aromatic hydrocarbons through hydrotreatment. The Ni–Mo catalyst demonstrates the best performance in such a transformation (Zhang et al., 2017).

7.3.3 Shale Oil Refining Technologies

Shale oils can be also processed in the FCC (Chen et al., 2015; García et al., 2022; de Graaf et al., 2015; Sullivan and Stangeland, 1978), but hydrotreatment is recommended as a preliminary refining step to decrease the level of heteroatom content of shale oil and then be cracked in the FCC unit. The behavior of hydrotreated shale oil in the FCC plant is reported to be similar to that of vacuum gas oils derived from conventional petroleum feedstocks (Chen et al., 2015; García et al., 2022; Sullivan and Stangeland, 1978).

Sullivan and Stangeland (1978) investigated three distinct processing routes for shale oil: (1) hydrotreating followed by hydrocracking; (2) hydrotreating followed by FCC; and (3) delayed coking followed by hydrotreating. Figures 7.36–7.38 present simplified flow diagrams of the three routes to process shale oil. Table 7.13 summarizes properties of the shale oil investigated by Sullivan and Stangeland

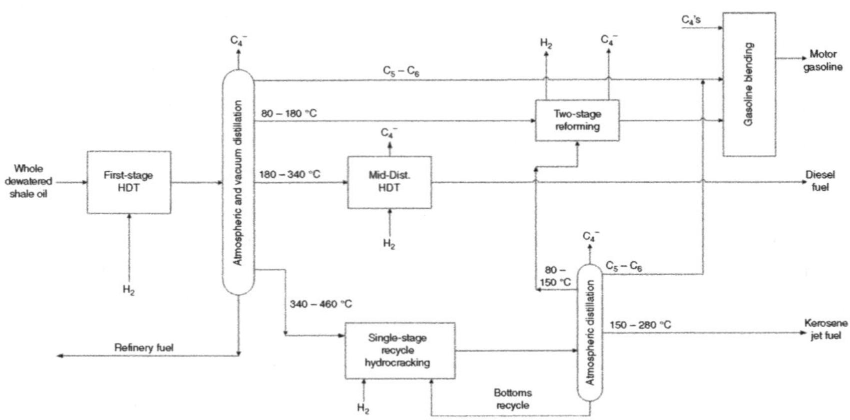

Figure 7.36 Simplified flow diagram of shale oil refining by hydrotreating followed by hydrocracking (Case 1). Notes: 1. HDT, hydrotreating; 2. C_4^- is butanes and lighter used for H_2 plant feed, gasoline blending, and refinery fuel; 3. Foul gas and water treated to recover NH_3 and sulfur; 4. Some naphtha from first-stage HDT also used for H_2 plant feed. *Source*: Sullivan and Stangeland (1978)/U.S. Department of Energy/Public Domain.

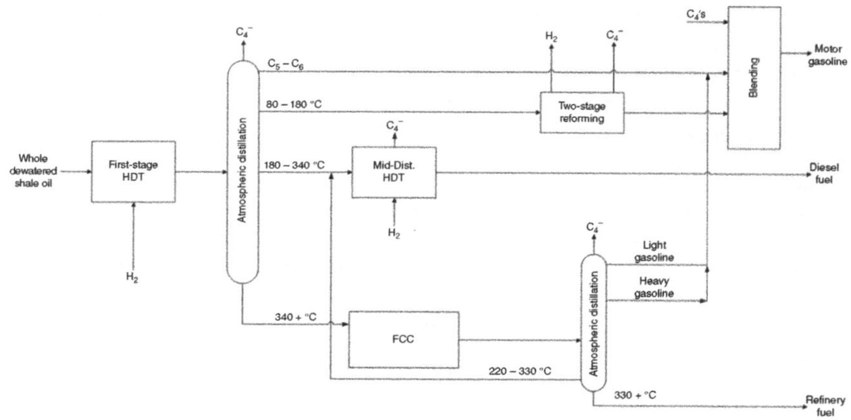

Figure 7.37 Simplified flow diagram of shale oil refining by hydrotreating followed by fluid catalytic cracking (Case 2). Notes: 1. C_4^- is butanes and lighter used for H_2 plant feed, gasoline blending, and refinery fuel; 2. Foul gas and water treated to recover NH_3 and sulfur; 3. HDT, hydrotreating; FCC, fluid catalytic cracking. *Source*: Sullivan and Stangeland (1978)/U.S. Department of Energy/Public Domain.

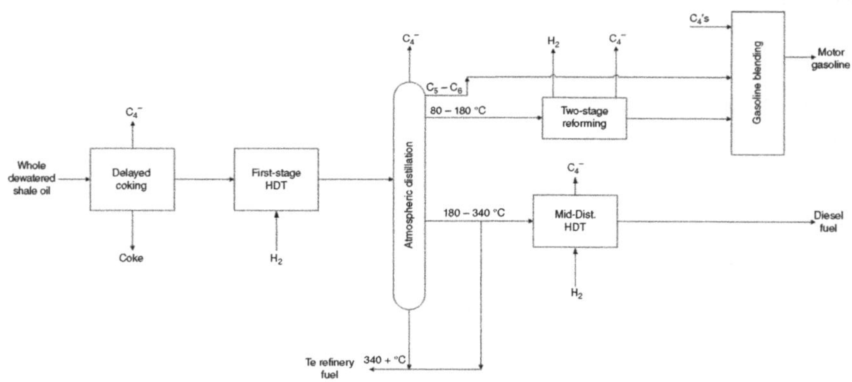

Figure 7.38 Simplified flow diagram of shale oil refining by delayed coking followed by hydrotreating (Case 3). Notes: 1. C_4^- is butanes and lighter used for H_2 plant feed, gasoline blending, refinery fuel, and liquid petroleum gas (LPG); 2. Foul gas and water treated to recover NH_3 and sulfur; 3. HDT, hydrotreating. *Source*: Sullivan and Stangeland 1978/U.S. Department of Energy/Public Domain.

(1978) along with the product yields obtained by their processing by the three routes mentioned above. It is evident from this data set that the maximum amount of finished salable transportation fuels is obtained by route 2, followed by route 1, and the least quantity of transportation fuels is produced by route 3. The chemical consumption of hydrogen was about 50 Nm^3/m^3.

Table 7.13 Shale oil properties and product yields derived from the three processing routes: (1) hydrotreating followed by hydrocracking; (2) hydrotreating followed by FCC; and (3) delayed coking followed by hydrotreating.

Properties of shale oil, and yields of products for the three routes of processing of shale oil		**Routes for shale oil processing**		
Specific gravity 15 °C/15 °C	0.933	Hydrotreatment followed by hydrocracking	Hydrotreatment followed by FCC	Delayed coking followed by hydrotreatment
Pour point, °C	32			
Ash content, wt%	0.03			
Sediment, vol%	0.1			
Average molecular weight, g/mol	326			
Kinematic viscosity at 50 °C, mm^2/s	25.45			
Kinematic viscosity at 99 °C, mm^2/s	5.54			
Carbon, wt%	84.3			
Hydrogen, wt%	11.29			
Sulfur, wt%	0.66			
Total nitrogen, wt%	2.18			
Oxygen, wt%	1.16			
Arsenic, ppm	28			
Iron, ppm	70			
C_7 – asphaltenes, wt%	0.17			
Ramsbottom carbon, wt%	1.85			
TBP distillation, %/°C				
Initial boiling point	84			
10	234			
50	391			
70	450			
95	567			

Properties of shale oil, and yields of products for the three routes of processing of shale oil	Routes for shale oil processing		
Product liquid volumetric yields, vol%			
LPG	0	0	2
Motor gasoline	17	38	17
Jet fuel	20	0	0
Diesel fuel	54	57	59
Total transportation fuels	91	95	78

Source: Adapted from Sullivan and Stangeland (1978).

Due to the very high nitrogen content of the investigated shale oil (2.2 wt%), the hydrotreatment step operated at high severity (hydrogen pressure of 12.7 MPa; liquid hourly space velocity of 0.6 h^{-1}; weight average bed temperature in the range of 407–418 °C; hydrogen to oil ratio of 199. HDN was found to obey first-order reaction kinetics with activation energy of 35 Kcal/mol (Sullivan and Stangeland, 1978).

The selection of processing scheme of shale oil will depend on product yields and flexibility desired.

7.3.4 Environmental Impact of Shale Oil

In general, shale oil production and refining have diverse impacts on the surrounding environment, generating various pollutants (Yang et al., 2020; Jia and Su, 2023; Speight, 2012). It affects greenhouse gas output, water consumption and pollution, surface disturbance, and disposal of solid waste (Speight, 2012a). Figure 7.39 (Yang et al., 2020) indicates that in comparison with the oil refining, the process of oil shale retorting (pyrolysis) and subsequent shale oil refining (shale to liquid (STL) conversion) gives 4.7 times higher CO_2-equivalent emissions per GJ (CO_2-eq/GJ). The combustion of coke gas obtained from the oil shale retorting used to provide the required heat for the oil shale pyrolysis reaction along with low fuel yields is the reason for the larger CO_2-eq/GJ fuel values.

Strict groundwater quality monitoring programs are needed to prevent water pollution. Key pollutants must be regularly measured to guarantee compliance with maximum acceptable limits. The best available practices must be applied to minimize the environmental impact of oil shale extraction, retorting, and refining.

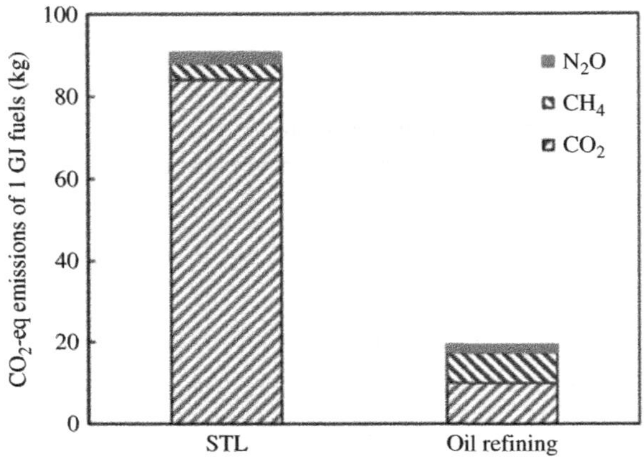

Figure 7.39 CO_2-equivalent emissions per GJ liquid fuel for STL and oil refining processes. *Source*: Yang et al. (2020)/with permission of Elsevier.

In addition, a comprehensive environmental impact assessment must be carried out before any oil shale operations are undertaken. These assessments should evaluate potential contamination risks and outline appropriate preventive measures to be implemented throughout the life cycle of the project (Jia and Su, 2023).

The management and disposal of solid waste, especially the rock that remains after shale oil extraction, are the most serious environmental problems in the process of conversion of shale to liquids (Speight, 2012a). The product generated in the process of retorting significant quantities of spent shale wastes, containing soluble salts, organic compounds, trace concentrations of various heavy metals, carcinogens, mutagens, and other hazardous substances, must be protected from leaching by snowmelt, rainfall, and groundwater, because leached salts and toxins can contaminate both aquifers and surface streams (Speight, 2012b). Before starting operations of conversion of shale to liquid, the developers are advised to turn to the lessons learned by those who have rich experience in this field of human activity such as the development of Estonian oil shale tracts and Alberta tar sand development that has been commercialized since 1967 (Speight, 2012b).

References

Abdi-Khanghah, M., Wu, K.W., Soleimani, A. et al. (2024). Experimental investigation, non-isothermal kinetic study and optimization of oil shale pyrolysis using two-step reaction network: maximization of shale oil and shale gas production. *Fuel* 371: 131828.

Alboudwarej, H., Beck, J., Svrcek, W.Y. et al. (2002). Sensitivity of asphaltene properties to separation techniques. *Energy & Fuels* 16 (2): 462–469.

Amiri, J. (2024). Fyfe, hydrogen research. https://fyfe.com.au/sectors/hydrogen/ (accessed 13 December 2024).

Bello, S.S., Wang, C., Zhang, M. et al. (2020). Recycling the CoMo/Al2O3 catalyst for effectively hydro-upgrading shale oil with high sulfur content and viscosity. *RSC Advances* 10 (61): 37287–37298.

Bello, S.S., Wang, C., Zhang, M. et al. (2021). A review on the reaction mechanism of hydrodesulfurization and hydrodenitrogenation in heavy oil upgrading fuel. *Energy & Fuels* 35 (14): 10998–11016. https://doi.org/10.1021/acs.energyfuels.1c01015 (accessed 9 July 2024).

Benyamna, A., Bennouna, C., Moreau, C. et al. (1991). Upgrading of distillate fractions of Timahdit Moroccan shale oil over a sulphided NiO-MoO_3/γ-Al_2O_3 catalyst. *Fuel* 70: 845–848.

Bhargava, S., Awaja, F., and Subasinghe, N.D. (2005). Characterisation of some Australian oil shale using thermal. X-ray and IR techniques. *Fuel* 84: 707–715.

Bhutada, G. (2023). How big is the market for crude oil? Visual Capitalist. https://www.visualcapitalist.com

Brelsford, R. (2023). Texas operator selects cushing for planned decarbonized refinery. *OGJ* (May 30).

Carp, S. and Keith, C. (2021). Honeywell UOP, "Advances in crude and chemicals". https://cdn.digitalrefining.com/data/articles/file/1002610-honeyweGBPGBP-apr-21-white.pdf (accessed 11 July 2024).

Chaudhuri, S. and Singh, R. (2021). Fuels and chemicals: finding the right refinery configuration for a less predictable world. *Hydrocarbon Processing* 3: 23–32.

Chen, X., Li, N., Yang, Y. et al. (2015). Novel propylene production route: utilizing hydrotreated shale oil as feedstock via two-stage riser catalytic cracking. *Energy & Fuels* 29 (11): 7190–7195.

Chen, X., Shen, B., Sun, J. et al. (2012). Characterization and comparison of nitrogen compounds in hydrotreated and untreated shale oil by electrospray ionization (ESI) Fourier transform ion cyclotron resonance mass spectrometry (FT-ICR MS). *Energy & Fuels* 26 (3): 1707–1714.

Chian, S.S.L., Lim, S.K., Low, S.Y. et al. (2016). A review of shale oil production methodologies and its impact to global energy and economy. *Research Journal of Applied Sciences, Engineering and Technology* 13 (7): 555–568.

Chishti, H.M. and Williams, P.T. (1990). Aromatic and hetero-aromatic compositional changes during catalytic hydrotreatment of shale oil. *Fuel* 78 (15): 1805–1815.

Chishti, H.M. and Williams, P.T. (1999). Aromatic and hetero-aromatic compositional changes during catalytic hydrotreatment of shale oil. *Fuel* 78 (15): 1805–1815. https://doi.org/10.1016/S0016-2361(99)00089 (accessed 11 July 2024).

Crawford, P.M. and Killen, J.C. (2010). New challenges and directions in oil shale development technologies. In: *Oil Shale: A Solution to the Liquid Fuel Dilemma* (ed. O.I. Ogunsola, A.M. Hartstein, and O. Ogunsola), 21–60. American Chemical Society.

da Silva, M.W. (2022). *Crude Oil Refining: A Simplified Approach*. CRC Press.

de Graaf, B., Radcliffe, C., Martin Evans, M. et al. (2015). Processing shale feedstocks, 1–6. www.digitalrefining.com/article/1001071 (accessed 11 July 2024).

DNV-GL (2017). Oil and Gas Forecast to 2050, Energy Transition Outlook 2017. https://www.ourenergypolicy.org/wp-content/uploads/2017/09/DNV-GL_Energy-Transistion-Outlook-2017_oil-gas_lowres-single_3108_3.pdf (accessed 28 June 2024).

Drane, A. (2022). Southern rock energy proposes a $5.6 billion 'Next Generation' refinery outside Victoria. *Houston Chronicle*, August 8. https://www.houstonchronicle.com/business/energy/article/Southern-Rock-Energy-proposes-5-6B-next-17359136.php (accessed 28 June 2024).

Dyni, J.R. (2006). Geology and resources of some world oil-shale deposits. U.S. Geological Survey Scientific Investigations Report 2005–5294. 1–42. http://www.usgs.gov/sir/2006/5294 (accessed 20 September 2024).

EI (2024). Energy Institute Statistical Review of World Energy 2024 (accessed 28 June 2024).

EIA (2023). Annual coal reports, 2023. U.S. Energy Information Administration (EIA). https://www.eia.gov/coal/annual/ (accessed 28 June 2024).

EIA (2024a). Monthly Energy Review, Table 1.3 and 10.1, April 2024, Preliminary Data. U.S. Energy Information Administration (accessed 28 June 2024). U.S. energy facts explained - consumption and production - U.S. Energy Information Administration (EIA).

EIA (2024b). Short-term energy outlook. U.S. Energy Information Administration, International Energy Statistics; and International Energy Outlook. https://www.eia.gov/analysis/globalrefining/outlookglobalrefining.pdf (accessed 28 January 2025).

EIA (2024c). Most US petroleum coke is exported. US Energy Information Administration. https://www.eia.gov/todayinenergy/detail.php?id=63705 (accessed 14 November 2024).

EIA (2021). International Energy Outlook 2021. https://www.eia.gov/outlooks/ieo/ (accessed 28 June 2024).

EIA (2015). World shale resource assessments. https://www.eia.gov/analysis/studies/worldshalegas/ (accessed 23 February 2025).

IEA (2024a). Oil market report – April 2024-analysis. www.iea.org/reports/oil-market-report-april-2024 (accessed 23 February 2025).

IEA (2024b). Heavy oil market. https://www.iea.org/reports/oil-market-report-may-2024 (accessed 23 February 2025).

IEA (2024c). IEA oil market report. https://www.iea.org/reports/oil-market-report-may-2024 (accessed 23 February 2025).

IEA (2023). *World Energy Outlook*, International Energy Agency (IEA). https://iea.blob.core.windows.net/assets/86ede39e-4436-42d7-ba2a-edf61467e070/WorldEnergyOutlook2023.pdf (accessed 20 June 2024).

IEA (2019). The future of hydrogen Seizing today's opportunities. Report prepared by the IEA for the G20. Japan: International Energy Agency (IEA). https://www.oecd.org/content/dam/oecd/en/publications/reports/2019/06/the-future-of-hydrogen_2d59d5dd/1e0514c4-en.pdf (accessed 20 March 2025).

Eser, S. and Guitian, J. (2013). Heavy-oil processing. In: *Petroleum Refining and Natural Gas Processing* (ed. M.R. Riazi, S. Eser, S.S. Agrawal et al.), 177–195. ASTM International: West Conshohocken, PA.

Eser, S. and Riazi, M.R. (2013). Crude oil refining processes. In: *Petroleum Refining and Natural Gas Processing* (ed. M.R. Riazi, S. Eser, S. Agrawal, and J.L. Peña), 103–104. West Conshohocken, PA: ASTM International.

Felix, G., Djimasbe, R., Varfolomeev, M.A. et al. (2024). 4-lump kinetic model for non-catalytic oil shale upgrading at sub- and supercritical water conditions. *Fuel* 357: 129987.

Ferrari, A. (2024). The unavoidable truth: the world's growing need for oil. *Journal of Petroleum Technology* 76 (06): 10–11.

Gao, S., Han, D.Y., Hou, D.D. et al. (2019). Utilization of phenolic compounds in light fraction of Fushun shale oil. *Petrol Proc Prtrochemicals* 50: 53–57 (In Chinese).

García, J.R., Fals, J., Dietta, L.E. et al. (2022). VGO from shale oil. FCC processability and co-processing with conventional VGO. *Fuel* 328: 125327.

GlobalNewsWire (2024). Global refinery fluid catalytic cracking units (FCCU) industry outlook 2024-2028: capacity and capital expenditure outlook with details of all operating and planned fluid catalytic cracking units. (GlobalNewsWire, 2024). www.globenewswire.com/news-release/2024/07/10/2911231/28124/en/Global-Refinery-Fluid-Catalytic-Cracking-Units-FCCU-Industry-Outlook-2024-2028-Capacity-and-Capital-Expenditure-Outlook-with-Details-of-All-Operating-and-Planned-Fluid-Catalytic-Cr.html (accessed 20 June 2024).

Gillis, D.B. and VanTine, F.M. (2003). UOP/FW USA solvent deasphalting process. *Handbook of Petroleum Refining Processes*, 3. UOP/FWUSA solvent deasphalting process. USA: UOP LLC a Honeywell Co-Des Plaines. www.uop.com, UOP Home (honeywell.com)

GlobalData (2022). Refinery hydrotreater units capacity and capital expenditure outlook with details of all operating and planned units, 2022-2026. www.globaldata.com (July 14).

Gomaa, S., Salem, K.G., and El-hoshoudy, A.N. (2023). Recovery of heavy oil and extra heavy oil; current status, new trends, and enhancement techniques. *Petroleum* 10: 15–25.

Goosens, A.G. (1996). Prediction of molecular weight of petroleum fractions. *Industrial & Engineering Chemistry Research* 35 (3): 985–988.

Guo, S.H. (2009). The chemistry of shale oil and its refining. In: *Coal, Oil Shale, Natural Bitumen, Heavy Oil and Peat*. Vol. II, Encyclopedia of Life Support Systems Publications, UNESCO.

Harvey, R. (2024). Fossil fuel use, emissions hit records in 2023. *Reuters* (June 20). https://www.reuters.com/business/environment/fossil-fuel-use-emissions-hit-records-2023-report-says-2024-06-19/ (accessed 10 August 2024).

Hermesmann, M. and Müller, T.E. (2022). Green, turquoise, blue, or grey? environmentally friendly hydrogen production in transforming energy systems. *Progress in Energy and Combustion Science* 90: 100996.

Hutton, A.C. (1987). Petrographic classification of oil shales. *International Journal of Coal Geology* 8 (3): 203–231.

Hutton, A.C. (1991). Classification, organic petrography and geochemistry of oil shale. *Proceedings 1990 Eastern Oil Shale Symposium*, 163–172. Lexington: University of Kentucky Institute for Mining and Minerals Research.

Järvik, O., Baird, Z.S., Rannaveski, R. et al. (2021). Properties of kukersite shale oil. *Oil Shale* 38 (4): 265–294.

Jenifer, A.C., Sharon, P., Prakash, A. et al. (2015). A review of the unconventional methods used for the demetallization of petroleum fractions over the past decade. *Energy & Fuels* 29 (12): 7743–7752.

Jia, B. and Su, J. (2023). Advancements and environmental implications in oil shale exploration and processing. *Applied Sciences* 13: 7657.

Jiang, Z., Zhang, W., Liang, C. et al. (2016). Basic characteristics and evaluation of shale oil reservoirs. *Petroleum Research* 1 (2): 149–163. https://doi.org/10.1016/S2096-2495(17)30039-X.

Jin, J.M., Kim, S., and Birdwell, J.E. (2012). Molecular characterization and comparison of shale oils generated by different pyrolysis methods. *Energy & Fuels* 26 (2): 1054–1062.

Johannes, I., Luik, H., and Kruusement, K. (2006). Mathematical models for kinetics of batchwise hydrogenation of shale oil. *Fuel Processing Technology* 87 (8): 711–716.

Kaiser, M.J., de Klerk, A., Gary, J.H. et al. (2019). *Petroleum Refining: Technology, Economics, and Markets*. CRC Press.

Kang, Z., Zhao, Y., and Yang, D. (2020). Review of oil shale in-situ conversion technology. *Applied Energy* 269 (115121): 1–18. https://doi.org/10.1016/j.apenergy.2020.115121.

Klemme, H.D. and Ulmishek, G.F. (1991). Effective petroleum source rocks of the world: stratigraphic distribution and controlling depositional factors. *AAPG Bulletin* 75: 1809–1851.

Kollmuss, A., Zink, H., and Polycarp, C. (2008). A comparison of carbon offset standards. Stokholm Environment Institute (SEI), Tricorona. https://www.globalcarbonproject.org/global/pdf/WWF_2008_A%20comparison%20of%20C%20offset%20Standards.pdf (accessed 14 February 2025).

Lababidi, H.M.S., Chedadeh, D., Riazi, M.R. et al. (2011). Prediction of product quality for catalytic hydrocracking of vacuum gas oil. *Fuel Journal* 90: 719–727.

Landau, M.V., Herskowitz, M., Givoni, D. et al. (1998). Medium severity hydrotreating and hydrocracking of Israeli shale oil. *Fuel* 77 (14): 1589–1597.

Landau, M.V. (1997). Deep hydrotreating of middle distillates from crude and shale oils. *Catalysis Today* 36 (4): 393–429.

LeClerc, H.O., Erythropel, H.C., and Backhaus, A. (2024). The CO_2 tree: the potential for carbon dioxide utilization pathways. *ACS Sustainable Chemistry & Engineering* 13 (1): 5–29.

Li, F., Liu, D., Sun, K. et al. (2024a). Towards a future hydrogen supply chain: a review of technologies and challenges. *Sustainability* 16 (5): 1890.

Li, C., Jin, Z., Zhang, L. et al. (2024b). Advancing the application of sub- and supercritical water in the in-situ conversion of immature and low-maturity shale. *Fuel* 359: 129891.

Li, N., Chen, C., Wang, B. et al. (2020). Distribution of nitrogen and oxygen compounds in shale oil distillates and their catalytic cracking performance. *Petroleum Science* 17: 1764–1778.

Li, T., Long, J., Du, W. et al. (2023). Three pathways towards elimination of CO_2 emissions from industrial plants that use hydrocarbon fuels. *Journal of Cleaner Production* 391: 136159.

Luik, H., Lindaru, E., Vink, N. et al. (1999a). Upgrading of Estonia shale oil distillation fractions. 1. hydrogenation of the "diesel fraction". *Oil Shale* 16: 141–148.

Luik, H., Vink, N., Lindaru, E. et al. (1999b). Upgrading of Estonia shale oil distillation fractions. 2. The effect of time and hydrogen pressure on the yield and composition of "diesel fraction" hydrogenation products. *Oil Shale* 16: 249–256.

Luik, H., Maripuu, L., Vink, N. et al. (1999c). Upgrading of Estonia shale oil distillation fractions. 3. hydrogenation of light Mazute. *Oil Shale* 16: 331–336.

Madugula, A.C.S., Sachde, D., Hovorka, S.D. et al. (2021). Estimation of CO_2 emissions from petroleum refineries based on the total operable capacity for carbon capture applications. *Chemical Engineering Journal Advances* 8: 100162.

Meyer, R.F. and Attanasi, E.D. (2003). Heavy oil and natural bitumen-strategic petroleum resources. *World* 434: 650–657.

Muñoz, J.A.D., Aguilar, R., Castañeda, L.C. et al. (2013). Comparison of correlations for estimating product yields from delayed coking. *Energy & Fuels* 27 (11): 7179–7190.

Navarro, U., Ni, M., and Orlicki, D. (2015). Understanding the potential for FCC feed to generate valuable products and how this knowledge can benefit refinery operation. www.digitalrefining.com/article/1001054 (March 2015).

Newborough, M. and Cooley, G. (2020). Developments in the global hydrogen market: the spectrum of hydrogen colours. *Fuel Cells Bulletin* 2020: 16–22.

Oil Sands Magazine (2024). Unconventional oil explained. January 18. www.oilsandsmagazine.com/news/2024/1/18/unconventional-oil-explained (accessed 20 June 2024).

Oni, A.O., Anaya, K., Giwa, T. et al. (2022). Comparative assessment of blue hydrogen from steam methane reforming, autothermal reforming, and natural gas decomposition technologies for natural gas-producing regions. *Energy Conversion and Management* 254: 115245.

Ortega, E. (2021). An overview of hydrotreating. *Chemical Engineering Progress* 117 (10): 29–33.

OSHA (2024). Occupational safety and health administration. US Government, Department of Labor, retrieved in July 2024 from www.osha.gov

Outlook, A.E. (2010). Energy information administration. *Department of Energy* 92010 (9): 1–15.

Plain, C. (2004). Improve economics of existing AR/VRDS units with the PRS Permutable Reactor System, Cécile Plain, Antwerp: BBTC October 2004. www.axens.net

Pope, G.J. (2000). An application of sequence stratigraphy in modeling oil yield distribution the Stuart oil shale deposit. MSC thesis. School of Natural Resource Sciences, Queensland, Australia. 1–121.

Rana, M.S., Ancheyta, J., Riazi, M.R. et al. (2013). Future directions in petroleum and natural gas refining. In: *Petroleum Refining and Natural Gas Processing* (ed. M.R. Riazi, S. Eser, S. Suresh et al.), 769–799. ASTM International.

Raja, M.A., Hao, C., Zhao, Y. et al. (2017a). Process simulation and assessment of hydrogen and high valued hydrocarbon fuels products from oil shale. *International Journal of Hydrogen Energy* 42 (8): 4922–4934.

Raja, M.A., Hao, C., Wang, H. et al. (2017b). Process analysis of hydrogen and liquid fuels production by oil shale. *Energy Technology* 5 (11): 1963–1978.

RefComm (2023). Refinery Community Inc., Cranberry, Pennsylvania, USA Inc. https://refiningcommunity.com/sda-wood-foster-wheeler-uop/ (accessed 20 June 2024).

Research Reports World (RRW) (2024). Global heavy oil market research report 2024. March 4. www.researchreportsworld.com/global-heavy-oil-market-26914641

Research and Market (2024). Refinery hydrocracking units capacity and capital expenditure (CapEx) forecast by region and countries with details of all operating and planned hydrocracking units to 2027. New York, USA. https://www.researchandmarkets.com/r/xo2hph (accessed 10 September 2024).

Riazi, M.R. (1989). Distribution model for properties of hydrocarbon-plus fractions. *Industrial and Engineering Chemistry Research* 28: 1731–1735.

Riazi, M.R. (1997). A continuous model for C7+ characterization of petroleum fractions. *Industrial and Engineering Chemistry Research* 36: 4299–4307.

Riazi, M.R. (2007). ASTM MNL50: characterization of petroleum fractions. ASTM International. http://www.astm.org/DIGITAL_LIBRARY/MNL/SOURCE_PAGES/MNL50.htm

Riazi, M.R., Eser, S., Agrawal, S. et al. (2013). ASTM MNL58: petroleum refining and natural gas processing. ASTM International. http://www.astm.org/DIGITAL_LIBRARY/MNL/SOURCE_PAGES/MNL58.htm

Riazi, M.R. and Gupta, R. (eds.) (2015). *Coal Production and Processing Technology*. CRC press.

Romero, S. (2020). The role of solvent deasphalting (SDA). In: *Refining*, Becht, Liberty Corner, NJ, October 6th. http://becht.com. https://refiningcommunity.com/wp-content/uploads/2020/12/The-Role-of-Solvent-De-Asphalting-SDA-in-Refining-2.pdf

Roy, R., Antonini, G., Hayibo, K.S. et al. (2025). Comparative techno-environmental analysis of grey, blue, green/yellow and pale-blue hydrogen production. *International Journal of Hydrogen Energy* 116: 200–210.

Saberi, F. and Hosseini-Barzi, M. (2024). Effect of thermal maturation and organic matter content on oil shale fracturing. *International Journal of Coal Science & Technology* 11 (16): 11–19.

Scheffer, B., Van Koten, M.A., Röbschläger, K.W. et al. (1998). The shell residue hydroconversion process: development and achievements. *Catalysis Today* 43 (3-4): 217–224.

Shekarifard, A. and Taheri-Shakib, J. (2016). Technical and scientific review on oil shale upgrading. *International Journal of Petrochemistry Science and Engineering* 1 (3): 78–83.

Shishkova, I., Stratiev, D., Kolev, I.V. et al. (2022). Challenges in petroleum characterization—a Review. *Energies* 15: 7765.

Solomon, P.R. and Miknis, F.P. (1980). Use of Fourier transform infrared spectroscopy for determining oil shale properties. *Fuel* 59 (12): 893–896.

Speight, J. (2012a). *Shale Oil Production Processes*. Waltham, USA: Elsevier Inc.

Speight, J.G. (2012b). Kerogen. In: *Shale Oil Production Processes*, 75–92. Waltham, USA: Elsevier Inc.

Speight, J. (2020). *Shale Oil and Gas Production Processes*, 873–912. Waltham, USA: Elsevier Inc.

Statista (2024a). Demand for crude oil worldwide from 2005 to 2023 with a forecast for 2024. Statista Research Department, May 22. www.statista.com/statistics/271823/global-crude-oil-demand/

Statista (2024b). Crude Oil Energy Production Outlook in the U.S. from 2022 to 2050. Statista Research Department, May 20. www.statista.com/statistics/264026/projections-of-us-oil-production/

Stratiev, D., Shishkova, I., Tankov, I. et al. (2019). Challenges in characterization of residual oils. A review. *Journal of Petroleum Science and Engineering* 178: 227–250.

Stratiev, D., Shishkova, I., Georgiev, G. et al. (2025). The incompatibility pitfall in refining opportunity crude oils. *Processes* 13 (593): 1–26.

Stratiev, D.S., Shishkova, I., Mitkova, M. et al. (2013). Method provides alternative for predicting RFCC yields. *Oil & Gas Journal* 111 (10): 22–28.

Su, Z.S., Liu, P., Cai, L. et al. (2008). Hydrotreating of Fushun shale oil. *Contemporary Chemical Industry* 37: 246–248.

Sullivan, R.F. and Stangeland. B. E. (1978). Converting green river shale oil to transportation fuels. *Annual Oil Shale Symposium, Colorado School of Mines*, April 12–14, 1–42.

Tasleem, S. and Alsharaeh, E.H. (2025). Role of green, yellow, blue, white and gold hydrogen in fuelling the path to net zero and sustainable future- a review. *Energy Conversion and Management* 326: 119500.

TotalEnergies (2024). Decarbonization of European refineries: a first agreement signed between TotalEnergies and air products for the delivery of green hydrogen, TotalEnergies press release, June 7, 2024. https://totalenergies.com/news/press-releases/decarbonization-european-refineries-first-agreement-signed-between

UOP (2011). A two-stage hydrocracking process. Modified from the diagram in following webpage. http://www.uop.com/refining/1071.html (accessed 20 February 2011).

van den Berg, F.G.A., Heijnis, R.M.A., Stamps, P.A. et al. (2003). A geochemical framework for understanding residue properties. *Petroleum Science and Technology* 21 (3–4): 449–460.

Vandenbroucke, M. (2003). Kerogen: from types to models of chemical structure. *Oil & Gas Science and Technology - Revue d'IFP Energies nouvelles* 58 (2): 243–269.

Volk, M., Wisecarver, K., and Sheppard, C. (2002). Fundamentals of Delayed Coking Joint Industry Project. Tulsa, OK: Department of Chemical Engineering, University of Tulsa.

Voolapalli, R.K., Thota, C., Gokak, D.T. et al. (2013). Fluid catalytic cracking. In: *Petroleum Refining and Natural Gas Processing* (ed. M.R. Riazi, S. Eser, and S.S. Agrawal), 127–158. West Conshohocken, PA: ASTM International.

Wang, H., Dai, F., Yang, Y. et al. (2017). Catalyst grading optimization and kinetic simulation of the shale oil hydrotreating process. *Energy & Fuels* 31 (4): 4353–4360.

Wang, L., Guo, J-X., Zhang, X-J. et al. (2023a). Nanoscale MOF-catalyzed pyrolysis of oil shale and kinetic analysis. *Journal of Analytical and Applied Pyrolysis* 174: 106149.

Wang, R., Wang, G., Zhao, G. et al. (2023b). Geological characteristics and resources potential of shale oil in Chang 7 member of upper triassic Yanchang formation in fuxian area, southern ordos basin, western China. *Unconventional Resources* 3: 237–247.

Wang, L., Gao, C-H., Xiong, R-Y. et al. (2024). Development review and the prospect of oil shale in-situ catalysis conversion technology. *Petroleum Science* 21: 1385–1395.

Williams, P.T. and Chishti, H.M. (2001). Reaction of nitrogen and sulphur compounds during catalytic hydrotreatment of shale oil. *Fuel* 80 (7): 957–63.

Xu, C. and Bell-Hammer, L. (2023). Global oil and gas reserves increase in 2022. *Oil & Gas Journal*, https://www.ogj.cCom/exploration-development/reserves/article/14286688/global-oil-and-gas-reserves-increase-in-2022 (accessed 10 December 2023).

Xu, C., Gao, J., Zhao, S. et al. (2005). Correlation between feedstock SARA components and FCC product yields. *Fuel* 84: 669–674.

Yan, Y., Gao, H., Sui, Y. et al. (2022). The catalytic effect of inner minerals on the combustion of oil shale char: characteristics and mechanism. *Fuel Processing Technology* 236: 107426.

Yang, Q., Yang, Q., Man, Y. et al. (2020). Technoeconomic and environmental evaluation of oil shale to liquid fuels process in comparison with conventional oil refining process. *Journal of Cleaner Production* 255: 120198.

Yoshida, R., Miyazawa, M., Yoshida, T. et al. (1996). Chemical structure changes in Condor shale oil and catalytic activities during catalytic hydrotreatment. *Fuel* 75 (1): 99–102.

Yu, H., Li, S., and Jin, G. (2010). Catalytic hydrotreating of the diesel distillate from fushun shale oil for the production of clean fuel. *Energy & Fuels* 24 (8): 4419–4424.

Zhang, K., Yu, J., Gao, S. et al. (2017). Understanding shale oil hydrotreatment with composition analysis using positive-ion mode atmospheric pressure photoionization Fourier transform ion cyclotron resonance mass spectrometry. *Energy & Fuels* 31 (2): 1362–1369.

Zhang, M., Wang, C., Wang, K. et al. (2020). Gentle hydrotreatment of shale oil in fixed bed over Ni-Mo/Al_2O_3 for upgrading. *Fuel* 281: 118495.

Zhao, G.F., Su, Z.S., and Liu, H. (2007). Study on diesel production by shale oil hydrocracking. *Contemporary Chemical Industry* 36: 361–366.

Zhou, H., Zeng, S., Yang, S. et al. (2018). Modeling and analysis of oil shale refinery process with the indirectly heated moving bed. *Carbon Resources Conversion* 1: 260–265. https://www.mapquest.com/us/illinois/uop-llc-a-honeywell-co-353251224

8

Petroleum Refining with Deep Conversion and Deep Desulfurization

George Hoekstra[1], *Marcio Wagner da Silva*[2], *and Thomas Murphy*[3]

[1] *Hoekstra Trading LLC, Bolingbrook, IL, USA*
[2] *Process Engineering, Petrobras, Sao Paulo, Brazil*
[3] *Valuation Risk & Strategy LLC, Skaneateles, NY, USA*

8.1 Heavy Oil Process Overview

Crude oil is a chemical soup containing tens of thousands of chemical compounds ranging from methane with one carbon atom to asphaltenes, which are large, complex chemical structures best used in asphalt. In a refinery, the crude oil is separated by distillation into light gas, gasoline, jet fuel, diesel fuel, and heavy oil.

The atmospheric crude distillation process separates the crude into groups of small and larger compounds with different boiling points. To be economically viable, a modern refinery must then chemically crack the largest compounds into smaller, lighter fragments that can then be added to the gasoline, diesel, and jet fuel streams.

The heaviest compounds exist in the bottom product stream from the atmospheric distillation tower, which contains what's called heavy oil or atmospheric resid. It contains compounds with normal boiling points above 343 °C (650 °F). This 343 °C+ heavy oil is distilled under vacuum into an intermediate fraction called vacuum gas oil which contains compounds boiling from 343 to 538 °C (650–1000 °F), and a bottom fraction called vacuum resid (similar to asphalt) containing the largest compounds boiling above 538 °C (1000+ °F) This vacuum distillation sets the stage to chemically convert the heavy oil and produce much more high-value fuel than the crude oil naturally contains.

This distillation sequence is shown in Figure 8.1. This chapter focuses on the heavy oil section highlighted by the dashed box.

Cleaner Petroleum Production and Refining Technologies, First Edition, Edited by M. R. Riazi and Harvey W. Yarranton.

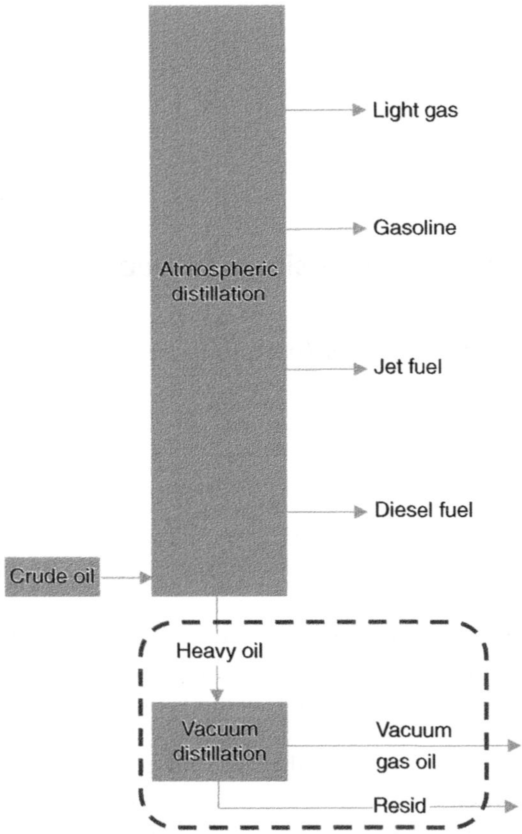

Figure 8.1 Crude oil distillation separates crude oil molecules into fractions containing compounds with low, intermediate, and high boiling point ranges. The bottom stream (heavy oil) is vacuum distilled into vacuum gas oil and resid.

8.1.1 Fluid Catalytic Cracking

Vacuum gas oil is chemically cracked in the Fluid Catalytic Cracker (FCC), producing lighter, broken compounds that are then distilled into the same boiling ranges as the streams coming from the atmospheric distillation process (that is, light gas, gasoline, jet fuel, diesel fuel, and unconverted heavy oil). Because these cracked streams are composed of fragments of heavy, high-boiling compounds, they contain higher levels of sulfur than those in the virgin (natural) streams.

A distinguishing feature of the FCC process is the solid catalyst, which circulates in a fluidized state, flowing from the cracking reactor where the oil is cracked, to the regeneration section where carbon formed on the catalyst during cracking is burned off the catalyst in air to regenerate the catalyst for recirculation to the cracking reactor.

8.1.2 Vacuum Gas Oil Hydrotreating

Some FCC process trains include a vacuum gas oil hydrotreater upstream of the FCC. This process feeds a mixed stream of vacuum gas oil and hydrogen through a fixed bed catalytic reactor to desulfurize and saturate aromatic compounds via hydrogenation reactions before the FCC feed is cracked. It allows feeding heavier, higher sulfur (lower value) feedstocks to the FCC train and gives big benefits in yield and quality of the FCC products.

8.1.3 Vacuum Gas Oil Hydrocracking

An alternative to the FCC is to crack vacuum gas oil in a hydrocracker, which both cracks and desulfurizes the vacuum gas oil in a flowing stream of hydrogen over a fixed bed of catalyst, yielding product streams that are lower in sulfur than those coming from the FCC. Vacuum gas oil hydrocracking is done in a high-pressure reactor operating at around 139 bar (2000 psig) pressure. Here, in addition to cracking the heavy molecules into smaller pieces, hydrogenation reactions remove sulfur and nitrogen atoms from the heavy oil molecules and saturate hydrogen-deficient compounds, producing higher yields of lower sulfur liquid products, especially diesel fuel, compared to catalytic cracking.

These vacuum gas oil cracking and desulfurization processes are depicted in the highlighted section of Figure 8.2.

8.1.4 Resid Upgrading

Most modern refineries also use chemical conversion processes to clean and crack the heaviest molecules in the resid stream into products suitable for use in fuels. This is a more challenging job for many reasons. The heavy oil molecules in resid contain lots of sulfur, nitrogen, and metal atoms (especially vanadium and nickel), which are deeply embedded in large aromatic molecules. To crack resid requires high temperatures, which increases carbon deposition and deactivation of the catalysts. Resids are also prone to form solid precipitates when processed, which increases catalyst deactivation and causes other problems like reactor plugging.

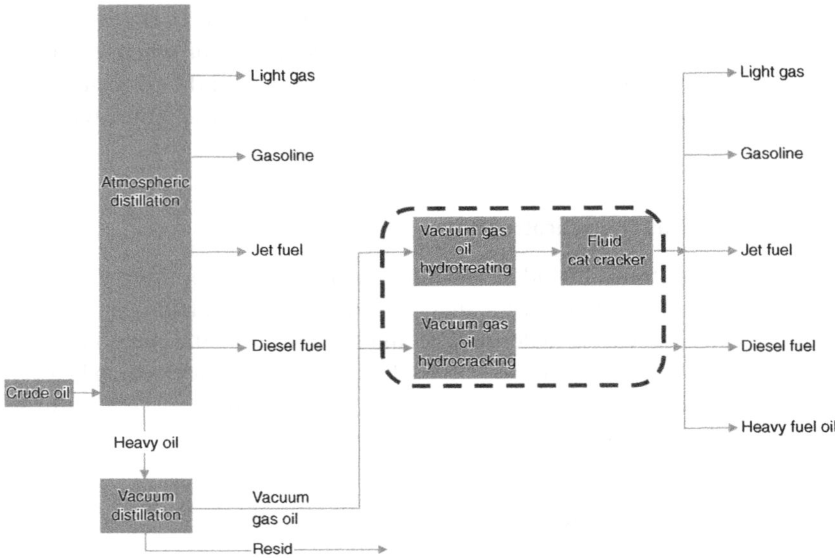

Figure 8.2 Vacuum gas oil cracking processes chemically crack and desulfurize intermediate heavy oil molecules into valuable fuels.

8.1.5 Resid Coking

Coking is a non-catalytic, thermal conversion process. Resid is fed to drums operating at near atmospheric pressure and 427 to 596 °C (800–925 °F). In the coke drum, side chains crack off large molecules and are released to a fractionator, yielding light gas and liquid products that are further refined into liquid fuels. At such high temperatures, in the absence of hydrogen or catalyst, carbon-rich molecular cores combine into even higher molecular weight compounds, producing more light hydrocarbons and eventually forming solid coke that contains most of the metals, sulfur, and nitrogen originally in the resid. The coke drums operate in a cyclic fashion – the coking cycle is followed by a decoking cycle to remove the coke from the drum.

Coking is a relatively inexpensive process that gives low yields of liquid products and typically 30 wt% relatively low-value coke. The liquid products from cokers, known as coker naphtha (gasoline) and coker gas oils (diesel and heavier), are highly olefinic, high in sulfur and nitrogen, and often contain silicon from additives used at the coker, which poison the hydrotreating catalyst; hence, they are among the most challenging streams in the refinery to process into finished fuels.

8.1.6 Resid Desulfurization

Starting in 1970, refiners started investing in resid hydrodesulfurization and resid hydrocracking processes around the world. This was driven by the demand for lower sulfur fuels and growing incentives to refine less expensive, heavy high-sulfur crude oils into high yields of cleaner liquid fuels. Desulfurization of resid was first commercialized in Japan for the production of lower sulfur heavy fuel oil in 1970. Since then, resid hydroprocessing has been a major focus for refining technology, catalyst development, and capital investment.

Referring again to Figure 8.1, the heavy oil from the bottom of an atmospheric crude distillation unit (also known as atmospheric resid) can be effectively desulfurized using the same basic process used to desulfurize gasoline, diesel, and vacuum gas oil streams, that is, fixed bed catalytic hydrodesulfurization in which feed and hydrogen are passed over a bed of solid catalyst at high pressure and temperature. Resid service is more demanding because the sulfur atom is less accessible, embedded deeply in the large heavy oil molecules, hence requiring better catalysts, larger reactors, higher pressures, and higher temperatures than those used for processing gasoline, diesel, and vacuum gas oil streams.

The nickel and vanadium atoms in resid molecules are also removed by hydrotreating; they deposit on the catalyst, causing faster deactivation and plugging of catalyst pores.

Desulfurization of vacuum resid brings more severe challenges because heavier feedstocks, combined with higher process severity, cause precipitation of the heaviest components (called asphaltenes) as solids on the catalyst and in process equipment. When asphaltene precipitation is a serious problem, the fixed-bed process becomes effectively inoperable.

8.1.7 Resid Hydrocracking

The difficulties in processing vacuum resid at high severity in fixed-bed reactors led to the development of new reactor technologies employing moving catalyst beds, slurry reactors, and ebullated bed reactors that enable fluidized and flowing catalyst and on-line catalyst replacement. These reactors enabled the successful application of catalytic hydrogenation to the most challenging service, hydrocracking of vacuum resid.

In this chapter, we will consider two broad categories of resid hydrogenation processes: resid hydrotreating, which is done mostly in fixed-bed reactors for desulfurization, and resid hydrocracking, which is mostly done in ebullated bed reactors for cracking (Figure 8.3).

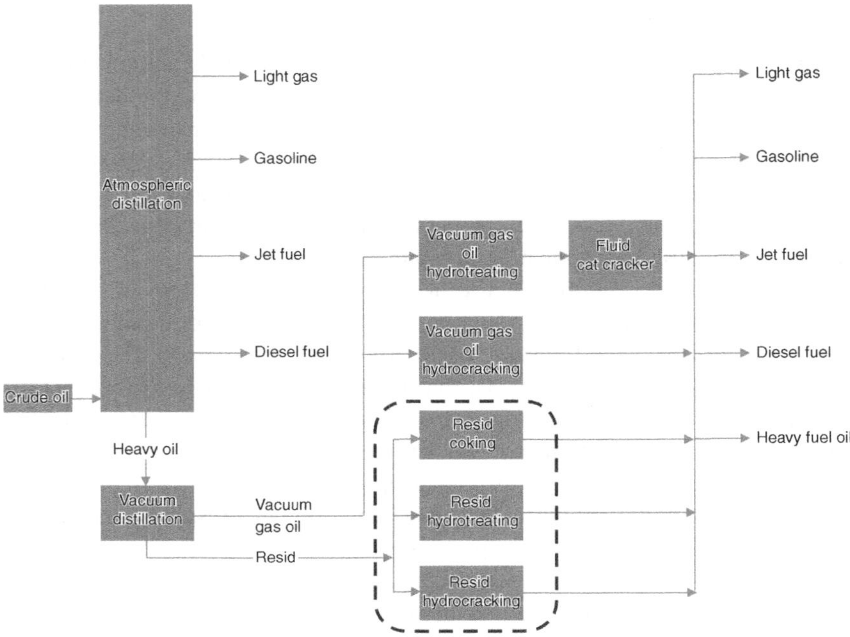

Figure 8.3 Resid cracking and desulfurization processes convert resid to high-value fuels.

8.2 A Closer Look at Heavy Oil Upgrading Reactors and Catalysts

8.2.1 Vacuum Gas Oil Hydrotreating

8.2.1.1 Vacuum Gas Oil Hydrotreating Process

Vacuum gas oil hydrotreating is done in a fixed-bed catalytic reactor containing multiple beds of catalysts as shown in Figure 8.4. Vacuum gas oil feed and recycle gas are fed to the reactor and flow through the catalyst beds.

Because the hydrogenation reactions are exothermic, the flowing liquid/vapor mixture increases in temperature as it moves down the reactor. This heat release is an important aspect of reactor control and unit performance. Cold recycle gas is fed to the reactor between the catalyst beds to quench the reaction mixture.

The reactions consume hydrogen and produce hydrogen sulfide (from desulfurization) and ammonia (from denitrogenation). The recycle gas system is designed and operated to maintain the desired composition of the reactor feed gas, including the makeup hydrogen supply.

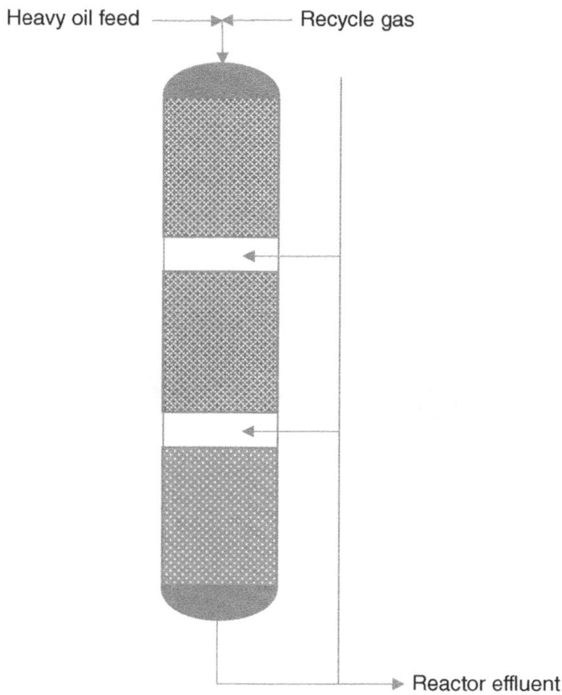

Figure 8.4 Reactor configuration for fixed-bed hydrotreating.

8.2.1.2 Vacuum Gas Oil Hydrotreating Catalyst

The catalyst is a particulate solid, usually in extruded form of 10 millimeter diameter, shown as the smallest particles in Figure 8.5. The top part of the first bed is loaded with larger particles of various sizes and shapes with high void space, like rings and wheels, as shown. These larger particles have little or no activity for reaction, but provide void space to trap particulate solids and contaminants coming from the feed or upstream operations, and solid reaction products that may form, especially at the top of the reactor.

Plugging of reactors has been a common problem, though, over the years, catalyst suppliers have developed better products and methods to load the top of the reactor with these grading materials to reduce that problem.

The extent of product value upgrade is determined mostly by the space velocity (oil volumetric feed rate/catalyst volume), reactor temperature, and hydrogen partial pressure.

Over time, the catalyst activity declines from coke deposition on the catalyst and, with some feeds, from deposition of metals from the feed on the catalyst. Reactor temperature is increased to compensate for catalyst deactivation and maintain

The Youth Science Association

Figure 8.5 The smallest particles are active main bed catalysts, and the larger ones are top bed grading materials.

constant product quality specifications. Refiners usually aim for cycle lives of several years before deactivation causes the maximum reactor temperature to be reached, at which time the unit is shut down to change and replace the catalyst. Occasionally, cycles are as short as several months, though that is not the norm.

Vacuum gas oil hydrotreating catalysts consist of extruded particles composed of a porous alumina base on which nickel, cobalt, and molybdenum are deposited. Advances in hydrotreating catalyst science and engineering have consistently brought billions of dollars/yr of innovation to the refining industry.

The primary reactions driving cat feed hydrotreater product value are desulfurization, denitrogenation, and aromatics saturation. When desulfurization is the predominant objective, cobalt molybdenum catalysts are usually chosen for their high activity and selectivity for desulfurization reactions. When aromatics saturation is the predominant objective, nickel molybdenum catalysts are usually chosen for their high activity and selectivity for aromatic saturation reactions.

8.2.2 Fluid Catalytic Cracking

8.2.2.1 Fluid Catalytic Cracking Process

Fluid catalytic crackers come in a wide variety of designs. The essential elements are shown in Figure 8.6. In this diagram, fluidized catalyst flows

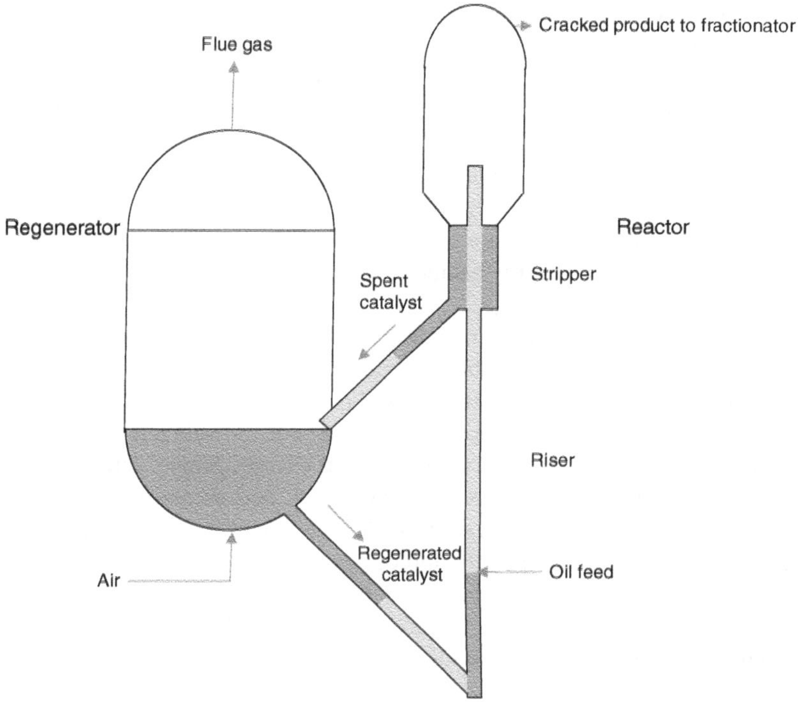

Figure 8.6 The essential components of a fluid catalytic cracker (FCC) are the reactor section (right) and the regenerator section (left). In this diagram, feed flows up through the reactor section and products flow out of the top, while fluidized catalyst circulates counterclockwise.

counterclockwise while oil vapors flow upward through the reactor section and out the top to a product fractionator. The reactor section consists of a riser in which oil vapors and hot catalyst flow upward to a disengaging section where cyclones (not shown) separate the catalyst particles and drop them back down to a section of the reactor called the stripper, which is a dense phase fluidized bed where upward flowing stripping gas helps remove cracked products as they continue to form in the stripper to be carried out the top of the reactor. Cracking reactions occur mostly in the riser, in which the hydrocarbon residence time is typically a few seconds.

The cracking reactions involve breaking carbon-carbon bonds to break down the high molecular weight hydrocarbons. This causes coke formation on the catalyst particles, which then flow downward through the stripper and standpipe through a slide valve, which controls their flow and release to the top of a fluidized bed in the regenerator section.

Air is fed to the bottom of the regenerator to burn coke off the catalyst, regenerating and heating the catalyst by combustion, with flue gas released at the top of the regenerator. The hot regenerated catalyst flows down through another standpipe and slide valve to the bottom of the riser, completing the catalyst flow circuit.

The FCC fractionator splits the cracked product into C1-C4 hydrocarbons, FCC gasoline, FCC diesel (commonly called light catalytic cycle oil (LCO)), and heavy catalytic cycle oil.

Because the cracking reactions are endothermic and coke combustion is exothermic, the integrated unit energy balance is an important aspect of unit design and operation.

Feeds to a fluid catalytic cracker include vacuum gas oil, resid, coker gas oils, and other intermediate streams suitable for cracking into fuels. Because the feed is cracked in the absence of hydrogen, the cracked products are highly olefinic compared to hydrocracked products. Propylene and butylene are valuable light olefin products. The high octane olefinic FCC gasoline is the largest single component of the gasoline pool and the largest source of octane barrels. Because the cracked molecules are broken pieces of heavy, high sulfur molecules, the FCC gasoline and diesel products must be hydrotreated or further processed to make clean fuels.

With low-value feeds, and potentially very high-value products, the FCC is a center of attention for those wanting to maximize refinery profit margins by optimizing feed, operating conditions, catalyst, product yields, and product quality.

8.2.2.2 Fluid Catalytic Cracking Catalysts

FCC catalysts consist of fine particles of zeolites dispersed in an amorphous solid matrix. Higher zeolite-to-matrix ratios lead to higher cracking activity. Because they operate in a fluidized state, the fluidization and attrition properties are especially important, in addition to the catalytic properties.

Catalysts are selected to optimize unit yields, product qualities, and to provide long-term stability. In addition to cracking activity, catalyst design affects important performance factors like coke selectivity, bottoms conversion, light olefins production, and tolerance of feed metals. Special solid additives are used to promote particular reactions like gasoline sulfur reduction, octane enhancement, combustion, and process emission control.

8.2.3 Vacuum Gas Oil Hydrocracking

8.2.3.1 Vacuum Gas Oil Hydrocracking Process

Vacuum gas oil hydrocracking achieves desulfurization, denitrogenation, aromatics saturation, and cracking of the feed. The addition of the cracking function gives rise to a wide variety of configurations and process designs for hydrocrackers.

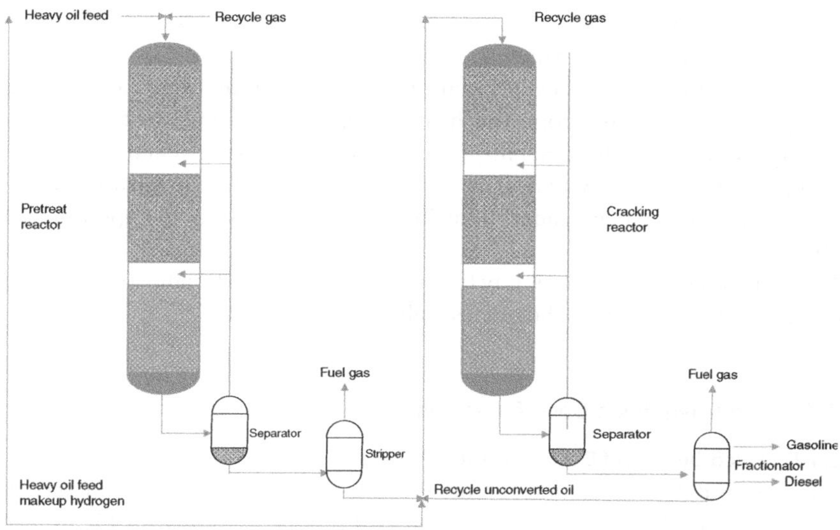

Figure 8.7 This recycle hydrocracker reactor configuration has two stages and includes recycle of unconverted oil to the second stage.

The configuration shown in Figure 8.7 has two reactor stages. The first reactor is equivalent to a vacuum gas oil desulfurizer as described in Section 8.2.1 and the second is a hydrocracking reactor. In this configuration, unconverted oil in the second-stage product is recycled back to the cracking reactor. The products include low-sulfur gasoline and low-sulfur diesel.

The feed to hydrocrackers can vary widely. Some hydrocrackers are dedicated to cracking diesel-range intermediates to gasoline. Others feed a combination of straight-run vacuum gas oil with other low-value intermediates to produce both gasoline and diesel in optimized quantities. Some hydrocrackers produce hydrogen-rich. low sulfur heavy oils for use in lubricant products.

8.2.3.2 Vacuum Gas Oil Hydrocracking Catalyst

The catalyst in the first stage of a hydrocracker is the same as the vacuum gas oil hydrotreating catalyst, except there is extra emphasis on having high denitrogenation activity. That is because the downstream hydrocracking catalyst performance is inhibited by the presence of nitrogen.

The hydrocracking catalyst contains a zeolitic component to provide the acidic cracking activity needed to crack the feed molecules and produce the desired gasoline and diesel products. The amount and design of the zeolite affect the extent of cracking and the yield distribution between gasoline and diesel. Higher zeolite content gives higher cracking activity with higher yields of gasoline and lighter products compared to diesel.

As in vacuum gas oil desulfurization, feed contamination, catalyst bed grading, catalyst deactivation, and heat release must be managed in hydrocracking. Because of the cracking function, hydrocracking reactions can be accompanied by runaway reactions with unusually high heat release; hence, there are special safety considerations involved in the design and operation of hydrocrackers.

Hydrocracking provides flexibility to optimize profitability by adjusting gasoline and diesel yields to produce the highest value product mix.

8.2.4 Fixed-bed Resid Desulfurization

8.2.4.1 Fixed-bed Resid Desulfurization Process

Fixed-bed resid desulfurization is a more challenging version of vacuum gas oil desulfurization. The reactor configuration is the same as shown in Figure 8.3. The challenges come in the form of slower desulfurization reactions, faster coke deactivation, much higher rates of vanadium and nickel deposition on the catalyst (which accelerates catalyst deactivation and plugs catalyst pores) and more ways to deposit more contaminants and other solids that cause physical plugging of the catalyst pores and catalyst beds.

8.2.4.2 Fixed-bed Resid Desulfurization Catalyst

In resid hydroprocessing, the removal of vanadium and nickel from high molecular weight feed components often becomes a critical technical challenge. Beds of demetallation catalyst and/or separate demetallation reactors are used with catalysts specially designed to remove vanadium and nickel in a manner that controls the metals deposition within the catalyst pores and along the length of the reactor.

Demetallation catalyst design typically puts more emphasis on the pore structure of the base material, as opposed to optimizing the forms of cobalt, nickel, and molybdenum species as in desulfurization catalyst design. Demetallation catalysts have high pore volumes and pore size distributions designed to facilitate the accumulation of large amounts of deposited metals. In some cases, the quantity of vanadium and nickel on a demetallation catalyst at the end of a cycle can exceed 100% of the original weight of the catalyst.

In addition to custom-designed catalysts, fixed-bed resid desulfurization requires more reactor volume, more catalyst beds, higher reactor pressure, higher reactor temperature, and more reactor quench points.

8.2.5 Ebullated Bed Resid Hydrocracking

8.2.5.1 Ebullated Bed Resid Hydrocracking Process

The continual push to make cleaner fuels from heavier, higher-sulfur, cheaper crude oils led to the commercialization of ebullated bed reactor technology for upgrading vacuum resids.

The ebullated bed reactor shown in Figure 8.8 was invented and developed by Edgar Johanson working at Hydrocarbon Research Institute in the 1960s. In an ebullated bed reactor, liquid, gas, and catalyst exist in a fluidized state in the bottom part of the reactor. Feed liquid, gas, and recycle liquid are fed from the bottom of the reactor through bubble caps on a perforated distributor plate and upward through the bed of catalyst.

The bed consists of continuously flowing liquid and gas moving through an expanded catalyst bed in which catalyst circulates through a 3-phase fluidized mixture while remaining in the bottom part of the reactor. At the top of the ebullated bed (where there is ideally no catalyst), liquid product spills over into a concentric conduit, which directs it back down to an ebullating pump that recycles it back up into the bottom of the fluidized bed. This liquid circulation is set to maintain a fluid velocity that maintains the catalyst level in the three-phase bed. The fluidizing pump can be inside or outside the reactor vessel. Product exits the top of the reactor to a vapor–liquid separator.

The reactor design allows for intermittent addition of fresh catalyst through a catalyst addition line at the top of the reactor down into the 3-phase ebullated bed, and withdrawal of equilibrium catalyst through a catalyst withdrawal line at the bottom of the reactor.

In an ebullated bed, the reaction mixture is in a near-isothermal state. There is little pressure drop, minimal bed plugging, and no reactor quench. Thus, the ebullated bed eliminates many of the problems inherent in severe processing of vacuum resid in a fixed-bed reactor.

8.2.5.2 Ebullated Bed Resid Hydrocracking Catalyst

Ebullated bed resid hydroprocessing catalysts are small (10 mm) extrudates similar to fixed-bed resid desulfurization and hydrocracking catalysts with alumina bases containing cobalt, nickel, and molybdenum and tungsten, with large pore diameters, high pore volumes and tailored to manipulate physical and mechanical properties to make them suitable for the ebullated bed reactor environment.

8.2.6 Meeting Clean Fuels Challenges with Heavy Oil Feeds

For the last 30 years, the global refining industry has been continually challenged to meet more stringent regulations on refinery emissions, clean fuels, vehicle

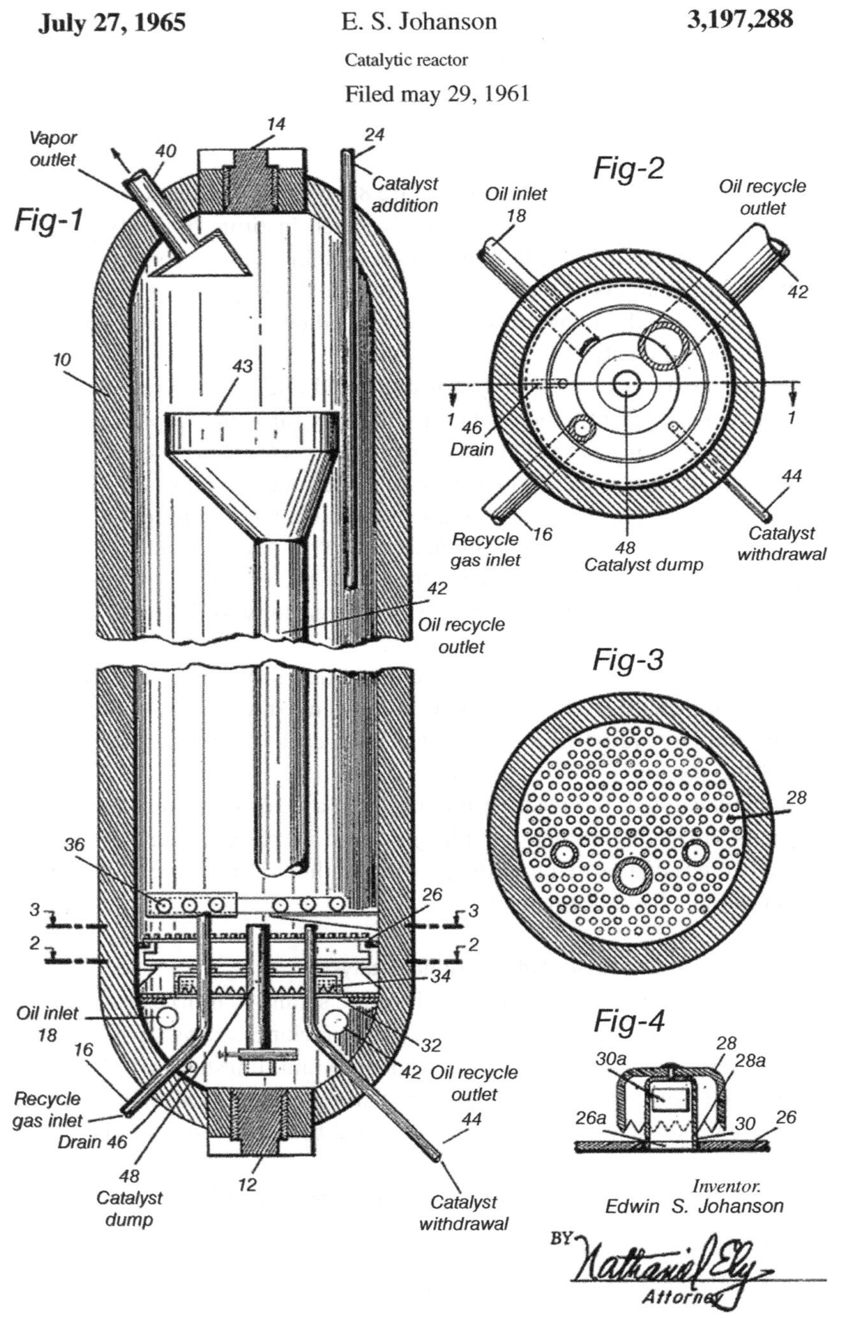

Figure 8.8 Ebullating bed reactor. U.S. patent 3,197,288 to Edgar Johanson/U.S. Patent.

exhaust emissions, waste disposal, and use of renewable fuels, with Europe, Japan, and the state of California leading the way.

In the United States, the Clean Air Act Amendments of 1990 set the groundwork for a decades-long effort to produce cleaner fuels. During the same time frame, Canada, Europe, and Japan followed similar paths. The U.S. Reformulated gasoline (RFG) program set stringent new specifications for gasoline sulfur, benzene, aromatics, olefins, volatility, and toxics to meet new air quality standards for control of carbon monoxide, lead, sulfur dioxide, nitrogen dioxide, ozone, and particulate matter in the air.

In the 1990s, the state of California took the lead in the development of fuel specifications in the United States with a specific focus on helping to reduce ground-level ozone in the Los Angeles Basin area. The new clean gasoline formulations were seasonal, and specific for different local regions, which increased the demands and placed new constraints on fuel blending and fuel supply chains.

With the U.S. Environmental Protection Agency (EPA)'s Tier 2 and Tier 3 gasoline sulfur standards, the sulfur in gasoline was reduced in a stepwise fashion from 350 ppm in 1990 to 10 ppm in 2020. Figure 8.9 shows the profile of gasoline sulfur reduction in the United States.

Figure 8.10 shows the current state of worldwide gasoline sulfur specifications, which will continue to tighten.

Ultra-low sulfur diesel fuel specifications have a similar history. In the United States, sulfur in diesel fuel has been reduced in steps from 5,000 ppm in 1990 to the "ultra-low sulfur diesel" (ulsd) specification of 15 ppm sulfur. The current U.S.

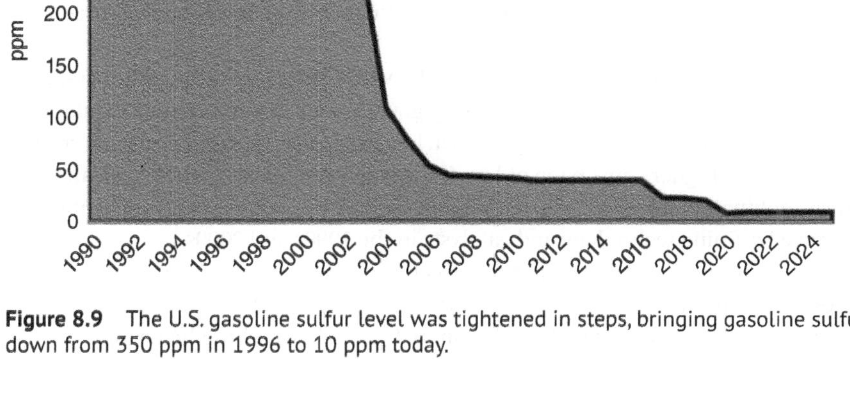

Figure 8.9 The U.S. gasoline sulfur level was tightened in steps, bringing gasoline sulfur down from 350 ppm in 1996 to 10 ppm today.

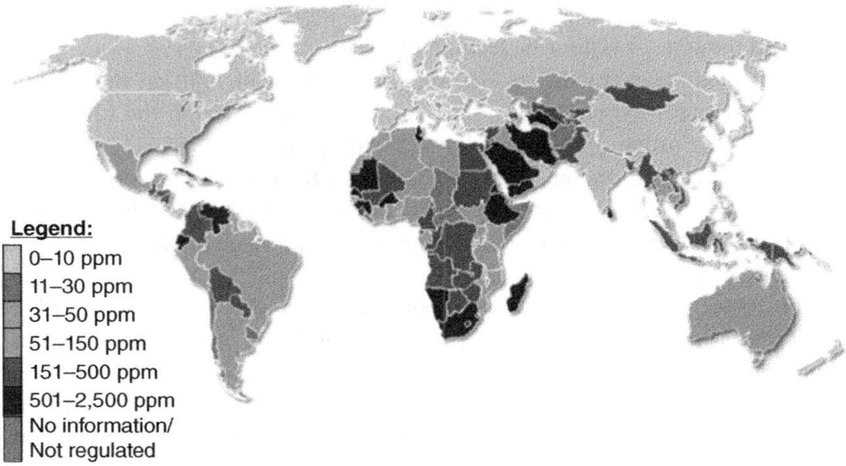

Figure 8.10 Global gasoline sulfur specifications as of 2022. *Source*: Hydrocarbon Engineering, March (2022).

specification of 15 ppm sulfur was phased in, first for on-road diesel fuel from 2006 to 2010, then for nonroad, locomotive, and marine diesel fuel to 500 ppm from 2007 to 2014, and to the current 15 ppm sulfur since 2014.

These clean gasoline and diesel fuel requirements were accompanied by the development of improved vehicle emissions control technologies that work in concert to reduce harmful emissions from vehicles, and by stringent fuel economy standards that reduced fuel consumption and emissions per mile travelled.

In 2020, a new sulfur specification of 0.5 wt% maximum was implemented by the International Maritime Organization (IMO), from the previous 3.5 wt%. This measure is reinforced in the Sulphur Emission Control Areas (SECA) with a 0.1 wt% specification. This is an international standard aimed at reducing harmful emissions from fuel oil used by ships that affect air quality in coastal and port areas. Participating countries implement their own legislation and enforcement policies to enforce the standard.

Starting in the 1990s, refineries around the world began investing in processes to comply with these ever-more demanding government environmental regulations, while, at the same time, processing heavier, less expensive, and more variable crudes, and achieving higher conversion of low-value heavy fractions to valuable clean fuels. This combination of needs drove refining capital investment and application of improved refining technology and catalysts in the 1990s and 2000s.

As an example of those investments, Section 5 presents an economic analysis of investment options for meeting more stringent gasoline sulfur specifications in a gasoline-producing refinery that uses an FCC unit for cracking heavy oils.

8.3 Economic Analysis of Gasoline Sulfur Reduction Investment

For the last 30 years, refiners have faced an ongoing need to invest capital to enable production of lower sulfur fuels from heavier, higher sulfur crude oils. These investments take many forms, depending on the specific needs of the markets and capabilities of the refineries. Investments for low-sulfur gasoline have focused on reducing the sulfur in the FCC gasoline stream, which accounts for 40% of the gasoline volume and over 95% of the sulfur in the U.S. gasoline pool.

This need to desulfurize FCC gasoline triggered growth of a relatively new process called cracked naphtha hydrodesulfurization, which has played an important role in enabling refineries to meet this challenge while sustaining and improving refining profitability. The number of commercial cracked naphtha hydrodesulfurization units has grown from near zero in the year 2000 to over 200 commercial units today. Before proceeding with investment analysis, we must introduce this relatively new member of the family of 21st-century refining processes.

8.3.1 Cracked Naphtha Hydrodesulfurization Process

8.3.1.1 Process Description

The hydrodesulfurization of cracked naphtha represents a great technology challenge because it is now necessary to remove over 95% of the sulfur components from this FCC gasoline stream without saturating olefins, which are the components that give it a high octane number for gasoline. Processes for doing this are described in the literature (Wagner, 2022).

In chemical engineering terms, a cracked naphtha desulfurizer is called a "selective catalytic" process because it must selectively catalyze chemical reactions of sulfur-containing molecules while simultaneously minimizing the reaction of high-octane olefins.

Nowadays, the most applied FCC naphtha hydrotreating process is the PRIME G+™ process, developed by the Axens Company, shown in Figure 8.11. This process exploits the fact that sulfur tends to concentrate in the heavier fractions of the FCC naphtha, while the olefins tend to concentrate in the lighter fraction; therefore, a fractionation step is carried out (in the light grey distillation tower in Figure 8.11) to remove the olefin-rich light naphtha before the heavier part of the FCC naphtha is exposed to deep hydrodesulfurization in the selective desulfurization step (as in Figure 8.11).

The cracked naphtha is fed to a diolefins reactor, which promotes hydrogenation of diolefins, and then the stream is separated into low-sulfur light and higher-sulfur heavy fractions in a distillation tower. While the high-octane light

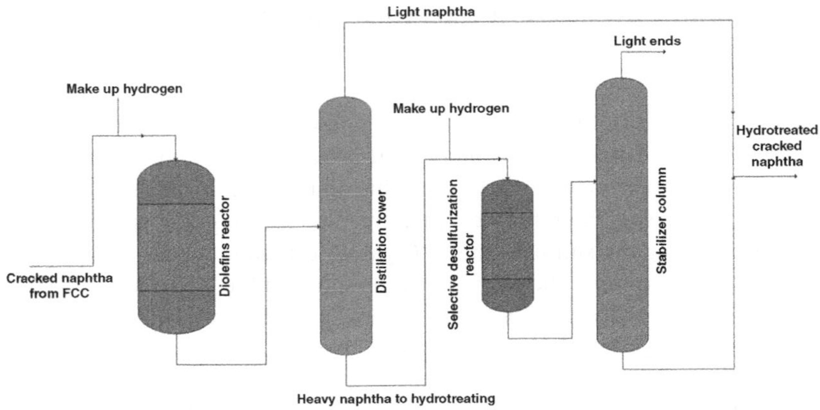

Figure 8.11 Axens PRIME G+ gasoline desulfurization process achieves deep desulfurization of cracked naphtha with minimal octane loss.

naphtha is recovered in the top, the heavy fraction is removed from the bottom of the column and sent to a selective hydrotreating section to remove the bulk of the sulfur with minimum olefin saturation and minimum octane loss. In the sequence, the hydrotreated naphtha is separated in a stabilizer column to remove light compounds, the bottom product is mixed with the light fraction, and the final product is directed to the refinery gasoline pool.

Other technologies that apply selective hydrotreating to reduce the sulfur content in cracked naphtha are the Scanfining™ process developed by ExxonMobil Company and the HyOctane™ process developed by Haldor Topsoe.

8.3.1.2 A Special Challenge – Dienes (Diolefins) Control

A key parameter in cracked naphtha hydrodesulfurization is the conjugated dienes content; these compounds are chemically unstable and tend to suffer oxidation reactions that lead to gum formation in the reactors, high pressure drop, and short operation campaigns. The selective desulfurization processes described above usually use a low-temperature hydrogenation step (as in Figure 8.11) as the first step to convert these dienes to more stable olefins.

8.3.1.3 Modern Tools for Optimization

Recently, refiners have been applying 21st-century lab analytical methods like ASTM D6730 (detailed hydrocarbon analysis) and ASTM D5623 (sulfur speciation) to measure the dienes, olefins, and sulfur compounds in cracked naphtha streams at the molecular level. These modern analytical methods, combined with process models, enable unit optimization and have recently been streamlined and automated for easy use in a refinery for better control and optimization of FCC naphtha hydrodesulfurization feed, distillation cut points, and reactor operating conditions (Hoekstra, 2020).

As one example, measuring olefins and sulfur compounds at the molecular level is essential for optimal control of naphtha fractionator cut points to maximize capture of high-octane olefins in light naphtha while simultaneously pushing the sulfur compounds down into the selective desulfurization reactor, resulting in lower product sulfur, less olefin saturation, and less octane loss.

8.3.2 Economic Analysis of Investment Options for Tier 3 Gasoline

8.3.2.1 Three Capital Investment Alternatives

The reduction of gasoline sulfur, by investment in the fluid catalytic cracking process train, is a good example to show the refining costs and benefits of clean fuel investments. Here, we provide an evaluation of the economics of three options for investing capital in an existing FCC process train to meet clean gasoline requirements.

The analysis considers the FCC train in a hypothetical refinery with an existing 55,000 Barrel/d FCC in a market that will soon adopt a new 30 ppm sulfur clean gasoline specification. The refinery's current FCC naphtha, which was previously suitable for use in gasoline, will then be much too high in sulfur for that market. In addition, there is a possibility that the 30 ppm sulfur specification will be tightened to 10 ppm in 7 years. Without investment, this refinery would likely be forced to shut down.

The investment options being considered involve pretreating the FCC feed using a 30,000 barrel/day vacuum gas oil desulfurization unit (also called a "cat feed hydrotreater" or "CFHT"), and/or desulfurizing the high-sulfur cracked FCC naphtha with an FCC gasoline hydrodesulfurization unit (or "HDS").

We assume that 40% of the refinery's gasoline pool consists of cracked gasoline from the FCC process train. Because FCC gasoline consists of cracked fragments of heavy, high-sulfur oil molecules, that stream contains up to 1,000 ppm sulfur and is responsible for 98% of the sulfur in the gasoline pool. To make a 30 ppm sulfur gasoline pool, the refinery considers three options for a new FCC train configuration, which are shown, with the capital costs and benefits, in Table 8.1.

The benefits in the "Years 1–6" column will apply for all years *if the sulfur mandate remains at 30 ppm. If the sulfur mandate is tightened to 10 ppm in year 7*, more severe processing will be required, which will cause reduced benefits for years 7–20, as shown in the "years 7–20" column.

With just an FCC unit, the refinery will not be able to make gasoline meeting the 30 ppm sulfur specification and will have to shut down.

Its first ("red") investment option adds a $90 million FCC gasoline hydrodesulfurization unit (HDS), as described in Section 8.3.1, to desulfurize the gasoline

Table 8.1 Investment alternatives, capital costs, and benefits.

		Benefit, $MM/year	
Investment alternatives	**Capital cost $MM**	**Years 1–6**	**Years 7–20 (if 10 ppm spec comes)**
Red HDS only	$90	$0/yr	–$25/yr
→ FCC → HDS → G			
Yellow CFHT only	$300	$30/yr	$10/yr
→ CFHT → FCC → G			
Green CFHT and HDS	$390	$56/yr	$50/yr
→ CFHT → FCC → HDS → G			

fraction of the FCC product, which enables making 30 ppm sulfur gasoline while maintaining current profitability.

In the second ("yellow") option, a $300 million cat feed hydrotreater (CFHT) is added to desulfurize and improve the crackability of the FCC feed. This is a vacuum gas oil desulfurization unit as described in Section 8.1.2. It is a high-horsepower hydrotreater with sufficient kinetic capacity to desulfurize the large heavy oil compounds before they are cracked. It also removes nitrogen and saturates aromatics, which improves the crackability of the FCC feed, allowing higher conversion and flexibility to optimize the yields and product value from the FCC unit.

The third ("green") investment adds both the CFHT and an HDS unit for a cost of $390 million.

8.3.2.2 Benefits in Cleaner Fuels, Increased Margin, and Octane Retention

In the case of the red investment, the capital cost is $90 million for the new HDS unit, which enables making 30 ppm sulfur gasoline. *This is the minimum investment required for continued operation supplying a market with a 30 ppm gasoline sulfur mandate.* This is the base case and is assigned a benefit of zero. In years 7–20, if the sulfur mandate is tightened to 10 ppm, there will be a $25 million/yr penalty versus the base case because the HDS unit will have to be run at higher severity, beyond its intended design, causing high octane loss. The refinery could still make 10 ppm sulfur gasoline, but the HDS unit will be over-stressed in years 7–20 and will saturate more olefins with an octane penalty of $25 million/yr.

In the case of the yellow investment, the capital cost is $300 million for the new CFHT, which enables production of 10 ppm sulfur gasoline because the CFHT removes enough sulfur from the FCC feed to enable production of FCC gasoline that is low enough in sulfur to blend a 30 ppm sulfur gasoline pool. It also improves the FCC feed crackability, which allows feeding lower-cost (for example, heavier, higher sulfur) feed to the FCC train with the same clean product yields, giving upgrade margin benefits of $30 million/yr in years 1–6 versus the base case. In years 7–20, if the sulfur mandate is tightened to 10 ppm, the upgrade margin benefit will fall from $30 million/yr to $10 million/yr because the 10 ppm gasoline sulfur mandate constrains flexibility to feed lower cost (for example, heavier, higher sulfur) feed to the FCC train with the same product yields and quality versus the base case.

In the case of the green investment, the capital cost is $390 million for a new CFHT and a new HDS unit, which delivers upgrade margin benefits of $50 million/yr. The upgrade margin benefit is higher than the yellow case because the existence of the HDS unit increases flexibility to feed even lower cost feeds (for example, coker products) to the FCC train while producing the same clean product yields and product sulfur. In addition, the HDS unit is designed to meet the 10 ppm sulfur mandate even starting in years 1–6, before 10 ppm is needed. This "over-design" for years 1–6 enables running it at lower severity, providing an octane benefit of $6 million/yr in years 1–6 versus the base case. This adds to a total benefit of $56 million/yr in years 1–6 versus the base case. In years 7–20, if the sulfur mandate is tightened to 10 ppm, the HDS unit severity will have to be increased to meet the 10 ppm sulfur mandate, which will eliminate the $6 million/yr octane benefit realized in years 1–6, reducing the total benefit to $50 million/yr for years 7–20.

8.3.2.3 Present Values

The net present values for a 20-year time horizon with a discount rate of 10% are shown in Table 8.2.

The red investment is a $90 million compliance investment that is necessary just to stay in business at the current level of profitability, with a net present value of −$90 million. In the scenario where the sulfur mandate is tightened to 10 ppm in year 7, the present value goes down by $104 million (from −$90 million to −$194 million) because of the large octane penalty that comes with making 10 ppm sulfur with a highly stressed HDS unit designed for 30 ppm sulfur (at this point this refinery might have to shut down).

For the yellow investment, the upgrade margin benefits bring the present value from −$90 million in the red minimum compliance case to $40 million. In the scenario where the sulfur mandate is tightened to 10 ppm, the present value goes down by $40 million (from $40 million to $0) because the FCC train must run a

Table 8.2 Net present values of three investment alternatives for two product sulfur scenarios.

	Net present value, $MM	
Investment alternatives	**30 ppm for 20 years**	**30 ppm for 6 years, then 10 ppm for 14 years**
Red HDS only → FCC → HDS → G	−\$90	−\$194
Yellow CFHT only → CFHT → FCC → G	\$40	\$0
Green CFHT and HDS → CFHT → FCC → HDS → G	\$86	\$62

more restricted feed slate when making 10 ppm S gasoline in years 7–20, which reduces upgrade margin in those years.

For the green investment alternative, the upgrade margin benefits plus gasoline octane benefits bring the present value up to \$86 million. If the sulfur mandate is tightened to 10 ppm in year 7, the present value goes down by \$24 million (from \$86 million to \$62 million) because the HDS unit no longer realizes the \$6 million/yr gasoline octane benefit versus the red base case.

8.3.3 Interpreting Tier 3 Investments with a Real Options Perspective

8.3.3.1 Derivatives

Options, futures, and other derivative securities are important elements of modern finance and can reduce risk when properly applied. As described in published literature (Foote and Murphy, 2001; Murphy, 2010), derivatives are embedded in financial assets, tangible and intangible property, structured products, trading and hedging strategies, and contracts that have flexible terms and conditions. Real options occur naturally as flexibility and growth opportunities that can be exercised over time in an environment of uncertainty.

Insurance can be modeled as a put option having an exercise price equal to the face value of the policy. Any loss in value of the asset is covered by an increase in value of the put option. Reinsurance contracts, commonly known as excess-of-loss contracts, are combinations of long and short call option positions on layers of potential losses. Manufacturing companies often rely on the futures market to lock in raw material costs, and real options exist in the flexible operation of physical equipment.

An oil refinery having the flexibility to switch inputs and outputs has an embedded real option to switch when heating oil is in greater demand than gasoline. Flexibility increases the value of every asset under conditions of uncertainty. Long-term contracts contain option-like features, including volumetric swing components when demand is uncertain. Contract prices can be tied to an index or a basket of commodities. The value of these derivatives is determined using complex quantitative models.

8.3.3.2 Model Risk

Models are merely abstractions of reality and are imperfect. The assumptions behind these models often don't hold up in reality. In the financial industry, quantitative models are used to measure value and risk to help managers make better decisions. In every case, it is critical that management understand both the asset being modeled and the theory and assumptions behind the quantitative model. Assessing model risk is a key element in preventing upsets and providing litigation support.

Engineering students learn the laws of physics and chemistry in an idealized setting (in a vacuum) before adding in real-world imperfections (like an atmosphere or friction). This is how we move from the basics of Newton's laws to chemical reactor kinetics or heat and mass transfer. Models are imperfect representations of reality. This is especially true in the social sciences, which depend on human behavior not scientific law. Unfortunately, we often forget this once we begin applying an equation, like Black-Scholes 1973, to value an option or a more complicated quantitative model in the case of a CDS or CMO.

8.3.3.3 Real Options on Future Sulfur Regulations

This real options concept certainly applies to investment and optimization decisions in refining, and it can bring important insights to capital investment analysis, even when applied in a simple, semi-quantitative way (Hoekstra et al., 2023a).

In our Tier 3 investment analysis, in the scenario where the sulfur mandate stays at 30 ppm for the full 20-year life of the project, the capital investments and present values are as shown in Table 8.3.

Looking just at this scenario, the green investment has the highest present value of $86 million and so, on paper, would be the best alternative. But, for a capital-constrained company, even the red alternative, a $90 million, might be attractive versus the alternative of going out of business.

In the second scenario, where the mandate is tightened to 10 ppm for years 7–20, the payoffs on the investment change to those shown in Table 8.4:

Table 8.3 Net present values of three investment alternatives for the 30 ppm sulfur scenario.

	Investment and present values, \$MM 30 ppm for 20 years	
Investment alternatives	**Capital investment**	**Present value**
Red HDS only ——→ FCC → HDS ——→ G	\$90	−\$90
Yellow CFHT only → CFHT → FCC ——→ G	\$300	\$40
Green CFHT and HDS → CFHT → FCC → HDS ——→ G	\$390	\$86

Table 8.4 Net present values of three investment alternatives for 30 ppm sulfur for 6 years, then 10 ppm sulfur for 14 years scenario.

	Investment and present values, \$MM 30 ppm for 6 years, then 10 ppm for 14 years	
Investment alternatives	**Capital investment**	**Present value**
Red HDS only ——→ FCC → HDS ——→ G	\$90	−\$194
Yellow CFHT only → CFHT → FCC ——→ G	\$300	\$0
Green CFHT and HDS → CFHT → FCC → HDS ——→ G	\$390	\$62

This scenario makes the green alternative look much more attractive – it provides sustainable, profitable operation over the full 20-year time horizon even after the 10 ppm mandate kicks in, with a positive present value of \$62 million.

By contrast, the red alternative suffers a lethal blow in this scenario with a present value of −\$194 million because of the high-octane loss on the over-stressed HDS unit in years 7–20 after the 10 ppm mandate kicks in.

Analysis of the investment alternatives for both scenarios exposes the high risk of the minimum compliance, stay in business strategy, which is that the 10-ppm requirement might be a lethal blow that will bankrupt the refinery if it kicks in.

The real options perspective helps interpret this result. The second scenario with 10 ppm sulfur reduces the green investment's present value by only \$24 million (from \$86 million to \$62 million), while the red investment present value falls by \$104 million (from minus \$90 million to minus \$194 million). The difference of \$80 million (\$104 − \$24 million) is a measure of the *payoff on a real option embedded in the green investment to keep octane loss low even if the 10-ppm scenario occurs*. The green investment alternative is more robust, as well as more profitable, because it provides flexibility to sustain profitability through future circumstances that would otherwise kill the refinery's profitability.

The real options perspective brings critical insight by quantifying the impact of a known coming contingency in a simple, semi-quantitative analysis.

8.3.3.4 Real Options on Future Margins

The above analysis focused on how changing sulfur specifications affects the economics of the red and green investments, and especially how those investments differ in their flexibility to respond to an uncertain future specification change. The analysis does not consider other uncertainties that would affect the green versus red investment decision.

Perhaps the largest uncertainty is the overall level of gasoline margins in the market. The above "fixed benefit" analysis essentially assumes the overall level of gasoline margins will stay constant over time. The green investment would deliver a benefit of \$56 million/yr, which corresponds to \$2.80/barrel of FCC feed, over what would otherwise occur. But to an investor who is bearish on gasoline margins, that \$2.80/barrel benefit could easily be wiped out by declining gasoline market margins, which would work against the higher cost green alternative. Similarly, an investor who is bullish on gasoline margins would be even more inclined to make the green investment.

U.S. gasoline margins have, in fact, been unusually high and volatile in recent years (Hoekstra et al., 2023b). Figure 8.12 shows probability distributions derived from historical data on the price spread between West Texas Intermediate (WTI) crude oil and New York Mercantile Exchange (NYMEX) front-month RBOB gasoline futures (RBOB stands for reformulated blend stock for oxygenate blending, which is a gasoline product produced by refineries for blending with ethanol to make finished gasoline).

We will use this margin between RBOB and WTI crude price to represent the behavior of gasoline margins in the United States. Noteworthy points about Figure 8.12 are as follows:

- The shaded distribution shows that the margin during 2012–2020 was centered around \$16.00/barrel and very rarely went as high as \$30.00/barrel during that 8-year period.

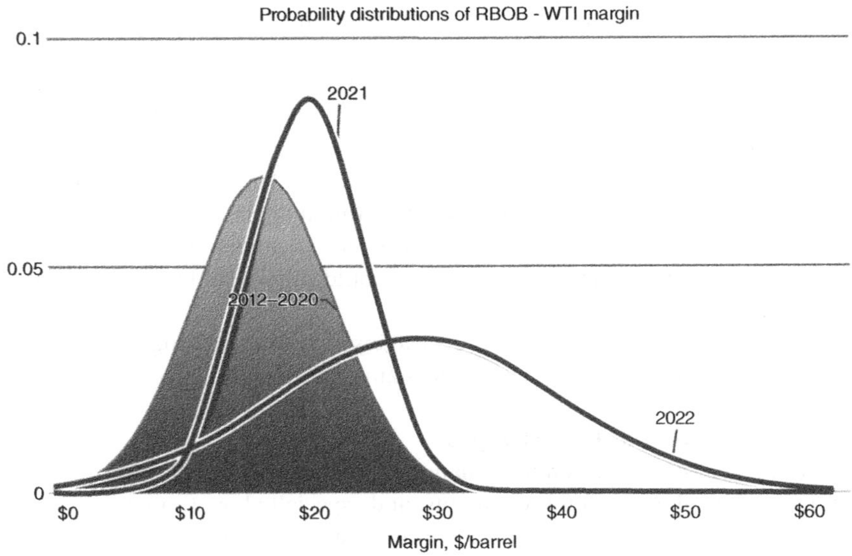

Figure 8.12 Probability distributions for RBOB gasoline-WTI crude price differential show the increasing level and volatility of gasoline margins in the United States.

- The 2021 distribution curve is centered on a mean of $21.60 per barrel, with a 1-year volatility similar to that of the previous 8 years combined.
- 2022 saw a spectacular increase in both the level and volatility of gasoline margin, with a mean of $30.40/barrel and a much broader range than all previous years.

This picture of increasing gasoline margin and margin volatility heightens the interest in considering margin volatility in refining capital investment analysis.

8.3.3.5 The Value of Flexibility

Margin volatility is important because refineries with flexibility can capture much higher margins in more volatile market environments. That is a key principle of real options – *flexibility provides managers with real options to adjust operations in volatile markets as changes occur, and the higher the volatility, the higher the value of those options.*

What does this margin volatility mean for our example case of the green versus red Tier 3 investment decision?

Suppose some combination of events causes gasoline market margins to increase. A red refinery, handcuffed by the lack of flexibility of its FCC train, cannot adjust operations to capture higher margins when opportunities arise. By contrast, the green refinery has many ways to adjust. It can adjust CFHT severity to boost the FCC's yield of low sulfur gasoline; it can feed higher sulfur feedstocks, coker gas oils, metal-containing feeds, and low-value intermediates to the FCC train when those feedstocks are cheap and crack them into fully valued Tier 3 gasoline; or it can make sub-10 ppm sulfur gasoline and generate sulfur credits at low cost for sale to refiners who have high incremental cost for sulfur reduction.

To factor margin volatility and flexibility into our investment analysis, we will treat the annual benefit of the green investment as an uncertain variable rather than a known constant. Consider the $56 million/yr or $2.80/barrel fixed benefit estimate as the expected value (mean) of a probability distribution that describes the green investment payoff in an uncertain margin environment. If we apply the variance rate of the 2012–2020 WTI-RBOB margin to represent the effect of other uncertainties on the market margin, the annual benefit from the green investment can be viewed as a normal probability distribution centered at $2.80/barrel (Figure 8.13).

This indicates that margin volatility provides an opportunity for upside benefits 4 times higher than the fixed $2.80/barrel estimated in the fixed-benefit analysis. Those upside benefits can only be captured by refineries with flexible FCC trains that make the necessary operational adjustments to move to the right of the curve and capture the high market margins when they occur.

The left side of Figure 8.13 shows the potential downside from falling gasoline margins, which could cause the investment payoffs to be less than $2.80/barrel, or even turn negative. But the green refinery has the option to throttle down the CFHT, or even idle it at times when net benefits are low. Notice that when the CFHT is idled, the green investment reduces to the red investment. So its payoff distribution is never less than the red investment.

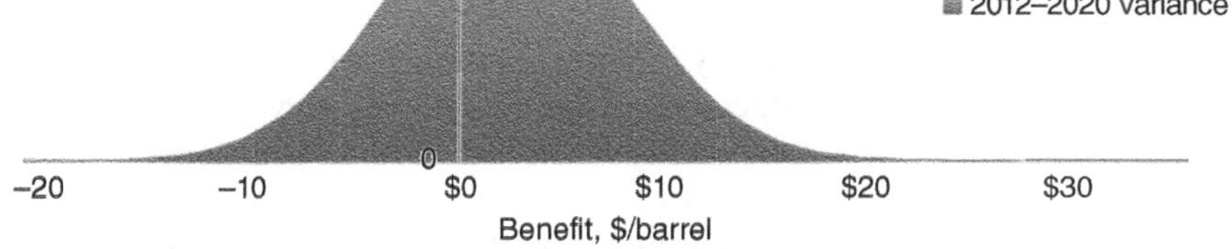

Figure 8.13 Probability distribution of the green investment benefit.

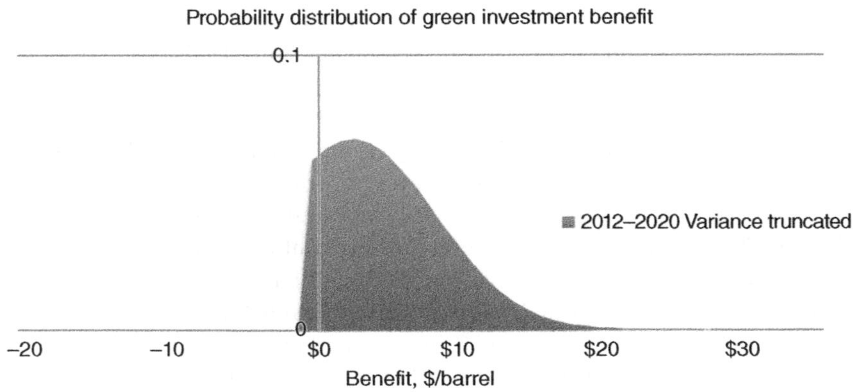

Figure 8.14 Probability distribution of the green investment benefit with throttled-down CFHT in low margin periods, which effectively truncates the probability distribution of investment benefits.

Assuming the refiner exercises its real options to flex CFHT operations as gasoline margins change, including throttling down or idling the unit when called for, the actual payoff distribution for the green investment can be viewed as the truncated distribution in Figure 8.14.

This is equivalent to the payoff distribution of a stock option – the owner of an option is positioned to capture the upside benefits of a volatile stock with strictly limited exposure to its downside risk. Similarly, the flexibility of the green refinery positions it to capture the upside potential of a volatile gasoline margin environment with limited exposure to its downside risk.

The value of this real option can be quantified for any particular refinery using option pricing theory, and that option value should be added to the fixed benefit estimate to give a more accurate picture of the payoff from more flexible FCC train investments.

An implication of the real options perspective is that higher gasoline margin volatility substantially increases the value of the green investment, in the same way that higher stock price volatility increases the value of a stock option. Hence, the view that margins will be more volatile is a strong incentive for the green versus the red investment.

We have seen these factors at play recently in refining operations and refining company earnings. Refining operations have flexed over very wide ranges in response to unpredictable events that have shocked fuel markets in the last few years. And unusually high differences in fuel margin capture rates emerged in refiners' 2021, 2022, and 2023 earnings reports. If you look closely, you will see that refiners with modern, flexible (green) FCC trains are delivering higher refining margin capture rates than those with bottlenecked (red) FCC trains in today's volatile fuel market (Hoekstra, 2022, 2023).

For the purpose of clarity in this paper, we are showing smooth, symmetric price curves. Real options work in all three economic regimes of risk, uncertainty, and extreme events (Foote and Murphy, 2001).

8.3.3.6 Takeaways

This economic analysis shows that an FCC train configuration with both a CFHT and an HDS unit is the most profitable configuration for the Tier 3 world because of its capability to consistently make high yields of ultra-low sulfur gasoline from low-value feedstocks, especially during periods of tightening product specifications and volatile fuel margins.

Gasoline margins and margin volatility have increased in the United States in recent years to an extent that suggests a flexible CFHT-FCC-HDS configuration can capture margins 4 times higher than that indicated by a fixed benefit analysis. Refiners who have this flexibility and use it proactively are the ones who are selectively capturing high market margins today.

This example demonstrates how margin volatility data and the theory of real options can be used in refining investment analysis to bring insight that informs decision-makers by showing the value of flexibility over a project's life, an approach that leads to better decisions in an industry with an uncertain regulatory environment and increasingly volatile input and output prices.

References

Foote, M. and Murphy, T. (2001). Real options: a flexible strategy. *Energy Risk International* (August), 46–49.

Hoekstra, G. (2020). *Gasoline Desulfurization for Tier 3 Compliance*. Chicago: Lulu.com.

Hoekstra, G. (2022). *Clean Fuels and the FCC. RefComm Galveston 2022*. Galveston.

Hoekstra, G. (2023). Big changes in the U.S. octane market. *Petroleum Technology Quarterly* Q4: 79–83.

Hoekstra, G., Murphy, T., and Silva, M. (2023a). Tier 3 gasoline production: challenges and opportunities for refiners. *Hydrocarbon Processing* (February).

Hoekstra, G., Murphy, T., and Silva, M. (2023b). Tier 3 gasoline production: the value of flexibility in volatile fuel markets. *Hydrocarbon Processing* (August).

Murphy, T. (2010). *Derivatives and Model Risk*. New Orleans: Association of American Law Schools Annual Meeting.

Wagner, M. (2022). *Crude Oil Refining – A Simplified Approach*. Boca Raton: CRC Press.

9

Hydrogen Production Using Renewable Feedstock

Gabriela Nascimento da Silva[1], *Clarissa Bergman-Fonte*[1], *Pedro R. R. Rochedo*[1,2], *Frédéric Lantz*[3], *and Alexandre Szklo*[1]

[1] *Centre for Energy and Environmental Economics (Cenergia), Energy Planning Program (PPE), COPPE, Universidade Federal do Rio de Janeiro, Brazil*
[2] *Research and Innovation Center on CO_2 and Hydrogen (RICH Center) and Management Science and Engineering Department, Khalifa University, Abu Dhabi, United Arab Emirates*
[3] *Economic, Energy and Environment, IFP Energies Nouvelles, IFP School, Rueil-Malmaison, France*

9.1 Introduction

In many sectors, hydrogen use is still a technological frontier and/or is not yet consolidated on a commercial scale, such as in transportation and in the electricity sector. However, hydrogen produced from fossil sources has been used for decades in oil refining and in industrial processes, such as the production of margarine, the production of nitrogen fertilizers, the production of methanol, etc. The industrial sector used around 56% of the world hydrogen demand in 2022, while oil refineries accounted for 43% (IEA, 2023) and have been the main producer of hydrogen for decades, with already amortized infrastructure, such as reactors and utility plants (IEA, 2023).

In an oil refinery, two units require significant amounts of hydrogen: the hydrocracking (HCC) and the hydrotreating (HDT) units.[1] HCC converts heavy molecules, which are difficult to fractionate in fluid catalytic cracking (FCC) units,

1 Hydrocracking units operate with the following hydrogen amount: 850–1,700 Nm^3/m^3; for hydrotreating, the values range from 50 to 675 Nm^3/m^3 (the lower limit corresponds to the minimum operating conditions for naphtha hydrotreating and the higher value refers to the condition for hydrotreating atmospheric residues) (Jones and Pujadó, 2006).

Cleaner Petroleum Production and Refining Technologies, First Edition, Edited by M. R. Riazi and Harvey W. Yarranton.

into high value-added lighter compounds, in the presence of a catalyst and at high pressures[2] (Parkash, 2003; Thybaut and Marin, 2016). It is one of the refinery's most versatile processes, converting feeds from many origins and qualities, from gas oils to vacuum residues, into products such as refinery gas, gasoline, naphtha, and diesel. Simultaneously, hydrogenation reactions and molecular breaks occur. Moreover, hydrogen facilitates some reactions, such as cracking, isomerization, heteroatom removal, and hydrogenation (Thybaut and Marin, 2016).

HDT refers to the processes of treating oil derivatives to remove pollutants that cause local effects from emissions and that can deactivate catalysts in refinery units, or to the processes for stabilizing intermediate streams, through double bond saturation (e.g., light cycle oil from the FCC unit). It operates under high pressure,[3] with the addition of hydrogen, and occurs in the presence of a catalyst (Stanislaus et al., 2010; Szklo and Schaeffer, 2007). Its main difference from HCC is the catalyst, which is selective for removing sulfur, nitrogen, and metals. HDT can be divided according to the category of reactions: hydrodesulfurization (HDS), olefin saturation, aromatic saturation (hydrodearomatization – HDA), and hydrodenitrogenation (HDN) – see Figure 9.1.

The continuous decline in the quality of some world's crude oils, both in terms of API grade and sulfur content (Szklo and Schaeffer, 2007), makes it increasingly necessary to use HDT units. At the same time, the stricter emission legislation for local pollutants, such as nitrogen oxides (NOx), particulate matter (PM), and sulfur oxides (SOx) (EPA, 2023; European Parliament and EU Council, 2016; IEA, 2016; IMO, 2021), and the consequent need for more severe HDT to adequate fuels also indicate a trend of increasing hydrogen consumption in refineries.

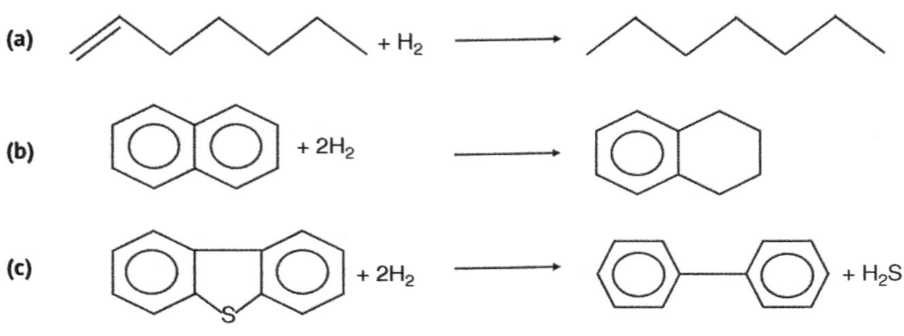

Figure 9.1 Typical hydrotreatment reactions: (a) saturation of an olefin (1-heptene); (b) saturation of an aromatic (naphthalene); (c) desulfurization of a thiophene compound. *Source*: Prepared by the authors.

2 The range of operating pressures for hydrocracking is around 70–140 bar (Speight, 2017).
3 For a typical catalytic hydrodesulfurization unit, the feedstock is fed under pressure of up to 70 bar (Speight, 2017).

Several countries, such as the United States, China, and European countries, have developed legislation[4] to control these contaminants, as a way to improve the quality of the finished fuels and minimize air pollution (Ranzan et al., 2015). To meet emission limits, sulfur-sensitive after-treatment systems are used in vehicles. Therefore, ultra low-sulfur diesel (ULSD) is required so as not to harm the after-treatment systems. This fuel, associated with new treatment technologies in engines, allows a drastic reduction in the emission of atmospheric pollutants. Even when used in older vehicles, ULSD has advantages, such as reduced particulate matter emissions and less deterioration of lubricating oil. Thus, the use of fuels with reduced pollutant content is a way to enable advanced technologies for the engine and for the gas after-treatment systems, so that significant emission reductions can be achieved (Blumberg et al., 2003).

The final product specification determines the required severity levels of the HDT unit. Mild HDT is generally used to remove only sulfur and olefins. Severe HDT, in addition to sulfur and olefins, also removes nitrogen compounds, difficult-to-remove sulfur compounds, and aromatic rings (Szklo and Schaeffer, 2007). The ultra-severe removal of sulfur from oil products involves complex technical issues, because, by evolving from more reactive compounds, such as mercaptans, to less reactive compounds, such as 4,6-dimethyl-dibenzothiophene (called DMDBTs), the difficulty of removing these pollutants increases. The lighter fractions of atmospheric distillation typically contain aliphatic compounds, such as mercaptans, which are very reactive and are easily removed in mild HDS. For the heavier fractions, the predominant compounds are less reactive, more difficult to remove: thiophenes, benzothiophenes, dibenzothiophenes, and other sulfur-containing polyaromatic compounds (Szklo and Schaeffer, 2007). Therefore, heavier oils have

4 The reduction of the sulfur emissions from fuels has been a long process, having started before the 1980s:

- Protocol to the 1979 Convention on Long-Range Transboundary Air Pollution on further Reduction of Sulfur Emissions: the objective was to reduce 30 % of the sulfur emissions by 1993 regarding 1980 emissions (United Nations, 1985);
- The Acid Rain Program (ARP), a market-based initiative of the Environmental Protection Agency (EPA), from 1990: the objective was to obtain reductions of sulfur dioxide (SO_2) and nitrogen oxides (NOx) from the power sector (EPA, 1990);
- The 1999 Göteborg Protocol: the objective was to reduce emissions of air pollutants: sulfur (mainly sulfur dioxide), nitrogen oxides, ammonia and volatile organic compounds (VOC) (UNECE, 1999);
- Directive n°2001/81/CE of the European Parliament: the objective was to limit emissions of acidifying and eutrophying pollutants and ozone precursors (European Parliament, 2001);
- The European regulatory pathway for vehicle emissions limits: for LDVs are ranged from Euro 1 to Euro 6, while for HDVs, from Euro I to Euro VI (European Commission, 2011);
- IMO 2020 Regulation: reduction of the sulfur content in the fuel oil used on board ships (IMO, 2019).

on average higher nitrogen and aromatic contents, which influence the reaction severity of HDT and, consequently, imply a higher demand for hydrogen per processed feed (Szklo and Schaeffer, 2007). The relationship is non-linear due to the chemistry of HDT (IEA, 2005).

The hydrogen used by HDT and/or HCC units is usually produced in the refinery's battery limits. An example is the catalytic reforming unit, in which hydrogen is a byproduct of a series of endothermic reactions that aim to convert naphtha into gasoline with a high-octane rating (or nowadays into BTXs to be used as petrochemical feedstock). This process lowers the content of paraffinic and naphthenic hydrocarbons and increases the content of aromatic compounds. Other byproducts of the catalytic reforming are refinery gas, LPG, isobutane, and n-butane. Benzene, toluene, and xylenes (BTX) can be diverted toward the petrochemical industry (Szklo and Schaeffer, 2007). However, when available in a refinery facility, hydrogen production from the catalytic reforming unit cannot meet, in quantity and quality, the high demands of the HDT and HCC. Therefore, a hydrogen generating unit (HGU) is necessary to complement this production (Gomes et al., 2009; Guedes et al., 2019; Meerman et al., 2012), typically through processes such as steam reforming light hydrocarbons (e.g., methane, ethane, naphtha...) or partial oxidation of heavy hydrocarbons. Usually, HGU processes natural gas or light streams from refinery processes (e.g., refinery gases found in fractioning columns). In HGUs, CO_2 emissions derive not only from fuel combustion to provide steam and heat, but also from the chemical reactions that convert hydrocarbons into CO_2 and H_2 (API, 2021).

Thus, the global refining industry faces challenges associated with the trade-off between emissions of pollutants with global impacts from its operation and emissions of pollutants with local impacts due to the use of oil products (Szklo and Schaeffer, 2007). This trade-off refers to the fact that the production of fuels with extremely low contaminant content requires more energy and hydrogen than the normally conducted process and thus results in higher CO_2 emissions by refineries (Szklo and Schaeffer, 2007).

In 2022, less than 1% of global hydrogen was produced via low GHG emissions routes, with no relevant example of this low-GHG-related-hydrogen supply in oil refineries. In the same period, natural gas was the most common feedstock processed (62%), followed by coal (21%) (IEA, 2023). Steam reforming of natural gas was the preferred technological production route, being the selected route for oil refineries. As mentioned before, steam reforming generates CO_2 emissions from energy use and from the chemical reaction itself, totaling around 9.5 kg of CO_2/kg of H_2 produced (IRENA, 2019).

However, to produce low-emission fuels in refineries, it is worth using low-carbon hydrogen as well. This becomes even more relevant with the intended use of co-processing of biomass, which also increases the hydrogen demand of hydrotreating units (see Chapter 11), and with the more stringent specification of fuels. Actually, the co-processing of lipid biomass in oil refining improves the quality of fuels and reduces raw material costs if waste is used, in addition to reducing GHG emissions (Bezergianni et al., 2018). However, this co-processing also results in higher hydrogen consumption in HDT, compared to oil derivatives: an increase of more than 40% in the hydrogen consumption, when rising the lipids/oil derivatives fractions in HDT from 0/100 to 30/70 (Bezergianni et al., 2018).

There are many ways to produce low-carbon hydrogen, from options that involve fossil hydrocarbon feeds with carbon capture[5] to options that use renewable energy (Hanley et al., 2018; IEA, 2023; Incer-Valverde et al., 2023; Pleshivtseva et al., 2023). In the context of the current energy transition, these low-carbon hydrogen production processes have gained great importance. This chapter describes and discusses routes for the production of hydrogen with low GHG emissions associated with oil refineries, starting with the most widely used route today, steam reforming, which can process fossil or renewable hydrocarbon inputs and be even associated with carbon capture. Subsequently, processes that have gained great importance recently are described, such as water electrolysis, oxyfuel combustion, and carbon capture technologies, which allow not only to avoid emissions into the atmosphere, but also to generate inputs for the production of synthetic fuels.[6]

9.2 Hydrogen Production in Oil Refineries

Steam reforming is a catalytic and endothermic process in which steam and a hydrocarbon or an alcohol are converted into hydrogen and CO_2. The most used input for hydrogen production via steam reforming is dry natural gas, due to its higher methane content (or C:H proportion) (Saeidi et al., 2023). The chemical reaction begins with methane reacting with steam at high temperatures in the presence of a catalyst, usually Ni-based supported on ceramic oxides, to produce syngas (basically H_2 and CO):

$$CH_4 + H_2O \rightarrow CO + 3H_2 \qquad \Delta H^0_{298} = +206\ \text{kJ/mol} \tag{9.1}$$

5 Lindsay et al. (2009) compared three configurations for producing low-carbon hydrogen via steam reforming associated with carbon capture. The authors showed that it is possible to reach 81% of emissions reductions at the cost of 35 USD/tCO_2.

6 Low TRL routes to produce H_2, such as photolysis, thermolysis with concentrating solar energy, among others, will not be the focus of this chapter.

In the following step, syngas reacts with more steam at high temperatures in *Water–Gas Shift* (WGS) reactors,[7] and CO is converted into CO_2. The hydrogen purification system, *Pressure Swing Adsorption* (PSA), removes CO_2 more easily than CO, so it is important to convert CO into CO_2 before the gas is sent to the PSA unit. This conversion also produces an additional amount of hydrogen, which increases the process yield (Meerman et al., 2012).

$$CO + H_2O \rightarrow CO_2 + H_2 \qquad \Delta H^0_{298} = -41\ \text{kJ/mol} \tag{9.2}$$

PSA consists of containers filled with specific adsorbents for this separation. The process brings hydrogen to purities higher than 99.9% by volume.

Figure 9.2 shows a simplified block diagram of methane steam reforming.

Due to the endothermic nature of the process, an energy source is required to generate the heat needed for the reactions. Part of the steam is produced using heat recovery from the effluent of the reforming furnace and the flue gases (Anzelmo et al., 2017). The rest must be supplied by burning additional fuels.

In steam reforming, three streams with potential for implementation of CO_2 capture can be identified[8]: (1) the stream that leaves the WGS reactor, upstream of the hydrogen purification unit (PSA); (2) the so-called *tail gas*, downstream of the PSA; and (3) the flue gas, at the outlet of the reformer (Soltani et al., 2014).

For cases in which a feedstock derived from biomass is used, such as biomethane, bionaphtha, or even bioethanol, this carbon is biogenic, and, therefore,

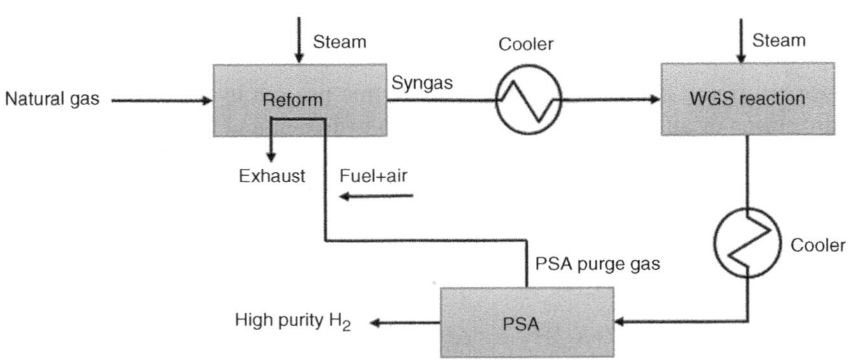

Figure 9.2 Simplified block diagram of *methane* steam reforming. *Source*: Prepared by the authors.

7 The most common configuration is based on two reactors: one at high temperatures (340–530 °C), to favor reaction kinetics; and another at low temperatures (180–230 °C) for a balance with higher yield.

8 This topic will be discussed in more detail in Section 9.3.1.

emissions are part of the carbon cycle. If CO_2 is captured in this case, the process has potential to remove CO_2 from the atmosphere in its life cycle (Bui et al., 2021; Rosa and Mazzotti, 2022; Wu et al., 2023). For example, the steam reforming of bioethanol is a very similar process to the steam reforming of fossil methane. However, in this case, the process is less endothermic (Silva, 2010).

$$C_2H_5OH + 3H_2O \rightarrow 2CO_2 + 6H_2 \qquad \Delta H^0_{298} = +173\ \text{kJ/mol} \tag{9.3}$$

Regarding biomethane as feedstock for steam reforming, in addition to the utilization of a renewable and low-cost input, providing an adequate destination to a waste stream is also a positive side effect. Moreover, biomethane (or the methane from biogenic sources) takes advantage of the existing infrastructure of steam reforming, thus avoiding higher investment costs.

The most common process for biogas production is anaerobic biodigestion (Fuess et al., 2021; Marafon et al., 2019). The main reactors used in this process are UASB (*Upflow Anaerobic Sludge Blanket*), covered lagoon digester, and the conventional anaerobic reactor. Anaerobic biodigestion is a sequential process that occurs in the absence of oxygen and relies on the activity of microorganisms to promote the fermentation of organic matter. The process results in biogas, a mixture of gases, composed mainly of methane (CH_4) and carbon dioxide (CO_2). The group of microorganisms active in biodigestion can be divided into three categories: acidogenic, acetogenic, and methanogenic bacteria. Each one acts on a step in the anaerobic biodigestion process (Marafon et al., 2019). In order to obtain biomethane, it is necessary to remove hydrogen sulfide (H_2S) and CO_2, depending on the feedstock used for producing biogas (Fuess et al., 2021). CO_2 can be compressed and stored in geological formations or used in the production of synthetic fuels. Likewise, the steam reforming of biomethane also has potential for carbon capture,[8] mainly by chemical absorption, with the use of solvents such as monoethanolamine (MEA). This removal and storage of biogenic CO_2 would make the hydrogen production process at the refinery potentially negative in terms of carbon dioxide emissions.

For instance, a study from Nascimento da Silva et al. (2024b) evaluated the potential for hydrogen production in oil refineries processing biomethane derived from a residue of the sugar and alcohol industry, the vinasse. The authors developed a case study for large refineries in Brazil and observed that two refineries (of a total of nine evaluated) have the potential to take advantage of the vinasse produced in distilleries within a radius of 100 km away. The seasonality of sugarcane and the dispersed location of distilleries are challenges for this option. Even so, the economic and logistical analysis of this study showed that a carbon price starting at 38 USD/tCO_2 could make it economically viable to use vinasse for hydrogen production in these refineries.

9.3 Hydrogen Produced from Water Electrolysis

Hydrogen from water electrolysis associated with renewable energy sources, the so-called "green hydrogen," has gained great international attention in recent years (IEA, 2023). Green hydrogen is considered a key energy carrier in the decarbonization strategy of several nations, such as the European Union member States,[9] Chile,[10] Australia,[11] Japan,[12] Canada,[13] South Africa,[14] among others. This process should become interesting in regions with great availability of natural resources, such as fresh water, solar irradiation throughout the year, and regular wind speeds.

However, data from the IEA (2023) estimates that, in 2022, water electrolysis represented less than 0.1% of the world's dedicated hydrogen production, from a total of 95 Mt. This process consists of an electrochemical reaction in which a water molecule is broken into H_2 and O_2,[15] with the application of an electric current. Electrolyzers, the equipment where electrolysis occurs, are large consumers of energy, thus they need to be powered with renewable electricity so that the process results in zero direct CO_2 emissions (Busch et al., 2023). The declining prices and advanced technological maturity of solar photovoltaics and wind energy make them interesting options for the production of green hydrogen (IRENA, 2022). Electricity produced from biomass waste associated with carbon capture[16] is also an interesting option, due to the advantage of not having the intermittency of wind and solar resources.

In addition to electricity, electrolysis requires a large amount of water: the ratio is about 11 L/kg of H_2 (Argonne, 2015). This water needs to be previously treated due to the high degree of purity required by the electrolyzers. The production of hydrogen occurs according to the reaction:

$$H_2O \rightarrow ½O_2 + H_2 \qquad \Delta H^0_{298} = 285{,}83 \text{ kJ/mol} \tag{9.4}$$

The main constituent parts of an electrolyzer are two electrodes, cathode and anode, where hydrogen and oxygen are formed, respectively; the electrolyte, which

9 Available at: A hydrogen strategy for a climate-neutral Europe, European Commission (2020).
10 Available at: National Green Hydrogen Strategy, Chile (2020).
11 Available at: National Hydrogen Strategy, Australia (2019).
12 Available at: Basic Hydrogen Strategy, Japan (2017).
13 Available at: The Hydrogen Strategy, Canada (2020).
14 Available at: Hydrogen society roadmap for South Africa 2021.
15 For each kg of hydrogen produced, 8 kg of oxygen is co-produced. This O_2 can be used in applications such as combustion processes in refining or in petrochemical processes in refinery-associated plants.
16 In this case, carbon capture via oxycombustion can be used, with the use of the oxygen co-produced in electrolysis. This topic will be covered in more detail in Section 9.3.1.

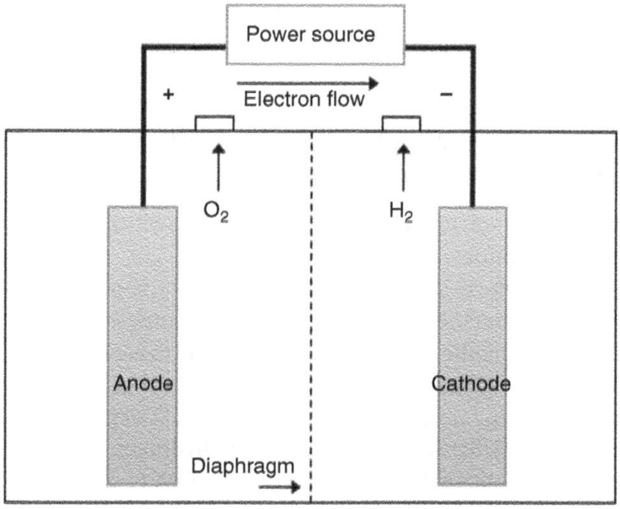

Figure 9.3 Basic schematic of water electrolysis. *Source*: Prepared by the authors.

assists in the conduction of ions; and a diaphragm, which is an electrical insulator that keeps the produced gases separate (Schiebahn et al., 2015). Figure 9.3 illustrates a basic schematic of water electrolysis:

There are several types of electrolyzers, however the most developed and currently used technologies, classified according to the type of electrolyte, are described as follows:

- **AEL, *Alkaline Electrolysis***: This is the most used technology, available commercially (IEA, 2023). Its electrolyte is composed of an aqueous electrolyte solution of potassium hydroxide or sodium hydroxide. This type of technology is not adequate for high level of flexibility in its operation, so it does not perform well with a variable electricity source.
- **PEM, *Polymer Electrolyte Membrane***: This technology works with a proton exchange membrane and a solid polymeric electrolyte. Hydrogen generation occurs in the same way as the alkaline electrolyzer. However, PEM performs better regarding operational flexibility than AEL, as the transport of protons across the membrane offers a rapid response to fluctuations in electricity supply (Palmer et al., 2021). Although this technology is the most suitable to operate with intermittent energy sources, some negative characteristics are the low lifetime and the use of noble metals (such as platinum and iridium) and complex materials in the polymeric membrane (nafion fluoropolymer-copolymer), which make the system more expensive (Kiemel et al., 2021).

- **SOEC, *Solid Oxide Electrolysis Cell*:** This technology applies steam at high temperatures. The electrolyte used is composed of a ceramic material, such as yttria-stabilized zirconia. Cathodes are usually composed of nickel, and anodes are strontium-doped lanthanum manganite. This electrolyzer is the one with the lowest technological maturity among the three, and its main disadvantage regards the degradation of materials due to high operating temperatures.

Table 9.1 Compares the main characteristics of AEL and PEM electrolyzers, the current commercially available options.

One challenge of electrolysis is the cost of electrolyzers. However, some authors and international agencies argue that this barrier tends to be overcome as the technology gains greater scale, expecting the decrease of unit cost due to economies of scale (Götz et al., 2016; IEA, 2019; IRENA, 2018). The levelized cost of hydrogen (LCOH) produced from wind energy ranges from 4.61 to 10.01 USD/kgH_2, while for solar PV this value is 7.1–14.87 USD/kgH_2. In comparison, hydrogen produced from steam reforming methane has an LCOH ranging from 1.03 to 2.16 USD/kgH_2 (Parkinson et al., 2019).

Besides, water electrolysis co-produces oxygen, and this gas can be used to capture carbon dioxide in the FCC via oxy-combustion in oil refineries. This possibility has been further explored in Nascimento da Silva et al. (2022) and will be discussed in Section 9.3.1.

In addition, Nascimento da Silva et al. (2024a) evaluated a case study for a refinery in Bahia, Brazil, in which solar and wind energy associated with water electrolysis

Table 9.1 Main features of AEL and PEM electrolyzers.

Characteristics	AEL	PEM
System Efficiency (LHV[a]) (%)	63–71	60–68
Lifetime (hours)	60,000–90,000	30,000–90,000
Scale of systems (MWe)	<6	<2
Current Densities (A/cm^2)	0,2–0,4	0,6–2
Operating Temperatures (°C)	60–80	50–80
Operating Pressures (bar)	<30	<200
Gas Purity (%)	99.5	99.99
Costs (USD/kWe[b])	500–1,400	1,100–1,800

a. Efficiency calculated based on hydrogen lower heating value (LHV).
b. Electrolysis-related characteristics are usually measured in electrical kW, i.e., based on the power received.

Source: Adapted from Buttler and Spliethoff (2018), IEA (2019a).

could be employed to produce hydrogen to supply the total refinery's demand. The economic evaluation addressed the whole hydrogen value chain through an optimization tool called HERA (Nascimento da Silva et al., 2023), including production, compression, and transportation via trucks or pipelines to the refinery, and showed costs for the hydrogen produced at an average of 5.6 USD/kgH_2. Consequently, the carbon price to make this option viable would be 448 USD/tCO_2, still much higher when compared to the case of vinasse biomethane, mentioned before. However, the availability of vinasse is very site-specific, while solar irradiation and high and constant wind speeds are more spread around the world and indicate greater potential for the water electrolysis route to supply H_2 to oil refineries.

9.3.1 Power-to-liquids

Hydrogen from electrolysis can also be used to produce syngas. This conversion occurs through the reaction with carbon dioxide (resulting from the capture of an emission source or from the atmosphere). This reaction, known as RWGS (*Reverse Water–Gas Shift*), leads to the production of water and carbon monoxide. Subsequent steps are employed to adjust the composition of the syngas formed, making it suitable for use as an input in a Fischer–Tropsch reactor. After treatment of the produced liquids, value-added products are obtained (Brynolf et al., 2018; Fonte, 2021). Figure 9.4 illustrates the possible transformations. The refining process units that can be used are indicated.

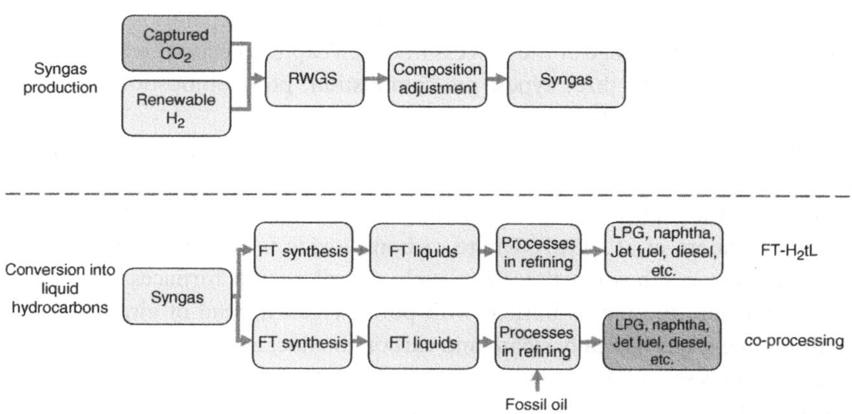

Figure 9.4 Transformation of renewable hydrogen into final products via different routes. Refining process units that can be used in each case are shown. *Source*: Prepared by the authors.

Similar to what occurs in the case of the FT-BtL Route (described in Chapter 11), but with a syngas of different origin, the Fischer–Tropsch liquids obtained from this synthesis consist of a mixture of hydrocarbons analogous to fossil oil. The syngas used in this case is purer than the one derived from the FT-BtL route. In this way, this stream can benefit from various refining processes, including distillation. The fractions obtained can be sent to different units, thus obtaining different final products, all similar to their corresponding fossil products. In this chapter, we used the name FT-H2tL (*Fischer–Tropsch Hydrogen-to-Liquids*) to describe this route.

In addition, co-processing is also an option in this case, which means that the Fischer–Tropsch liquids and fossil oils can be fed together in refining units. This strategy is more technologically mature compared to the complete redesign of refineries and falls under the ASTM D1655 standard for jet fuel production, employing up to 5% by volume of Fischer–Tropsch hydrocarbons (ICAO, 2021). The products obtained have a renewable carbon content.

However, in addition to hydrogen to compose the necessary inputs for the FT-H2tL route, a source of CO_2 is necessary. Petroleum refineries have several CO_2-emitting processes (Abdul-Manan et al., 2017; Mastropasqua et al., 2020), which can be captured not only to produce Fischer–Tropsch liquids but also to be stored, avoiding its release to the atmosphere. The main attributes necessary for a potential source for CO_2 capture in refineries are discussed in the next paragraphs.

Carbon capture is an alternative that involves separating carbon dioxide from gas streams. The higher its concentration in the stream, the easier it is to separate it (Rochedo and Szklo, 2013). In general, capture technologies can be classified according to the separation principle: chemical absorption, physical absorption, adsorption, chemical loop, membranes, and cryogenic separation (Concawe, 2018). Regarding the stage of the process in which capture occurs, carbon capture can be classified into three types: pre-combustion, post-combustion, and oxy-combustion (Yao et al., 2018).

In pre-combustion systems, fuels are converted into syngas, and thus carbon is removed prior to combustion, but this method is not often reported for refineries. In contrast, post-combustion systems are the most mature capture technology (Escudero et al., 2016), where carbon dioxide is captured from flue gas resulting from combustion processes such as boilers and furnaces. Finally, in oxy-combustion, the fuel is burned with pure oxygen instead of air, and so the resulting flue gas is basically steam and carbon dioxide, easily separated through cooling/compression (De Mello et al., 2013; Yao et al., 2018).

Although post-combustion is considered the most viable capture process for the refining and petrochemical industry, with abatement costs between 70 and 120 USD/tCO_2, the dispersed sources of carbon dioxide and the low concentration of this gas in the streams make capture in refining plants a challenge (Escudero et al., 2016). Carbon capture in refining is more often reported in steam reforming units (Yao et al., 2018), fluid catalytic cracking (FCC) (Güleç et al., 2023; Tang et al., 2023), and cogeneration systems for electricity and heat (Yáñez et al., 2022). The most consolidated process is the capture with solvent MEA,[17] with a capture efficiency of 90% even for low carbon dioxide concentrations in the exhaust (4–16%) (Yáñez et al., 2022). However, due to the high energy penalty of the solvent regeneration step, corrosion, and high costs of the solvent, other methods have been constantly researched – chemical looping, cryogenic distillation, adsorption, membrane, and the combination of some of them – as well as other solvents – phase change solvents, deep eutectic solvents (DESs), blends of chemical and physical solvents, blends of different chemical solvents, blending chemical solvents with ionic liquids (ILs), adding a surfactant to the solvent, etc. (Khan et al., 2023; Leonzio and Shah, 2024).

In addition, as mentioned before, FCC units present the possibility of capturing the carbon dioxide resulting from the regeneration of the catalyst via oxy-combustion (De Mello et al., 2013; Nascimento da Silva et al., 2022). This type of capture is also suitable for cogeneration systems of electricity and heat (Escudero et al., 2016). In the FCC unit, the thermal balance matters. Burning coke with pure oxygen reaches higher temperatures than burning with air, which can harm the materials in the unit. One way to limit these temperatures is to recirculate a stream of CO_2 in the system, so that this gas acts as a thermal diluent (De Mello et al., 2013). However, the challenge of oxy-combustion lies in obtaining pure oxygen, a step that results in a large energy penalty due to the high electricity consumption of an air separation unit (ASU) (Tang et al., 2023). Thus, the implementation of carbon capture via oxy-combustion in refining could be an opportunity to use the oxygen co-produced in the water electrolysis reaction (addressed in Section 9.2) without the energy penalty of an air separation unit. This idea was evaluated for an oil refinery by (Nascimento da Silva et al., 2022).

9.4 Conclusion

As the demand for hydrogen by oil refineries is expected to grow, either due to the stricter fuel specifications or due to the co-processing biomass, it is worth finding ways of producing hydrogen that are consistent with current global climate ambitions. Alternatives involve low-carbon, zero-carbon, or carbon-negative hydrogen. The first option includes the steam reforming of fossil methane with carbon capture and storage, or even the steam reforming of biomethane with the

17 Monoethanolamine (MEA) is a primary amine widely used industrially, with high rates of removal of acid gases, such as CO_2.

advantage of low costs and already existing partial infrastructure. Bioethanol is also an option. For the second option, hydrogen from electrolysis associated with renewable energy sources is interesting, mainly due to the spread availability of wind and solar resources worldwide. The third option refers mainly to the steam reforming of biomethane associated with carbon capture and storage: the advantages of this route include the low costs and environmental benefits, but the limited availability of feedstock is a barrier. It is also worth noting that electrolysis can provide not only hydrogen, but also oxygen for refining units, which can be used in oxy-combustion associated with carbon capture processes, especially in FCC units. The most appropriate strategy will depend on the characteristics of the refinery and the availability of resources nearby.

References

Abdul-Manan, A.F.N., Arfaj, A., and Babiker, H. (2017). Oil refining in a CO_2 constrained world: effects of carbon pricing on refineries globally. *Energy* 121: 264–275. https://doi.org/10.1016/j.energy.2017.01.046.

Anzelmo, B., Wilcox, J., and Liguori, S. (2017). Natural gas steam reforming reaction at low temperature and pressure conditions for hydrogen production via Pd/PSS membrane reactor. *Journal of Membrane Science* 522: 343–350. https://doi.org/10.1016/j.memsci.2016.09.029.

API (2021). *Compendium of Greenhouse Gas Emissions Methodologies for the Natural Gas and Oil Industry*. American Petroleum Institute.

Argonne (2015). *Development of a Life Cycle Inventory of Water Consumption Associated with the Production of Transportation Fuels*. Argonne: Argonne National Lab.

Bezergianni, S., Dimitriadis, A., Kikhtyanin, O. et al. (2018). Refinery co-processing of renewable feeds. *Progress in Energy and Combustion Science* 68: 29–64. https://doi.org/10.1016/j.pecs.2018.04.002.

Blumberg, K.O., Walsh, M.P., and Pera, C. (2003). Low-sulfur gasoline & diesel: the key to lower vehicle emissions. The International Council on Clean Transportation (ICCT).

Brynolf, S., Taljegard, M., Grahn, M. et al. (2018). Electrofuels for the transport sector: a review of production costs. *Renewable and Sustainable Energy Reviews* 81: 1887–1905. https://doi.org/10.1016/j.rser.2017.05.288.

Bui, M., Zhang, D., Fajardy, M. et al. (2021). Delivering carbon negative electricity, heat and hydrogen with BECCS – comparing the options. *International Journal of Hydrogen Energy* 46: 15298–15321. https://doi.org/10.1016/j.ijhydene.2021.02.042.

Busch, P., Kendall, A., and Lipman, T. (2023). A systematic review of life cycle greenhouse gas intensity values for hydrogen production pathways. *Renewable and Sustainable Energy Reviews* 184: 113588. https://doi.org/10.1016/j.rser.2023.113588.

Buttler, A. and Spliethoff, H. (2018). Current status of water electrolysis for energy storage, grid balancing and sector coupling via power-to-gas and power-to-liquids: a review. *Renewable and Sustainable Energy Reviews* 82: 2440–2454. https://doi.org/10.1016/j.rser.2017.09.003.

Concawe (2018). Concawe Review 27. Volume 27, Number 1. https://www.concawe.eu/wp-content/uploads/Concawe-Review-27-1_webfile.pdf (accessed May 2023).

De Mello, L.F., Gobbo, R., Moure, G.T. et al. (2013). Oxy-combustion technology development for fluid catalytic crackers (FCC) - large pilot scale demonstration. *Energy Procedia* 37: 7815–7824. https://doi.org/10.1016/j.egypro.2013.06.562.

EPA (1990). *The Acid Rain Program*. United States Environmental Protection Agency.

EPA (2023). Process of reviewing the national ambient air quality standards https://www.epa.gov/criteria-air-pollutants/process-reviewing-national-ambient-air-quality-standards (accessed 25 March 2024).

Escudero, A.I., Espatolero, S., and Romeo, L.M. (2016). Oxy-combustion power plant integration in an oil refinery to reduce CO_2 emissions. *International Journal of Greenhouse Gas Control* 45: 118–129. https://doi.org/10.1016/j.ijggc.2015.12.018.

European Commission (2011). Commission Regulation (EU) No 582/2011 (accessed on May 2023).

European Parliament (2001). DIRECTIVE 2001/81/EC of the European Parliament and of the Council. https://eur-lex.europa.eu/eli/dir/2001/81/oj/eng (accessed May 2023).

European Parliament, EU Council (2016). Directive (EU) 2016/2284 of the European Parliament and of the Council-of 14 December 2016-on the reduction of national emissions of certain atmospheric pollutants, amending Directive 2003/35/EC and repealing Directive 2001/81/EC. https://eur-lex.europa.eu/eli/dir/2016/2284/oj/eng (accessed May 2023).

Fonte, C.B. (2021). Avaliação de rota de produção de eletrocombustíveis para aviação utilizando eletricidade solar fotovoltaica e captura de dióxido de carbono diretamente do ar. Universidade Federal do Rio de Janeiro.

Fuess, L.T., Cruz, R.B.C.M., Zaiat, M. et al. (2021). Diversifying the portfolio of sugarcane biorefineries: anaerobic digestion as the core process for enhanced resource recovery. *Renewable and Sustainable Energy Reviews* 147: 111246. https://doi.org/10.1016/j.rser.2021.111246.

Gomes, G.L., Szklo, A., and Schaeffer, R. (2009). The impact of CO_2 taxation on the configuration of new refineries: an application to Brazil. *Energy Policy* 37: 5519–5529. https://doi.org/10.1016/j.enpol.2009.08.019.

Götz, M., Lefebvre, J., Mörs, F. et al. (2016). Renewable power-to-gas: a technological and economic review. *Renew Energy* 85: 1371–1390. https://doi.org/10.1016/j.renene.2015.07.066.

Guedes, F., Szklo, A., Rochedo, P. et al. (2019). Climate-energy-water nexus in Brazilian oil refineries. *International Journal of Greenhouse Gas Control* 90: 102815. https://doi.org/10.1016/j.ijggc.2019.102815.

Güleç, F., Meredith, W., and Snape, C.E. (2023). CO_2 capture from fluid catalytic crackers via chemical looping combustion: regeneration of coked catalysts with oxygen carriers. *Journal of the Energy Institute* 107: 101187. https://doi.org/10.1016/j.joei.2023.101187.

Hanley, E.S., Deane, J.P., and Gallachóir, B.P.Ó. (2018). The role of hydrogen in low carbon energy futures – a review of existing perspectives. *Renewable and Sustainable Energy Reviews* 82: 3027–3045. https://doi.org/10.1016/j.rser.2017.10.034.

ICAO (2021). Conversion processes. Environment. https://www.icao.int/environmental-protection/GFAAF/Pages/Conversion-processes.aspx (accessed 23 May 2023).

IEA (2005). The European Refinery Under the EU Emissions Trading Scheme. https://seors.unfccc.int/applications/seors/attachments/get_attachment?code=5RUQ0TNKB1L5YUFU90G5SD8SH1FBOINN (accessed May 2023).

IEA (2016). Energy and Air Pollution. IEA, Paris. https://www.iea.org/reports/energy-and-air-pollution, Licence: CC BY 4.0.

IEA (2019). The future of hydrogen: seizing today's opportunities. IEA, Paris. https://www.iea.org/reports/the-future-of-hydrogen, Licence: CC BY 4.0.

IEA (2023). Global Hydrogen Review 2023. IEA, Paris. https://www.iea.org/reports/global-hydrogen-review-2023, Licence: CC BY 4.0.

IMO (2019). IMO 2020: Consistent Implementation of MARPOL Annex VI. International Maritime Organization.

IMO (2021). Sulphur oxides (SOx) and particulate matter (PM) – Regulation 14 https://www.imo.org/en/OurWork/Environment/Pages/Sulphur-oxides-(SOx)-%E2%80%93-Regulation-14.aspx (accessed 25 March 2024).

Incer-Valverde, J., Korayem, A., Tsatsaronis, G. et al. (2023). "Colors" of hydrogen: definitions and carbon intensity. *Energy Conversion and Management* 291: 117294. https://doi.org/10.1016/j.enconman.2023.117294.

IRENA (2018). *Hydrogen From Renewable Power: Technology Outlook for the Energy Transition*. Abu Dhabi: International Renewable Energy Agency.

IRENA (2019). *Hydrogen: A Renewable Energy Perspective*. International Renewable Energy Agency.

IRENA (2022). *Renewable Technology Innovation Indicators: Mapping progress in Costs, Patents and Standards*. International Renewable Energy Agency.

Jones, D.S.J. and Pujadó, P.R. (2006). *Handbook of Petroleum Processing*. Springer.

Khan, U., Ogbaga, C.C., Abiodun, O.A.O. et al. (2023). Assessing absorption-based CO_2 capture: research progress and techno-economic assessment overview. *Carbon Capture Science and Technology* 8: 100125. https://doi.org/10.1016/j.ccst.2023.100125.

Kiemel, S., Smolinka, T., Lehner, F. et al. (2021). Critical materials for water electrolysers at the example of the energy transition in Germany. *International Journal of Energy Research* 45: 9914–9935. https://doi.org/10.1002/er.6487.

Leonzio, G. and Shah, N. (2024). Recent advancements and challenges in carbon capture, utilization and storage. *Current Opinion in Green and Sustainable Chemistry* 46: 100895. https://doi.org/10.1016/j.cogsc.2024.100895.

Lindsay, I., Lowe, C., Reddy, S. et al. (2009). Designing a climate friendly hydrogen plant. *Energy Procedia* 1: 4095–4102. https://doi.org/10.1016/j.egypro.2009.02.217.

Marafon, A.C., Salomon, K.R., Amorim, E.L.C. et al. (2019). Use of sugarcane vinasse to biogas, bioenergy, and biofertilizer production. *Sugarcane Biorefinery, Technology and Perspectives* 179–194. https://doi.org/10.1016/B978-0-12-814236-3.00010-X.

Mastropasqua, L., Pegorin, A. and Campanari, S. (2020). Low fuel utilisation solid oxide fuel cell system for CO_2-free hydrogen production in oil refineries. *Journal of Power Sources* 448: 227461. https://doi.org/10.1016/j.jpowsour.2019.227461.

Meerman, J.C., Hamborg, E.S., van Keulen, T. et al. (2012). Techno-economic assessment of CO_2 capture at steam methane reforming facilities using commercially available technology. *International Journal of Greenhouse Gas Control* 9: 160–171. https://doi.org/10.1016/j.ijggc.2012.02.018.

Nascimento da Silva, G., Lantz, F., Rochedo, P.R.R. et al. (2023). *HERA – Hydrogen Economics and Infrastructure Optimization Model*. IFPEN Economic Papers. IFP Energies nouvelles (IFPEN).

Nascimento da Silva, G., Lantz, F., Rochedo, P.R.R. et al. (2024a). Developing and applying the Hydrogen Economics and infRAstructure optimization model (HERA). *International Journal of Hydrogen Energy* 61: 1170–1186. https://doi.org/10.1016/j.ijhydene.2024.02.015.

Nascimento da Silva, G., Rochedo, P.R.R. and Szklo, A. (2022). Renewable hydrogen production to deal with wind power surpluses and mitigate carbon dioxide emissions from oil refineries Levelized Cost of Hydrogen. *Applied Energy* 311: 118631. https://doi.org/10.1016/j.apenergy.2022.118631.

Nascimento da Silva, G., Szklo, A. and Rochedo, P. (2024b). Potential and economic evaluation of hydrogen production processing vinasse biomethane derived from the sugar and alcohol industry for oil refineries. Paper under submission.

Palmer, G., Roberts, A., Hoadley, A. et al. (2021). Life-cycle greenhouse gas emissions and net energy assessment of large-scale hydrogen production via electrolysis and solar PV. *Energy & Environmental Science* 14: 5113–5131. https://doi.org/10.1039/D1EE01288F.

Parkash, S. (2003). Hydrocracking processes. In: *Refining Processes Handbook*, 62–108. https://doi.org/10.1016/B978-075067721-9/50003-7.

Parkinson, B., Balcombe, P., Speirs, J.F. et al. (2019). Levelized cost of CO_2 mitigation from hydrogen production routes. *Energy & Environmental Science* 12: 19–40. https://doi.org/10.1039/c8ee02079e.

Pleshivtseva, Y., Derevyanov, M., Pimenov, A. et al. (2023). Comprehensive review of low carbon hydrogen projects towards the decarbonization pathway. *International Journal of Hydrogen Energy* 48 (10): 3703–3724. https://doi.org/10.1016/j.ijhydene.2022.10.209.

Ranzan, C., Ranzan, L., Trierweiler, L.F. et al. (2015). Sulfur determination in diesel using 2D fluorescence spectroscopy and linear models. *IFAC-PapersOnLine* 48 (8): 415–420. https://doi.org/10.1016/j.ifacol.2015.09.003.

Rochedo, P.R.R. and Szklo, A. (2013). Designing learning curves for carbon capture based on chemical absorption according to the minimum work of separation. *Applied Energy* 108: 383–391. https://doi.org/10.1016/j.apenergy.2013.03.007.

Rosa, L. and Mazzotti, M. (2022). Potential for hydrogen production from sustainable biomass with carbon capture and storage. *Renewable and Sustainable Energy Reviews* 157: 112123. https://doi.org/10.1016/j.rser.2022.112123.

Saeidi, S., Sápi, A., Khoja, A.H. et al. (2023). Evolution paths from gray to turquoise hydrogen via catalytic steam methane reforming: current challenges and future developments. *Renewable and Sustainable Energy Reviews* 183: 113392. https://doi.org/10.1016/j.rser.2023.113392.

Schiebahn, S., Grube, T., Robinius, M. et al. (2015). Power to gas: technological overview, systems analysis and economic assessment for a case study in Germany. *International Journal of Hydrogen Energy* 40: 4285–4294. https://doi.org/10.1016/j.ijhydene.2015.01.123.

Silva, M.E. (2010). Análise experimental da reforma a vapor de etanol: aspectos técnicos, econômicos e ecológicos. Guaratinguetá: Universidade Estadual Paulista.

Soltani, R., Rosen, M.A., and Dincer, I. (2014). Assessment of CO_2 capture options from various points in steam methane reforming for hydrogen production. *International Journal of Hydrogen Energy* 39: 20266–20275. https://doi.org/10.1016/j.ijhydene.2014.09.161.

Speight, J.G. (2017). *Handbook of Petroleum Refining*. CRC Press.

Stanislaus, A., Marafi, A., and Rana, M.S. (2010). Recent advances in the science and technology of ultra low sulfur diesel (ULSD) production. *Catalysis Today* 153: 1–68. https://doi.org/10.1016/j.cattod.2010.05.011.

Szklo, A. and Schaeffer, R. (2007). Fuel specification, energy consumption and CO_2 emission in oil refineries. *Energy* 32: 1075–1092. https://doi.org/10.1016/j.energy.2006.08.008.

Tang, Y., Li, S., Liu, C. et al. (2023). Process simulation and techno-economic analysis on novel CO_2 capture technologies for fluid catalytic cracking units. *Fuel Processing Technology* 249: 107855. https://doi.org/10.1016/j.fuproc.2023.107855.

Thybaut, J.W. and Marin, G.B. (2016). Multiscale Aspects in hydrocracking: from reaction mechanism over catalysts to kinetics and industrial application. *Advances in Catalysis* 59: 109–238. https://doi.org/10.1016/bs.acat.2016.10.001.

UNECE (1999). Gothenburg Protocol. https://unece.org/environmental-policy/air/protocol-abate-acidification-eutrophication-and-ground-level-ozone (accessed May 2023).

United Nations (1985). Protocol to the 1979 Convention on Long-Range Transboundary Air Pollution on the Reduction of Sulphur Emissions or Their Transboundary Fluxes by at Least 30 Per Cent. https://unece.org/sites/default/files/2021-10/1985.Sulphur.e.pdf (accessed May 2023).

Wu, N., Lan, K., and Yao, Y. (2023). An integrated techno-economic and environmental assessment for carbon capture in hydrogen production by biomass gasification. *Resources, Conservation and Recycling* 188: 106693. https://doi.org/10.1016/j.resconrec.2022.106693.

Yáñez, É., Meerman, H., Ramírez, A. et al. (2022). Fully integrated CO_2 mitigation strategy for an existing refinery: a case study in Colombia. *Applied Energy* 313: 118771. https://doi.org/10.1016/j.apenergy.2022.118771.

Yao, Y., Marano, J., Morrow, W.R. et al. (2018). Quantifying carbon capture potential and cost of carbon capture technology application in the U.S. refining industry. *International Journal of Greenhouse Gas Control* 74: 87–98. https://doi.org/10.1016/j.ijggc.2018.04.020.

10

Closer Integration Between Refining and Petrochemical Assets and Their Role in the Future of the Downstream Industry

Marcio Wagner da Silva[1], *George Hoekstra*[2], *and Thomas Murphy*[3]

[1] *Process Engineering, Petrobras, Sao Paulo, Brazil*
[2] *Hoekstra Trading LLC, Bolingbrook, IL, USA*
[3] *Valuation Risk and Strategy LLC, Skaneateles, NY, USA*

10.1 Introduction and Context

The driving force of change in the downstream sector is a result of the hostile scenario imposed on fossil fuels and the need to reduce the environmental footprint in the Energetic Matrix. To maximize the added value of the processed crude oil, ensure lower environmental impact, and maximize the longevity of the refining industry. The downstream players are looking for alternatives and the non-energetic derivatives like lubricants and petrochemical intermediates like ethylene, propylene, and aromatics appear like attractive options due to their higher added value and circularity potential in comparison with the transportation fuels. Recent forecasts indicate a growing demand for petrochemicals over the coming years, as presented in Figure 10.1.

A simple analysis of Figure 10.1 reveals the power of the petrochemicals market in the downstream sector over the next years, causing these derivatives to surpass transportation fuels as the main driver of the global crude oil demand in the medium term.

Considering only the propylene market, it is expected to have an annual growth rate of 3.6% until 2032, and the market size can reach a total value of 150 USD billion, as presented in Figure 10.2.

On the other hand, the gasoline market tends to suffer contraction in the medium term as presented in Figure 10.3.

Cleaner Petroleum Production and Refining Technologies, First Edition, Edited by M. R. Riazi and Harvey W. Yarranton.

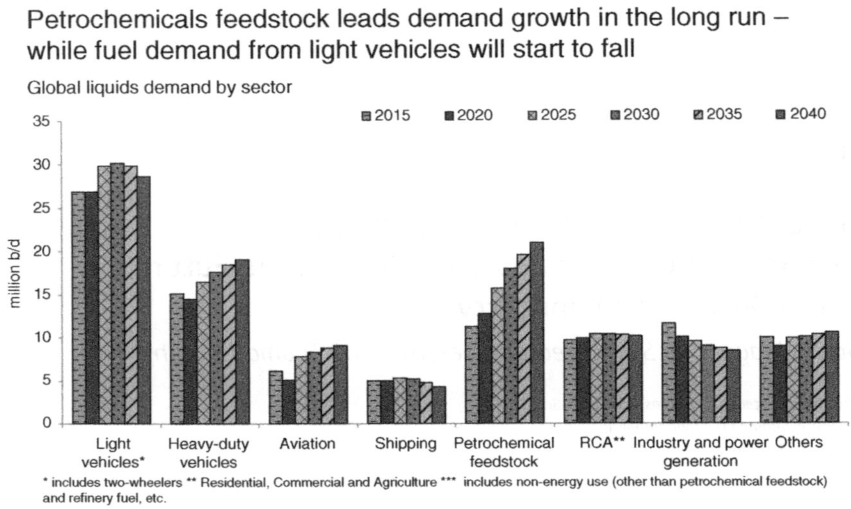

Figure 10.1 Growing trend in the demand by petrochemical intermediates. *Source*: Wood Mackenzie (2020)/GPCA.

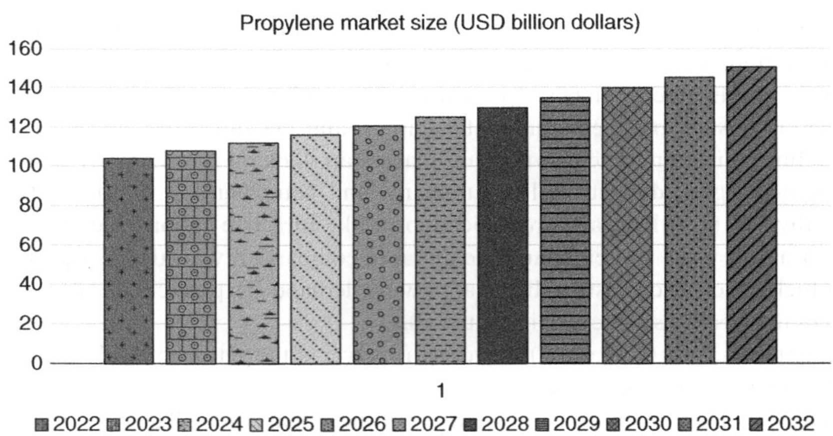

Figure 10.2 Evolution of propylene market size 2022–2032. *Source*: Precedence Research (2022)/Precedence Research.

It is interesting to note that the gasoline demand tends to be sustained by developing economies. This fact tends to restrict the consumer market, which tends to suffer with lower refining margins, reinforcing the competitive advantage for those refiners capable of directing the naphtha to petrochemicals against gasoline.

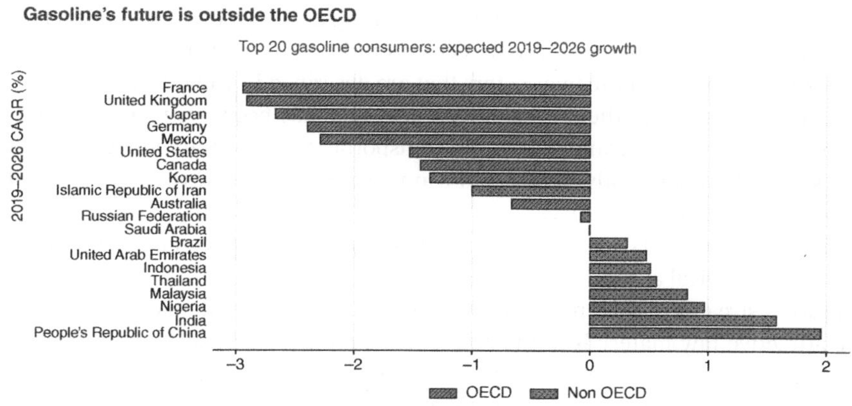

Figure 10.3 Growth of gasoline demand for the next years. *Source*: IEA (2021)/with permission of IEA.

Table 10.1 Refining and petrochemical industry characteristics and synergies opportunities.

Refining industry	Petrochemical industry
Large feedstock flexibility	Raw material from naphtha/NGL
High capacities	Higher operation margins
Self-sufficient in power/steam	High electricity consumption
High hydrogen consumption	High availability of hydrogen
Streams with low added Value (Unsaturated Gases & C2)	Streams with low added value (heavy aromatics, pyrolysis gasoline, C4's)
Strict regulations (benzene in gasoline, etc.)	Strict specifications (hard separation processes)
Transportation fuel demand is declining at the global level	High-demand products

10.2 What Is the Petrochemical Integration Concept?

Regarding the petrochemical intermediates, a closer integration between refining and petrochemical assets is the best way to maximize the yield of petrochemicals as it promotes synergies between refining and the petrochemical industry to minimize risks and costs and maximize the added value to each molecule of the processed crude. Table 10.1 summarizes the concept of petrochemical integration of refining assets.

The focus of closer integration between refining and petrochemical assets is to maximize the added value of the processed crude oil using available synergistic opportunities between the two sectors that are described in Table 10.1.

As aforementioned, the petrochemical industry has been growing at considerably higher rates as compared to the transportation fuels market over the last years, which, additionally, represents a noble destiny with less environmental impact for crude oil derivatives.

Figure 10.4 presents an example of available synergies between typical refining and petrochemical assets.

Process streams considered with low added value to refiners like fuel gas (C2) are attractive raw materials to the petrochemical industry. Also, streams considered residual to petrochemical industries (butanes, pyrolysis gasoline, and heavy aromatics) can be applied to refiners to produce high-quality transportation fuels. This can help the refining industry meet the environmental and quality regulations to derivatives. Furthermore, the concept of petrochemical integration considers value construction through the optimization of utilities and power consumption between the assets, leading to a significant reduction of operating costs once the energy represents more than 60% of the OPEX costs in a typical downstream processing unit.

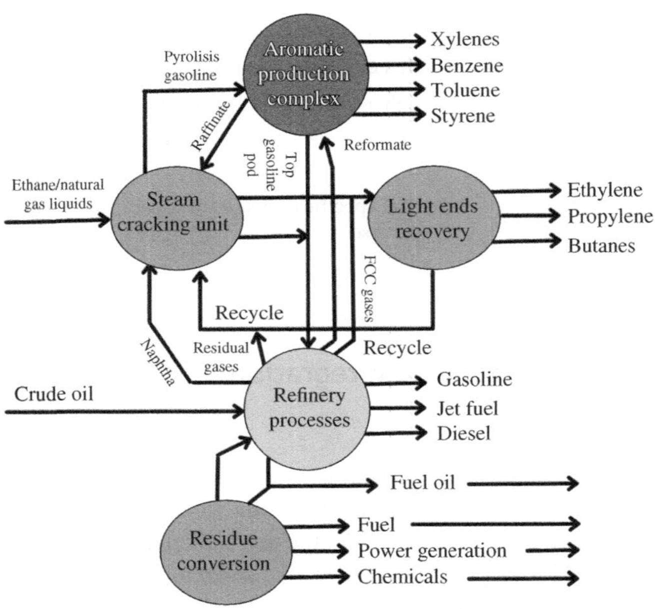

Figure 10.4 Example of synergies between refining and petrochemical assets.

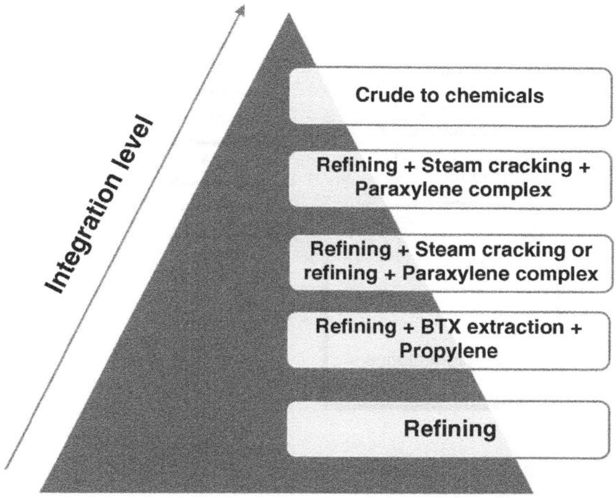

Figure 10.5 Petrochemical integration levels (based on IHS Markit, 2018).

The level and potential of petrochemical integration depend on the refining configuration of the refining asset. The more complex the refinery, the higher the integration potential with the petrochemical industry. Figure 10.5 presents a classification of the petrochemical integration grades based on IHS Markit Company criteria.

According to the classification proposed, the crude-to-chemicals refineries represent the maximum level of petrochemical integration, where the processed crude oil is totally converted into petrochemicals like ethylene, propylene, and BTX.

The closer integration between refining and petrochemical assets can be another tool to help the downstream players optimize energy consumption. The refining assets are normally self-sufficient in steam and power generation, while they are normally deficient in hydrogen generation. In turn, the petrochemical plants can export hydrogen but are deficient in power and steam generation. With these characteristics, the synergy between refining and petrochemical assets can help the players optimize their energy demand.

10.3 The Crude Oil to chemicals – Maximum Added Value to the Processed Crude

Due to the increasing market and higher added value as well as the trend of reduction in transportation fuel demand, some refiners and technology developers have dedicated their efforts to developing crude-to-chemicals refining

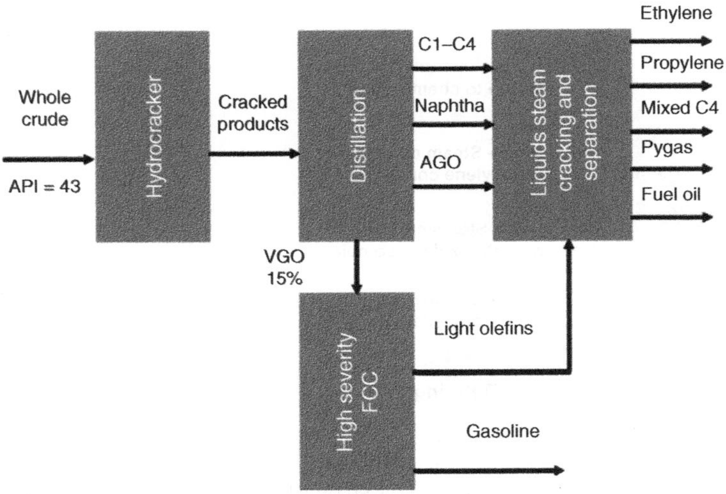

Figure 10.6 Saudi Aramco crude oil to chemicals concept (IHS Markit, 2017).

assets. One of the big players that has invested in this alternative is the Saudi Aramco Company. The concept is based on the direct conversion of crude oil to petrochemical intermediates, as presented in Figure 10.6.

The process presented in Figure 10.6 is based on the quality of the crude oil and deep conversion technologies like High Severity or petrochemical FCC units and deep hydrocracking technologies. The processed crude oil is light with low residual carbon, which is a common characteristic of the Middle East crude oils. The processing scheme involves a deep catalytic conversion process aiming to reach maximum conversion to light olefins. In this refining configuration, the petrochemical FCC units have a key role to ensure high added value to the processed crude oil. An example of FCC technology developed to maximize the production of petrochemical intermediates is the PetroFCC™ process by UOP Company. This process combines a petrochemical FCC with separation processes optimized to produce raw materials for the petrochemical process plants, as presented in Figure 10.7. Other available technologies are the HS-FCC™ process commercialized by Axens Company, and INDMAX™ process licensed by Lummus Company. The basic process flow diagram for HS-FCC technology is presented in Figure 10.8.

It is important to consider that both technologies presented in Figures 10.7 and 10.8 are based on Petrochemical FCC units that present special design due to the most severe operating conditions.

In petrochemical FCC units, the reaction temperature reaches 600 °C and higher catalyst circulation rate raises the gas production, which requires a scaling

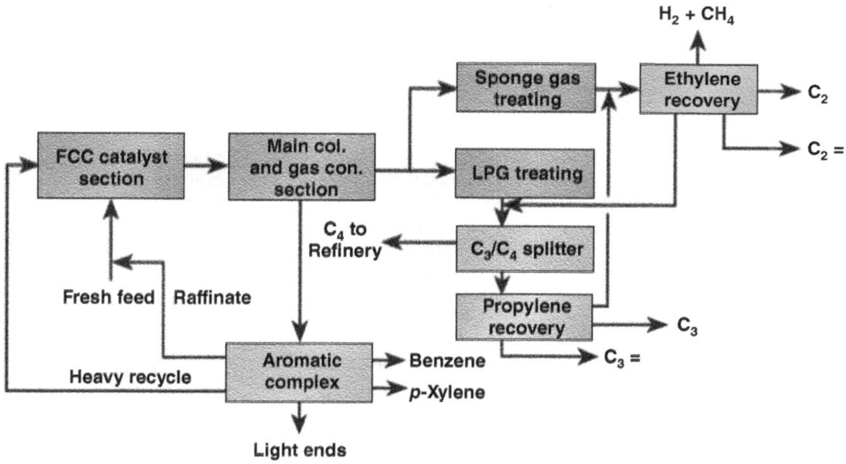

Figure 10.7 PetroFCC process technology by UOP company. *Source*: Honeywell UOP - www.uop.honeywell.com.

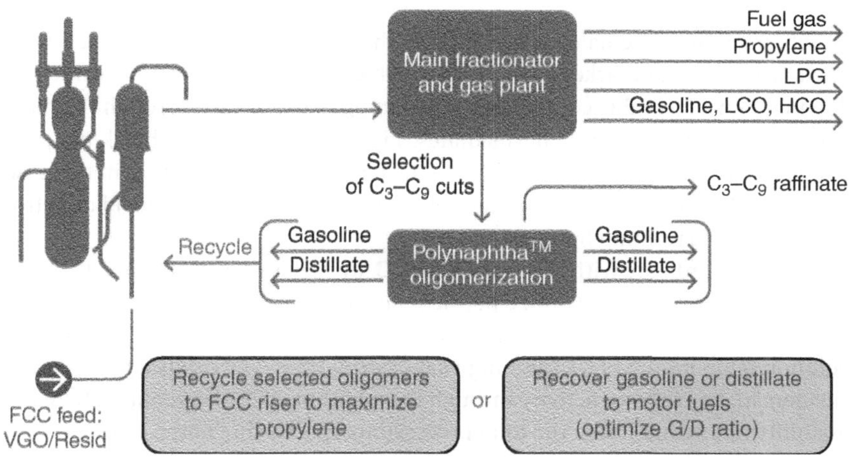

Figure 10.8 HS-FCC process technology by Axens Company.

up of the gas separation section. The higher thermal demand makes it advantageous to operate the catalyst regenerator in total combustion mode, leading to the necessity of installing a catalyst cooler system.

The installation of petrochemical catalytic cracking units requires a deep economic study considering the high capital investment and higher operational costs; however, some forecasts indicate growth of 4.0% per year in the market of

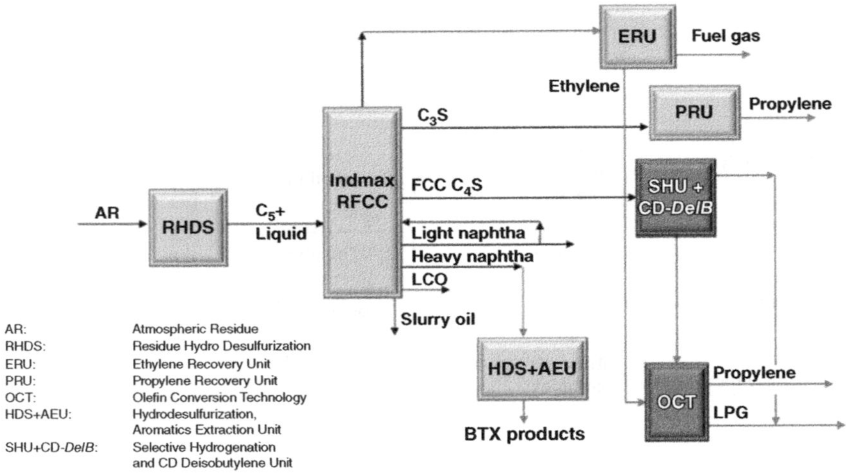

Figure 10.9 Olefins maximization in the refining hardware with INDMAX FCC technology by Chevron Lummus Global Company. *Source*: Sarin (2017)/Lummus Technology, LLC.

petrochemical intermediates until 2025. In this scenario, the capital investment aimed at raising the market share in the petrochemical sector can be attractive, allowing then a favorable competitive positioning to the refiner, through the maximization of petrochemical intermediates. Figure 10.9 presents a block diagram showing a case study demonstrating how the petrochemical FCC unit, in this case the INDMAX technology by Lummus Company, can maximize the yield of petrochemicals in the refining hardware.

In refining hardware with conventional FCC units, it is possible to operate under operating conditions to maximize petrochemicals (Ethylene, Propylene, C4's). In this operation mode, the FCC unit operates under high severity, translated to high operation temperature (TRX) and high catalyst/oil ratio. The catalyst formulation considers higher catalyst activity through the addition of ZSM-5 zeolite. There is a possibility of a reduction in the total processing capacity due to the limitations in blowers and cold area capacity.

An improvement in the octane number of cracked naphtha has been observed despite a lower yield, due to the higher aromatics concentration in the cracked naphtha. In some cases, the refiner can use the cracked naphtha recycle to improve the LPG yield even more.

In the maximum LPG operation mode, the main restrictions are the cold area processing capacity, metallurgical limits in the hot section of the unit, treating

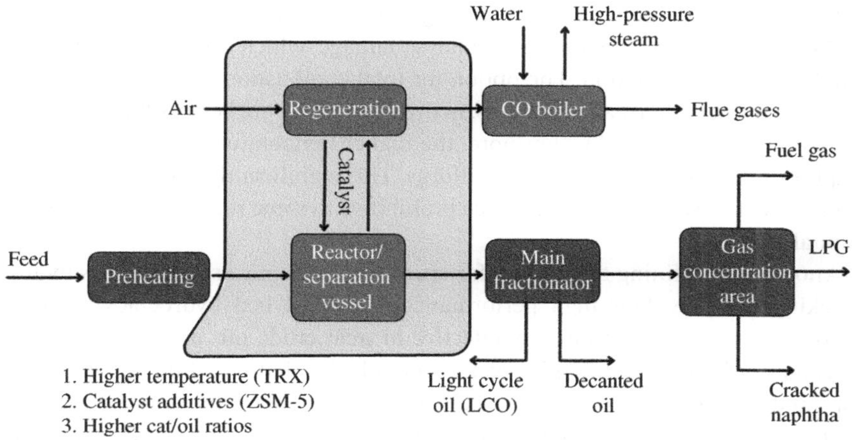

Figure 10.10 Optimization of process variables in FCC units to improve the yield of petrochemical intermediates.

section processing capacity, as well as the top systems of the main fractionating column. In markets with falling demand for transportation fuels, this is the most common FCC operation mode.

By changing the reaction severity, it is possible to maximize the production of petrochemical intermediates, mainly propylene, in conventional FCC units, as shown in Figure 10.10.

The use of FCC catalyst additives such as ZSM-5 can increase unit propylene production by up to 9.0%. Despite the higher operating costs, the higher revenues from the higher added value of derivatives should lead to a positive financial result for the refiner, according to current market projections. A relatively common strategy also applied to improve the yield of LPG and propylene in FCC units is the recycling of cracked naphtha, leading to an over-cracking of the gasoline range molecules.

These operating conditions require the installation of a catalyst cooler system, which raises the process unit profitability through the total conversion enhancement and selectivity to noblest products as propylene and naphtha against gases and coke production. The catalyst cooler is necessary when the unit is designed to operate under total combustion mode due to the higher heat release rate presented in the following text.

$$C + \frac{1}{2}O_2 \rightarrow CO \text{ (partial combustion)} \qquad \Delta H = -27 \text{ kcal/mol}$$

$$C + O_2 \rightarrow CO_2 \text{ (total combustion)} \qquad \Delta H = -94 \text{ kcal/mol}$$

In this case, the temperature of the regeneration vessel can reach values close to 760 °C, leading to higher risks of catalyst damage, which is minimized through catalyst cooler installation. The option for total combustion mode needs to consider the refinery thermal balance, as, in this case, it will not be possible to produce steam in the CO boiler. Furthermore, the higher temperature in the regenerator requires materials with noblest metallurgy. This significantly raises the installation costs of these units, which can be prohibitive to some refiners with restricted capital access.

Another key refining technology for crude oil-to-chemicals refineries is hydrocracking. Despite their high performance, the fixed bed hydrocracking technologies cannot be economically effective to treat crude oils directly due to the possibility of short operating lifecycle. Technologies that use ebullated bed reactors and continuum catalyst replacement allow higher campaign period and higher conversion rates. Among these technologies, the most known are the H-Oil and Hyvahl™ technologies developed by Axens Company, the LC-Fining Process by Chevron-Lummus, and the Hycon™ process by Shell Global Solutions. These reactors operate at temperatures above 450 °C and pressures up to 250 bar. Figure 10.11 presents a typical process flow diagram for a LC-Fining process unit, developed by Chevron Lummus Company, while the H-Oil™ process by Axens Company is presented in Figure 10.12.

Catalysts applied in hydrocracking processes can be amorphous (alumina and silica-alumina) and crystalline (zeolites) and have bifunctional characteristics once the cracking reactions (in the acid sites) and hydrogenation (in the metal sites) occur simultaneously.

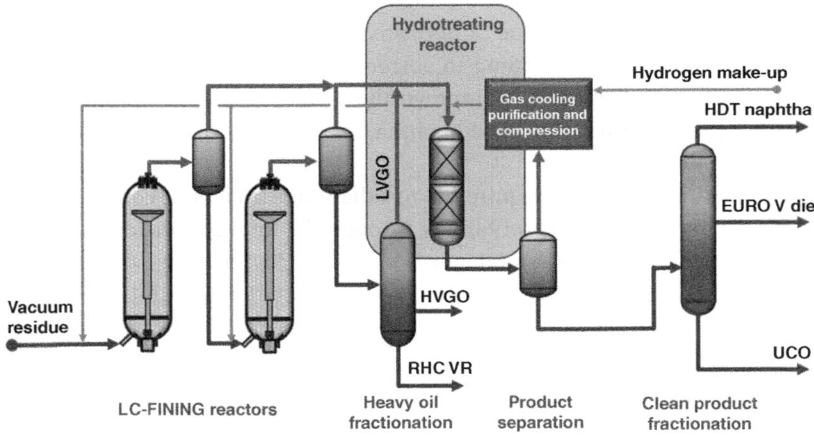

Figure 10.11 Process flow diagram for LC-Fining technology by CLG company.

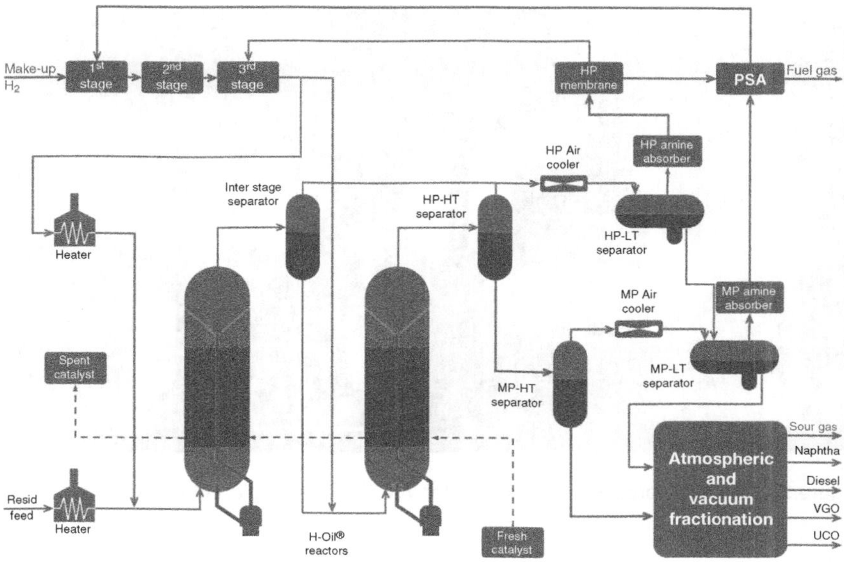

Figure 10.12 Process flow diagram for H-Oil process by Axens Company.

An improvement in relation to ebullated bed technologies is the slurry phase reactors, which can achieve conversions higher than 95%, despite some technology licensors reporting conversions higher than 93% using ebullated bed processes. In this case, the main available technologies are the HDH™ process (Hydrocracking-Distillation-Hydrotreatment), developed by PDVSA-Intevep, VEBA-Combicracking Process (VCC)™ commercialized by KBR Company, the EST™ process (Eni Slurry Technology) developed by the Italian state oil company ENI, and the Uniflex™ technology developed by UOP Company. Figure 10.13 presents a basic process flow diagram for the VCC technology by KBR Company.

In the slurry phase hydrocracking units, the catalysts are injected with the feedstock and activated in situ, while the reactions are carried out in slurry phase reactors, minimizing the reactivation issue, and ensuring higher conversions and operating lifecycle. Figure 10.14 presents a basic process flow diagram for the Uniflex slurry hydrocracking technology by UOP Company.

Other commercial technologies for slurry hydrocracking process are the LC-Slurry™ technology developed by Chevron Lummus Company and the Microcat-RC™ process by Exxon Mobil Company.

For this side, the steam cracking process has a fundamental role in the petrochemical industry. Nowadays, most of the light olefins, light ethylene, and propylene are produced through the steam cracking route. Steam cracking consists of a thermal cracking process that can use gas or naphtha to produce olefins.

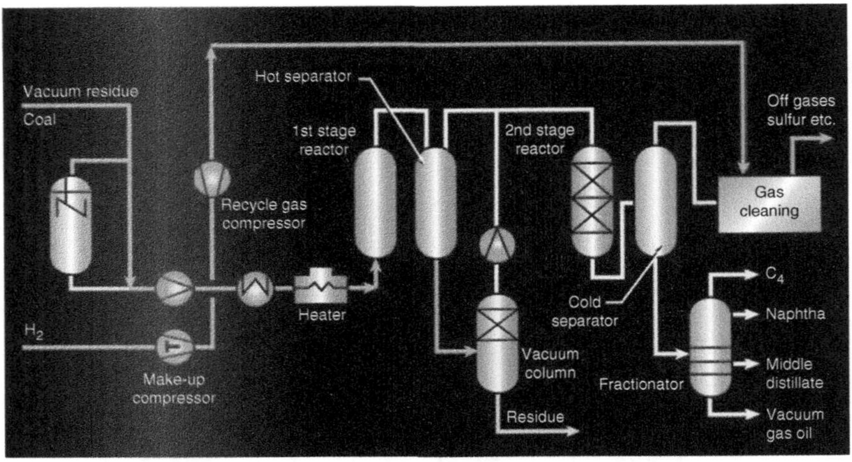

Figure 10.13 Basic process arrangement for VCC slurry hydrocracking by KBR Company.

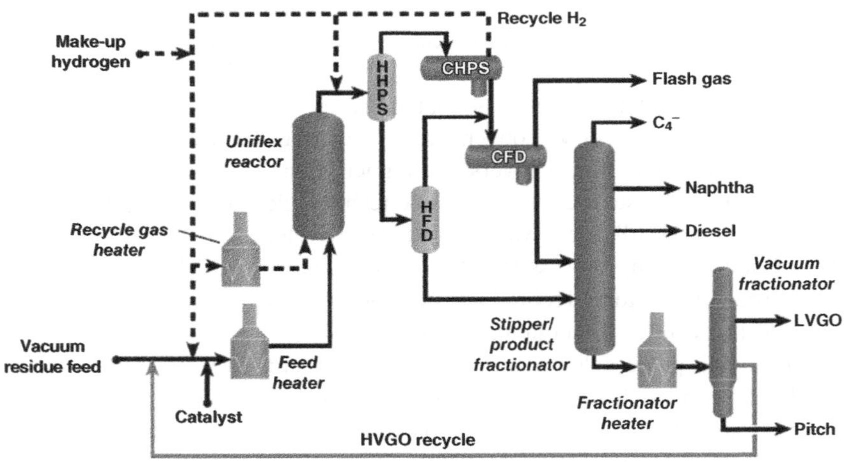

Figure 10.14 Process flow diagram for Uniflex slurry phase hydrocracking technology by UOP Company. *Source*: Adapted from UOP Company (2019).

The naphtha-to-steam cracking is composed basically of straight-run naphtha from crude oil distillation units. Normally, to meet the requirements as petrochemical naphtha, the stream needs to present high paraffin content (higher than 66%).

Due to its relevance, great technology developers have dedicated their efforts to improve steam cracking technologies over the years, especially related to the steam cracking furnaces. Companies like Stone & Webster, Lummus, KBR, Linde, and

Technip develop technologies for steam cracking processes. One of the most known steam cracking technologies is the SRT™ process (Short Residence Time), developed by Lummus Company, which applies a reduced residence time to minimize the coking process and ensure a higher operational lifecycle. Another commercial technology dedicated to optimizing the yield of ethylene is the SCORE™ technology developed by KBR and ExxonMobil Companies, which combines a selective steam cracking furnace with high-performance olefins recovery section.

The cracking reactions occur in the furnace tubes; the main concern and limitation to the operating lifecycle of steam cracking units is the coke formation in the furnace tubes. The reactions are conducted under high temperatures, between 500 °C and 700 °C, according to the characteristics of the feed. For heavier feeds like gas oil, lower temperature is applied to minimize the coke formation. The combination of high temperatures and low residence time is the main characteristic of the steam cracking process. Although it is possible to operate with naphtha, nowadays the steam cracking operators have chosen to operate with ethane or LPG against naphtha due to the competitive prices related to the new sources of natural gas liquid (NGL). Despite this trend over the last years, in markets where a gasoline surplus is observed, naphtha can still be an attractive alternative as feedstock to steam crackers.

According to some forecasts, the demand for propylene will rise from 130 million metric tons in 2020 to around 190 million metric tons in 2030. Due to the increasingly light feed used in refineries and steam cracking units, which tend to favor ethylene production to the detriment of propylene, the propylene demand tends to be supplied by on-purpose propylene production routes like propane dehydrogenation, methanol to olefins (MTO), and olefins metathesis.

As quoted above, some technology developers are dedicating their efforts to develop commercial crude-to-chemicals refineries. Figure 10.15 presents the concept of crude-to-chemicals refining scheme by Chevron Lummus Company.

Other great refining technology developers, such as UOP, Shell Global Solutions, ExxonMobil, Axens, and others, are developing crude-to-chemicals technologies, reinforcing that this is a trend in the downstream market. Figure 10.16 presents a highly integrated refining configuration capable of converting crude oil into petrochemicals developed by UOP Company.

As presented in Figure 10.16, the production focus changes to the maximum, adding value to the crude oil through the production of high added value petrochemical intermediates or chemicals for general purposes, leading to a minimum production of fuels. As aforementioned, big players such as Saudi Aramco Company have made great investments in COC technologies aiming to achieve even more integrated refineries and petrochemical plants, raising considerably their competitiveness in the downstream market. The major technology licensors, such as Axens, UOP, Lummus, Shell, ExxonMobil, etc., have been applying resources to develop technologies capable

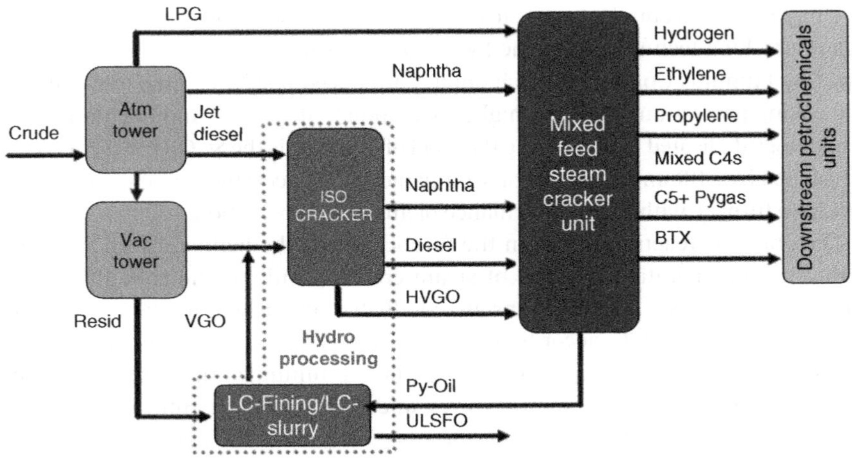

Figure 10.15 Crude to chemicals concept by Chevron Lummus company (Chevron Lummus Global Company, 2019). © Copyright 2023 by Lummus technology, LLC. All rights reserved. 2023.

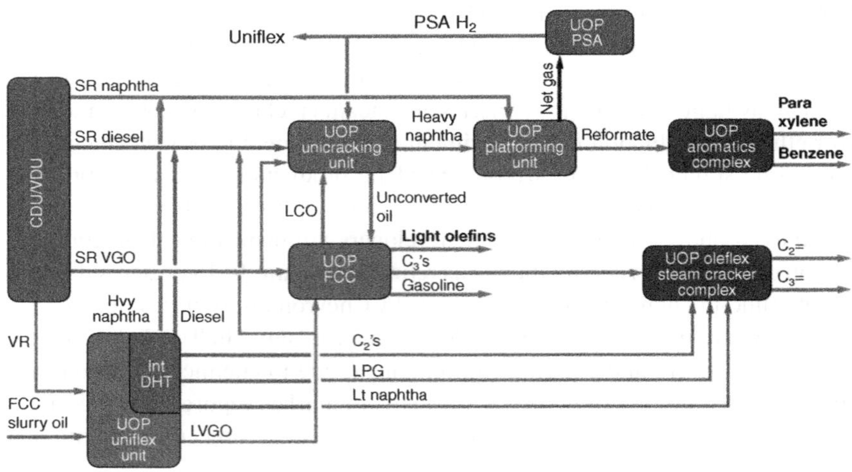

Figure 10.16 Integrated refining configuration based in crude-to-chemicals concept by UOP Company. *Source*: Adapted from Honeywell UOP.

of allowing a closer integration in the downstream sector, aiming to allow refiners to extract the maximum added value from the processed crude oil, an increasing necessity in a scenario where the refining margins are under pressure. Based on data from The Catalyst Group Company (TCGR) in 2019, there were some capital investments in crude-to-chemicals projects as presented in Table 10.2.

Table 10.2 Crude oil to chemicals investments based on The Catalyst Group (2019) with permission.

Company	Location	Capital spending (US$ Billion)	Enterprise type
Zhejiang Petroleum and Chemical	China	26	Greenfield
Hengli Petrochemical	China	11	Greenfield
Shengong Petrochemical	China	11.84	Greenfield
Ningbo Zhongjin Petrochemical	China	5.0	Revamp
Huajin Aramco Petrochemical	China	10 +	Greenfield
SABIC/Fuhaichuang Petrochemical	China	NA	Greenfield
SINOPEC/SABIC (Tianjin Petrochemical	China	45	Revamp
Petrochina	China	-	Revamp
Petrochina	China	-	Revamp
CNOOC	China	-	Revamp
SINOPEC	China	2.8	Greenfield
SINOPEC	China	4.2	Greenfield
SINOPEC	China	4.26	Greenfield
Total Capital Investments in China		**120.1**	
Other Asia			
Hengyi Group	Brunei	20	Greenfield
Saudi Aramco/ADNOC/India Consortium	India	44	Greenfield
Petronas/Saudi Aramco (RAPID)	Malaysia	2.7	Greenfield
ExxonMobil (Singapore Chemical Plant)	Singapore	< 1	Revamp
Pertamina/Rosneft	Indonesia	15	Greenfield
Total Capital Investments in Other Asia		**82.7**	
Middle East			
ADNOC	UAE	45	Revamp
Saudi Aramco/SABIC	Saudi Arabia	30	Greenfield
Saudi Aramco/Total	Saudi Arabia	5	Greenfield

(Continued)

Table 10.2 (Continued)

Company	Location	Capital spending (US$ Billion)	Enterprise type
KNPC/KIPIC (Al-Zour Refinery)	Kuwait	13	Greenfield
Oman Oil Company/Kuwait Petroleum International (Duqm Refinery)	Oman	15	Greenfield
Total Capital Investments in the Middle East		**108**	
Europe			
MOL Group	Hungary	4.5	Revamp
Total Capital Investments		**315**	

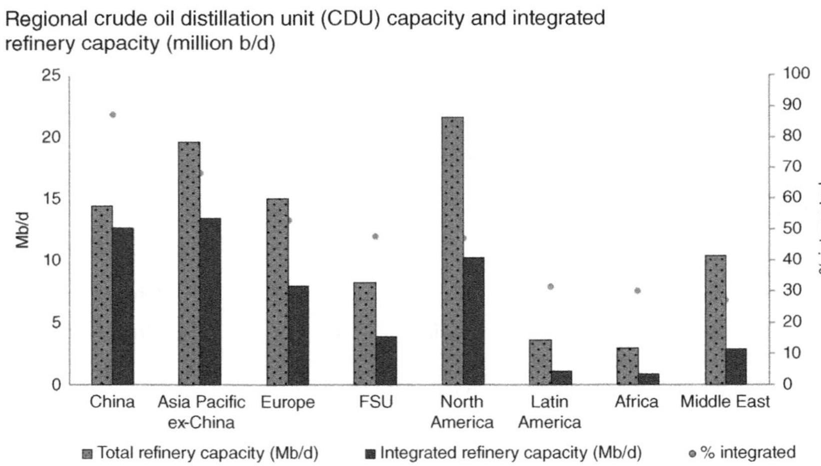

Figure 10.17 Crude oil distillation capacity and integrated refinery capacity for each continent. *Source*: Wood Mackenzie (2023)/with permission of Wood Mackenzie.

Some of these capital investments were postponed due to the economic crisis provoked by the COVID-19 pandemic, but these data reinforce the crude-to-chemicals trend in the market. It is interesting to note that close to 64% of the global crude-to-chemicals investments are made by Asian players.

To reinforce this, Figure 10.17 presents a comparison between the relation of crude oil distillation capacity and the integrated refinery capacity for each continent.

Figure 10.17 shows that the Asian players have a superior integration capacity of their refining assets in comparison with other continents. As mentioned above, this can be translated into a significant competitive advantage to the Asian players and a great potential for competitive imbalance of the downstream market, considering the recent forecasts that indicate growing demand for petrochemicals. Furthermore, it is possible to see the power of China in the Asian and global downstream market.

It is important to consider that the cost composition involves several factors, and this scenario can be different according to the local business environment. Figure 10.18 presents a comparison between petrochemical yields of traditional refineries, a benchmark integrated refinery and Hengli crude-to-chemicals complex, according to data from IHS Markit.

Analyzing Figure 10.18, it is possible to note the higher added value reached in crude-to-chemicals refineries when compared even with highly integrated refineries.

As aforementioned, facing the current trend of reduction in transportation fuel demand at the global level, the capacity of maximum adding value to crude oil can be a competitive differential to refiners. Due to the high capital investment needed for the implementation that allows the conventional refinery to achieve the maximization of chemicals, capital efficiency also becomes an extremely important factor in the current competitive scenario, as well as the operational flexibility, related to the processed crude oil slate.

The dynamics of the downstream industry tends to show a significant migration of players from transportation fuels to non-energetic derivatives, especially petrochemicals, due to their higher added value and the hostile scenario imposed on the fossil fuels in order to minimize the environmental footprint of the energetic matrix. We are seeing these changes in the profile of investments in the downstream sector nowadays. According to International Energy Agency (IEA), the capital investments in petrochemical plants grow from US$ 15 billion in 2014

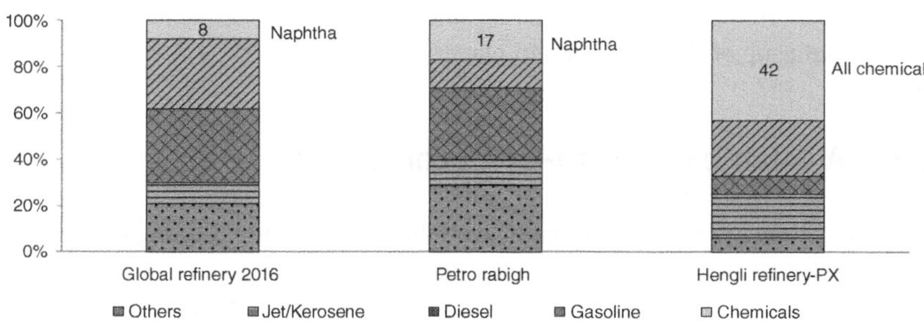

Figure 10.18 Petrochemicals yield comparison (IHS Markit, 2018).

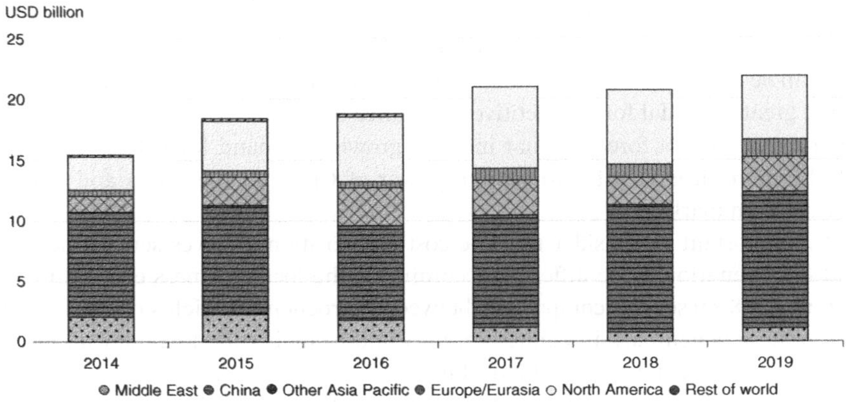

Figure 10.19 Capital investments in petrochemical plants between 2014 and 2019. *Source*: IEA (2022)/with permission of IEA.

to US$ 22 billion in 2019. Again, most of these investments were made by Asian players as presented in Figure 10.19.

Analyzing deeply Figure 10.19, it is possible to verify the relevance of the China investments in petrochemical plants over the years. Considering the total investments in 2019, the Chinese market was responsible for 52% of the capital investments in petrochemical plants and the Asian market was responsible for 67% of the total investments, while the North American players were responsible for close to 24%, reinforcing the high capacity of integration between refining and petrochemicals assets in the Asian market.

Despite their relatively low participation in crude-to-chemicals assets and petrochemical plants (only 5 % in Figure 10.19), the Middle East market has a great potential for petrochemical integration. In the end of 2022, the Company SATORP (a Joint Venture between Saudi Aramco and Total Energies) announced a capital investment of US$ 11 billion in the Amiral crude-to-chemicals plant in the Middle East to maximize the added value to the processed crude through petrochemicals maximization, reinforcing again the relevance of crude-to-chemicals trend in the strategic planning of the downstream players.

10.4 Available Crude-to-chemicals Routes

Nowadays, there are three technically available routes that are being considered for capital investments to crude-to-chemicals refining complexes. Figure 10.20 presents the concepts based on the information of IHS Markit Company.

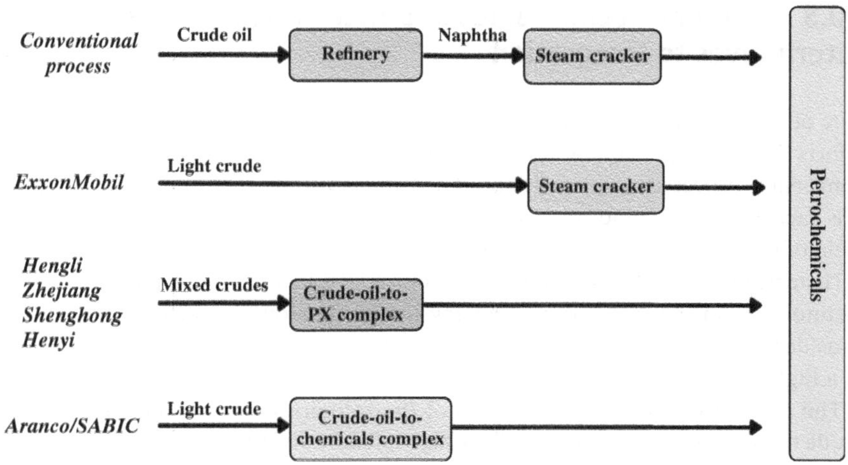

Figure 10.20 Crude-to-chemicals concepts. *Source*: Adapted from IHS Markit (2019).

The conventional routes consider the processing of crude oil in a conventional crude oil refinery, producing petrochemical intermediates like naphtha, which is supplied to a petrochemical asset like a steam cracking unit. The ExxonMobil route is based on the direct feed of selected crude oils, normally light and low-contaminant crudes, to petrochemical assets, while the Chinese enterprise Hengli Zhejiang Shenghong Henyi project considers the feed of mixed crude oil slate to a crude to PX (Para-Xylene) complex to ensure the domestic Chinese market, which presents high demand for light aromatics (BTX). A conventional, highly integrated refining hardware is capable of achieving 15 to 20% of petrochemicals yield, while a crude-to-chemicals refinery can reach up to 40% as presented in Figure 10.18.

As aforementioned, the Aramco/Sabic concept is based on a high complexity refining hardware to convert selected crude oil (light) to maximize the yield of petrochemical intermediates, mainly light olefins.

Although the advantages presented by closer integration between refining and petrochemical assets, it is important to understand that the players of the downstream industry are facing a transitional period where, as presented in Figure 10.1, the transportation fuels are responsible for a great part of the revenues. In this business scenario, it is necessary to define a transition strategy where the economic sustainability achieved by the current status (transportation fuels) needs to be invested to build the future (maximize petrochemicals).

10.5 Refining Technologies and Relatively Low-cost Alternatives to Integration

One of the main restrictions to the implementation of petrochemical integration efforts is the high capital spending demanded for the installation of deep conversion processes, especially the crude-to-chemicals refining assets. Despite this fact, there are some relatively low-cost options capable of maximizing the yield of petrochemicals in brownfield refining assets.

Historically, the refining industry has optimized its operations and developed technologies in the sense of maximizing the production of transportation fuels, considering the high demand. Figure 10.21 shows a typical refining configuration of a high-complexity refinery focused on the production of transportation fuels.

The current refining configurations of the refineries aim to add value to the crude oil through the maximization of high added value fuels, mainly middle distillates (Kerosene and Diesel).

The trend of reduction in transportation fuel demand associated with the growing petrochemical market has led the refiners to search optimize the refining

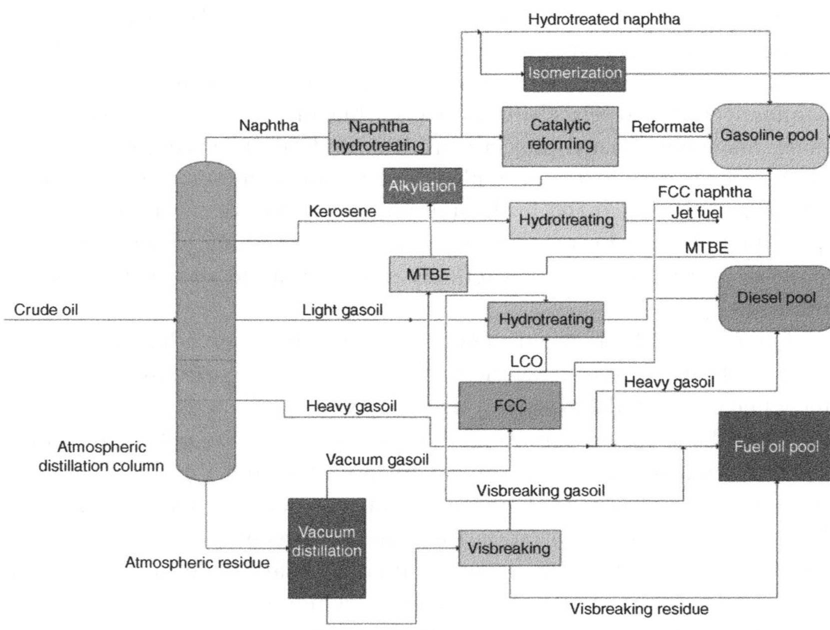

Figure 10.21 Refining arrangement to a refinery under coking/hydrocracking configuration.

hardware to raise the yield of petrochemicals to the detriment of fuels, promoting a closer integration with the petrochemical sector.

In this sense, flexible refining technologies as fluid catalytic cracking (FCC) and catalytic reforming have gained prominence in the downstream industry, as these are capable of maximizing the production of high added value petrochemical intermediates (olefins and aromatics, respectively). However, some refiners have come across the high cost of capital as a barrier to further integration with the petrochemical industry in view of the greater need for investments associated with petrochemical maximizing units. Usually, the installation of units dedicated to the production of petrochemical intermediates requires high capital investment in high reaction severity units, such as the FCC petrochemical and high complexity separation units, as in the case of catalytic reforming units dedicated to the production of light aromatics.

Despite capital investment restrictions, there are some alternatives to maximize petrochemical intermediates in the refining hardware with relatively low capital investment. The fractionating of straight-run naphtha can be an attractive alternative, as presented in Figure 10.22.

In this case, the light fraction of naphtha may be directed to the market for higher value-added petrochemical intermediates, while the heavier fraction may comprise the gasoline, diesel, or jet fuel pool, according to local market demand.

Due to its process characteristics, as shown in Figure 10.23, the conventional FCC unit can have its variables optimized to produce petrochemical intermediates.

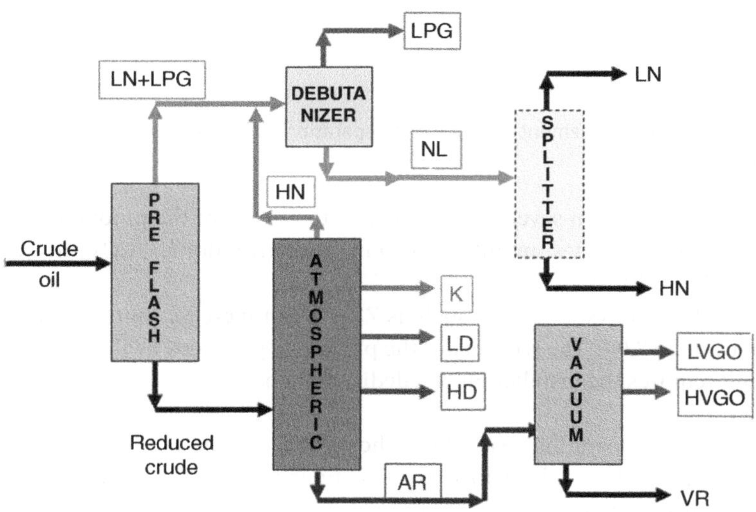

Figure 10.22 Crude oil distillation scheme considering the fractionation of straight-run naphtha.

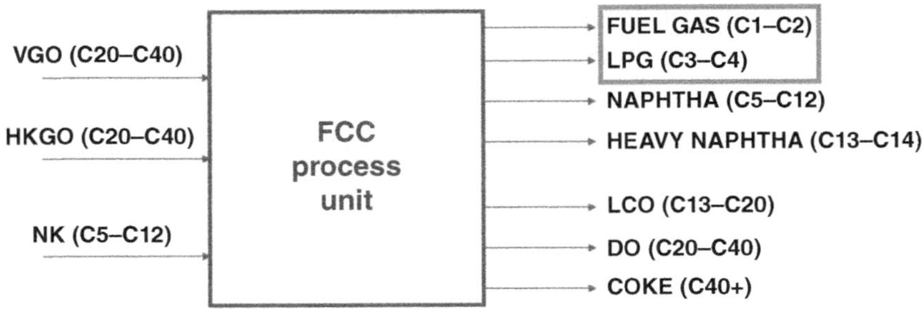

Figure 10.23 Possible feedstocks and derivatives produced by FCC process units.

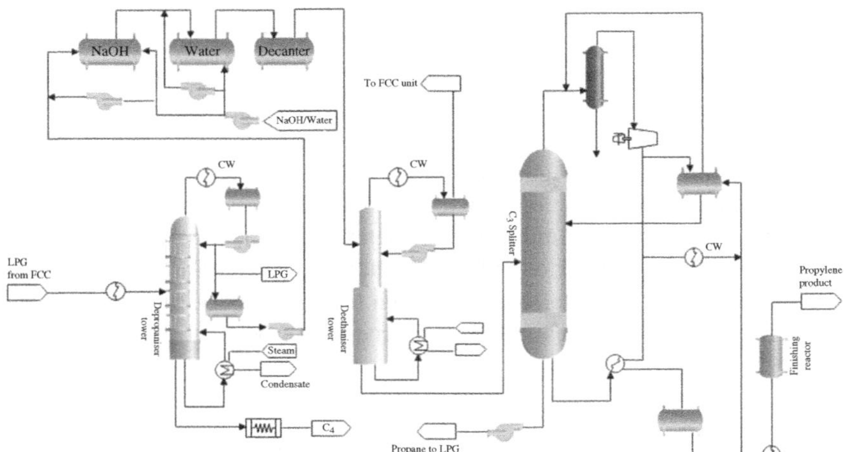

Figure 10.24 Typical arrangement of propylene separation unit based on the heat pump concept.

By changing the reaction severity, it is possible to maximize the production of petrochemical intermediates, mainly propylene, in conventional FCC units, as presented in Figure 10.10.

The use of FCC catalyst additives such as ZSM-5 can increase unit propylene production by up to 8.0%. The recovery of the propylene produced by the catalytic cracking unit requires the installation of a dedicated separation unit, as shown in Figure 10.24.

Despite the higher operating costs, the higher revenues from the higher added value of derivatives should lead to a positive financial result for the refiner, according to current market projections. A relatively common strategy also applied to improve the yield of LPG and propylene in FCC units is the recycling of cracked naphtha, leading to an over-cracking of the gasoline range molecules.

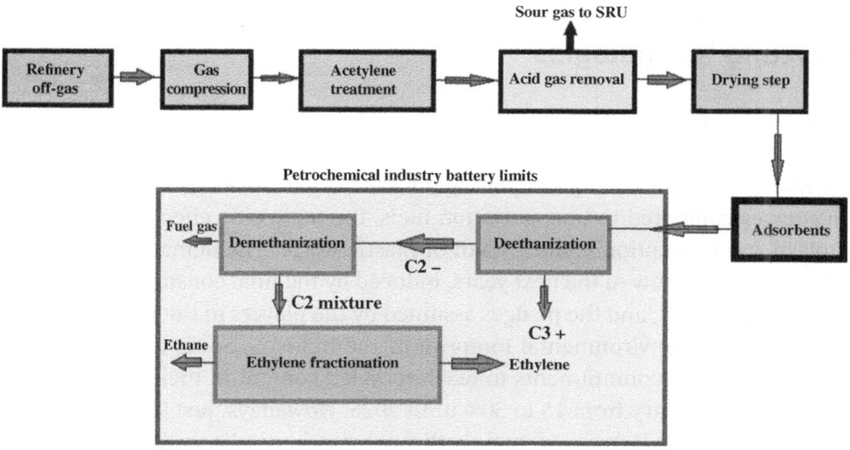

Figure 10.25 Typical arrangement of the olefins recovery process from refinery off-gases.

Another alternative for refiners with low capital availability is the installation of units dedicated to the recovery of olefins from refinery off-gases, as shown in Figure 10.25.

The off gases from deep conversion units such as FCC and delayed coking have high olefin content (>20% in some cases). In markets with high availability of natural gas, the installation of olefin recovery units tends to be economically viable to the refiner.

The off gases from thermal or catalytic cracking (FCC, Delayed Coking, Visbreaking, etc.) are sent to a compression unit and subsequently passed through a nickel catalyst bed, where the acetylene is hydrogenated to ethane. The presence of acetylene contributes to coke deposition in the petrochemical reactors. Due to low olefin content, the gases from separation or treatment units as distillation and hydrotreating aren't used as petrochemical intermediates and are normally directed to the fuel gas refinery pool with the natural gas supply.

After the acetylene removal, the off-gas stream is directed to an amine treatment aiming to remove sour gases such as H_2S and CO_2. These gases are poisonous to the catalysts applied in the petrochemical conversion steps and raise the corrosion rate in the transfer pipelines. Other common contaminants in the refinery residual gases, which mainly stem from Fluid Catalytic Cracking units (FCC), are nitriles, Mercury, arsine/phosphine, oxygen, ammonia, and nitrogen oxides. These contaminants need to be removed, as they act as poisons to the petrochemical conversion catalysts, leading to a reduction in the operational campaign of these units, consequently raising the operational costs.

10.6 Closing the Sustainability Cycle – Plastics Recycling Technologies

As described above, we are facing a continuously growing demand for petrochemicals and a great part of these crude oil derivatives has been applied to produce common-use plastics. Despite their higher added value and significant economic advantages compared to transportation fuels, the main side effect of the growth of plastics consumption is the growth of plastic waste. The demand for recycled plastics tends to grow in the next years, induced by the final consumer awareness, stricter regulations, and the pledges assumed by the players in the plastics market to minimize the environmental footprint of the industry. Some of these players have made public commitments to reach recycled content in their production in percentages that vary from 15 to 50% until 2025. Nowadays, just 9% of the global plastics production is recycled, mainly through mechanical recycling processes.

The mechanical recycling technologies have been applied with successful results, especially for polyethylene terephthalate (PET) and polyethylene (PE), which are used to produce bottles. Although this implies success, there are some concerns and restrictions related to the use of mechanically recycled plastics for noblest purposes such as food-grade packages due to the contamination risk.

Despite the efforts related to the mechanical recycling of plastics, the increasing volumes of plastic waste call for the most effective recycling routes to ensure the sustainability of the petrochemical industry through the regeneration of the raw material. In this sense, some technology developers have been making dedicated investments and efforts to develop competitive and efficient chemical recycling technologies for plastics.

Called advanced recycling technologies, these technologies involve basically two processing routes: the conversion or the feedstock recycling and the decomposition that recovers the monomer of the polymer.

Among the conversion routes (plastics to feedstock), one of the most applied technologies for plastics recycling is the catalytic pyrolysis, where the long polymeric chain is converted into smaller hydrocarbon molecules that can be fed to steam cracking units to reach a real circular petrochemical industry. Another route is the thermal pyrolysis of plastics, which already relies on some commercial technologies available, as well as the gasification process.

Another promising chemical recycling route for plastics based on conversion (plastics to feedstock) is the hydrocracking of plastic waste. In this case, the chemical principle involves the cracking of carbon-carbon bonds of the polymer under high hydrogen pressure, which leads to the production of stable, low-boiling-point hydrocarbons. The hydrocracking route presents some advantages in comparison with thermal or catalytic pyrolysis, once the number of aromatics or unsaturated molecules is lower than that achieved in the pyrolysis processes, leading to a more stable feedstock for steam cracking or other downstream processes. Moreover, it is more selective, producing gasoline range hydrocarbons, which can be easily applied in the highly integrated refining hardware.

Among the recycling routes involving the decomposition of plastic waste (monomer recycling), it is possible to quote the solvolysis and the microwave processes.

Chemical recycling of plastics offers a great opportunity for technology developers and scientists, especially related to the development of effective catalysts to promote depolymerization reactions, which can ensure the recovery of high-added-value molecules such as BTX. More than that, the chemical recycling of plastics is an urgent necessity to close the sustainability cycle of an essential industry in our society.

It is fundamental to understand that the plastic recycling is necessary to the achievement of a real circular economy as required by the society, especially considering the growing demand for petrochemical intermediates in the coming years.

These data reinforce the relevance of the plastics waste recycling technologies for the sustainability of the downstream industry at the same level as the energy management efforts of the players of this industry.

10.7 Conclusion

Nowadays, it is still difficult to imagine the global energetic matrix free of fossil transportation fuels, especially in developing economies. Despite this fact, recent forecasts, growing demand for petrochemicals, and the pressure to minimize the environmental impact produced by fossil fuels create a positive scenario and act as the main driving force to closer integration between refining and petrochemical assets. In the extreme scenario, the zero-fuel refineries tend to grow in the medium term, especially in developed economies.

The synergy between refining and petrochemical processes raises the availability of raw material to petrochemical plants and makes the supply of energy to these processes more reliable. At the same time, it ensures a better refining margin to refiners due to the high added value of petrochemical intermediates compared to transportation fuels. The development of crude-to-chemicals technologies reinforces the necessity for closer integration of refining and petrochemical assets by the brownfield refineries, aiming to face the new market that tends to focus on petrochemicals over transportation fuels. It is important to note the competitive advantage of the refiners from the Middle East that have easy access to light crude oils, which can be easily applied in crude-to-chemicals refineries. As presented above, crude oil-to-chemicals refineries are based on deep conversion processes that require high capital spending. This fact can put pressure on the refiners with restricted access to capital, again reinforcing the necessity to look for close integration with the petrochemical sector, aiming to achieve competitiveness.

In the extreme side of the petrochemical integration trend, there are the zero fuels refineries. As quoted above, it is still difficult to imagine the downstream market without transportation fuels. However, it seems a serious trend and the players of the downstream sector need to consider a focus shift in their strategic plans, like opportunity and threat, mainly considering the pressure on transportation fuels due to the decarbonization necessity and new technologies.

Despite the benefits of petrochemical integration, it is fundamental to take into account the necessity to reach a circular economy in the downstream industry. To achieve this goal, the chemical recycling of plastics is essential. Despite this concern, there are promising technologies that can ensure the closing of the sustainability cycle of the petrochemical industry.

Considering the higher added value of the petrochemicals compared to transportation fuels, the growing market, and the necessity to reduce the carbon intensity of the global energetic matrix, petrochemical integration tends to be the way to follow in the downstream industry in the upcoming years. The current scenario of the downstream industry indicates a few pillars to competitive operations in the market such as high bottom barrel conversion capacity, closer integration between refining and petrochemical assets, energy efficiency, and low-carbon emission routes for hydrogen generation. Despite the advantages, as described above, the sustainability of the downstream industry is still dependent on effective routes to plastics recycling technologies, which need to be developed with the support of the downstream players, governments, and technology developers to reach a real circular economy.

References

Beccari, M. and Romano, U. (eds.) (2006). *Encyclopedia of Hydrocarbons, Volume II – Refining and Petrochemicals*. ENI.

Chang, R.J. (2018). Crude oil to chemicals – industry developments and strategic implications. Presented at Global Refining & Petrochemicals Congress Houston, USA.

Charlesworth, R. (2019). Crude oil to Chemicals (COTC) – an industry game Changer? IHS Markit Company. Presented in 14th GPCA Forum.

Couch, K. (2019). The refinery of the future – a flexible approach to petrochemicals integration. Honeywell UOP company. Presented in 12th Asian Downstream Summit.

Cui, K. (2019). Why crude to chemicals is the obvious way forward. *Wood Mackenzie*.

Deloitte Company (2019). The future of petrochemicals: growth surrounded by uncertainties.

Frecon, J., Le Bars, D., and Rault, J. (2019). Flexible upgrading of heavy feedstocks. *PTQ Magazine*.

Gelder, A. (2023). Refinery-petrochemical integration disrupts gas-based cracker feedstock advantage. Wood Mackenzie.

Gelder, A. and Bailey, G. (2020). The future of petrochemicals: a tale of two transitions. Wood Mackenzie.

International Energy Agency (IEA) (2021). *Oil 2021: Analysis and Forecast to 2026*. IEA.

IEA (International Energy Agency) (2022). Investments in petrochemical plants by region, 2014-2019. https://www.iea.org/data-and-statistics/charts/investment-in-petrochemical-plants-by-region-2014-2019

Kim, W.C. and Mauborge, R. (2004). *Blue Ocean Strategy*. Harvard Business Review.

Lambert, N., Ogasawara, I., and Abba, I. et al. (2014). HS-FCC for propylene: concept to commercial operation. *PTQ Magazine*.

Liu, S., Kots, P.A., Vance, B.C. et al. (2021). Plastic waste to fuels by hydrocracking at mild conditions. *Science Advances* 7 (17): eabf8283.

Maller, A. and Gbordzoe, E. (2016). High severity fluidized catalytic cracking (HS-FCC™): from concept to commercialization. Technip stone & webster technical presentation to REFCOMM™.

Marsh, M. and Wery, J. (2018). *Filling the Propylene Gap – Shaping the Future with On-Purpose Technologies*. UOP Company.

Mukherjee, U. and Gillis, D. (2019). Advances in Residue Hydrocracking. *PTQ Magazine*.

Muldoon, B.S. (2019). Profit pivot points in a crude to chemicals integrated complex. Presented at Ethylene Middle East Technology Conference.

Murphy, J.J. and Payn, C.F. (2019). Oil to chemicals: new approaches. Catalyst Group Report. *PTQ Magazine*.

Oyekan, S.O. (2019). *Catalytic Naphtha Reforming Process*, 1e. CRC Press.

Precedence Research (2022). Propylene market size 2022-2032. https://www.precedenceresearch.com/propylene-market#:~:text=The%20global%20propylene%20market%20size,3.76%25%20between%202023%20and%202032.

Sarin, A.K. (2017). Integrating refinery with petrochemicals: advanced technological solutions for synergy and improved profitability. Presented at Global Refining & Petrochemicals Congress, Mumbai, India.

Silva, M.W. (2020). More petrochemicals with less capital spending. *PTQ Magazine*.

Silva, M.W. (2022). *Crude Oil Refining: A Simplified Approach*, 1e. CRC Press.

Silva, M.W. (2023). Closing the sustainability cycle. *PTQ Magazine*.

The Catalyst Group (TCGR) (2021). *Advances in Catalysis for Plastic Conversion to Hydrocarbons.*

Vu, T., Ritchie, J. (2019). *Naphtha Complex Optimization for Petrochemical Production*, UOP Company.

Wood Mackenzie Presentation (2019). Refinery-Petrochemical Integration (Downstream SME Knowledge Share).

Zhou, T. and Baars, F. (2010). Catalytic reforming options and practices. *PTQ Magazine.*

11

Greening Existing Petroleum Refineries via Biomass Processing

Clarissa Bergman-Fonte[1]*, Aline T. Carvalho*[1]*,*
Gabriela Nascimento da Silva[1]*, Pedro R. R. Rochedo*[1,2]*,*
Frédéric Lantz[3]*, and Alexandre Szklo*[1]

[1] *Centre for Energy and Environmental Economics (Cenergia), Energy Planning Program (PPE), COPPE, Universidade Federal do Rio de Janeiro, Brazil*
[2] *Research and Innovation Center on CO_2 and Hydrogen (RICH Center) and Management Science and Engineering Department, Khalifa University, Abu Dhabi, United Arab Emirates*
[3] *Economic, Energy and Environment, IFP Energies Nouvelles, IFP School, Rueil-Malmaison, France*

11.1 Introduction

The Paris Agreement aims to limit global warming to "well below 2 °C, preferably to 1.5 °C" compared to pre-industrial levels (UNFCCC, 2020). According to mitigation scenarios evaluated by the Working Group III contribution to the Sixth Assessment Report from the IPCC (IPCC, 2022), achieving the temperature restriction of 1.5 °C with limited or no overshoot means reaching net zero CO_2 emissions around 2035 to 2070. In such mitigation scenarios, the consumption of fossil oil is anticipated to decrease around 30–78% by 2050, relative to 2020 levels. This means that, in the decarbonization context, oil refineries are subject to the risk of stranded assets, an expression that refers to the possibility of assets becoming no longer able to earn their economic return prior to the end of their lifetime (IRENA, 2017). Global estimates of potential stranded oil and gas assets amount to around US$ 1.4 trillion, with the potential to induce destabilizing effects on the macroeconomic and financial systems (Semieniuk et al., 2022). This includes assets within the oil refining sector.

Cleaner Petroleum Production and Refining Technologies, First Edition, Edited by M. R. Riazi and Harvey W. Yarranton.

Fortunately, the refining sector has particularities that might enable its integration to this energy transition context, thus reducing the risks of asset stranding. Refineries consist of complex systems with multiple operations that depend on the properties of the crude oil to be processed and the desired final products. Among this diversity of processes, there are process units that are more resilient to changes toward decarbonization or, in other words, are more prone to revamps toward processing renewable feeds (Bergman-Fonte et al., 2023). In addition, refineries have an entire logistics infrastructure already in place and a skilled labor force.

Furthermore, hydrogen production, which typically occurs in refineries from fossil light hydrocarbons (e.g., methane, ethane, or light naphtha streams), can be generated from biomethane or even water electrolysis using renewable energy sources, with this latter also providing pure oxygen that can be used in oxyfuel combustion (e.g., in FCC units, see Chapter 9).

In addition, for the expansion of the greenfield facilities, modular mini-refineries – that is, small-scale refineries composed of independently designed units – may become attractive in a context of uncertain long-term demand, investment constraints, and short-term implementation time (Teixeira et al., 2021). These types of processing facilities can also be planned to be revamped to process renewable feedstocks (Bergman-Fonte et al., 2023).

Therefore, the present chapter provides a comprehensive review on how refineries can produce value-added products from bio-based raw materials such as oily crops, used cooking oils, residual fats, lignocellulosic biomass, and sugary and starchy biomass. Chapter 9 relates to the present chapter, as it describes opportunities for refineries to produce and use renewable hydrogen.

11.2 General Concepts

The use of biofuels such as ethanol and biodiesel (Fatty Acid Methyl Esters–FAMEs–or Fatty Acid Ethyl Esters–FAEEs) is already consolidated in various regions (IEA, 2023). Ethanol can be produced from sugar, starch, or lignocellulosic feedstocks (Pechstein and Kaltschmitt, 2018) and is applied as a substitute and/or an octane enhancer for gasoline (Anderson, 2012), while biodiesel is produced from triglyceride-based biomass (Karmakar et al., 2010) and can be blended with fossil diesel (USDoE, 2023). However, biofuels are not limited to these already consolidated options. This chapter focuses on liquid hydrocarbons produced from biogenic sources, as described in the paragraphs below, since their synthesis can take advantage of existing oil refineries. These hydrocarbons can compose not only fuels (for example, LPG, gasoline, jet fuel, and diesel) but also feedstock for the petrochemical industry (such as naphtha and basic petrochemicals – ethylene, propylene, benzene, toluene, and xylene).

Alternatively, value-added products composed of liquid hydrocarbons can be obtained from biomass instead of from crude oil (Holden and Gilpin, 2013), particularly from oily crops, used cooking oils, residual fats, sugary biomass, and lignocellulosic materials. Different production pathways lead to the conversion of these biogenic feedstocks into final products, which are similar to their fossil counterparts (and/or to premium components of the blend of finished products). Thus, they are drop-in products, that is, they can take advantage of the infrastructure already in place for their fossil correspondents (van Dyk et al., 2019). The production of these alternative products involves at least one process unit commonly found in oil refineries, although modifications might be needed, for example, regarding catalysts and operating conditions (Bergman-Fonte et al., 2023).

Regarding production facilities, there are three types of strategies that are possible for processing biomass feedstocks. They are illustrated in Figure 11.1. The construction of entirely new plants, or *greenfield biorefineries*, makes current oil refineries more susceptible to asset stranding. The complete conversion of traditional oil refineries, leading to *grayfield biorefineries*, is an alternative that results in complete asset utilization. In this case, the term *repurpose* can be applied, meaning that there is a change in the main concept of the plant. As an intermediate option, there is the partial conversion of traditional refineries, leading to *brownfield biorefineries*, which thus receive both bio- and fossil-based feedstocks. Here, the term *revamp* might be used since it refers to a change or upgrade in equipment. Moreover, the concept of biomass *co-processing* refers to the strategy, present in this last case, in which bio- and fossil-based oils are processed together in the same refinery unit. This way, the final products contain carbon from both renewable and fossil sources (van Dyk et al., 2019).

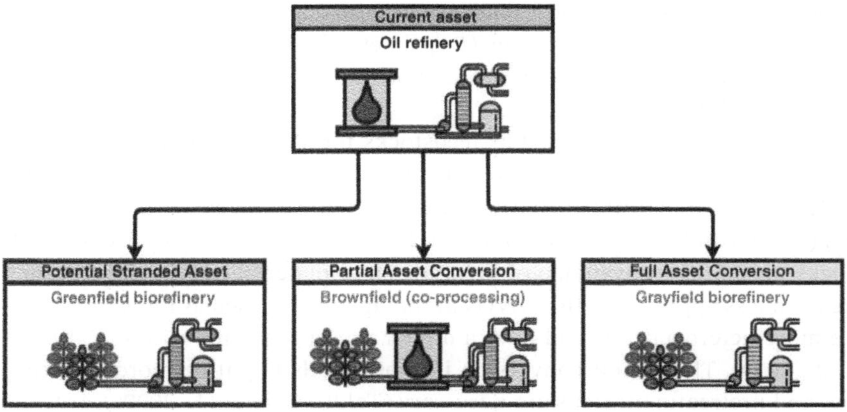

Figure 11.1 Strategies for producing liquid hydrocarbons from bio-based feedstocks.

The partial or total adaptation of conventional refineries allows for the utilization not only of process units, which compose the ISBL (Inside Battery Limits) portion of the plant, but also of OSBL (Outside Battery Limits) structures, including utilities, storage tanks, distribution systems, and administrative buildings. Moreover, the existing specialized workforce in the petroleum refining industry can also be employed. The partial conversion of refineries via co-processing, specifically, is considered an option that can be readily applied, utilizing existing facilities with reduced extra capital investment (Al-Sabawi et al., 2012; Han et al., 2020). According to Yáñez et al. (2021), co-processing has the potential to accelerate the transition towards more sustainable fuels. According to van Dyk et al. (2019), co-processing is a key strategy for expanding biofuel production.

The following sections present a review of the state of the art of the most feasible technological options available for each type of feedstock, with a focus on co-processing. It is important to emphasize that there are multiple possible production routes, and this description does not exhaust all the possibilities.

11.3 Oily Crops, Used Cooking Oils, Residual Fats

11.3.1 Feedstock Description

Oily crops include crops such as soybeans, rapeseed, canola, and palm, among others. Through extraction and/or pressing techniques, it is possible to obtain vegetable oils from these feedstocks (Al-Sabawi et al., 2012; Han et al., 2020). Also known as straight vegetable oils (SVOs), these liquids are mainly composed of triglycerides (Hammond, 2003). These oils, in general, exhibit chemical and thermal instability, high viscosity, and a tendency to form deposits, properties that make their direct use as fuels challenging (Dupain et al., 2007). Thus, their conversion into other oils such as biodiesel (FAMEs or FAEEs) or Hydrotreated Vegetable Oils (HVOs) is necessary.

Many of the oily crops used as feedstocks for SVO production are edible, which represents competition for their use for fuel production (Prasad and Ingle, 2019). Thus, other triglyceride-based biomass such as used cooking oils (UCOs) and animal fats might also be employed (No, 2014). These feedstocks present the advantage of being residual materials. Nevertheless, they are even less appropriate than SVOs for being directly used as fuels since they are more heterogeneous and have a higher presence of impurities (Carvalho et al., 2019). Thus, their conversion into other oils might be more challenging, and pretreatment steps are necessary (Toldra-Reig et al., 2020; Cárdenas et al., 2021).

11.3.2 Fuel Production Pathways in Refineries

11.3.2.1 Co-processing in Hydrotreaters for Diesel and for Unstable Feeds

Hydrotreaters (HDTs) promote the stabilization of their input stream through a catalytic process that utilizes hydrogen as a reactant. This process, which can be applied to different feedstocks, converts double and triple bonds into single bonds via hydrocarbon saturation reactions. In addition, it removes undesirable elements. In this sense, reactions that may occur include hydrodesulfurization (HDS), hydrodenitrogenation (HDN), and hydrodeoxygenation (HDO), which remove sulfur, nitrogen, and oxygen, leading to the formation of byproducts such as H_2S, NH_3, and water, respectively (Gary and Handwerk, 2001; Parkash, 2003). Decarbonylation (DCO) and decarboxylation (DCO_2) reactions, which eliminate oxygen as CO and CO_2, respectively, might also take place. In the case of co-processing, since biomass-derived compounds have high oxygen content, HDO, DCO, and DCO_2 reactions are particularly important (Srifa et al., 2014).

In refineries, diesel hydrotreaters, hereafter referred to as D-HDTs, and hydrotreaters that deal with unstable feeds (mainly heavy feedstocks, such as gas-oils and vacuum residues), hereafter referred to as U-HDTs, can receive SVOs, UCOs, or residual fats in their feed streams. U-HDTs operate under higher severity conditions of pressure and temperature, which are more adequate for treating the blend of fossil feed with SVOs, UCOs, or residual fats. In both cases, the product presents a content of renewable carbon (Al-Sabawi et al., 2012; Carmona et al., 2019) and can be incorporated into the diesel pool.

After the D-HDT or U-HDT unit, isomerization and hydrocracking (HCK) processes might be employed, respectively, to branch and shorten hydrocarbon chains, ensuring the suitability of the obtained liquid to be used as aviation fuel (Carvalho et al., 2019). This process also yields, as co-products, fractions of LPG, naphtha, and diesel.

It is worth noting that the ASTM D1655 rule (ASTM, 2022) already allows the production of aviation kerosene through co-processing by employing up to 5% (volume) of feedstock such as SVOs or animal fats. However, the feedstocks eligible for sustainable aviation fuel (SAF) production change by jurisdiction. A summary of allowed types of feedstocks according to different regulations can be found in (Perez et al., 2024).

Figure 11.2 shows a general scheme of these pathways.

As detailed in Section 11.3.3.1, studies show that it is possible to work with different percentages of biomass material in the input of D-HDT or U-HDT units, potentially achieving 100% of bio-feedstock. In the case of a fully biogenic feedstock, the main product of hydrotreating SVOs, UCOs, or animal fats is known as HVO. The principal product of the hydrotreating, isomerization, and HCK process

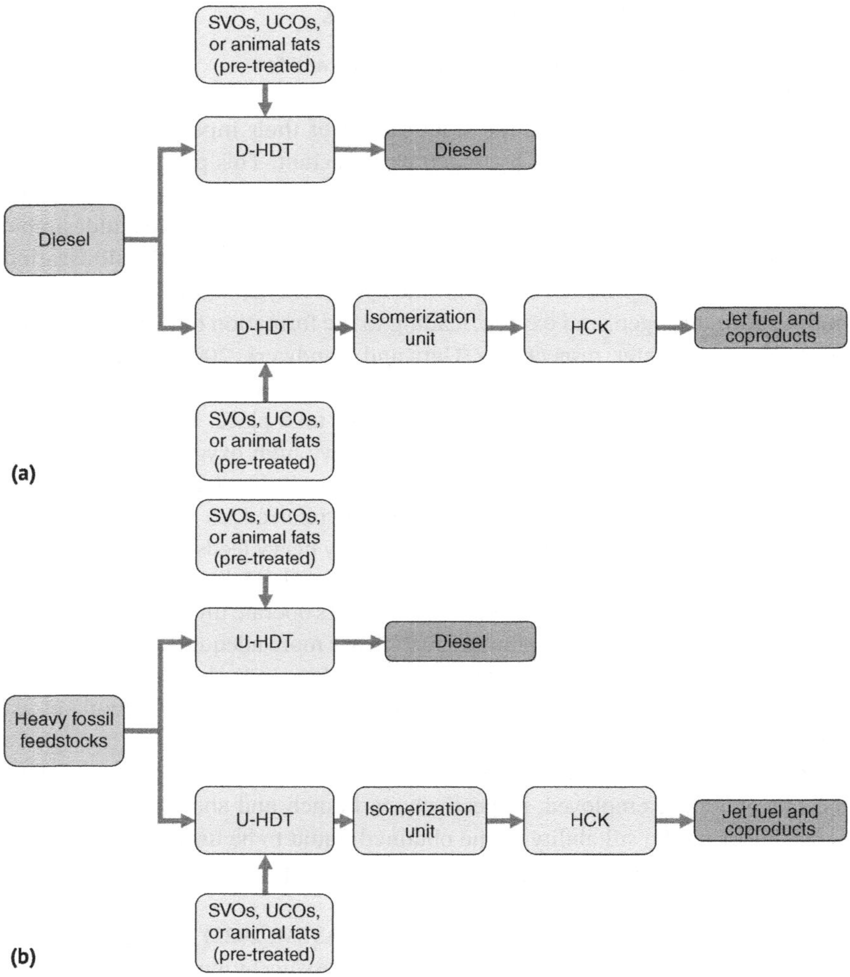

Figure 11.2 General scheme for SVO, UCOs, or animal fats co-processing in D-HDTs (a) or U-HDTs (b).

sequence is referred to as hydroprocessed ester and fatty acid (HEFA). They are drop-in alternatives to fossil diesel and jet fuel, respectively.

It is worth pointing out that, although FAME/FAEE biodiesel is a more established option for use in blends with fossil diesel (Lorne, 2023), blending fossil diesel with HVO presents important advantages. Indeed, HVO can be produced in refineries themselves and is a drop-in fuel, unlike biodiesel, whose synthesis takes place in separate facilities and whose composition differs from fossil diesel, presenting lower quality (Tiwari et al., 2011). Thus, the choice for HVO might be beneficial for the refining sector because it prevents assets from becoming stranded and allows the production of a less specified fossil diesel, since the blend is made with a bioproduct with better properties.[1]

11.3.2.2 Co-processing in Hydrotreaters for Kerosene

Kerosene hydrotreaters (K-HDTs) can also receive SVOs, UCOs, or residual fats in their feed streams, resulting in a product with bio-carbon content that can be added to the jet fuel pool. Again, it is important to emphasize that ASTM D1655 (ASTM, 2022) permits up to 5% (volume) of such feedstocks in co-processing for aviation fuels; however, feedstocks eligible for SAF production change by jurisdiction (Perez et al., 2024), as pointed out in Section 11.3.2.1.

Figure 11.3 shows a general scheme of the pathway.

Since jet fuel has strict specifications, its production by co-processing in K-HDT units faces some important challenges, such as the fact that additions of renewable feedstock can cause freezing point increase. This can be addressed through separate isomerization and HCK units, as shown in the route presented in Section 11.3.2.1. In the case of co-processing in the K-HDT unit only, it is essential to guarantee the employment of catalysts that promote selective deoxygenation, desulfurization, and isomerization. K-HDT co-processing is recommended up to 5% (volume) of renewable feedstocks in the feed stream in order to be done without major modifications, aside from the catalyst change (Perez et al., 2024).

11.3.2.3 Co-processing in Fluid Catalytic Crackers

Fluid Catalytic Cracking units (FCCs) convert heavy feedstocks (such as gas-oils from different units) into higher value-added products, including LPG, naphtha, LCO (Light Cycle Oil, a product that can be part of the diesel pool after

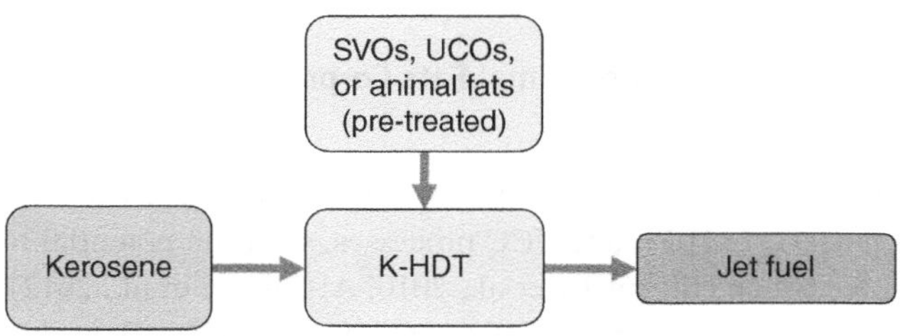

Figure 11.3 General scheme for SVO, UCOs, or animal fats co-processing in K-HDTs.

1 HVO has better chemical stability if compared to FAME/FAEE biodiesel. In addition, in general, HVO exhibits lower sulfur content and density values (Zuniga et al., 2011; Vignesh et al., 2021; Longanesi et al., 2022).

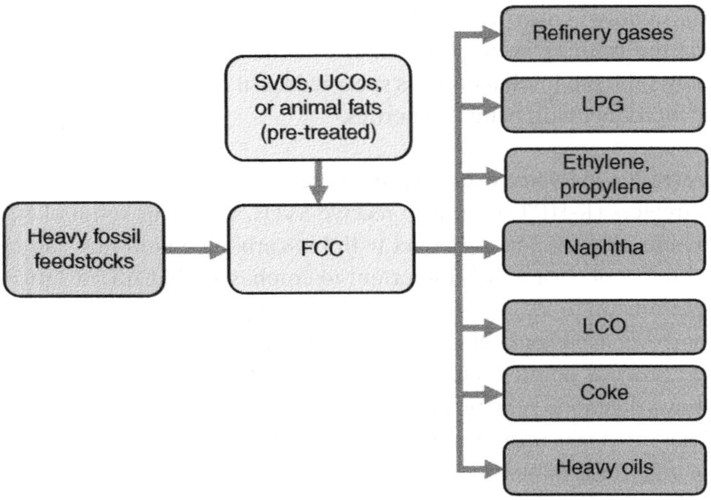

Figure 11.4 General scheme for SVO, UCOs, or animal fats co-processing in FCCs.

being processed in U-HDT units), and basic petrochemicals. FCCs can receive, in addition to their usual fossil feed, a fraction of SVOs, UCOs, or residual fats (Dupain et al., 2007; Bielansky et al., 2010), resulting in products with renewable carbon content that can be incorporated into their respective final product pools. Figure 11.4 shows a general scheme of the route.

According to the choice of operating conditions, the process can focus more on producing naphtha or other products such as basic petrochemicals (ethylene and propylene) (Szklo et al., 2012). Also, process and catalyst innovations make it possible to generate fully renewable petrochemical products, including aromatics, as reported by Petrobras (Brelsford, 2023).

11.3.3 Key Aspects of SVO, UCOs, and Animal Fats Co-processing

11.3.3.1 Operating Conditions

Studies indicate the possibility of working with different percentages of biomass feedstock in the D-HDT, U-HDT, and FCC processes, with the potential to reach 100% renewable feedstock (Bielansky et al., 2010; Al-Sabawi et al., 2012). However, it is relevant to note that different levels of renewable feedstock lead to different conversions and yields in terms of final products. In K-HDT units, up to 5% in volume of the feed stream shall be composed of bio-feedstocks.

For co-processing in D-HDT or U-HDT units, conventional catalysts can be applied, but there are also various studies on new catalysts, as shown in the literature review found in Al-Sabawi et al. (2012). In K-HDT co-processing, catalyst change is key to guarantee the required properties for jet fuel (Perez et al., 2024). In the case of co-processing in FCCs, the predominant application is of commercial equilibrium catalysts (Ecat), similar to those used in conventional operations. Catalyst innovations can also enable specific applications. Regarding reaction temperature conditions, generally, no significant changes are observed compared to the conventional use of the units (Al-Sabawi et al., 2012). As for hydrogen pressure, a relevant variable only in the case of hydrotreaters, D-HDTs, and U-HDTs need to employ higher pressures compared to processing fossil feeds alone. According to Szklo et al. (2012), the removal of unwanted elements via hydrotreating becomes more difficult when comparing oxygen with sulfur, being more challenging in terms of catalysts and the required pressure. Thus, biomass-derived compounds, which have a high oxygen content in their composition, require more severe conditions in this process. Pressure ranges between 5 and 15 MPa are found in the literature (ETIP Bioenergy, 2023) for SVO co-processing, tending toward the higher values. In the case of K-HDTs, with the right catalyst, low-pressure operation is possible, in the range of 2.5–3.0 MPa (Perez et al., 2024).

11.3.3.2 Strengths and Challenges

The processing of SVOs, UCOs, and residual fats in refinery units has a high level of technological maturity, with existing examples of applications in the refining industry, as described in Section 11.3.3.3. Another advantage is that the products obtained are drop-in, meaning they are chemically similar to fossil products currently used, and thus can take advantage of existing infrastructure. Challenges are related to competition with the biodiesel (FAME or FAEE) sector, which uses the same kind of feedstocks, and with the food sector (in the particular case of SVOs from edible sources). The need for logistics for the supply of these feedstocks is also a challenge. Given that units in existing refineries were initially designed to deal with fossil oil, their large scales might require the collection of biomass raw materials over large distances. Regarding UCOs and residual fats, an additional challenge is the fact that they must be collected from dispersed locations. Kristiana et al. (2022) show that the collection of UCOs as a percentage of the estimated potential varies significantly among countries as a result of each country's collection capabilities. In this sense, modular mini-refineries might be appropriate for co-processing in terms of scale. Section 11.6 shows a case study that approaches these issues.

11.3.3.3 Existing Initiatives

A list of existing initiatives in Europe and the United States regarding the co-processing of renewable feedstocks in refineries can be found in Vertz and Sayal (2020).

For example, the BP refinery in Lingen, Germany, co-processes UCOs together with fossil oil in hydrotreaters, focusing on producing aviation fuel (bp, 2022). This strategy is in line with ASTM D1655 standards and can help meet initial mandates for blending renewable and fossil jet fuel (Energy Intelligence, 2022).

In France, the TotalEnergies refinery La Mède has been undergoing transformation since 2015 to process bio-feedstocks only. In 2019, a production capacity of 500,000 tonnes per year of renewable diesel was established, using a variety of raw materials: SVOs, UCOs, and animal fats (TOTAL, 2020).

In the United States, a study shows that eight refineries had been converted to produce renewable diesel and/or jet fuel by the end of 2022, including full and partial conversions (Fitzgibbon et al., 2023).

In Brazil, in 2006, Petrobras patented a technology for co-processing vegetable oils or animal fats with fossil diesel in hydrotreaters, naming it H-BIO. In 2020, industrial-scale tests were completed using soybean oil (Petrobras, 2020), and some Brazilian refineries are currently authorized to operate with SVO feedstock in hydrotreating units (ANP, 2015, 2021, 2022). In 2023, with technology developed by Petrobras, the Riograndense Refinery processed 100% SVO in FCC, this being the first test of this kind in the world (Petrobras, 2023). Finally, it is worth noting that the new investments related to co-processing are currently foreseen in the country (MME, 2023). Nevertheless, despite the mature technology and planned investments, Brazil still lacks regulations allowing the commercialization of co-processing products.

11.4 Lignocellulosic Biomass

Lignocellulosic biomass is widely available as a residue from agricultural and forestry industries and can be obtained through planted forests, contributing to increasing the global potential for biofuel production (van Dyk et al., 2019). Furthermore, these are non-edible raw materials that do not face the issue of competition with food. On the other hand, planted forestry, especially when it is monospecific, does not promote biodiversity (Carnus et al., 2006; Messier et al., 2022). Agricultural residues can become a problem when they do not have proper disposal, with the possibility of GHG emission during their decomposition process (Portugal-Pereira et al., 2015). Their conversion process to fuels is also more complex compared to that of oilseeds (Bezergianni et al., 2018). Lignocellulosic biomass is mainly composed of cellulose, hemicellulose, and lignin, which include functional groups such as hydroxyl, carboxyl, aldehyde, ketone, ester, and ether (van Dyk et al., 2019). These groups are responsible for the higher oxygen content, lower heat capacity, water miscibility, and increased corrosiveness compared to fossil and vegetable oils (van Dyk and Saddler, 2021). This section will present technological alternatives to process this raw material into fuels using existing petroleum refinery assets.

11.4.1 Fischer–Tropsch Biomass to Liquid (FT-BtL) Pathway

The Fischer–Tropsch synthesis is preceded by an important process called gasification. Gasification is a thermochemical conversion, whose desired product is synthesis gas (mainly CO and H_2) (Cortazar et al., 2023). Different sources of carbon can be used in this process, such as coal, biomass, pyrolysis oil, and captured CO_2 (Cortazar et al., 2023; Nascimento da Silva et al., 2024). Depending on the catalysts, operational conditions, and downstream processes, this route allows the production of liquid hydrocarbons equivalent to aviation kerosene, gasoline, naphtha, and diesel (Tagomori et al., 2019). The use of biomass as a raw material for this process introduces significant challenges compared to coal or natural gas. The purity of synthesis gases is an important issue for the Fischer–Tropsch synthesis (Tagomori et al., 2019). Due to biomass characteristics such as high oxygen concentration and other contaminants (sulfur, nitrogen compounds, and alkali metals), if the traditional gasification process is used, more steps for cleaning synthesis gases are required (Tagomori et al., 2019; Cortazar et al., 2023). Alternative technologies such as plasma gasification contribute to the production of gases with higher purity but also entail higher costs (Richardson et al., 2012). There is also the option of biomass entrained flow gasification using pyrolysis oil as an intermediate product (Leijenhorst et al., 2015), which helps simplify and reduce the cost of biomass transportation and may allow dealing with large-scale entrained-bed gasifiers (Leijenhorst et al., 2015).

In the case of direct gasification, biomass must undergo a pretreatment involving drying, adjusting of pellet or chip sizes, and acidity conditioning (Hoffmann and Szklo, 2011). Gasification begins with solid-gas reactions and is succeeded by gas-phase reactions (Kimura et al., 2006; Zhang and Pang, 2019; Cortazar et al., 2023). These reactions are presented below (Hoffmann, 2010) apud (Kaltschmitt and Streicher, 2009). The heating rate, oxidant, carbon source, steam-to-feed ratio, and residence time will influence the products obtained in the reaction, indicating the complexity of controlling the process. The use of pure steam as a gas agent, for example, increases H_2 production and gas product heat value (Cortazar et al., 2023). This technology has been used on an industrial scale for fuel production using coal as a raw material (van Dyk and Saddler, 2021), classifying this route as Technology Readiness Level (TRL) between 7 and 8. However, considering the challenges in cleaning the synthesis gas when dealing with biomass gasification as opposed to coal, the technological readiness is considered between 6 and 7.

Gasification Gas–solid Reactions:

$$\text{Complete Combustion:} \quad C + O_2 \leftrightarrow CO_2 \quad \Delta H = -393.5\,\text{kJ/mol} \tag{11.1}$$

$$\text{Partial Combustion:} \quad C + 1/2\ O_2 \leftrightarrow CO \quad \Delta H = -123.1\,\text{kJ/mol} \tag{11.2}$$

$$\text{Methanation:} \quad C + 2H_2 \leftrightarrow CH_4 \quad \Delta H = -87.5\,\text{kJ/mol} \tag{11.3}$$

$$\text{Vapor–carbon Reaction:} \quad C + H_2O \leftrightarrow CO + H_2 \quad \Delta H = +118.5\,\text{kJ/mol} \tag{11.4}$$

$$\text{Boudard Reaction:} \quad C + CO_2 \leftrightarrow 2CO \quad \Delta H = +159.9\,\text{kJ/mol} \tag{11.5}$$

Gas-phase Reactions:

$$\text{Water–gas Shift:} \quad CO + H_2O \leftrightarrow CO_2 + H_2 \quad \Delta H = -40.9\,\text{kJ/mol} \tag{11.6}$$

$$\text{Methanation:} \quad CO + 3H_2 \leftrightarrow CH_4 + H_2O \quad \Delta H = -203.0\,\text{kJ/mol} \tag{11.7}$$

After gasification, several steps are necessary for gas cleaning, removal of acidic gases, and adjustment of composition. The cleaning step removes ash, sludge, particulate matter, and contaminants such as ammonia (NH_3) and hydrogen chloride (HCl) (Tagomori et al., 2019). The removal of acidic gases is necessary because the catalyst (usually based on iron and cobalt) used in the next step can be damaged by halogens, sulfur, and particulate contaminants[2] (Tagomori et al., 2019). It is also necessary to adjust the H_2:CO ratio to around 2:1, which is the condition indicated for the Fischer–Tropsch synthesis. Then, the synthesis gases are ready for the Fischer–Tropsch reaction, presented below. This process involves the adsorption and dissociation of reagents (H_2 and CO) on the catalyst surface to form chain initiators that lead to the reaction and promote carbon chain growth (Tagomori et al., 2019). The yield of each size of alkanes and alkenes is a function of the reaction conditions and catalysts used.

Fischer–Tropsch Reactions:

$$\text{Paraffins Production:} \quad nCO + (2n+1)H_2 \leftrightarrow C_nH_{2n+2} + nH_2O \tag{11.8}$$

$$\text{Olefins Production:} \quad nCO + (2n)H_2 \leftrightarrow C_nH_{2n} + nH_2O \tag{11.9}$$

The FT-BtL product is a mixture of different sizes and types of carbon chains. Therefore, it is necessary to process this product in some well-known refining units before its final use. First, it is necessary to separate different carbon chain ranges (with a greater interest in naphtha, gasoline, kerosene, and diesel) through distillation. Then, hydroprocessing (HDT and HCK) units can be used to promote the saturation of olefins and some long carbon chain breaking to

---

2 Maximum allowable concentration for FT process of sulfur and nitrogen compounds is 1 ppm (volume); of halogen compounds is 10 ppb (volume); and dust shall be completely removed (Meerman et al., 2011).

maximize the production of medium and light distillates. In the case of aviation bio-kerosene (bio-jet) production, hydroisomerization units are also used to tailor the desired product (Carvalho et al., 2019).

11.4.2 Hydrotreated Pyrolysis Oil (HDPO) Pathway

Pyrolysis oil (or bio-oil) is obtained through a thermal decomposition process of biomass at high temperatures (400 °C–500 °C), in the absence of oxygen (Bezergianni et al., 2018). In the case of its use as fuel, fast pyrolysis (see Figure 11.5) is applied, with typical heating rate of 1000 K/s, short residence time, and high conversion into a liquid product (Talmadge et al., 2014). This reaction achieves a conversion rate of around 70–80% (on a mass basis) (Bezergianni et al., 2018). The raw material for the reaction needs to be pre-dried, ground, and homogenized to avoid density variations in the reactor. Pyrolysis oil has a heating value ranging between 23 and 34 MJ/kg (Vecino Mantilla et al., 2014) and is rich in oxygenated compounds (around 40%), as heteroatoms from different functional groups (van Dyk et al., 2019). Its composition varies according to the raw material used. Pyrolysis oil exhibits higher acidity (typical total acid number value over 100 KOH mg/g (Talmadge et al., 2014)) and lower chemical stability compared to fossil derivatives and even SVOs. The byproducts of this reaction are non-condensable gases (5–10% w/w) and biochar (20–30% w/w) (Bezergianni et al., 2018; Ximenes et al., 2022). The production of pyrolysis oil is considered TRL 8, as there are at least four industrial plants in operation (BTG Bioliquids, 2023; Ensyn, 2023).

According to Bezergianni et al. (2018), two processes are most studied for the co-processing of pyrolysis oil in refineries: hydroprocessing and fluidized catalytic cracking (FCC). Some concerns about co-processing bio-oil in hydroprocessing units are related to its high instability and its immiscibility with petroleum feeds (gasoils, for example) (Sousa-Aguiar et al., 2018). Bezergianni et al. (2018) report

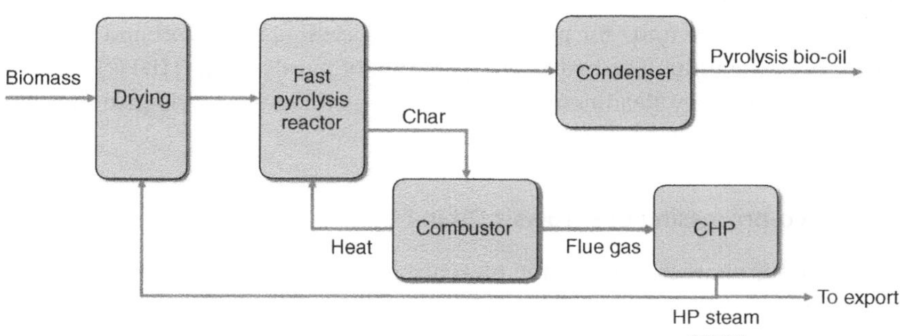

Figure 11.5 Simplified fast pyrolysis scheme. Adapted from Lammens (2022), Oliveira and Schure (2020).

that the most used catalysts in this process are $CoMo/Al_2O_3$ and $NiMo/Al_2O_3$, also used in existent hydroprocessing units. However, the main issue of bio-oil co-processing is catalyst deactivation (Sousa-Aguiar et al., 2018). Bezergianni et al. (2018) conclude that at least one pretreatment step of bio-oil is still necessary for it to be co-processed in HDTs, indicating that alternative steps and catalysts need to be studied to increase the feasibility of this route. Sousa-Aguiar et al. (2018) cite some alternative catalysts studies for bio-oil hydrotreatment using noble metals or alternative bases for catalysts such as TiO_2, CeO_2, and ZrO_2. Some of those experiments presented good results, but these alternatives are still more expensive and in initial development phases (Sousa-Aguiar et al., 2018).

Additionally, the route through hydrothermal liquefaction (HTL) of biomass has also been extensively studied, as presented by Lindfors et al. (2023). According to their review, HTL bio-oil is more deoxygenated and consequently more hydrophobic and viscous, but less dense than pyrolysis oil. However, Sharma et al. (2022) and Zhu et al. (2022) also highlight that some pretreatment steps are necessary before co-processing in HDTs.

The deoxygenation process is followed by cracking (breaking of carbon chains), DCO (removal of CO), decarboxylation (removal of CO_2), and HDO (removal of oxygen through the formation of H_2O) reactions. All reactions occur in parallel. The production of CO and CO_2 leads to a loss in carbon conversion, as the primary interest is to retain carbon in the final products. A higher conversion rate through increased removal of H_2O is achieved under specific conditions of pressure, temperature, and catalyst (Bezergianni et al., 2018).

The gases formed in these reactions result in a reactive mixture at the end of the process that is quite different from the conventional HDS process of fossil feedstock. The water produced as a byproduct impacts the efficiency of alumina, and the increased gas production makes pressure control more difficult. Temperature control is also important as the reaction is exothermic, and the catalyst loses efficiency at high temperatures (Bezergianni et al., 2018).

The most suitable units for pyrolysis oil co-processing are diesel and unstable hydroprocessing units. The hydrodeoxygenation of pyrolysis oil (HDPO) is a process with Technology Readiness Level (TRL) 5, as there is no evidence that it is done on an industrial scale.

11.4.3 Co-processing of Pyrolysis Oil in FCCs

Pyrolysis oil is characterized by the presence of aromatic compounds, phenolics, and other oxygenated functional groups (van Dyk et al., 2019). This makes the cracking process a more suitable option to form shorter hydrocarbon chains closer to the required specifications for higher-value fuels. In this case, available refining assets include FCCs, HCKs, and Delayed Coking units (DCUs) (Sousa-Aguiar et al., 2018; van Dyk et al., 2019).

Many studies and demonstration-scale tests have been conducted for the application of pyrolysis oil co-processing in FCC units (Wang et al., 2014; Pinho et al., 2017; Talmadge et al., 2021; Lindfors et al., 2023). The FCC unit is widely present in petroleum refineries,[3] primarily used for producing high-octane gasoline. In addition to promoting the breaking of carbon chains, it increases the olefinicity and aromaticity of molecules. FCC is a catalytic conversion process with a significant reliance on catalysts, primarily zeolites (Pinho et al., 2017). This unit offers broad flexibility and can be adjusted to maximize the production of, for example, medium distillates and propene (Szklo et al., 2012).

According to Pinho et al. (2022), pyrolysis oil co-processing has been successfully tested on a demonstration scale for concentrations up to 20% (by mass) concerning the total load of the FCC unit, without prior hydrotreatment. The fossil fraction of the FCC load is mostly composed of vacuum gasoil (VGO). For this process, an independent feed of pyrolysis oil is used at a point axially below the fossil VGO entry nozzle at the base of the riser, thus avoiding immiscibility issues. As the flow in the riser is upward, the fossil gasoil entry occurs first, allowing this product to reach the catalyst before the bio-oil, creating a protective layer that prevents unwanted adhesion of bio-oil coke to the catalyst. Figure 11.6 illustrates this process. Cyclones are used to separate hydrocarbons from the catalyst. An important advantage of this process is that minimal changes are needed in the existing units, representing a low

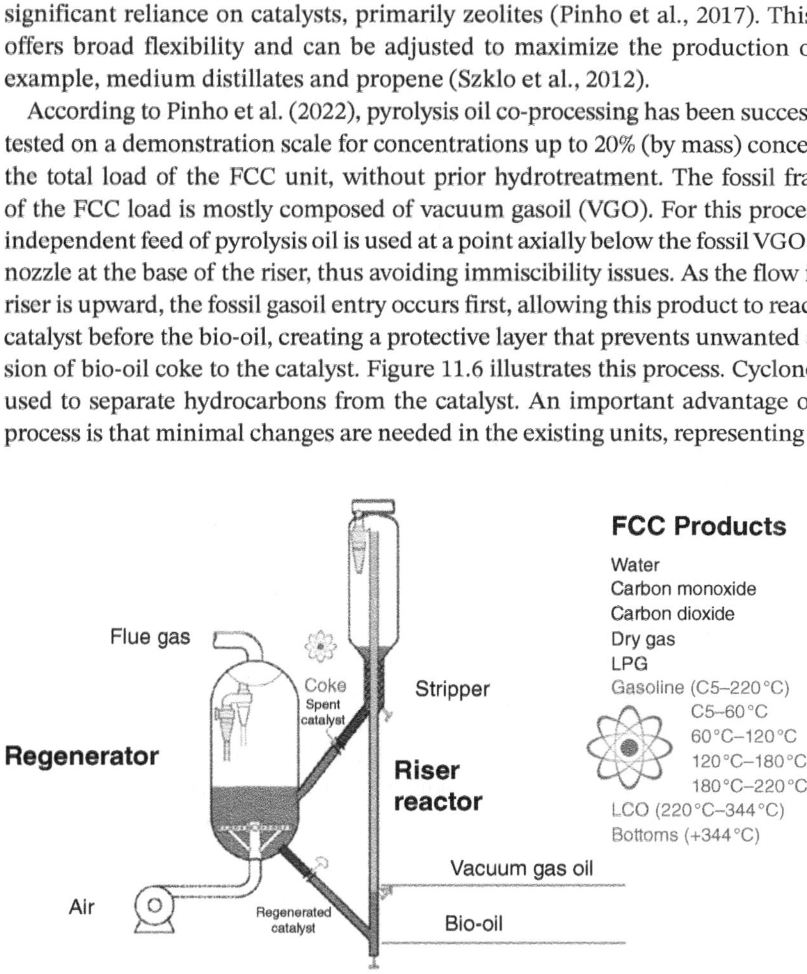

Figure 11.6 Co-processing in FCC unit and its products. *Source*: Pinho (2022)/with permission of Elsevier.

3 FCC installed capacity in the world is around 15.3 million bpd, which corresponds to 17.7% of the world's crude distillation refining capacity (Oil and Gas Journal, 2023).

investment option for the adaptation of existing plants. In addition to a new nozzle, a pyrolysis oil tank and all interconnecting piping and control systems for this new feed must be included (Oliveira and Schure, 2020).

In terms of the FCC yields, Pinho et al. (2022) indicate that between 40% and 54% of the renewable carbon concentrates in the formed coke, 31–34% in liquid fractions (gasoline, diesel, and gasoil), and the remaining is distributed among the formed gases CO and CO_2.[4] These percentages vary depending on the concentration of pyrolysis oil at the unit inlet and the specific characteristics of pyrolysis oil and the fossil feed used. The fraction of renewable carbon present in the produced liquid fuels will result in emissions savings in Scope 3, corresponding to greenhouse gas emissions indirectly associated with the company's activities in this case. On the other hand, the renewable fraction in the FCC coke represents savings in emissions that occur during the refining process itself and is considered Scope 1 (WRI and WBSCD, 2013). The coke formed in the FCC provides heat for the reaction when oxidized during the catalyst regeneration.

Despite promising results, the level of technological readiness associated with this technology (TRL) is still considered as between 6 and 7, since there is no record of industrial-scale operation.

11.4.4 Co-processing of Pyrolysis Oil in Coking and Deasphalting Units

As mentioned before, studies are being conducted considering the use of lignocellulosic biomass in delayed coking units (Siskin et al., 2011), a unit widely present in various Coking or Hycon refineries worldwide[5] (Oil and Gas Journal, 2023). According to Siskin et al. (2011), pyrolysis oil in a fraction of up to 20%, preferably up to 10% (on a mass basis), has been co-processed in delayed coking and fluidized bed coking units. The temperature and pressure conditions used in the furnace (between 480 °C and 520 °C and between 44 psig and 580 psig) and in the coke drums (between 410 °C and 475 °C and between 15 psig and 44 psig) are similar to the conditions used for fossil feedstock only. Siskin et al. (2011) state that the unit's operation is facilitated in the case of co-processing, as it improves the drying rate due to

4 The mass yield reported by Pinho et al. (2022) for one of the FCC processing 100% fossil HVO feed is: 3.9% of natural gas; 13.8% of liquefied petroleum gas (LPG); 39.9% of gasoline; 20.3% of diesel; 16.1% of fuel oil (i.e., 76.3% in total of liquids); and 6.4% of coke. Yields in all lighter fractions showed a slight decrease in the case of co-processing of bio-oil. The coke fraction remained the same in the case of 10% bio-oil co-processing and increased in the case of 20% bio-oil co-processing. The decrease in yields for gaseous and liquid fractions is attributed to the loss of carbon to the byproducts CO and CO_2 formed during co-processing.

5 Delayed Coking Unit installed capacity in the world is around 5.4 million bpd, which corresponds to 6.3% of the world's crude distillation refining capacity (Oil and Gas Journal, 2023).

the free radicals present in lignin, reducing coke handling problems and increasing coke quality. In the case of fluidized bed coking, the addition of alkali metal salts to the catalyst, preferably cesium or potassium carbonate on an alumina base (Siskin et al., 2011), enhances the gasifier's efficiency. However, this technology is considered TRL 5 or 6, as it has only reached the stage of isolated tests.

In addition, a patent filed by Petrobras (Ximenes et al., 2022) presents the option of introducing the bio-oil feed into deasphalting units, with very promising results. The deasphalting process is an extraction process that maximizes the concentration of higher-value-added products, increasing the recovery of lighter fractions concentrated in the so-called deasphalted oil stream. This stream is subsequently processed in the bottom of the barrel conversion sections, such as FCC units. Two types of bio-oil were tested: pyrolysis oil and catalytic pyrolysis oil. For both cases, bio-oils were co-processed at different mass fractions, from 10% to 75%. The solvent used was pentane, the temperature was maintained at 65 °C, and the system was pressurized to 1379 kPa to keep the solvent in the liquid phase (Ximenes et al., 2022). The results showed a significant linearity between the properties of the products and the concentrations of bio-oil in the feed, confirming the incorporation of bio-oil into the deasphalted stream (Pontes et al., 2021). Ximenes et al. (2022) state that their main motivation was to identify "new integration points for bio-oil in the refining scheme to bring cost-effectiveness to second-generation biofuel production processes." Despite being very promising, the use of this technology is still in early stages of study and needs further development to achieve higher technological readiness. Due to the limited existing studies, this route can be considered to have a Technology Readiness Level (TRL) between 3 and 4.

Therefore, this section presents several points of insertion of bio-oil into existing refining units: Hydrotreatment, Catalytic Cracking, Coking, and Deasphalting. Among them, the option of co-processing in FCC currently has the highest technological readiness. In the case of the deasphalting process, it should be seen as a complementary treatment process that possibly would allow the use of a larger fractions of bio-oil in the FCC feed.

11.5 Sugary and Starchy Biomass

Sugary biomass (such as sugarcane and sugar beet) and starchy biomass (like maize and wheat), as well as lignocellulosic materials, can be used to produce ethanol through fermentative processes. Ethanol can be used directly as a fuel or converted to liquid hydrocarbons, mainly jet fuel, through the process known as Alcohol-to-Jet (AtJ) (Pechstein and Kaltschmitt, 2018). In addition, it can be dehydrated to form ethylene, a basic petrochemical (Zhang and Yu, 2013).

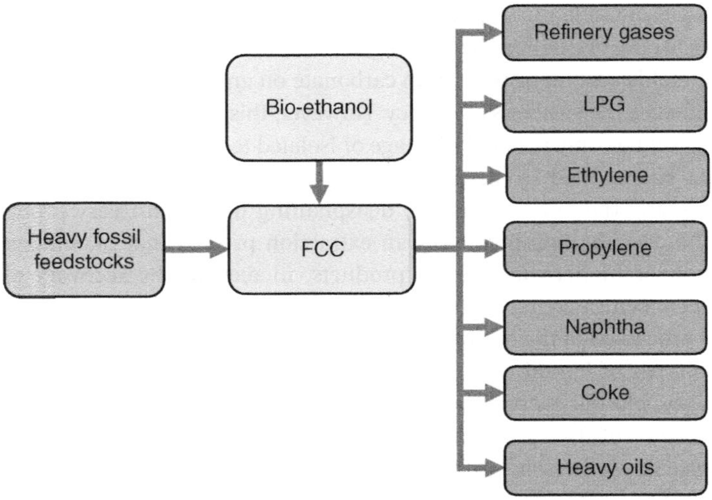

Figure 11.7 General scheme for ethanol co-processing in FCCs.

In refineries, ethylene can be obtained by co-processing ethanol in FCC units with their usual feedstock (heavy fossil oils) (Pinho et al., 2007). This process produces ethylene by ethanol dehydration as the main product, as well as lighter hydrocarbons from the heavy fossil feedstock. Figure 11.7 shows a general scheme of the route.

Ethanol shall be introduced to a maximum amount of 20% (mass basis) of the FCC feed stream. A typical FCC catalyst can be used, having zeolites in its constitution (Pinho et al., 2007).

In usual ethanol dehydration processes, ethanol has to be superheated to feed the reactor, while in FCC co-processing, the catalysts from the regenerator of the unit provide the necessary heat for both ethanol dehydration and hydrocarbon cracking reactions. The ethylene and coproducts are drop-in, which is also an advantage of the process. Nevertheless, the process still needs to reach higher maturity levels. Currently, existing initiatives include tests in companies such as Petrobras (Pinho et al., 2007) and ExxonMobil (DAKKA et al., 2024).

11.6 Opportunities and Challenges for Biomass Co-processing – A Case Study Application

Since existing units in refineries were projected to deal with fossil resources and not biomass-derived feedstocks, it is important to understand if the collection of the required amounts of biomass for co-processing might represent a logistics challenge. To evaluate this issue, this section presents a case study for Brazil in which the availability of biomass close to existing refineries is assessed. In addition, suitable locations for modular mini-refineries, in order to ensure the supply of these resources, are also evaluated.

11.6.1 Methodology

The possibility of co-processing SVOs, UCOs, residual fats, and pyrolysis oils (POs) in refining units was evaluated. For the first three renewable feedstocks, co-processing was considered in D-HDT and U-HDT units, despite the fact that it is also possible to co-process them in other units. For POs, co-processing was considered only in FCC units, since it is a process with higher readiness level when compared to the other pathways, as mentioned in Section 11.4. The analysis was carried out for each Brazilian refinery with processing capacity greater than 100,000 barrels/day. Their names and the Brazilian state where they are located are shown in the map in Figure 11.8.

Figure 11.8 Map of Brazil showing refineries with processing capacity greater than 100,000 barrels/day.

For each evaluated refinery, the total capacity of D-HDT and U-HDT units was accounted for, and the possibility of feeding of 25% by mass of SVOs, UCOs, or residual fats in a blend with fossil feedstock was assumed. It was also considered a yield of 20% in the transformation of oily crops into SVOs (Huo et al., 2009) and 40% in the transformation of crude animal fat into pre-treated fat (Soares Dias et al., 2022).

Furthermore, the total capacity of FCC units was accounted for, and an input of 15% by mass of POs was assumed. This lower input percentage is due to the fact that this is a lower-quality oil. A yield of 50% in the transformation of lignocellulosic biomass into pyrolysis oil was also considered (Matos et al., 2020).

These hypothetical demands, in order of magnitude, were compared to the availability of biomass close to the refineries mentioned before. For oily crops and lignocellulosic biomass, data from Tagomori (2022) were used, which estimated the potential of soybeans and woody biomass in Brazil until 2050. In the case of UCOs and animal fats, data from the Brazilian Institute of Geography and Statistics for residential consumption of soybean oil (IBGE, 2018) and for cattle and swine farming (IBGE, 2021) were used, respectively.[6]

Furthermore, a hypothetical modular mini-refinery with a processing capacity of 15,000 barrels/day was considered, with U-HDT capacity of 8,100 barrels/day and FCC capacity of 5,600 barrels/day. These values were estimated using the CAESAR tool, which simulates the production profile, utility consumption, and GHG emissions of refineries (Guedes et al., 2019). Similar assumptions to those described above were also made regarding the amount of renewable feedstock at the unit feed and the conversion of raw materials.

These hypothetical demands were compared to the availability of biomass in Brazilian cities. Georeferencing tools were applied to the data from Tagomori (2022), IBGE (2018), and IBGE (2021) to determine the best locations, in the national territory, in terms of availability of soybeans, used soybean oil, animal fat, and woody biomass.

11.6.2 Results and Discussion

The charts in Figure 11.9 compare the availability of raw materials close to existing refineries with the hypothetical demand for co-processing. These charts and paragraphs below refer to Brazilian refineries. Please see the map in Figure 11.8 for their names and locations.

---

6 It is important to highlight that there is a methodological difference between the data used for the availability of soybeans/woody biomass, which are estimates for 2050, and for UCOs/animal fats, which are actual data for the referred years.

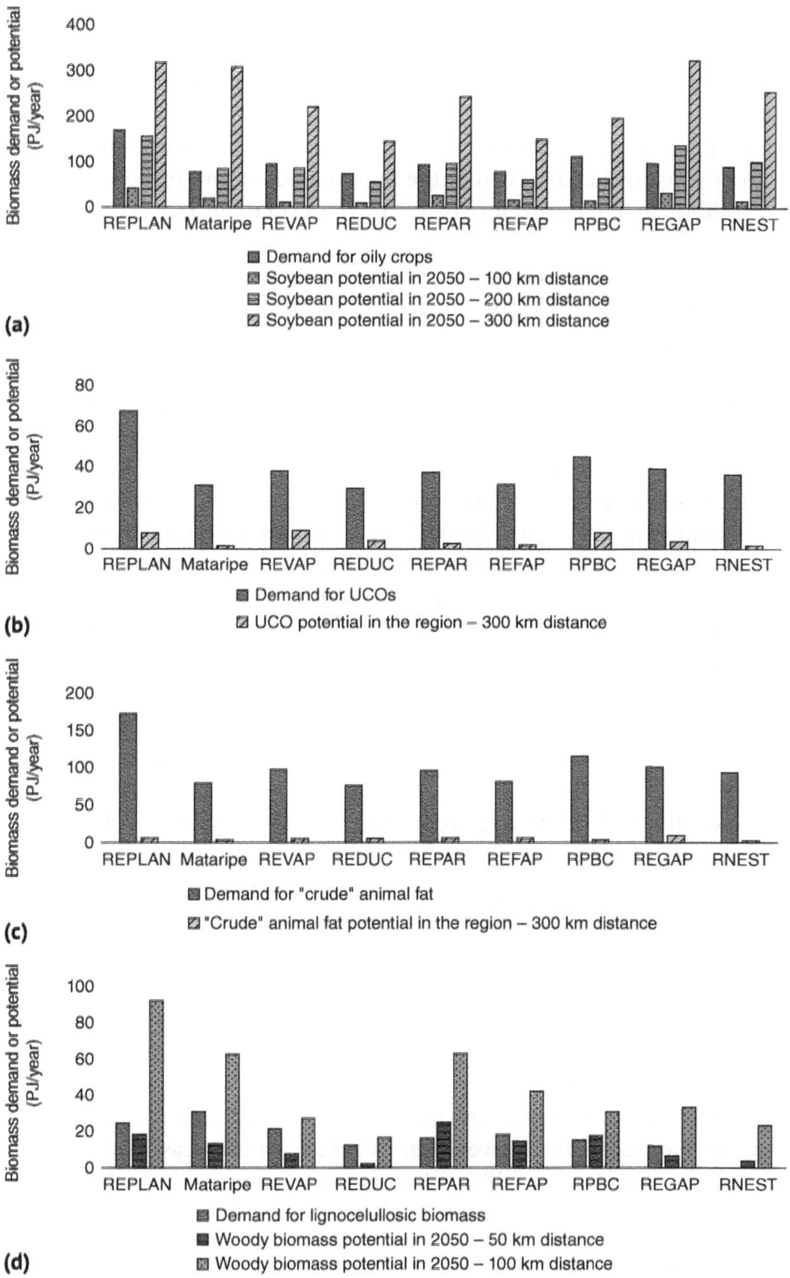

Figure 11.9 Availability of raw materials and hypothetical demand from existing refineries for co-processing.

Figure 11.9a shows that, for the Mataripe, REPAR, REGAP, and RNEST refineries, a distance of 200 km is sufficient to supply, with soybean oil, the calculated demand for co-processing. For the other refineries, the availability of soybeans within a distance of 300 km meets this demand. Figures 11.9b, c show that, even at a distance of 300 km, the availability of UCOs and animal fat is not sufficient to meet the co-processing demand, within the limit defined in this exercise. Finally, Figure 11.9d shows that, for a distance of 50 km, the availability of woody biomass near REPAR and RPBC is sufficient to meet the co-processing demand. For the other refineries, this demand can be met within a distance of 100 km. The shorter distances evaluated in this last case are due to the fact that pyrolysis oil is less stable than other materials (Han et al., 2020), making its transportation over long distances more challenging.

The graphs in Figure 11.10 display Brazilian cities[7] with the highest availability of the evaluated raw materials and the demand for co-processing from the hypothetical modular mini-refinery.

Figure 11.10a shows that a distance of 100 km around the analyzed cities is sufficient to supply the calculated demand for co-processing in the hypothetical mini-refinery, even if the decision is made to feed 100% of SVO into the HDTs unit eventually. Figure 11.10b shows that a distance of at least 200 km around the evaluated cities would be necessary to supply the co-processing demand with used cooking oil. Figure 11.10c shows that, in general, a radius of 300 km is necessary around the evaluated cities to meet the demand for co-processing with animal fat. Finally, Figure 11.10d shows that, for a distance of 50 km, it is already possible to meet the co-processing demand with woody biomass in the hypothetical mini-refinery.

This exercise shows that, for large refineries, due to the fact that their units were initially projected to deal with fossil oil and not with biomass, scale can be a challenge for co-processing: in the case of oily crops and lignocellulosic biomass, it is necessary to collect raw materials in large radii around the refineries. The use of residual materials (UCOs or animal fats) in relevant blends is prohibitive. For modular mini-refineries, the availability of oily crops and lignocellulosic biomass is consistent with the scale of the processing units, allowing for co-processing. The use of residual materials is still more challenging; however, it could be explored in particular cases.

It is important to highlight some limitations of the analyses conducted. Actually, the demand for biomass will be lower than calculated in each case, since not all HDT or FCC units in refineries will be allocated for this purpose; the utilization factor of the units shall be less than 100%; and the renewable feed into the units

---

7 For each city, it was chosen to display in the graphs the Brazilian state where it is located (the location of each state can be seen in the map in Figure 11.8).

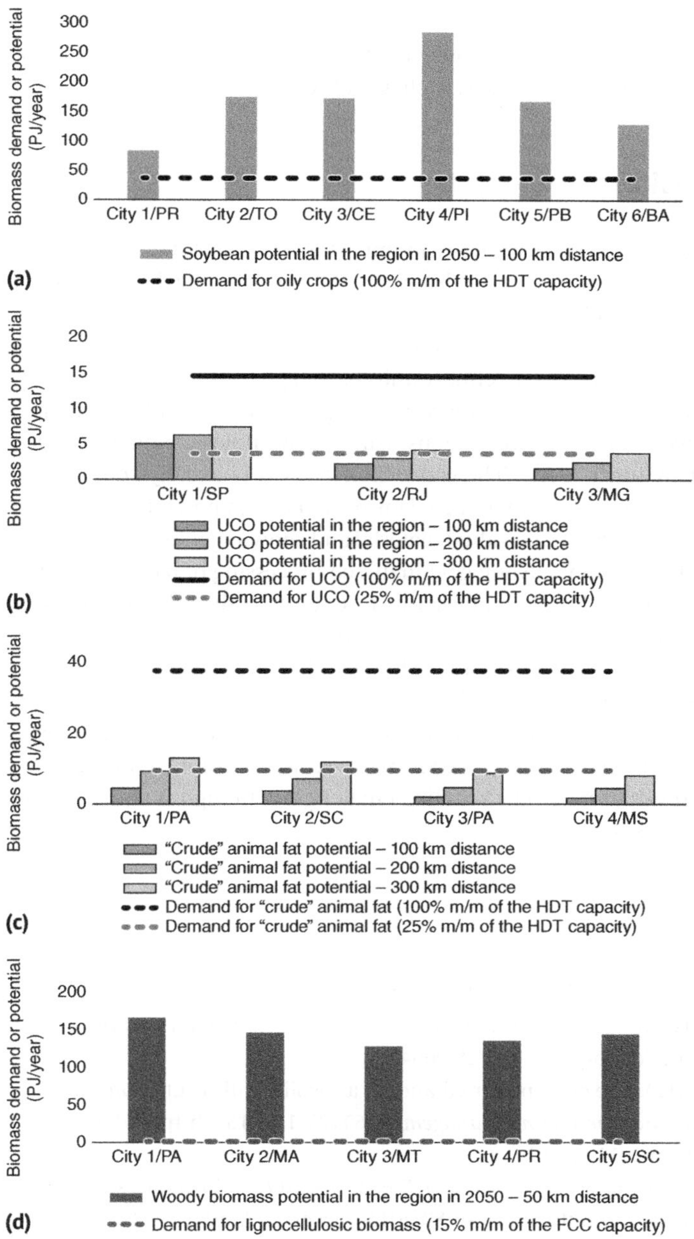

Figure 11.10 Cities with the highest availability of the evaluated raw materials and demand for co-processing from hypothetical mini-refinery.

may be lower than the percentages assumed in this study. Likewise, the availability of raw materials may also be lower, especially in the case of oily crops, which face competition for other uses, such as in the food sector.

11.7 Conclusion

Biomass co-processing in oil refineries is an important strategy for expanding production of drop-in biofuels and bio-feedstocks to the petrochemical sector. In general, it can be readily applied with little extra capital investment.

This chapter described several biomass co-processing options. The co-processing of SVOs, UCOs, and residual fats in refinery units has a high readiness level, although competition with other sectors and supply chain logistics might impose challenges to these alternatives. Co-processing options using lignocellulosic biomass, in general, show lower maturity levels. The conversion process is usually complex and requires a significant amount of hydrogen. However, the high potential of this kind of biomass, widely available as a residual material and showing no competition with other sectors, is an important advantage. Finally, co-processing of bioethanol (from sugary, starch, or lignocellulosic biomass) in FCC units, although still in testing phases, is an interesting alternative to use this kind of biomass in refineries.

This chapter also presented a case study showing that large refineries, originally designed for fossil oil, might face challenges in co-processing biomass due to the need for raw material collection over large distances. For modular mini-refineries, the availability of biomass aligns better with scale, enabling co-processing. However, using residual materials remains challenging, though it could be considered in specific cases.

References

Al-Sabawi, M., Chen, J., and Ng, S. (2012). Hydroprocessing of biomass-derived oils and their blends with petroleum feedstocks: a review. *Energy and Fuels* 26 (9): 5355–5372. https://doi.org/10.1021/ef3006417.

Anderson, S.T. (2012). The demand for ethanol as a gasoline substitute. *Journal of Environmental Economics and Management* 63 (2): 151–168. https://doi.org/10.1016/j.jeem.2011.08.002.

ANP (2015). *AUTORIZAÇÃO Nº 80, DE 23 DE FEVEREIRO DE 2015*. https://atosoficiais.com.br/anp/autorizacao-n-80-2015- (accessed January 2023).

ANP (2021). *AUTORIZAÇÃO Nº 810, DE 20 DE DEZEMBRO DE 2021*. https://atosoficiais.com.br/anp/autorizacao-n-810-2021- (accessed January 2023).

ANP (2022). *AUTORIZAÇÃO SPC-ANP Nº 196, DE 4 DE ABRIL DE 2022*. https://atosoficiais.com.br/anp/autorizacao-n-196-2022- (accessed January 2023).

ASTM (2022). ASTM D1655-22: standard specification for aviation turbine fuels. https://www.astm.org/d1655-22.html (accessed 3 January 2023).

Bergman-Fonte, C., Nascimento da Silva, G., Império, M. et al. (2023). Repurposing, co-processing and greenhouse gas mitigation – the Brazilian refining sector under deep decarbonization scenarios: a case study using integrated assessment modeling. *Energy* 282: 128435. https://doi.org/10.1016/j.energy.2023.128435.

Bezergianni, S., Dimitriadis, A., Kikhtyanin, O. et al. (2018). Refinery co-processing of renewable feeds. *Progress in Energy and Combustion Science* 68: 29–64. https://doi.org/10.1016/j.pecs.2018.04.002.

Bielansky, P., Reichhold, A., and Schonberger, C. (2010). Processing of pure vegetable oils in a continuous FCC pilot plant. In: *Refereed Proceedings The 13th International Conference on Fluidization - New Paradigm in Fluidization Engineering* (ed. S.D. Kim, Y. Kang, J.K. Lee, et al.). Engineering Conferences International.

bp (2022). Bp refinery in Lingen starts production of sustainable aviation fuel from used cooking oil. *Press Releases*. https://www.bp.com/en/global/air-bp/news-and-views/press-releases/bp-refinery-in-lingen-starts-production-of-saf.html (accessed 22 May 2022).

Brelsford, R. (2023). Petrobras processes 100% renewable feed in Riograndense FCC. *Oil & Gas Journal*. https://www.ogj.com/refining-processing/petrochemicals/article/14301266/petrobras-partners-process-100-renewable-feedstock-in-riograndense-refinerys-fcc (accessed 5 March 2025).

BTG Bioliquids (2023). *Commercial-Scale Production Plants in Operation*. https://www.btg-bioliquids.com/plants/ (accessed March 2025).

Cárdenas, J., Orjuela, A., Sánchez, D.L. et al. (2021). Pre-treatment of used cooking oils for the production of green chemicals: a review. *Journal of Cleaner Production* 289: 125129. https://doi.org/10.1016/j.jclepro.2020.125129.

Carmona, H.D.P., Torre Alfaro, O.D.L., Alayón, A.B. et al. (2019). Co-processing of straight run gas oil with used cooking oil and animal fats. *Fuel* 254: 115583. https://doi.org/10.1016/j.fuel.2019.05.166.

Carnus, J.-M., Parrotta, J., Brockerhoff, E. et al. (2006). Planted forests and biodiversity. *Journal of Forestry* 104 (2): 65–77. https://doi.org/10.1093/jof/104.2.65.

Carvalho, F., da Silva, F.T.F., Szklo, A. et al. (2019). Potential for biojet production from different biomass feedstocks and consolidated technological routes: a georeferencing and spatial analysis in Brazil. *Biofuels, Bioproducts and Biorefining* 13 (6): 1454–1475. https://doi.org/10.1002/bbb.2041.

Cortazar, M., Santamaria, L., Lopez, G. et al. (2023). A comprehensive review of primary strategies for tar removal in biomass gasification. *Energy Conversion and Management* 276: 116496. https://doi.org/10.1016/j.enconman.2022.116496.

Dakka, J.M., Sharma, A.K., Xu, X. et al. (2024). WO2025038172 - Ethanol conversion using FCC System. https://patents.google.com/patent/WO2025038172A1/en (accessed March 2025).

Dupain, X., Costa, D.J., Schaverien, C.J. et al. (2007). Cracking of a rapeseed vegetable oil under realistic FCC conditions. *Applied Catalysis B: Environmental* 72 (1–2): 44–61. https://doi.org/10.1016/j.apcatb.2006.10.005.

Energy Intelligence (2022). Refinery co-processing may help meet early SAF mandates, alternative fuels. https://www.energyintel.com/0000017d-9a80-d04d-a9fd-bf9d4bd80000 (accessed 22 May 1987).

Ensyn (2023). *Refinery Feedstocks: Low Carbon Transportation Fuels.* https://www.ensyn.com/refinery-feedstocks.html (accessed March 2025).

ETIP Bioenergy (2023). EVC2: hydrotreatment to HVO, value chains. https://www.etipbioenergy.eu/value-chains/conversion-technologies/conventional-technologies/hydrotreatment-to-hvo (accessed 17 December 2023).

Fitzgibbon, T., Nariman, K., and Roth, B. (2023). Converting refineries to renewable fuels: no simple switch, Our Insights, McKinsey. https://www.mckinsey.com/industries/oil-and-gas/our-insights/converting-refineries-to-renewable-fuels-no-simple-switch (accessed 12 February 2024).

Gary, J.H. and Handwerk, G.E. (2001). *Petroleum Refining: Technology and Economics, Chemical Engineering Science.* Golden, Colorado, USA: Marcel Dekker, Inc. https://doi.org/10.1016/0009-2509(94)87025-x.

Guedes, F., Szklo, A., Rochedo, P. et al. (2019). Climate-energy-water nexus in Brazilian oil refineries. *International Journal of Greenhouse Gas Control* 90: 102815. https://doi.org/10.1016/j.ijggc.2019.102815.

Hammond, E.W. (2003). VEGETABLE OILS | Composition and Analysis. In: *Encyclopedia of Food Sciences and Nutrition*, 2e (ed. B. Caballero), 5916–5921. Oxford: Academic Press. https://doi.org/10.1016/B0-12-227055-X/01227-X.

Han, X., Wang, H., Zeng, Y. et al. (2020). Advancing the application of bio-oils by co-processing with petroleum intermediates: a review. *Energy Conversion and Management: X* 10: 100069. https://doi.org/10.1016/j.ecmx.2020.100069.

Hoffmann, B.S. (2010). *O ciclo combinado com gaseificação integrada e a captura de CO_2: Uma solução para mitigar as emissões de CO_2 em termelétricas a carvão em larga escala no curto prazo*? Mestrado: Universidade Federal do Rio de Janeiro.

Hoffmann, B.S. and Szklo, A. (2011). Integrated gasification combined cycle and carbon capture: a risky option to mitigate CO_2 emissions of coal-fired power plants. *Applied Energy* 88 (11): 3917–3929. https://doi.org/10.1016/j.apenergy.2011.04.002.

Holden, E. and Gilpin, G. (2013). Biofuels and sustainable transport: a conceptual discussion. *Sustainability (Switzerland)* 5 (7): 3129–3149. https://doi.org/10.3390/su5073129.

Huo, H., Wang, M., Bloyd, C. et al. (2009). Life-cycle assessment of energy and greenhouse gas effects of soybean-derived biodiesel and renewable fuels. In: *Environmental Science & Technology*, 43 (3): 750–756. https://doi.org/10.1021/es8011436.

IBGE (2018). Pesquisa de orçamentos familiares (POF). https://www.ibge.gov.br/estatisticas/sociais/saude/24786-pesquisa-de-orcamentos-familiares-2.html (accessed 31 May 2023).

IBGE (2021). Pesquisa Pecuária Municipal. https://www.ibge.gov.br/estatisticas/economicas/agricultura-e-pecuaria/2041-np-producao-da-pecuaria-municipal.html (accessed 31 March 2023).

IEA (2023). Renewables 2023. www.iea.org (accessed April 2024).

IPCC (2022). *Climate Change 2022: Mitigation of Climate Change. Contribution of Working Group III to the Sixth Assessment Report of the Intergovernmental Panel on Climate Change* (ed. P.R. Shukla, J. Skea, R. Slade, et al.). Cambridge, UK and New York, NY, USA: Cambridge University Press. https://doi.org/10.1017/9781009157926.

IRENA (2017). *Stranded Assets and Renewables: How the Energy Transition Affects the Value of Energy Reserves, Buildings and Capital Stock.* Abu Dhabi. International Renewable Energy Agency.

Kaltschmitt, M. and Streicher, W. (2009). Energie aus biomasse. In: *Regenerative Energien in Österreich: Grundlagen, Systemtechnik, Umweltaspekte, Kostenanalysen, Potenziale, Nutzung* (ed. M. Kaltschmitt and W. Streicher), 339–532. Wiesbaden: Vieweg+Teubner. https://doi.org/10.1007/978-3-8348-9327-7_9.

Karmakar, A., Karmakar, S., and Mukherjee, S. (2010). Properties of various plants and animals feedstocks for biodiesel production. *Bioresource Technology* 101 (19): 7201–7210. https://doi.org/10.1016/j.biortech.2010.04.079.

Kimura, T., Miyazawa, T., Nishikawa, J. et al. (2006). Development of Ni catalysts for tar removal by steam gasification of biomass. *Applied Catalysis B: Environmental* 68 (3): 160–170. https://doi.org/10.1016/j.apcatb.2006.08.007.

Kristiana, T., Baldino, C., and Searle, S. (2022). An estimate of current collection and potential collection of used cooking oil from major Asian exporting countries. www.theicct.org (accessed April 2024).

Lammens, T.M. (2022). Effect of various green carbon tracking methods on life cycle assessment results for fluid catalytic cracker co-processing of fast pyrolysis bio-oil. *Energy & Fuels* 36 (20): 12617–12627. https://doi.org/10.1021/acs.energyfuels.2c01676.

Leijenhorst, E.J., Assink, D., van de Beld, L. et al. (2015). Entrained flow gasification of straw- and wood-derived pyrolysis oil in a pressurized oxygen blown gasifier. *Biomass and Bioenergy* 79: 166–176. https://doi.org/10.1016/j.biombioe.2014.11.020.

Lindfors, C., Elliott, D.C., Prins, W. et al. (2023). Co-processing of biocrudes in oil refineries. *Energy & Fuels* 37 (2): 799–804. https://doi.org/10.1021/acs.energyfuels.2c04238.

Longanesi, L., Pereira, A.P., Johnston, N. et al. (2022). Oxidative stability of biodiesel: recent insights. *Biofuels, Bioproducts and Biorefining* 16 (1): 265–289. https://doi.org/10.1002/bbb.2306.

Lorne, D. (2023). Biofuels in the road transport sector, IFPEN economic studies. https://www.ifpenergiesnouvelles.com/article/biofuels-dashboard-2022 (accessed 18 February 2024).

Matos, M., Mattos, B.D., de Cademartori, P.H.G. et al. (2020). Pilot-scaled fast-pyrolysis conversion of eucalyptus wood fines into products: discussion toward possible applications in biofuels, materials, and precursors. *Bioenergy Research* 13 (2): 411–422. https://doi.org/10.1007/s12155-020-10094-y.

Meerman, J.C., Ramírez, A., Turkenburg, W.C. et al. (2011). Performance of simulated flexible integrated gasification polygeneration facilities. Part A: a technical-energetic assessment. *Renewable and Sustainable Energy Reviews* 15 (6): 2563–2587. https://doi.org/10.1016/j.rser.2011.03.018.

Messier, C., Bauhus, J., Sousa-Silva, R. et al. (2022). For the sake of resilience and multifunctionality, let's diversify planted forests! *Conservation Letters* 15 (1): e12829.

MME (2023). Novo PAC mira em combustíveis de baixo carbono para a promoção da transição energética, Notícias do Novo PAC. https://www.gov.br/casacivil/pt-br/assuntos/noticias-do-novo-pac/novo-pac-mira-em-combustiveis-de-baixo-carbono-para-a-promocao-da-transicao-energetica (accessed 11 February 2024).

Nascimento da Silva, G., Lantz, F., Rochedo, P.R.R. et al. (2024). Developing and applying the hydrogen economics and infrastructure optimization model (HERA). *International Journal of Hydrogen Energy* 61: 1170–1186. https://doi.org/10.1016/j.ijhydene.2024.02.015.

No, S.-Y. (2014). Application of hydrotreated vegetable oil from triglyceride based biomass to CI engines – a review. *Fuel* 115: 88–96. https://doi.org/10.1016/j.fuel.2013.07.001.

Oil and Gas Journal (2023). Worldwide refining survey. https://digital.ogj.com/ (accessed January 2024).

Oliveira, C. and Schure, K.M. (2020). *Decarbonisation options for the Dutch Refinery Sector*. PBL Netherlands Environmental Assessment Agency.

Parkash, S. (2003). *Refining Processes Handbook, Refining Processes Handbook*. Burlington, MA, USA: Gulf Professional Publishing. https://doi.org/10.1016/B978-0-7506-7721-9.X5000-2.

Pechstein, J. and Kaltschmitt, M. (2018). Liquid hydrocarbon fuels derived from alcohols. In: *Energy from Organic Materials (Biomass)*, 1023–1046. https://doi.org/10.1007/978-1-4939-7813-7_1039.

Perez, M.J.L., Nygaard, G.T. and Verdier, S. (2024). *SAF production via co-processing in the kerosene hydrotreater*. Digital Refining. https://www.digitalrefining.com/article/1003094/saf-production-via-co-processing-in-the-kerosene-hydrotreater (accessed June 2024).

Petrobras (2020). Concluímos testes para produção de diesel renovável, Fatos e dados. https://petrobras.com.br/fatos-e-dados/concluimos-testes-para-producao-de-diesel-renovavel.htm (accessed 11 May 2022).

Petrobras (2023). Com tecnologia da Petrobras, Refinaria Riograndense processa carga com matéria-prima 100% renovável, Energia. https://agenciagov.ebc.com.br/noticias/202311/com-tecnologia-da-petrobras-refinaria-riograndense-processa-carga-com-materia-prima-100-renovavel (accessed 12 December 2023).

Pinho, A.R., Cabral, J.A.R., and Leite, L.F. (2007). EP 1 942 089 A1 - Process for converting ethanol and hydrocarbons in a fluidized catalytic cracking unit. Brazil. https://patentimages.storage.googleapis.com/72/17/f4/f8af3519ebb8b4/EP1942089A1.pdf (accessed 4 March 2025).

Pinho, A.R., de Almeida, M.B.B. and Rochedo, P.R.R. (2022). Renewable-carbon recovery in the co-processing of vacuum gas oil and bio-oil in the FCC process – where does the renewable carbon go? *Fuel Processing Technology* 229: 107176. https://doi.org/10.1016/j.fuproc.2022.107176.

Pinho, A.R., de Almeida, M.B.B., Mendes, F.L. et al. (2017). Fast pyrolysis oil from pinewood chips co-processing with vacuum gas oil in an FCC unit for second generation fuel production. *Fuel* 188: 462–473. https://doi.org/10.1016/j.fuel.2016.10.032.

Pontes, N.S., Silva, R.V.S., Ximenes, V.L. et al. (2021). Chemical speciation of petroleum and bio-oil coprocessing products: investigating the introduction of renewable molecules in refining processes. *Fuel* 288: 119654. https://doi.org/10.1016/j.fuel.2020.119654.

Portugal-Pereira, J., Soria, R., Rathmann, R. et al. (2015). Agricultural and agro-industrial residues-to-energy: techno-economic and environmental assessment in Brazil. *Biomass and Bioenergy* 81: 521–533. https://doi.org/10.1016/j.biombioe.2015.08.010.

Prasad, S. and Ingle, A.P. (2019). Impacts of sustainable biofuels production from biomass. In: *Sustainable Bioenergy: Advances and Impacts* (ed. M. Rai and Ingle, A.P.), 327–346. Elsevier Inc. https://doi.org/10.1016/B978-0-12-817654-2.00012-5.

Richardson, Y., Blin, J., and Julbe, A. (2012). A short overview on purification and conditioning of syngas produced by biomass gasification: catalytic strategies, process intensification and new concepts. *Progress in Energy and Combustion Science* 38 (6): 765–781. https://doi.org/10.1016/j.pecs.2011.12.001.

Semieniuk, G., Holden, P.B., Mercure, J.-F. et al. (2022). Stranded fossil-fuel assets translate to major losses for investors in advanced economies. *Nature Climate Change* 12 (6): 532–538. https://doi.org/10.1038/s41558-022-01356-y.

Sharma, K., Haider, M.S., Castello, D. et al. (2022). Understanding the miscibility and co-feeding potential of hydrothermal liquefaction biocrude in refinery streams: role of hydrodeoxygenation. *Energy & Fuels* 36 (21): 13105–13114. https://doi.org/10.1021/acs.energyfuels.2c02692.

Siskin, M., Phillips, G.E., Kelemen, S.R. et al. (2011). *Biomass Conversion Process.* United States of America, United States. US20110233042A1. https://patents.google.com/patent/US20110233042 (accessed January 2024).

Soares Dias, A.P., Ramos, M., and Rijo, B. (2022). Rendering of beef tallow for biodiesel production: microwave versus boiling water and acetone fat extraction. *Processes* 10 (4): 666. https://doi.org/10.3390/pr10040666.

Sousa-Aguiar, E.F., Ximenes, V.L., Araujo Ribeiro de Almeida, J.M. et al. (2018). Catalysts for co-processing biomass in oil refining industry. In: *Sustainable Catalysis for Biorefineries* (ed. F. Frusteri, D. Aranda, and G. Bonura). The Royal Society of Chemistry. https://doi.org/10.1039/9781788013567-00001.

Srifa, A., Faungnawakij, K., Itthibenchapong, V. et al. (2014). Production of bio-hydrogenated diesel by catalytic hydrotreating of palm oil over NiMoS2/γ-Al_2O_3 catalyst. *Bioresource Technology* 158: 81–90. https://doi.org/10.1016/j.biortech.2014.01.100.

Szklo, A.S., Uller, V.C., and Bonfá, M.H.P. (2012). *Fundamentos do Refino de Petróleo: Tecnologia e Economia*. 3a edição. (ed. Editora Interciência). Rio de Janeiro.

Tagomori, I.S. (2022). *Bioenergy and Carbon Capture and Storage Spatially Explicit Modelling in Brazil*. Universidade Federal do Rio de Janeiro.

Tagomori, I.S., Rochedo, P., and Szklo, A. (2019). Techno-economic and georeferenced analysis of forestry residues-based Fischer-Tropsch diesel with carbon capture in Brazil. *Biomass and Bioenergy* 123: 134–148. https://doi.org/10.1016/j.biombioe.2019.02.018.

Talmadge, M., Kinchin, C., Li Chum, H. et al. (2021). Techno-economic analysis for co-processing fast pyrolysis liquid with vacuum gasoil in FCC units for second-generation biofuel production. *Fuel* 293: 119960. https://doi.org/10.1016/j.fuel.2020.119960.

Talmadge, M.S., Baldwin, R.M., Biddy, M.J. et al. (2014). A perspective on oxygenated species in the refinery integration of pyrolysis oil. *Green Chemistry* 16 (2): 407–453. https://doi.org/10.1039/C3GC41951G.

Teixeira, R.C., Szklo, A.S., and Branco, D.C. (2021). Adding flexibility to petroleum refining through the introduction of modular plants – a case study for Brazil. *Energy Sources, Part B: Economics, Planning, and Policy* 16 (7): 617–637. https://doi.org/10.1080/15567249.2021.1952495.

Tiwari, R., Rana, B.S., Kumar, R. et al. (2011). Hydrotreating and hydrocracking catalysts for processing of waste soya-oil and refinery-oil mixtures. *Catalysis Communications* 12 (6): 559–562. https://doi.org/10.1016/j.catcom.2010.12.008.

Toldra-Reig, F., Mora, L., and Toldra, F. (2020). Trends in biodiesel production from animal fat waste. *Applied Science* 10 (3644): 1–17.

TOTAL (2020). La Mède: A facility for the energies of tomorrow. https://www.total.com/energy-expertise/projects/bioenergies/la-mede-a-forward-looking-facility (accessed 24 January 2021).

UNFCCC (2020). The Paris agreement, process and meetings. https://unfccc.int/process-and-meetings/the-paris-agreement/the-paris-agreement (accessed 27 January 2021).

USDoE (2023). Biodiesel blends, alternative fuels data center. https://afdc.energy.gov/fuels/biodiesel_blends.html (accessed 12 February 2024).

van Dyk, S. and Saddler, J. (2021). *Progress in the Commercialization of Biojet / Sustainable Aviation Fuels: Technologies, Potential and Challenges*. IEA Bioenergy. Task 39.

van Dyk, S., Su, J., Mcmillan, J.D. et al. (2019). Potential synergies of drop-in biofuel production with further co-processing at oil refineries. *Biofuels, Bioproducts and Biorefining* 13 (3): 760–775. https://doi.org/10.1002/bbb.1974.

Vecino Mantilla, S., Gauthier-Maradei, P., Álvarez Gil, P. et al. (2014). Comparative study of bio-oil production from sugarcane bagasse and palm empty fruit bunch: yield optimization and bio-oil characterization. *Journal of Analytical and Applied Pyrolysis* 108: 284–294. https://doi.org/10.1016/j.jaap.2014.04.003.

Vertz, L. and Sayal, S. (2020). Reduce, repurpose, reinvent: Long-term refinery outlook defined by diverging regional imperatives, IHS Markit - Energy & Natural Resources Research & Analysis. https://ihsmarkit.com/research-analysis/reduce-repurpose-reinvent-longterm-refinery-outlook.html (accessed 24 January 2021).

Vignesh, P., Pradeep Kumar, A.R., Ganesh, N.S. et al. (2021). Biodiesel and green diesel generation: an overview. *Oil & Gas Science and Technology – Revue d'IFP Energies nouvelles* 76: 6. https://doi.org/10.2516/ogst/2020088.

Wang, B.G., Gao, B.J., and Bi, B.Y. (2014). Process for catalytic cracking a pyrolysis oil. United States of America. US9469815B2. http://patents.google.com/patent/US9469815B2/en (accessed June 2024).

WRI and WBSCD (2013). Technical guidance for calculating scope 3 emissions. https://ghgprotocol.org/sites/default/files/2023-03/Scope3_Calculation_Guidance_0%5B1%5D.pdf (accessed 22 December 2023).

Ximenes, V.L., Bugueta, P.C.P., De Almeida, M.B.B. et al. (2022). Process of coprocessing a lignocellulosic liquid stream and an intermediate fossil stream in an oil refining process and a process for producing fuel from a deasphalted oil stream. United States Patent.

Yáñez, É., Meerman, H., Ramírez, A. et al. (2021). Assessing bio-oil co-processing routes as CO_2 mitigation strategies in oil refineries. *Biofuels, Bioproducts and Biorefining* 15 (1): 305–333. https://doi.org/10.1002/bbb.2163.

Zhang, M. and Yu, Y. (2013). Dehydration of ethanol to ethylene. *Industrial & Engineering Chemistry Research* 52 (28): 9505–9514. https://doi.org/10.1021/ie401157c.

Zhang, Z. and Pang, S. (2019). Experimental investigation of tar formation and producer gas composition in biomass steam gasification in a 100 kW dual fluidised bed gasifier. *Renewable Energy* 132: 416–424. https://doi.org/10.1016/j.renene.2018.07.144.

Zhu, C., Gutiérrez, O.Y., Santosa, D.M. et al. (2022). Impact of coprocessing biocrude with petroleum stream on hydrotreating catalyst stability. *Energy & Fuels* 36 (16): 9133–9146. https://doi.org/10.1021/acs.energyfuels.2c01748.

Zuniga, A.D.G., Paula, M.M., Coimbra, J.S. dos R., et al. (2011). Revisão: propriedades físico-químicas do biodiesel. *Pesticidas: Revista de Ecotoxicologia e Meio Ambiente* 21: 55–72. https://revistas.ufpr.br/pesticidas/article/viewFile/25939/17311 (accessed 24 March 2024).

12

Liquid Fuels from Syngas and CO_2

Chike George Okoye-Chine[1,2] *and B. Frank Gupton*[2]

[1] *Global R&D, Avery Dennison, Mentor, OH, USA*
[2] *Department of Chemical & Life Science Engineering, College of Engineering, Virginia Commonwealth University, Richmond, VA, USA*

12.1 Introduction

Liquid fuels obtained from syngas and CO_2 are sustainable liquid fuels that are environmentally friendly compared to fossil-based liquid fuels. The production of liquid fuels from syngas (i.e., a mixture of carbon monoxide and hydrogen) is the catalytic transformation of syngas obtained from natural gas, coal, and biomass into longer-chain hydrocarbons via Fischer–Tropsch (FT) synthesis, which are upgraded to liquid fuels such as gasoline, jet fuel, and diesel (Dry, 1996; Schulz and Claeys, 1999; Anders Blekkan et al., 2007; Khodakov, 2009; Bhatelia et al., 2011). The liquid fuels obtained from FT synthesis are normally termed clean fuels because of the very low sulfur content compared to fossil-based liquid fuels (Khodakov, 2009). The production of liquid fuels from syngas via FT synthesis is typically carried out at a temperature range of 200–300 °C and a pressure range of 1.01–6.06 MPa (Rao et al., 1992). Most times, the H_2/CO ratio of the syngas in the feed stream influences the choice of using either an iron-based or cobalt-based catalyst for the FT synthesis. Cobalt-based catalysts are most suitable for syngas with a high H_2/CO ratio (i.e., $H_2/CO \geq 2$), which is mostly obtained from natural gas (Davis, 2007). Conversely, iron-based catalysts are more suitable for syngas with a low H_2/CO ratio (i.e., $H_2/CO \leq 2$), which is normally obtained from coal, because they adjust the H_2/CO ratio due to their high intrinsic selectivity for the water–gas shift (WGS) reaction (Jacobs et al., 2004).

Cleaner Petroleum Production and Refining Technologies, First Edition, Edited by M. R. Riazi and Harvey W. Yarranton.

One of the primary aims of the Paris Agreement under the United Nations Framework Convention on Climate Change (UNFCCC) is to substantially reduce the emission of greenhouse gases such as CO_2 into the atmosphere to drastically minimize the risk and impact of climate change in the world (Paris Agreement, 2015). This action is essential because our environment has suffered from natural disasters such as flooding, tsunamis, extreme weather conditions, etc. These natural disasters have been attributed to climate change caused by the high emission of CO_2 and other greenhouse gases into the atmosphere. There are technologies used for capturing CO_2 from flue gas produced by process plants and the atmosphere. Most times, the captured CO_2 is stored in an underground storage facility to prevent the emission of a large amount of CO_2 into the atmosphere and mitigate climate change. Also, considering the high demand for transportation fuels globally, converting CO_2 into valuable liquid fuels will help meet the increasing global demand for fuels and minimize the excessive dependence on fossil fuels (Gao et al., 2020). Meanwhile, the transformation of CO_2 into liquid fuels is gaining lots of attention because it is one of the promising ways of meeting the high demand for liquid fuels and an effective way of minimizing the emission of CO_2 into the atmosphere, hence, mitigating climate change and global warming (Olah et al., 2011; Genovese et al., 2013; Banerjee et al., 2016; Bahri et al., 2019). One of the key challenges of converting CO_2 into liquid fuels is that CO_2 is thermodynamically stable and chemically inert (Gao et al., 2020; Reddy et al., 2022). Thus, a high-energy substance (such as H_2) and an effective catalytic system are required to convert CO_2 into liquid fuels (Sakakura et al., 2007; Benson et al., 2009; Zhou et al., 2019).

This chapter will discuss the various methods of syngas production and the further transformation of the produced syngas into long-chain hydrocarbons via FT synthesis and upgrading to liquid fuels such as gasoline, jet fuel, and diesel. Also, we will examine the various chemical and biological conversion techniques for the transformation of CO_2 into liquid fuels.

12.2 Syngas Production

Syngas is produced from a carbonaceous raw material (Steynberg and Dry, 2004; Steynberg, 2004). The word "carbonaceous" means that the raw material contains carbon. Also, hydrogen should be present in the raw material to boost the production of syngas, especially in situations where oxygen is used alone as the gasifying agent for the gasification of the carbonaceous feedstock. Syngas production is achieved by the following methods: steam reforming of CH_4, autothermal reforming of CH_4, partial oxidation of CH_4, dry reforming of CH_4, gasification of asphaltenes, coal gasification, and biomass gasification. This section will focus on the various methods through which syngas is produced.

12.2.1 Syngas Production from Methane

12.2.1.1 Steam Reforming of Methane

CH_4 is the main component of natural gas, and it is a greenhouse gas, which when emitted into the atmosphere causes global warming. In a bid to minimize the emission of CH_4, it is important to utilize CH_4 by converting it into syngas. One of the ways of transforming CH_4 into syngas is by steam methane reforming. Steam methane reforming is the most widely used technique for the production of syngas from natural gas (Salhi et al., 2011; Gangadharan et al., 2012). Steam reforming of methane is an endothermic reaction between steam and methane at a high temperature (usually 750–950 °C) and a high pressure (usually 1.42–2.03 MPa) (Zhang et al., 2021), as shown in Equation 12.1, in the presence of a metal-based catalyst for syngas production. The typical composition of products formed from steam reforming of methane consists of around 70–75% H_2, 7–10% CO, 6–14% CO_2, and a small amount of residual methane 2–6% (Singh et al., 2015; Ji and Wang, 2021). Nickel-based catalysts are commonly used for the industrial application of the steam reforming of methane for the production of syngas due to their low cost and high activity as a result of their strongest breaking ability of the C—H bond (Matsumura and Nakamori, 2004; Han et al., 2020). Due to steam reforming of methane is a highly endothermic reaction, it requires a highly efficient external energy supply from the reformer furnace. For steam reforming of methane, the molar steam/carbon (S/C) ratio typically exceeds 2.5, and the excess steam is necessary to aid the completion of the reaction and minimize coke formation; however, additional heat must be introduced (Maier et al., 2011). Furthermore, the high temperature requirement of steam reforming of methane always results in the severe sintering of the nickel catalyst, which swiftly deactivates Ni catalysts and makes them shortlived. In a bid to address this challenge, there is ongoing research to develop a sintering-resistant catalyst for syngas production from the steam reforming of methane. One of the potential techniques is the fabrication of a core-shell structure to confine nickel nanoparticles to preserve their original sizes (Han et al., 2020).

Meanwhile, during the steam reforming of methane, a side reaction called the WGS reaction, as shown in Equation 12.2, could take place to further aid the formation of H_2, leading to an adjustment in the composition of the syngas (Rahbari et al., 2018).

$$CH_4 + H_2O \longrightarrow CO + H_2 \quad \Delta H_R = +206\,\text{kJ/mol} \tag{12.1}$$

$$CO + H_2O \longrightarrow CO_2 + H_2 \quad \Delta H_R = -41\,\text{kJ/mol} \tag{12.2}$$

Figure 12.1 shows the flow diagram for the production of syngas through the method of steam reforming of methane. Methane and steam are mixed in a mixer before being flowed into the multitubular fixed bed reactor containing

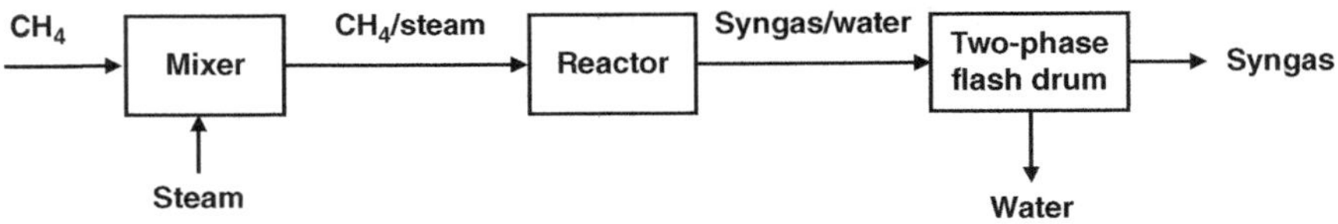

Figure 12.1 Schematic flow diagram of synthesis gas production via steam reforming of methane.

catalysts (Gangadharan et al., 2012). The syngas and water formed are separated in a two-phase flash drum. Steam methane reforming produces syngas with an H_2/CO ratio of 4 or above, which is the reason why the method of steam reforming of methane is a preferred choice for hydrogen production. On the other hand, techniques such as partial oxidation of methane produce syngas with an H_2/CO ratio of 1, and autothermal reforming of methane produces syngas with an H_2/CO ratio of 2. The production of diesel from FT synthesis would require syngas with an H_2/CO ratio from less than 1 to about 2, depending on the catalyst used (Raju et al., 2009). It has to be noted that the production of H_2-rich syngas is highly favored during steam reforming of methane due to the inadequate distribution of external heat in the tubular reactors packed with Ni-based catalysts, which causes heat transfer limitation. This leads to the conversion of some fraction of the produced CO to CO_2 and H_2 via an exothermic WGS reaction. This is not favorable for the production of syngas, where the desired H_2/CO ratios are to be less than 3. However, the use of wall-coated microchannel reactors for steam reforming of methane could provide a possible solution by providing quick distribution of external heat to the catalyst at a suitable temperature, catalyst, and steam-to-methane feed ratios, allowing steam reforming of methane technique to produce syngas with an appropriate H_2/CO ratio ($H_2/CO \leq 3$) for the production of synthetic fuel, which is usually achieved via autothermal reforming (Simsek et al., 2011).

12.2.1.2 Autothermal Reforming of Methane

Autothermal reforming is the combination of partial oxidation with steam reforming (Nezhad et al., 2009; Faheem et al., 2021; Osman et al., 2021; Aryafard et al., 2023; Zahid et al., 2023). In other words, steam, O_2 (or air), and methane react over a nickel catalyst in an autothermal reactor to produce syngas (Oni et al., 2022). Figure 12.2 shows the schematic diagram of autothermal reforming of methane for syngas production. Apart from nickel-based catalysts, platinum-, palladium-, or ruthenium-based catalysts are also used for autothermal reforming of methane. However, nickel-based catalysts are widely used due to their low cost compared to the other active metals (Murmura et al., 2016). Autothermal reforming can achieve high thermal efficiency and adjustable H_2/CO ratio and reduce its operational costs because it does not need an external energy supply (Carapellucci and Giordano,

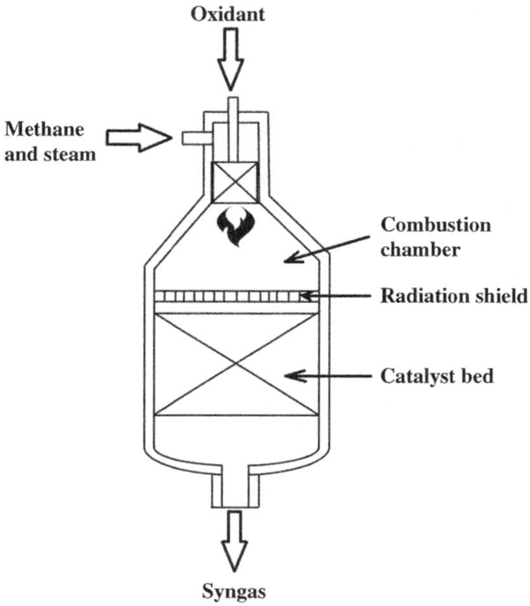

Figure 12.2 Schematic diagram of an autothermal reformer for syngas production.

2020; Fan et al., 2021). The energy efficiency of autothermal reforming is 60–75%, while that of steam reforming is 70–85% (He and De Wilde, 2021). Autothermal reforming of methane is the most suitable and cost-effective method for the production of syngas for Gas-to-Liquid (GTL) plants. The Oryx GTL plant in Qatar is based on autothermal reforming technology at a low steam/carbon ratio (S/C = 0.6), which produces syngas with an H_2/CO ratio equal to 2, which is suitable for the FT synthesis (Ruiz et al., 2008). H_2/CO ratio of 2 can be accomplished via the recirculation of carbon dioxide or a carbon dioxide-rich off-gas and by the reduction of the quantity of steam in the feed. There are two types of autothermal reforming systems, namely: (1) It has a catalyst bed in which combustion and steam reforming reactions occur simultaneously, and it is compact and useful for fuel cell applications; (2) It has two different sections – the first section is the combustion section where non-catalytic partial oxidation takes place using a burner, while the second section is the catalyst bed section where the steam reforming reaction occurs in the catalyst bed (Nezhad et al., 2009). The latter is preferably suitable for GTL plants.

12.2.1.3 Dry Reforming of Methane

Considering the rising concern of global warming, it is vital to reduce the emission of greenhouse gases such as CO_2 and CH_4 in the atmosphere (Zhang et al., 2021).

CH_4 is very much more effective at trapping heat in the atmosphere than CO_2, and thus, even a small amount of CH_4 emission into the atmosphere can have a huge influence on the greenhouse gas footprints (Howarth, 2014). For the benefit of the environment, the use of dry reforming of methane helps to reduce the emission of CH_4 and CO_2 into the atmosphere, which helps to minimize global warming. Dry reforming of methane is considered to be the technology with the highest potential to effectively utilize CH_4 and CO_2 (Liang et al., 2023). Figure 12.3 displays the schematic diagram of the dry reforming of methane for the production of syngas. Dry reforming of methane transforms CH_4 and CO_2 into syngas with an H_2/CO ratio of approximately 1 (Ranjekar and Yadav, 2021). Due to the high endothermicity of the dry reforming of methane as shown in Equation 12.3, it favors a reaction temperature of 800 °C or above for higher feed conversions (Naidu et al., 2022). Coke formation is very evident in the dry reforming of methane between the temperatures of 550–700 °C due to the Boudouard reaction (Palmer et al., 2020), as shown in Equation 12.4.

$$CH_4 + CO_2 \longrightarrow 2CO + 2H_2 \quad \Delta H_R = +247 \text{ kJ/mol} \tag{12.3}$$

$$2CO \longrightarrow C + CO_2 \quad \Delta H_R = -171 \text{ kJ/mol} \tag{12.4}$$

The major challenges in the commercialization of dry reforming of methane for the production of syngas are the high energy needed because it is a highly endothermic reaction and catalyst deactivation due to the coke deposition on the catalyst's surface (Han et al., 2017; Pham et al., 2020; He et al., 2022; Jahangiri and Saidi, 2022). The rapid and immense deactivation that occurs with catalysts used in dry reforming of methane, especially Ni-based catalysts, is still a huge barrier toward its industrialization and commercialization for the production of

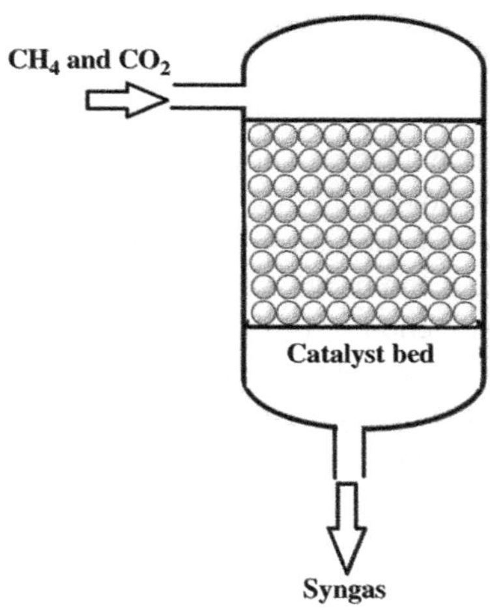

Figure 12.3 Schematic diagram of dry reformer for syngas production.

syngas (Abdulrasheed et al., 2019). Bian et al. reported the use of mesoporous alumina with high porosity and surface area as a support for Ni catalyst could help to increase activity and reduce coke deposition during dry reforming of methane (Bian et al., 2021). Bitters et al. (2022), in their review, stated the great prospects of using bimetallic combinations of earth-abundant transition metals, such as Ni—Co, Ni—Fe, and Co—Mo, toward minimizing coke deposition on the catalysts during dry reforming of methane. Yentekakis et al. (2021) buttressed this point by stating that bimetallic catalysts provide synergistic effects via metal-to-metal interactions that make dry reforming of methane catalysts sintering-and-coking resistant. Movasati et al. (2019) doped Ni supported on CeO_2-$ZnAl_2O_4$ catalyst with Co, which enhanced the catalyst's surface area, improved reduction in the active particle size, and strengthened the active phase/support interaction, which led to coke deposit resistance. Chein and Fung (2019) reported that the modification of Ni/Al_2O_3 by the addition of CeO_2 helped with the suppression of carbon deposition and enhancement of the catalyst activity. This is attributed to the formation of $CeAlO_3$ in the catalyst. Ceria-based catalysts are notable for having good resistance to coke, which is vital for the dry reforming of methane (Teh et al., 2021). Luisetto et al. (2017) reported that the Ru-promoted Ni catalyst was remarkably active and stable, while the unpromoted Ni catalyst was deactivated as a result of the formation of Ni^{2+} comprising inactive phases. Mao et al. (2023) reported that the incorporation of a small amount of noble metal rhodium into a Ni-based catalyst enhanced the low-temperature stability and anti-coking performance of the catalyst. This is ascribed to the inhibition of the direct decomposition of CH_x*, enhancement of the adsorption and activation of CO_2, metal dispersion, and metal oxide reduction. Song et al. (2020) reported that a Mo-doped Ni nanocatalyst that is stabilized at the edges of a single-crystalline MgO support was tested for the production of synthesis gas from dry reforming of methane, and no coking was detected after a catalyst run of more than 850 hours of continuous operation under 60 L per unit mass of catalyst per hour of reactive gas flow.

Nickel-based catalysts are promising for dry reforming of methane due to their reactivity, availability, and cost-effectiveness. However, it has been established that sintering and coking can limit their industrial use for dry reforming of methane. Thus, researchers are exploring various catalyst supports to enhance the dispersion of nickel particles and change the reaction process as well as utilize various noble metals as promoters for nickel-based catalysts to regulate the chemical properties of nickel particles to prevent coking and sintering (Liang et al., 2024). There are new varieties of catalysts utilizing noble metals, structured silica foams, and zeolites that are being investigated for dry reforming of methane for the production of syngas (Ranjekar and Yadav, 2021).

12.2.1.4 Partial Oxidation of CH_4

Generally, partial oxidation of methane can be conducted at a high temperature (1,200–1,500 °C) without a catalyst (see Figure 12.4); however, the reaction can be carried out at a significantly lower temperature (800–900 °C) by the use of a suitable catalyst (see Figure 12.5) (Elbadawi et al., 2021). Partial oxidation of CH_4 for the production of syngas is a mildly exothermic reaction (Loktev et al., 2022), as shown in Equation 12.5. Thus, partial oxidation of CH_4 does not require an external heat source to sustain the process, which makes this process more economical to heat. Also, partial oxidation can be combined with endothermic reactions, such as steam reforming or dry reforming, to make these processes more energy efficient (York et al., 2003). Some of the advantages that the partial oxidation of CH_4 has over steam reforming of methane and dry reforming of methane are that high conversion and selectivity at lower process temperature are achieved, and partial oxidation of methane can be conducted under high gas hourly space velocity (GHSV), which is an indication of less investment and less production scale to achieve the same capacity (Fazlikeshteli et al., 2023a).

$$CH_4 + \tfrac{1}{2}O_2 \longrightarrow CO + 2H_2 \quad \Delta H_R = -36\ \text{kJ/mol} \tag{12.5}$$

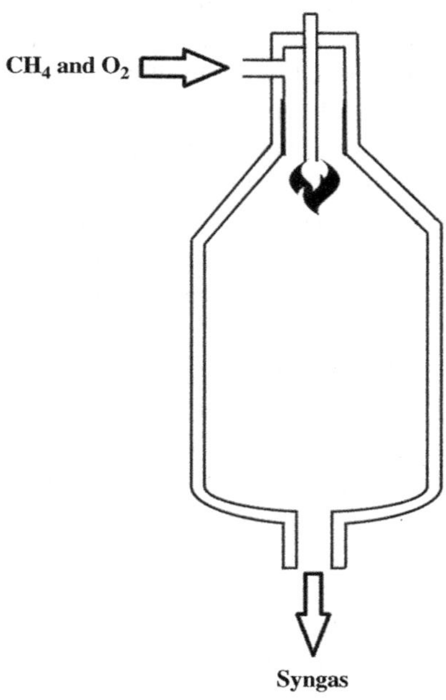

Figure 12.4 Schematic diagram of a reactor for the partial oxidation of methane without a catalyst for syngas production.

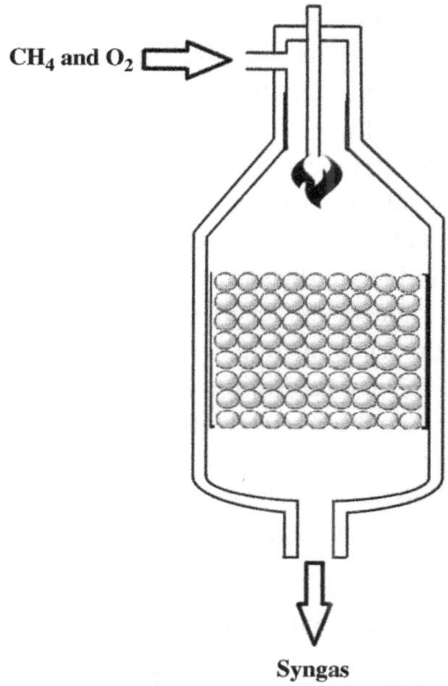

Figure 12.5 Schematic diagram of a reactor for the partial oxidation of methane with a catalyst for syngas production.

Partial oxidation of CH_4 produces H_2/CO ratios that are suitable for FT synthesis and methanol synthesis and operates at reaction temperatures where no NOx emissions are produced (Jin et al., 2000; Requies et al., 2005). Partial oxidation of CH_4 typically produces syngas consisting of a mixture of H_2 and CO in a ratio of approximately 2:1 (Ding et al., 2024). The partial oxidation of CH_4 technology does not require huge amounts of expensive superheated steam like steam reforming of methane, which is a highly energy-consuming and capital-intensive process (Tang et al., 1998). Nevertheless, an O_2 separation plant, which is also expensive, may be needed in situations where N_2 from the air is undesirable in high-pressure downstream processes (York et al., 2003).

Usually, two types of metal-based catalysts are used for the catalytic partial oxidation of methane to achieve high conversion of methane and high selectivity to syngas. They are supported noble metal-based catalysts such as Rh, Ru, Ir, Pt, and Pd and supported transition metal-based catalysts such as Ni, Co, and Cu (Fazlikeshteli et al., 2023a). Also, Al_2O_3 and CeO_2 are the commonly used catalyst supports. Even though noble metal-based catalysts are much more expensive than transition metal-based catalysts, they are commonly used for the partial oxidation of CH_4 because they are highly active and less susceptible to coke formation. Meanwhile, using bimetallic catalysts would perform much better than using monometallic catalysts for the partial oxidation of methane for the production of syngas. This was observed as bimetallic Ru–Pd/CeO_2 catalysts exhibited superior catalytic activity and stability during the partial oxidation of CH_4 than monometallic Ru–CeO_2 and Pd–CeO_2 (Fazlikeshteli et al., 2023a). Similarly, bimetallic Pd–CeO_2 catalysts exhibited considerably higher methane conversion and syngas yield with better stability than monometallic Ni/CeO_2 and Pd/CeO_2 (Fazlikeshteli et al., 2023b). This was attributed to the strong synergistic interaction between Pd and Ni on the surface of the CeO_2.

One of the recent developments in the transformation of methane to syngas is the tri-reforming of methane (TRM), which is the combination of steam reforming, dry reforming, and partial oxidation in a single reactor. Tri-reforming of methane has a relatively lower carbon footprint and energy requirement and permits the direct use of flue gas in the process feedstock as well as allows for outstanding flexibility to regulate the H_2/CO ratio for downstream processes (Alli et al., 2023).

12.2.2 Syngas Production via Gasification

12.2.2.1 Gasification of Asphaltenes

Asphaltenes are often defined as the heaviest component in crude oil (Sheu, 2002; Hoepfner et al., 2013), which are mostly aromatics (Groenzin and Mullins, 1999). They are brown-to-black amorphous solids consisting of carbon, hydrogen, oxygen, nitrogen, and sulfur (Moschopedis et al., 1976). Wallace et al. (2001) pioneered the combination of gasifying asphaltenes by partial oxidation and the extraction of asphaltene with a solvent. This allowed the low heat from the gasification process to be applied to the retrieval of the solvent used for the extraction of asphaltenes from the asphaltene-containing substance. The extraction of asphaltenes is carried out from its hydrocarbon source by blending it with a solvent in adequate amounts that will cause at least some of the asphaltenes to precipitate. After being precipitated, the asphaltenes are gasified to produce syngas via a very exothermic gasification process. The low-level heat in the syngas is utilized in the removal and recovery of the solvent from both the deasphalted hydrocarbon substance and the asphaltenes before gasification (Wallace et al., 2001).

Moreno-Arciniegas et al. (2009) carried out a study by performing nine trials on the gasification of asphaltenes to produce syngas and evaluate their composition. The gasifying agent used for the study was oxygen, of which different amounts of oxygen (33%, 40%, and 47%) were used for the complete combustion of asphaltenes at temperatures of 900 °C, 950 °C, and 1,000 °C. Based on the results obtained from the study, the proportion of H_2 and CO increases with increasing temperature, while the amount of tars and wastes decreases. As a result of a larger breakdown of asphaltene chains and improved gasification, syngas mass yields of 77% were attained. At 1,000 °C, the molar percentage of CO tended to increase when the amount of oxygen was increased, whereas the molar percentage of H_2 increased when the amounts of oxygen of 33% and 47% were applied. Optimal results were obtained when the amount of oxygen utilized was 40%, as measured by the increase in H_2 and CO with the increase in temperature and oxygen. The amount of CH_4 produced in the tests was relatively very high but decreased from 54% to 38% as temperature increased from 900 °C to 1,000 °C. Moreno-Arciniegas et al. (2009) proposed that by using a mixture of oxygen and steam as the gasifying agent, the amount of CH_4 can be further reduced while the amount of CO and H_2 in the syngas is further increased. They further suggested that the operating temperature for the gasification of asphaltenes must be higher than 1,000 °C to achieve a carbon conversion rate greater than 75%. It is therefore convenient that recent studies take into account an operating temperature range of 1,200–1,500 °C.

12.2.2.2 Coal Gasification

The production of syngas from coal gasification involves the transformation of coal into syngas in the presence of gasifying medium such as O_2, air, steam, CO_2, or their mixtures in different proportions (Żogała, 2014). The composition of syngas from coal gasification varies depending on the feedstock and operating conditions of the process. However, according to the report by the National Energy Technology Laboratory (NETL) on the Wabash River Coal Gasification Repowering Project over the four-year demonstration period from 1996 through 1999, the typical composition of syngas from coal gasification is 30–60% CO, 25–30% H_2, 0–5% CH_4, 5–15% CO_2, small amounts of water vapor, and smaller amounts of H_2S, COS, NH_3, and other trace contaminants. Coal gasification has the capability of producing less expensive clean energy, minimizing greenhouse gas emissions, clean electricity, greener chemicals, and H_2-rich syngas (Matamba et al., 2022). Coal gasification is a known essential technology of clean coal utilization that shows substantial benefits in the production of H_2-rich syngas and the reduction of CO_2 emission (Dai et al., 2023). The efficiency of coal gasification and the composition of the syngas produced during the gasification process are intensely influenced by the feedstock and the operating conditions of the process (Corbetta et al., 2015). In other words, there are key factors such as coal type, temperature, pressure, gasifying agents, and catalysts that influence the performance of the gasification process (Dai et al., 2023). The moisture content of coal type is vital for the air gasification process. However, air-steam gasification is a more effective method for coal gasification than air gasification (Gungor et al., 2012). Coal gasification techniques are classified into the conventional coal gasification and the relatively new coal gasification. The conventional coal gasification technologies are grouped into fixed bed or moving bed, fluidized bed, and entrained bed gasification, while the relatively new coal gasification technologies are plasma gasification, supercritical water gasification, decoupling gasification, and chemical-looping gasification. While conventional coal gasification technologies are at a mature stage and used for industrial-scale application, the relatively new coal gasification technologies are still in their infancy stage.

Meanwhile, underground coal gasification is a promising technology that is gaining attention considering that it reduces the cost of syngas production from coal (Pei et al., 2016). In the underground coal gasification process, coal is burned and gasified underground, which may make it safer, cleaner, and cheaper than using a surface gasifier (Friedmann et al., 2009; Gür and Canbaz, 2020). Underground coal gasification combines mining, exploitation, and gasification, and it is applied to unmineable coal seams that are deep or steeply dipping without the need for mining (Bhutto et al., 2013). Underground coal gasification creates a robust approach to ease the tension between supplying energy and ensuring sustainable development (Pei et al., 2016). Underground coal gasification can be improved in the following ways: minimizing or eliminating subsidence by careful geological study and selective gasification; enhancing the diffusion of steam from the jet core to the coal surface by a swirl jet; using swirl flow injection to increase the surface contact between the gas and coal surface, which will boost the gasification reaction; managing leaching with the aid of extensive geological research and improved operating conditions; and increasing the amount of syngas by boosting the moisture content in the converting medium.

12.2.2.3 Biomass Gasification

Biomass is a renewable organic material from plants and animals. Biomass, as a source of energy, has the prospect of generating sustainable power and fuels and contributing toward a cleaner future (Lv et al., 2007). Utilizing biomass as a carbon-neutral organic source in an integrated system creates valuable products and minimizes waste and the consumption of non-renewable resources (AlNouss et al., 2020). Gasification is a preferred option for the conversion of biomass to syngas because it provides a higher electrical efficiency than combustion, which makes the generated syngas very useful in the generation of clean energy. Biomass gasification is a thermochemical conversion of organic materials, usually at high temperatures (>700 °C) without combustion, with a controlled quantity of oxygen and/or steam to produce syngas and solid residues (Chang et al., 2023). Figure 12.6 shows the flow diagram of the gasification of biomass for the production of syngas. Commonly, the heat for the gasification of biomass can be created within the reactor or produced outside the reactor and transferred to the feedstock. Air or O_2 is supplied into the gasifier to generate internal heat by the partial oxidation of the biomass, while the combustion of a fraction of the biomass feedstock or product may be combusted to generate heat externally from the gasifier (Maschio et al., 1994).

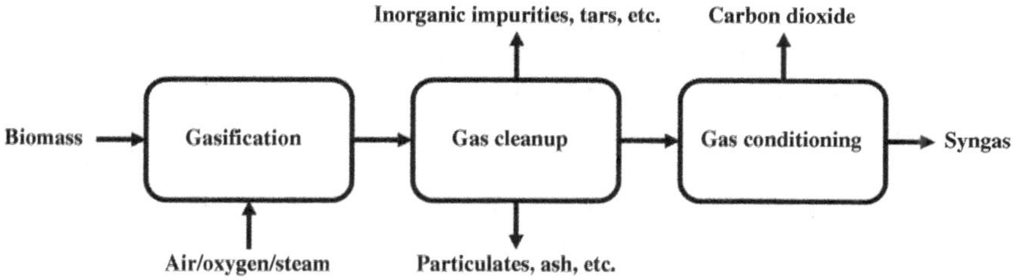

Figure 12.6 Schematic flow diagram of syngas production from biomass gasification.

The composition of the syngas is determined by the use of air, oxygen, or steam as an oxidizer for biomass gasification (Couto et al., 2013). Other factors that influence the final composition of syngas are the biomass conditions, type of gasifier, and pressure and temperature of the process. Typically, biomass gasification produces raw syngas that contains CO, H_2, CO_2, and CH_4 (AlNouss et al., 2020). It also contains some small quantities of undesirable impurities, which are called contaminants and are composed of tars, nitrogen-based compounds (NH_3, HCN, etc.), sulfur-based compounds (H_2S, COS, etc.), hydrogen halides (HCl, HF, etc.), and trace metals (Na, K, etc.) (Abdoulmoumine et al., 2015). Thus, raw syngas is normally subjected to cleanup or purification, which is an important step before the syngas is utilized for downstream applications.

Plasma technology is a promising technology that is applied to the gasification of biomass and has gained growing attention as an environmentally friendly and efficient method for the transformation of biomass to syngas (Kuo et al., 2020). Plasma gasification of biomass is a thermochemical process utilized for the conversion of biomass into syngas by subjecting the biomass to extremely high temperatures up to 10,000 °C in the presence of plasma (highly ionized gas containing charged particles) (Molino et al., 2016; Chanthakett et al., 2021). The most common plasma gases are air or nitrogen; steam or carbon dioxide is better for process chemistry; and water-stabilized plasma torches have been shown to have unique advantages (Hrabovsky, 2011). The high energy content of plasma can make it gasify treated materials directly without the need for oxidizing media, thereby resulting in carbon production at the expense of syngas production, which is the reason why a stoichiometric quantity of air, CO_2, steam, O_2, or the proper mixture is added to prevent this (Hlina et al., 2014). The benefits of using plasma technology for the gasification of biomass are high conversion efficiency, the ability to handle diverse biomass feedstocks, the production of clean syngas with the presence of minimal tar and char formation, and the potential for energy recovery and minimization of waste. However, the drawbacks of plasma technology for biomass gasification are the high energy demands and high capital costs (Tamošiūnas et al., 2023).

Being that plasma technology is considered a promising technology for biomass gasification to achieve a sustainable and circular economy, further research work must be carried out to optimize the process, improve energy efficiency, and reduce costs.

There are several new developments in the gasification technology. The U.S. Department of Energy's Gasification Systems Program is into the development of modular designs for the conversion of a variety of materials such as coal, biomass, municipal solid waste, and waste plastics into synthesis gas to enable the low-cost production of electricity, transportation fuels, high-value chemicals, hydrogen, and other useful products. Another recent development in gasification technology is the development of new burners that can operate over a range of turndown ratios and Wobbe's index ranges, which allows for the utilization of low calorific value fuels without compromising on emissions or temperatures (Deore et al., 2023). Furthermore, Fulcrum BioEnergy's Sierra BioFuels Plant, located in Reno, Nevada, the United States, developed a gasification process that converts municipal solid waste into a low-carbon synthetic crude oil, which can further be refined into usable transportation fuels such as diesel and jet fuel. In other words, it is a process of transforming trash into fuel. After the municipal solid waste has undergone drying and sorting, the municipal solid waste is heated in an oxygen-poor environment to produce a gas called syngas, which is then transformed into synthetic crude oil using a FT reactor. According to the report by Chemical & Engineering News (c&en), Fulcrum described Sierra BioFuels Plant as the world's first commercial-scale landfill-waste-to-fuel plant, and it was designed to make 42 million L of fuel per year (Bettenhausen, 2024).

12.3 Liquid Fuel Production from Syngas via FT Synthesis

12.3.1 Chemistry of FT Synthesis

FT synthesis is the catalytic transformation of syngas into ultraclean liquid fuels (Schanke et al., 1995; Yang et al., 2014; Eschemann and de Jong, 2015). Liquid fuels obtained from FT synthesis are called ultraclean fuels because they have a very low content of sulfur and aromatic compounds compared to liquid fuels derived from petroleum (Khodakov, 2009). FT synthesis is a condensation polymerization reaction, and it is highly exothermic (Dry and Hoogendoorn, 1981; Satterfield et al., 1986; Rao et al., 1992; Dalai and Davis, 2008). This is expressed in Equation 12.6.

$$CO + 2H_2 \longrightarrow -CH_2- + H_2O \quad \Delta H_R = -39.4\ \text{kcal/mol} \tag{12.6}$$

For the production of liquid fuels from syngas via FT synthesis, desired products such as paraffins (saturated hydrocarbons) and olefins (unsaturated hydrocarbons) are formed, as shown in Equations 12.7 and 12.8, respectively.

$$nCO + (2n + 1)H_2 \longrightarrow C_nH_{2n+2} + nH_2O \tag{12.7}$$

$$nCO + 2nH_2 \longrightarrow C_nH_{2n} + nH_2O \tag{12.8}$$

Small amounts of byproducts such as oxygenates (alcohols), carbon dioxide, and methane could be formed during the FT reaction, as shown in Equations 12.9–12.11, respectively. Meanwhile, water is a major byproduct because a large amount of it is formed during FT synthesis.

$$CO + 2nH_2 \longrightarrow C_nH_{2n+1}OH + (n - 1)H_2O \tag{12.9}$$

$$CO + H_2O \longleftrightarrow CO_2 + H_2 \tag{12.10}$$

$$CO + 3H_2 \longrightarrow CH_4 + H_2O \tag{12.11}$$

12.3.2 FT Synthesis Catalysts

The regularly used heterogeneous catalysts for FT synthesis are Ru, Ni, Fe, and Co. Ru is highly active for FT synthesis, but it is very expensive and insufficient in supply. Ni is not favorable for FT synthesis because it favors a high formation of methane, which is an undesirable product. Therefore, Fe and Co are the heterogeneous catalysts used for commercial FT synthesis. Iron-based catalysts are currently used industrially for the conversion of coal-derived syngas into FT liquids because iron-based catalysts possess inherent WGS capability, which increases the low H_2/CO ratio of coal-derived syngas, thereby improving hydrocarbon product yields in the FT synthesis. In FT synthesis, iron-based catalysts can be used for both high-temperature conditions (300–350 °C) and low-temperature conditions (200–240 °C) (Abelló and Montané, 2011), while cobalt-based catalysts are only suitable for low-temperature conditions because higher temperatures cause more methane formation with cobalt-based catalysts than iron-based catalysts. A recent metal price chart shows that cobalt is about 255 times more expensive than iron ore, but cobalt-based catalysts are a good alternative to iron-based catalysts for FT synthesis because they exhibit good activity at lower reaction pressures, which means that the higher catalyst costs of cobalt-based catalysts can be offset by lower operating costs.

At similar reaction conditions in FT synthesis, the product distribution for cobalt-based and iron-based catalysts is different. For instance, at reaction conditions of 3 MPa and 240 °C, a cobalt-based catalyst will exhibit somewhat higher selectivity for heavier hydrocarbons than an iron-based catalyst. The product distribution is mainly influenced by the choice of operating temperature. Higher temperature results in a gasoline/diesel ratio of 2:1, while lower temperature results in a gasoline/diesel ratio of 1:2, which is more or less regardless if the catalyst is an iron-based catalyst or cobalt-based catalyst. Higher temperatures normally shift selectivity toward lower carbon number products, more hydrogenated products, branching increases, and secondary products such as ketones and aromatics also increase.

12.3.3 FT Synthesis Reactors

The type and amount of desired products in FT synthesis are deciding factors for the reaction temperature that will be used. Low-temperature Fischer–Tropsch (LTFT) synthesis is conducted in either a tubular fixed bed reactor or slurry-phase reactor at operating conditions of 200–240 °C and 2.0–4.0 MPa (Yung et al., 2009). The LTFT synthesis is utilized in the Shell GTL facility at Bintulu, Malaysia, the Escravos GTL plant at Escravos, Nigeria, and the Oryx GTL plant at Ras Laffan, Qatar. LTFT synthesis produces more waxes, which are upgraded to diesel through hydrocracking. Both cobalt-based catalysts and iron-based catalysts are utilized for LTFT synthesis. Meanwhile, high-temperature Fischer–Tropsch (HTFT) synthesis is conducted in a circulating fluidized bed (CFB) reactor or fixed fluidized bed (FFB) reactor at the operating conditions of 300–350 °C and 2.5 MPa. Sasol in South Africa utilizes HTFT synthesis in their coal-to-liquid (CTL) plants. HTFT synthesis is utilized purposefully to produce olefins (C_3–C_{11}), which are upgraded to gasoline. Iron-based catalysts are predominantly utilized for HTFT synthesis. The schematic flow diagram of synthesis gas production and the FTS process is shown in Figure 12.7.

FT synthesis is a highly exothermic reaction, which makes the removal of heat from a reactor a vital factor in the design of a commercial reactor. Typically,

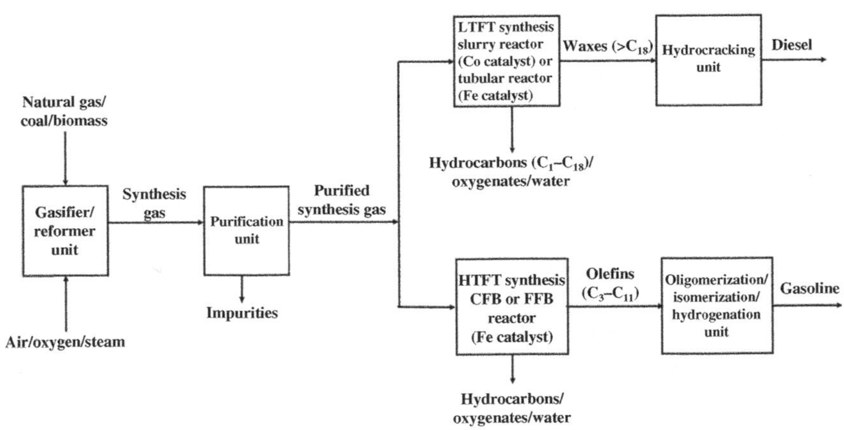

Figure 12.7 Schematic flow diagram of synthesis gas production and the FT synthesis process.

three different types of reactor design can be utilized commercially for FT synthesis, namely: fixed bed reactor, fluidized bed reactor, and slurry bed reactor. The multitubular fixed bed reactors, which are also known as Arge reactors, were jointly developed by Lurgi and Ruhrchemie and commissioned in 1955 (Espinoza et al., 1999). Sasol in Sasolburg, South Africa, utilized the Arge reactors for the production of heavy liquid hydrocarbons and waxes in an LTFT synthesis process targeted for liquid fuel production (Speight, 2020). Most of the Arge reactors utilized for LTFT synthesis in the past are now being replaced by slurry bed reactors, which are considered state-of-the-art technology for LTFT synthesis because they offer better temperature control and higher conversion. Other companies using slurry bed reactors for FT technology are Exxon and Statoil (Espinoza et al., 1999). On the other hand, fluidized bed reactors were developed for HTFT synthesis for the production of low molecular gaseous hydrocarbons and gasoline. Fluidized bed reactors were initially developed in a circulating mode, e.g., Sasol's Synthol reactors, and they have been upgraded by a FFB type of design called Advanced Synthol reactors, and they have high throughputs (Speight, 2020).

12.4 Conversion of CO_2 into Liquid Fuels

Captured CO_2 is mostly stored in an underground storage facility to avoid the release of an enormous amount of CO_2 into the atmosphere to mitigate global warming and climate change. Several technologies are commercially available for the capturing and storing of CO_2 underground, but there is a limitation in storage capacity (Do et al., 2022). Therefore, it is necessary to develop technologies that can be used in converting the captured CO_2, which is a key carbon source for synthesizing high-value materials such as liquid fuels. Considering the high demand for transportation fuels globally, transforming CO_2 into valuable liquid fuels will help meet the increasing global demand and lessen the too much dependence on fossil fuels (Gao et al., 2020). The chemical methods (catalytic hydrogenation, photocatalytic, and electrochemical) and biological methods (photosynthesis and non-photosynthesis) are the technologies utilized for the conversion of CO_2 into liquid fuels.

12.4.1 Chemical Conversion of CO_2 into Liquid Fuels

12.4.1.1 Hydrogenation of CO_2

Hydrogenation via heterogeneous catalysis is the most straightforward method for directly transforming CO_2 into valuable products (Fujiwara et al., 1995; Wang et al., 2016). The hydrogenation of CO_2 into liquid fuels such as gasoline, aviation fuel, diesel, methanol, ethanol, and higher alcohols through heterogeneous catalysis could help to minimize the environmental challenges caused by the enormous release of CO_2 into the atmosphere and help achieve sustainable liquid fuels that could be a good alternative to fossil fuels (Gao et al., 2020). The liquid fuels obtained from the catalytic hydrogenation of CO_2 can be called carbon-neutral fuels because CO_2 captured from the air is utilized for the production of liquid fuels and re-emitted into the atmosphere when the liquid fuels are combusted (Dowson and Styring, 2017).

Industrial-scale production of methanol is generally carried out through catalytic conversion of syngas (mixture of CO and H_2) over $Cu/ZnO/Al_2O_3$ catalyst, which is commonly known to be susceptible to sintering (Fichtl et al., 2015). A similar commercial $Cu/ZnO/Al_2O_3$ catalyst exhibits very low catalytic performance for CO_2 hydrogenation to methanol (Allam et al., 2019). In a bid to improve the catalytic performance of the Cu-based catalysts for the hydrogenation of CO_2 to methanol, the Cu-based catalysts are modified by adding various activators or other metals (Zr, Si, La, Ti, Cr, Ga, Ce, Fe, Nb, Pd, etc.). In terms of the maturity of the technique for the conversion of CO_2 to methanol, CO_2 hydrogenation via heterogeneous catalysis is the most developed of the existing methods. The production of methanol through a CO_2 hydrogenation reaction is thermodynamically exothermic (Tada et al., 2017; Tada and Satokawa, 2018; Rui et al., 2020). The reaction for the catalytic hydrogenation of CO_2 into methanol is shown in Equation 12.12 (Atsbha et al., 2021).

$$CO_2 + 3H_2 \longleftrightarrow CH_3OH + H_2O \ \Delta H_{298K} = -49.5 \text{ kJ/mol} \tag{12.12}$$

The hydrogenation of CO_2 to liquid fuels with two or more carbon atoms (C_{2+}) has encountered more challenges than methanol because CO_2 is highly thermodynamically stable and chemically inert (Gao et al., 2020), which leads to the poor adsorption of CO_2 on a catalyst surface, thereby causing a low C/H ratio at catalytic sites. Thus, it favors the formation of CH_4 and hinders the chain growth probability (Wang et al., 2011; Prieto, 2017). So far, fewer studies have been carried out on the catalytic hydrogenation of CO_2 into liquid hydrocarbon fuels (C_{5+}), which comprises gasoline (C_5–C_{11}), jet fuel (C_8–C_{16}), and diesel (C_{10}–C_{20}) (Gao et al., 2017, 2020; Wei et al., 2017; Shi et al., 2018; Wang et al., 2018; Yao et al., 2020). The synthesis of liquid hydrocarbon fuels from CO_2 hydrogenation could be accomplished via two pathways, namely: a methanol route, which involves the hydrogenation of CO_2 to methanol (see Equation 12.13) and followed by the conversion of methanol to liquid hydrocarbon fuels (see Equation 12.14); and the FTS route, which entails the hydrogenation of CO_2 to CO via the reverse water–gas shift (RWGS) reaction (see Equation 12.15) and followed by the hydrogenation of CO to liquid hydrocarbon fuels via FTS (see Equation 12.16) (Wei et al., 2016; Yao et al., 2017; Wang et al., 2018).

Hydrogenation of CO_2 to methanol:

$$CO_2 + 3H_2 \rightleftharpoons CH_3OH + H_2O \tag{12.13}$$

Methanol to hydrocarbon reaction:

$$CH_3OH \rightleftharpoons -CH_2- + H_2O \tag{12.14}$$

RWGS reaction:

$$CO_2 + H_2 \rightleftharpoons CO + H_2O \tag{12.15}$$

FTS reaction:

$$CO + 2H_2 \rightleftharpoons -CH_2- + H_2O \tag{12.16}$$

Iron-based catalyst is the most suitable catalyst used to catalyze both the hydrogenation of CO_2 into CO and the hydrogenation of CO via the FT reactions because it limits the formation of methane and is active for both the RWGS and the FT reactions. In the case of the methanol pathway, the hydrogenation of CO_2 to methanol requires a metallic catalyst (e.g., Cu—ZnO—Al_2O_3), and then the conversion of methanol to liquid hydrocarbon fuels requires an acidic zeolite catalyst. This means that if the methanol pathway is to be carried out in a single pot, it means a tandem catalyst with a favorable metallic and acidic property will be required. The products resulting from the methanol route contain mostly aromatics and branched aliphatics, which are appropriate for direct usage as gasoline due to the high octane number of 90–95. On the other hand, products from the FT route are a mixture of hydrocarbons of a wide range of carbon chain lengths, thereby making the selectivity toward gasoline and diesel range inadequate. Additionally, the energy requirement for producing liquid hydrocarbon fuels via FT is higher than that of the methanol route. However, the methanol route requires a higher cost of catalyst and is susceptible to catalyst deactivation than the FT route. The two-step method for both the FT route and methanol route has been extensively investigated, and more improvement is required. Nonetheless, the one-step method for both the FT route and methanol route for the conversion of CO_2 to liquid hydrocarbon fuels will be more cost-effective. The one-step method for both the FT route and methanol route is still in the infancy stage. However, good progress has been achieved in recent times in the production of liquid hydrocarbon fuels from CO_2 hydrogenation via the one-step methanol route using bifunctional catalysts (Panzone et al., 2020).

12.4.1.2 Photocatalysis of CO_2

Photocatalytic reduction of CO_2 to liquid fuels is a promising route to minimize global warming and boost the energy supply (Wang et al., 2020). The photocatalytic conversion of CO_2 to liquid fuels involves the production of liquid fuels, such as formic acids, methanol, and ethanol, from the reduction of CO_2 by using either artificial light or solar light (sunlight) with semiconductor photocatalysts (Corma and Garcia, 2013). Solar light is one of the sustainable, plentiful, and inexpensive energy sources, which has been used for the conversion of CO_2 to liquid fuels (Park et al., 2021). CO_2 is a linear molecule that is highly thermodynamically stable due to its high bond energy of 750 kJ/mol and usually requires high energy inputs to break the C=O bonds for photocatalytic CO_2 conversion (Li et al., 2016; Remiro-Buenamañana and García, 2019; Gong et al., 2022). While there is unlimited solar energy hitting the earth (1.3 × 105 TW) for CO_2 conversion, CO_2 is optically inert at either visible or UV radiation in the wavelengths of 200–900 nm, and thus, there is a need for photocatalysts with a suitable band structure, which can be excited by sunlight and then transfer the photogenerated electrons to CO_2 to successfully achieve CO_2 reduction (Li et al., 2016).

Photocatalytic conversion of CO_2 primarily comprises the following five steps (Xu et al., 2023):

1. Absorption of light.
2. Separation of photogenerated electrons and holes.
3. Adsorption of CO_2.
4. Redox reactions on the catalyst surface.
5. Desorption of the resulting products.

Table 12.1 shows the possible reactions of CO_2 reduction to different products and their corresponding standard redox potentials.

Table 12.1 CO_2 reduction to different products and their corresponding standard redox potentials.

Reaction	E^0 [V] vs. SHE at pH 7
$CO_2 + e^- \longrightarrow CO_2^-$	−1.9
$CO_2 + 2H^+ + 2e^- \longrightarrow HCOOH$	−0.61
$CO_2 + 2H^+ + 2e^- \longrightarrow CO + H_2O$	−0.52
$2CO_2 + 12H^+ + 12e^- \longrightarrow C_2H_4 + 4H_2O$	−0.34
$CO_2 + 4H^+ + 4e^- \longrightarrow HCHO + H_2O$	−0.48
$CO_2 + 6H^+ + 6e^- \longrightarrow CH_3OH + H_2O$	−0.38
$2CO_2 + 12H^+ + 12e^- \longrightarrow CH_3CH_2OH + 3H_2O$	−0.24
$2H^+ + 2e^- \longrightarrow H_2$	−0.42

Source: Xu et al. (2023)/with permission of Elsevier.

The most widely studied photocatalyst for CO_2 reduction to liquid fuels is TiO_2 because of its easy availability, cheap cost, nontoxicity, long-term stability, etc. It can be used in its pure form or doped with different metals and nonmetals to boost its photoactivity. Nevertheless, the quick recombination of the photogenerated electron–hole pairs through the microstructure of TiO_2 photocatalysts and their low adsorption capability for CO_2 have a negative impact on their catalytic efficiency in CO_2 reduction (Pomilla et al., 2018). Therefore, other materials have been developed for the photocatalytic conversion of CO_2 to liquid fuels. Materials such as metal–organic framework (MOF), copper oxides, metal oxynitrides, and graphitic carbon nitride have been utilized for the photocatalytic conversion of CO_2 to liquid fuels, and they look promising considering that they showed better light-harvesting properties than TiO_2. The photocatalytic conversion of CO_2 to liquid fuels is in its infancy and is still considered by the poor efficiency of the processes when compared to other methods. The photocatalytic conversion of CO_2 to liquid fuels is still faced with huge challenges because of the need to design photocatalysts with sites accomplishing the thermodynamic and kinetic requirements such as high reduction potential of photogenerated electrons and simultaneous transfer of electrons and protons.

12.4.1.3 Electrochemical Conversion of CO_2

In recent times, the technology for the electrochemical conversion of CO_2 has made good progress in achieving high selective catalysts, larger-scale reactors, specific process modeling, and mechanistic comprehension of the CO_2 reduction reaction, which makes the technology promising for commercial application for the electrocatalytic reduction of CO_2 to value-added products (Wang et al., 2021; Zhao et al., 2021; Overa et al., 2022). Electrochemical conversion of CO_2 via reduction reaction is a promising method for the conversion of CO_2 to liquid fuels and it can also store the electricity produced from green energy sources such as wind, solar, and tide into chemical energy (Feng et al., 2021). Also, water is a clean and inexpensive source of hydrogen or protons for CO_2 reduction. Electrochemical conversion or reduction of CO_2 encompasses a cathodic half-cell reaction of CO_2 reduction and an anodic half-cell reaction of water oxidation (Ganesh, 2016). It could also be defined as the concurrent reduction of CO_2 reduction and oxidation of water as shown in Equations 12.17 and 12.18.

$$\text{Cathode reaction: } CO_2 + 2H^+ + 2e^- \rightleftarrows CO + H_2O \tag{12.17}$$

$$\text{Anode reaction: } 2H_2O \rightleftarrows O_2 + 4H^+ + 4e^- \tag{12.18}$$

The electrical potential supplied to the electrochemical system results in water oxidation into oxygen molecules (O_2), protons (H^+), and electrons (e^-) at the anode. O_2 is given off at the anode, while electrons flow via the external circuit to the cathode, where they can combine either with protons travelled via the membrane/junction/salt bridge to form H_2 or with CO_2 to form different value-added products such as formic acids, methanol, ethanol, propanol, etc. (Ganesh, 2016), as shown in Equations 12.19–12.24. The specific product to be produced is dependent on reaction conditions and the catalyst in the cathodic system. Generally, the CO_2 reduction reaction competes with the hydrogen evolution reaction (HER), and if the appropriate reaction conditions and catalysts are not used in the CO_2 reduction reaction, the main reduced product would be H_2.

$$4H^+ + 4e^- \longrightarrow 2H_2\uparrow \quad (12.19)$$

$$CO_2 + 2H^+ + 2e^- \longrightarrow CO\uparrow + H_2O \quad (12.20)$$

$$CO_2 + 2H^+ + 2e^- \longrightarrow HCOOH \quad (12.21)$$

$$CO_2 + 6H^+ + 6e^- \longrightarrow CH_3OH + H_2O \quad (12.22)$$

$$2CO_2 + 12H^+ + 12e^- \longrightarrow C_2H_5OH + H_2O \quad (12.23)$$

$$3O_2 + 18H^+ + 18e^- \longrightarrow C_3H_7OH + H_2O \quad (12.24)$$

The electrochemical conversion of CO_2 can be carried out at ambient temperature and atmospheric pressure, which makes it a safe method of conversion of CO_2 into value-added products. Also, the electrochemical conversion of CO_2 can be driven by any source of electricity such as solar, geothermal, hydro, wind, tidal, and atomic power. In addition, the electrochemical conversion method can be utilized as an off-peak electricity or renewable energy storage vehicle through the conversion of electrical energy to chemical energy by producing fuels from CO_2, and the stored energy can later be utilized for end-use by the oxidization of the fuels through fuel cells or normal fuel-burning engines (Agarwal et al., 2022).

The important components for the electrochemical conversion of CO_2 to liquid fuels are electrochemical cells, electrolytes, and electrocatalysts. The electrochemical cells utilized for electrocatalytic reduction of CO_2 are categorized into H-type, polymer electrolyte membrane flow cell, solid oxide electrolysis cell, and differential electrochemical mass spectrometry cell. However, the H-type cell is the well-known cell for the electrocatalytic reduction of $CO_{2,}$ and it is commercially available. Meanwhile, the electrolytes used for the electrochemical conversion of CO_2 into liquid fuels are aqueous electrolytes (e.g., Na+, K+, OH^-, CO_3^{2-}, etc.), organic solvent-based electrolytes (e.g., acetone, ethyl acetate, toluene, etc.), and ionic liquid-based electrolytes. The low efficiencies of organic solvent-based electrolytes do not encourage their use. The use of aqueous electrolytes for the electrocatalytic reduction of CO_2 increases and diversifies selectivity toward liquid fuel products via C—C coupling, while the ionic liquid-based electrolytes can only reduce CO_2 to C_1-based liquid fuels for most of the electrocatalysts (Okoye-Chine et al., 2022). Generally, electrocatalytic reduction of CO_2 is dependent on the utilization of active transition metals such as Cu, Co, Pd, etc., to decrease the high overpotentials needed for the activation of CO_2 to the $CO_2^{\bullet}$ radical anion or other intermediates that can be transformed further (Kang et al., 2020). An efficient electrocatalyst must possess an outstanding quality mixture of high activity, high selectivity, and low overpotential (Okoye-Chine et al., 2022). Cu-based electrocatalysts have been mostly utilized for the electrochemical reduction of CO_2 to liquid fuels because they possess high Faraday efficiency and high selectivity.

One of the promising technologies for the conversion of CO_2 to liquid fuels is the photoelectrochemical method. This technology combines electrochemical conversion and photocatalytic conversion methods for the reduction of CO_2 into liquid fuels. Photoelectrochemical efficiency could be improved via the use of 2D/3D materials, MOFs, plasmonic metals, porous nanostructures, carbon-based catalysts, molecular catalysts, and biological molecules (Kumaravel et al., 2020).

Electrochemical CO_2 reduction reaction is a promising strategy for minimizing CO_2 emissions contributing to global warming due to its easy operating system, simple constructions, operation at neutral pH, ambient temperature and atmospheric pressure, and low energy utilization to form value-added products such as formic acid, methane, ethanol, and carbon using renewable electricity (Usman et al., 2021). However, the electrochemical CO_2 reduction reaction strategy is faced with numerous challenges that affect the commercialization of the technology due to high overpotentials, electrode instability, and low Faradic efficiencies of the desirable products.

12.4.2 Biological Conversion of CO_2 into Liquid Fuels

Autotrophs naturally convert CO_2 into biomass by receiving energy from sunlight or inorganic electron donors (Nisar et al., 2021). This phenomenon is utilized for biological CO_2 fixation for the conversion of CO_2 into liquid biofuels. Among the major methods for CO_2 fixation, biological CO_2 fixation is the most attractive for industrial scale due to its relatively mild reaction conditions, high yield, high selectivity, and good scale-up prospect (Shi et al., 2019). Plants and microorganisms with the ability to convert CO_2 into biofuels via enzymatic processes can be categorized into photosynthetic and non-photosynthetic systems. Algae through the energy obtained from the sunlight absorb CO_2 and other nutrients to form a high-lipid-content biomass, which is further processed to liquid biofuels. Usually, algae are grown in an open pond system or a closed photobioreactor for a high-lipid-content biomass (Do et al., 2022). Meanwhile, with cyanobacteria, liquid biofuels can be directly synthesized from CO_2 without going through lipid synthesis. Cyanobacteria are photosynthetic prokaryotic microorganisms, and they are more manageable for genetic engineering (Li and Liao, 2013). The major challenge of the photosynthesis approach for CO_2 fixation to liquid biofuels is the reliance on sunlight, which affects the upscalability, efficiency, and tractability. Due to the intrinsic limitation connected with the use of photosynthetic organisms, a non-photosynthetic biological approach is required for the biological conversion of CO_2 to liquid biofuels. The non-photosynthetic biological approach is a light-independent biological technique. The non-photosynthetic biological systems have higher efficiency than the photosynthetic biological systems, and the non-photosynthetic organisms can use inexpensive low electron emissions together with available gaseous carbon feedstock (made up of CO, CO_2, and H_2) from synthesis gas, biogas, and waste gases from industries (Okoye-Chine et al., 2022). Nonetheless, one of the major barriers to the advancement of the non-photosynthetic biological approach for CO_2 fixation to liquid biofuels is the inadequate metabolic understanding and genetic tools, as well as the under-explored number of non-photosynthetic strains.

Even though biological CO_2 fixation can reduce heat and electricity consumption, commercial feasibility is deficient because of the high land requirements and unstable resource management such as sunlight and algae growth. Since none of the biological methods for the conversion of CO_2 to liquid biofuels have achieved industrial success, further improvement and research work in this area are important. The evolving technologies for the biological conversion of CO_2 into biofuels are the following: the use of genetically modifying microbes like bacteria and algae to boost their ability to capture and fixate CO_2 to create pathways for the synthesis of desired liquid fuels; combining enzymes from different microorganisms to create effective metabolic pathways for the conversion of CO_2 to liquid fuels in which each enzyme catalyzes a specific step in the bioreaction process; the integration of CO_2 capture technologies with biological conversion methods for the direct utilization of atmospheric CO_2 as a feedstock for the production of biofuels.

12.5 Conclusion

Liquid fuels produced from syngas and CO_2 are sustainable liquid fuels that are environmentally friendly. Syngas is obtained from various methods: steam reforming of CH_4, autothermal reforming of CH_4, partial oxidation of CH_4, dry reforming of CH_4, gasification of asphaltenes, coal gasification, and biomass gasification. The produced syngas undergo purification to meet the desired H_2/CO ratio suitable for conversion to long-chain hydrocarbons through the process of

FT synthesis. LTFT synthesis produces more waxes, which are upgraded to diesel via the process of hydrocracking, while HTFT synthesis is used purposefully for the production of olefins (C_3–C_{11}), which are upgraded to gasoline. Both cobalt-based catalysts and iron-based catalysts are used for LTFT synthesis, while iron-based catalysts are primarily used for HTFT synthesis. LTFT synthesis and HTFT synthesis are operated at an industrial scale.

In a bid to achieve carbon neutrality, CO_2, which is a major greenhouse gas, is used as a feedstock for the production of liquid fuels. Technologies such as catalytic hydrogenation, photocatalytic conversion, electrochemical conversion, and biological conversion are used for the conversion of CO_2 to liquid fuels. Progress has been made in the production of liquid fuels from CO_2 hydrogenation, especially with the production of methanol from CO_2 hydrogenation. However, more research is needed to design and develop a highly active, stable, and cost-effective catalyst for the production of C_{2+} liquid fuels from the catalytic hydrogenation of CO_2.

The photocatalytic conversion of CO_2 to liquid fuels is the production of liquid fuels from the reduction of CO_2 by using either artificial light or solar light (sunlight) with semiconductor photocatalysts. The process of converting CO_2 into liquid fuels by photocatalysis is still in its infancy stage and has limited efficiency when compared to alternative approaches. Due to the requirement to design photocatalysts with sites accomplishing the thermodynamic and kinetic requirements, such as the high reduction potential of photogenerated electrons and simultaneous transfer of electrons and protons, the photocatalytic conversion of CO_2 to liquid fuels remains challenging and requires more investigation.

The electrochemical conversion of CO_2 to liquid fuels is a promising technique considering its high energy efficiency and mild reaction conditions. The key components for the electrochemical conversion of CO_2 to liquid fuels are electrochemical cells, electrolytes, and electrocatalysts, which influence the overall performance of the electrochemical process measured in terms of reaction rate, current density, selectivity, stability, and overpotential. Cu-based electrocatalysts are mostly used for the electrochemical reduction of CO_2 to liquid fuels due to their high Faraday efficiency and high selectivity. Despite the progress made so far, the electrochemical conversion of CO_2 to liquid fuels is still bedeviled with some challenges in selectivity toward desired products, Faraday efficiency, overpotential, and competition with the HER. Overcoming these challenges will require additional research on surface modifications of the electrode materials, exploiting the synergy between the support materials and active metal electrodes, and optimizing the electrolytes and cell design.

Biological CO_2 fixation is attractive for large-scale use due to its relatively mild reaction conditions, high yield, high selectivity, and good scale-up prospect. The biological conversion of CO_2 to liquid biofuels is achieved by photosynthetic method or non-photosynthetic method. The primary obstacle to the photosynthetic method for fixing CO_2 into liquid biofuels is its dependence on sunshine, which has an impact on its tractability, efficiency, and upscalability. The biological conversion of CO_2 into liquid biofuels requires a non-photosynthetic biological technique because of the inherent limitations associated with the employment of photosynthetic organisms. Non-photosynthetic biological approach is a light-independent biological technique. Compared to photosynthetic biological systems, non-photosynthetic organisms are more efficient and can utilize cheap low-emission sources of electrons in addition to readily available gaseous carbon feedstock (composed of CO, CO_2, and H_2) from biogas, synthesis gas, and industrial waste gases. However, the underutilization of non-photosynthetic strains and the lack of genetic resources and metabolic knowledge are key obstacles to the development of the non-photosynthetic biological strategy for CO_2 fixation to liquid biofuels.

References

Abdoulmoumine, N., Adhikari, S., Kulkarni, A. et al. (2015). A review on biomass gasification syngas cleanup. *Applied Energy* 155: 294–307. https://doi.org/10.1016/j.apenergy.2015.05.095.

Abdulrasheed, A., Jalil, A.A., Gambo, Y. et al. (2019). A review on catalyst development for dry reforming of methane to syngas: recent advances. *Renewable and Sustainable Energy Reviews* 108 (March): 175–193. https://doi.org/10.1016/j.rser.2019.03.054.

Abelló, S. and Montané, D. (2011). Exploring iron-based multifunctional catalysts for Fischer-Tropsch synthesis: a review. *ChemSusChem* 4 (11): 1538–1556. https://doi.org/10.1002/cssc.201100189.

Agarwal, A.S., Rode, E., Sridhar, N. et al. (2022). Conversion of CO_2 to value added chemicals: opportunities and challenges. In: *Handbook of Climate Change Mitigation and Adaptation*, 3e (ed. M. Lackner, B. Sajjadi and W.-Y. Chen), 1585–1623. Springer Nature. https://doi.org/10.1007/978-3-030-72579-2_86.

Allam, D., Bennici, S., Limousy, L. et al. (2019). Improved Cu- and Zn-based catalysts for CO_2 hydrogenation to methanol. *Comptes Rendus Chimie* 22 (2–3): 227–237. https://doi.org/10.1016/j.crci.2019.01.002.

Alli, R.D., De Souza, P.A.L., Mohamedali, M. et al. (2023). Tri-reforming of methane for syngas production using Ni catalysts: current status and future outlook. *Catalysis Today* 407: 107–124. https://doi.org/10.1016/j.cattod.2022.02.006.

AlNouss, A., McKay, G., and Al-Ansari, T. (2020). Production of syngas via gasification using optimum blends of biomass. *Journal of Cleaner Production* 242: 118499. https://doi.org/10.1016/j.jclepro.2019.118499.

Anders Blekkan, E., Borg, O., Froseth, V. et al. (2007). Fischer-Tropsch synthesis on cobalt catalysts: the effect of water. In: *Catalysis* (ed. J.J. Spivey and K.M. Dooley), 13–32. Cambridge: Royal Society of Chemistry. https://doi.org/10.1039/b601307b.

Aryafard, E., Farsi, M., Alipoor, A.R. et al. (2023). The effect of CO_2 feeding on syngas quality in autothermal reforming of methane over Ni-Al_2O_3 catalyst: possibility of CO_2 conversion to hydrogen. *Reaction Kinetics, Mechanisms and Catalysis* 136 (2): 671–687. https://doi.org/10.1007/s11144-023-02374-3.

Atsbha, T.A., Yoon, T., Seongho, P. et al. (2021). A review on the catalytic conversion of CO_2 using H_2 for synthesis of CO, methanol, and hydrocarbons. *Journal of CO_2 Utilization* 44 (December 2020): 101413. https://doi.org/10.1016/j.jcou.2020.101413.

Bahri, S., Venezia, A.M., and Upadhyayula, S. (2019). Utilization of greenhouse gas carbon dioxide for cleaner Fischer-Tropsch diesel production. *Journal of Cleaner Production* 228: 1013–1024. https://doi.org/10.1016/j.jclepro.2019.04.310.

Banerjee, A., Dick, G.R., Yoshino, T. et al. (2016). Carbon dioxide utilization via carbonate-promoted C–H carboxylation. *Nature* 531: 215–219. https://doi.org/10.1038/nature17185.

Benson, E.E., Kubiak, C.P., Sathrum, A.J. et al. (2009). Electrocatalytic and homogeneous approaches to conversion of CO_2 to liquid fuels. *Chemical Society Reviews* 38: 89–99. https://doi.org/10.1039/b804323j.

Bettenhausen, C. (2024). *Fulcrum BioEnergy Abandons Trash-to-fuel Plant in Nevada*. Chemical & Engineering News (c&en). https://cen.acs.org/energy/Fulcrum-BioEnergy-abandons-trashfuel-plant/102/web/2024/06

Bhatelia, T., Ma, W., Davis, B. et al. (2011). Kinetics of the Fischer-Tropsch reaction over a Ru-promoted Co/Al_2O_3 catalyst. *Chemical Engineering Transactions* 25: 707–712. https://doi.org/10.3303/CET1125118.

Bhutto, A.W., Bazmi, A.A., and Zahedi, G. (2013). Underground coal gasification: from fundamentals to applications. *Progress in Energy and Combustion Science* 39 (1): 189–214. https://doi.org/10.1016/j.pecs.2012.09.004.

Bian, Z., Zhong, W., Yu, Y. et al. (2021). Dry reforming of methane on Ni/mesoporous-Al_2O_3 catalysts: effect of calcination temperature. *International Journal of Hydrogen Energy* 46 (60): 31041–31053. https://doi.org/10.1016/j.ijhydene.2020.12.064.

Bitters, J.S., He, T., Nestler, E. et al. (2022). Utilizing bimetallic catalysts to mitigate coke formation in dry reforming of methane. *Journal of Energy Chemistry* 68: 124–142. https://doi.org/10.1016/j.jechem.2021.11.041.

Carapellucci, R. and Giordano, L. (2020). Steam, dry and autothermal methane reforming for hydrogen production: a thermodynamic equilibrium analysis. *Journal of Power Sources* 469 (May): 228391. https://doi.org/10.1016/j.jpowsour.2020.228391.

Chang, Y.J., Chang, J.S., and Lee, D.J. (2023). Gasification of biomass for syngas production: research update and stoichiometry diagram presentation. *Bioresource Technology* 387 (June): 129535. https://doi.org/10.1016/j.biortech.2023.129535.

Chanthakett, A., Arif, M.T., Khan, M.M.K. et al. (2021). Performance assessment of gasification reactors for sustainable management of municipal solid waste. *Journal of Environmental Management* 291 (April): 112661. https://doi.org/10.1016/j.jenvman.2021.112661.

Chein, R.Y. and Fung, W.Y. (2019). Syngas production via dry reforming of methane over CeO_2 modified Ni/Al_2O_3 catalysts. *International Journal of Hydrogen Energy* 44 (28): 14303–14315. https://doi.org/10.1016/j.ijhydene.2019.01.113.

Corbetta, M., Bassani, A., Manenti, F. et al. (2015). Multi-scale kinetic modeling and experimental investigation of syngas production from coal gasification in updraft gasifiers. *Energy and Fuels* 29 (6): 3972–3984. https://doi.org/10.1021/acs.energyfuels.5b00648.

Corma, A. and Garcia, H. (2013). Photocatalytic reduction of CO_2 for fuel production: possibilities and challenges. *Journal of Catalysis* 308: 168–175. https://doi.org/10.1016/j.jcat.2013.06.008.

Couto, N., Rouboa, A., Silva, V. et al. (2013). Influence of the biomass gasification processes on the final composition of syngas. *Energy Procedia* 36: 596–606. https://doi.org/10.1016/j.egypro.2013.07.068.

Dai, F., Zhang, S., Luo, Y. et al. (2023). Recent progress on hydrogen-rich syngas production from coal gasification. *Processes* 11 (6): 1765. https://doi.org/10.3390/pr11061765.

Dalai, A.K. and Davis, B.H. (2008). Fischer-Tropsch synthesis: a review of water effects on the performances of unsupported and supported Co catalysts. *Applied Catalysis A: General* 348 (1): 1–15. https://doi.org/10.1016/j.apcata.2008.06.021.

Davis, B.H. (2007). Fischer-Tropsch synthesis: comparison of performances of iron and cobalt catalysts. *Industrial and Engineering Chemistry Research* 46 (26): 8938–8945. https://doi.org/10.1021/ie0712434.

Deore, S.P., Gadkari, P., Mahajani, S.M. et al. (2023). Development of a new premixed burner for biomass gasifier generated low calorific value producer gas for industrial applications. *Energy* 279 (June): 128140. https://doi.org/10.1016/j.energy.2023.128140.

Ding, Z., Chen, S., Yang, T. et al. (2024). Atomically dispersed MoNi alloy catalyst for partial oxidation of methane. *Nature Communications* 15 (1): 1–10. https://doi.org/10.1038/s41467-024-49038-x.

Do, T. N., You, C., and Kim, J. (2022). A CO_2 utilization framework for liquid fuels and chemical production: techno-economic and environmental analysis. *Energy and Environmental Science* 15 (1): 169–184. https://doi.org/10.1039/d1ee01444g.

Dowson, G.R.M. and Styring, P. (2017). Demonstration of CO_2 conversion to synthetic transport fuel at flue gas concentrations. *Frontiers in Energy Research* 5 (OCT): 1–11. https://doi.org/10.3389/fenrg.2017.00026.

Dry, M.E. (1996). Practical and theoretical aspects of the catalytic Fischer-Tropsch process. *Applied Catalysis A: General* 138 (2): 319–344. https://doi.org/10.1016/0926-860X(95)00306-1.

Dry, M.E. and Hoogendoorn, J.C. (1981). Technology of the Fischer-Tropsch process. *Catalysis Reviews* 23 (1–2): 265–278. https://doi.org/10.1080/03602458108068078.

Elbadawi, A.H., Ge, L., Li, Z. et al. (2021). Catalytic partial oxidation of methane to syngas: review of perovskite catalysts and membrane reactors. *Catalysis Reviews - Science and Engineering* 63 (1): 1–67. https://doi.org/10.1080/01614940.2020.1743420.

Eschemann, T.O. and de Jong, K. P. (2015). Deactivation behavior of Co/TiO_2 catalysts during Fischer–Tropsch synthesis. *ACS Catalysis* 5 (6): 3181–3188. https://doi.org/10.1021/acscatal.5b00268.

Espinoza, R.L., Steynberg, A.P., Jager, B. et al. (1999). Low temperature Fischer-Tropsch synthesis from a Sasol perspective. *Applied Catalysis A: General* 186 (1–2): 13–26. https://doi.org/10.1016/S0926-860X(99)00161-1.

Faheem, H.H., Tanveer, H.U., Abbas, S.Z. et al. (2021). Comparative study of conventional steam-methane-reforming (SMR) and auto-thermal-reforming (ATR) with their hybrid sorption enhanced (SE-SMR & SE-ATR) and environmentally benign process models for the hydrogen production. *Fuel* 297 (May 2020): 120769. https://doi.org/10.1016/j.fuel.2021.120769.

Fan, D., Gao, Y., Liu, F. et al. (2021). Autothermal reforming of methane over an integrated solid oxide fuel cell reactor for power and syngas co-generation. *Journal of Power Sources* 513 (July): 230536. https://doi.org/10.1016/j.jpowsour.2021.230536.

Fazlikeshteli, S., Vendrell, X., and Llorca, J. (2023a). Bimetallic Ru–Pd supported on CeO_2 for the catalytic partial oxidation of methane into syngas. *Fuel* 334 (P2): 126799. https://doi.org/10.1016/j.fuel.2022.126799.

Fazlikeshteli, S., Vendrell, X., and Llorca, J. (2023b). Low-temperature partial oxidation of methane over Pd–Ni bimetallic catalysts supported on CeO_2. *International Journal of Hydrogen Energy* 48 (32): 12024–12035. https://doi.org/10.1016/j.ijhydene.2022.07.020.

Feng, J., Ni, J., and Pan, H. (2021). Synergistic carbon and hydrogen reactions in the electrochemical reduction of CO_2 to liquid fuels. *Journal of Materials Chemistry A* 9 (17): 10546–10561. https://doi.org/10.1039/d1ta00758k.

Fichtl, M.B., Schlereth, D., Jacobsen, N. et al. (2015). Kinetics of deactivation on Cu/ZnO/Al_2O_3 methanol synthesis catalysts. *Applied Catalysis A: General* 502: 262–270. https://doi.org/10.1016/j.apcata.2015.06.014.

Friedmann, S.J., Upadhye, R., and Kong, F.M. (2009). Prospects for underground coal gasification in carbon-constrained world. *Energy Procedia* 1 (1): 4551–4557. https://doi.org/10.1016/j.egypro.2009.02.274.

Fujiwara, M., Kieffer, R., Ando, H. et al. (1995). Development of composite catalysts made of Cu-Zn-Cr oxide/zeolite for the hydrogenation of carbon dioxide. *Applied Catalysis A, General* 121 (1): 113–124. https://doi.org/10.1016/0926-860X(95)85014-7.

Ganesh, I. (2016). Electrochemical conversion of carbon dioxide into renewable fuel chemicals – the role of nanomaterials and the commercialization. *Renewable and Sustainable Energy Reviews* 59: 1269–1297. https://doi.org/10.1016/j.rser.2016.01.026.

Gangadharan, P., Kanchi, K.C., and Lou, H.H. (2012). Evaluation of the economic and environmental impact of combining dry reforming with steam reforming of methane. *Chemical Engineering Research and Design* 90 (11): 1956–1968. https://doi.org/10.1016/j.cherd.2012.04.008.

Gao, P., Li, S., Bu, X. et al. (2017). Direct conversion of CO_2 into liquid fuels with high selectivity over a bifunctional catalyst. *Nature Chemistry* 9 (10): 1019–1024. https://doi.org/10.1038/nchem.2794.

Gao, P., Zhang, L., Li, S. et al. (2020). Novel heterogeneous catalysts for CO_2 hydrogenation to liquid fuels. *ACS Central Science* 6 (10): 1657–1670. https://doi.org/10.1021/acscentsci.0c00976.

Genovese, C., Ampelli, C., Perathoner, S. et al. (2013). Electrocatalytic conversion of CO_2 to liquid fuels using nanocarbon-based electrodes. *Journal of Energy Chemistry* 22 (2): 202–213. https://doi.org/10.1016/S2095-4956(13)60026-1.

Gong, E., Ali, S., Hiragond, C.B. et al. (2022). Solar fuels: research and development strategies to accelerate photocatalytic CO_2 conversion into hydrocarbon fuels. *Energy and Environmental Science* 15 (3): 880–937. https://doi.org/10.1039/d1ee02714j.

Groenzin, H. and Mullins, O.C. (1999). Asphaltene molecular size and structure. *The Journal of Physical Chemistry A* 103 (50): 11237–11245. https://doi.org/10.1021/jp992609w.

Gungor, A., Ozbayoglu, M., Kasnakoglu, C. et al. (2012). A parametric study on coal gasification for the production of syngas. *Chemical Papers* 66 (7): 677–683. https://doi.org/10.2478/s11696-012-0164-0.

Gür, M. and Canbaz, E.D. (2020). Analysis of syngas production and reaction zones in hydrogen oriented underground coal gasification. *Fuel* 269 (65): 117331. https://doi.org/10.1016/j.fuel.2020.117331.

Han, B., Wang, F., Zhang, L. et al. (2020). Syngas production from methane steam reforming and dry reforming reactions over sintering-resistant Ni@SiO_2 catalyst. *Research on Chemical Intermediates* 46 (3): 1735–1748. https://doi.org/10.1007/s11164-019-04060-3.

Han, J.W., Park, J.S., Choi, M.S. et al. (2017). Uncoupling the size and support effects of Ni catalysts for dry reforming of methane. *Applied Catalysis B: Environmental* 203: 625–632. https://doi.org/10.1016/j.apcatb.2016.10.069.

He, C., Wu, S., Wang, L. et al. (2022). Recent advances in photo-enhanced dry reforming of methane: a review. *Journal of Photochemistry and Photobiology C: Photochemistry Reviews* 51 (November 2021): 100468. https://doi.org/10.1016/j.jphotochemrev.2021.100468.

He, Z. and De Wilde, J. (2021). Numerical simulation of commercial scale autothermal chemical looping reforming and bi-reforming for syngas production.

Chemical Engineering Journal 417 (November 2020): 128088. https://doi.org/10.1016/j.cej.2020.128088.

Hlina, M., Hrabovský, M., Kavka, T. et al. (2014). Production of high quality syngas from argon/water plasma gasification of biomass and waste. *Waste Management* 34 (1): 63–66. https://doi.org/10.1016/j.wasman.2013.09.018.

Hoepfner, M.P., Limsakoune, V., Chuenmeechao, V. et al. (2013). A fundamental study of asphaltene deposition. *Energy and Fuels* 27 (2): 725–735. https://doi.org/10.1021/ef3017392.

Howarth, R.W. (2014). A bridge to nowhere: methane emissions and the greenhouse gas footprint of natural gas. *Energy Science and Engineering* 2 (2): 47–60. https://doi.org/10.1002/ese3.35.

Hrabovsky, M. (2011). Plasma aided gasification of biomass, organic waste and plastics. *Proceedings of the 30th ICPIG, Belfast, UK*, 28–31.

Jacobs, G., Chaney, J.A., Patterson, P.M. et al. (2004). Fischer-Tropsch synthesis: study of the promotion of Re on the reduction property of Co/Al_2O_3 catalysts by in situ EXAFS/XANES of Co K and Re L_{III} edges and XPS. *Applied Catalysis A: General* 264 (2): 203–212. https://doi.org/10.1016/j.apcata.2003.12.049.

Jahangiri, A. and Saidi, M. (2022). Syngas production via auto-thermal methane reforming using modified perovskite catalysts: performance evaluation of $La_{1-x}Gd_xNiO_{3-\delta}$ perovskite oxides. *Fuel* 317 (October 2021): 123495. https://doi.org/10.1016/j.fuel.2022.123495.

Ji, M. and Wang, J. (2021). Review and comparison of various hydrogen production methods based on costs and life cycle impact assessment indicators. *International Journal of Hydrogen Energy* 46 (78): 38612–38635. https://doi.org/10.1016/j.ijhydene.2021.09.142.

Jin, R., Chen, Y., Li, W. et al. (2000). Mechanism for catalytic partial oxidation of methane to syngas over a Ni/Al_2O_3 catalyst. *Applied Catalysis A: General* 201 (1): 71–80. https://doi.org/10.1016/S0926-860X(00)00424-5.

Kang, X., Wang, B., Hu, K. et al. (2020). Quantitative electro-reduction of CO_2 to liquid fuel over electro-synthesized metal-organic frameworks. *Journal of the American Chemical Society* 142 (41): 17384–17392. https://doi.org/10.1021/jacs.0c05913.

Khodakov, A.Y. (2009). Fischer-Tropsch synthesis: relations between structure of cobalt catalysts and their catalytic performance. *Catalysis Today* 144 (3–4): 251–257. https://doi.org/10.1016/j.cattod.2008.10.036.

Kumaravel, V., Bartlett, J., and Pillai, S.C. (2020). Photoelectrochemical conversion of carbon dioxide (CO_2) into fuels and value-added products. *ACS Energy Letters* 5 (2): 486–519. https://doi.org/10.1021/acsenergylett.9b02585.

Kuo, P.C., Illathukandy, B., Wu, W. et al. (2020). Plasma gasification performances of various raw and torrefied biomass materials using different gasifying agents. *Bioresource Technology* 314 (June): 123740. https://doi.org/10.1016/j.biortech.2020.123740.

Li, H. and Liao, J.C. (2013). Biological conversion of carbon dioxide to photosynthetic fuels and electrofuels. *Energy and Environmental Science* 6 (10): 2892–2899. https://doi.org/10.1039/c3ee41847b.

Li, K., Peng, B., and Peng, T. (2016). Recent advances in heterogeneous photocatalytic CO_2 conversion to solar fuels. *ACS Catalysis* 6 (11): 7485–7527. https://doi.org/10.1021/acscatal.6b02089.

Liang, D., Wang, Y., Wang, Y. et al. (2024). Dry reforming of methane for syngas production over noble metals modified M-Ni@S-1 catalysts (M = Pt, Pd, Ru, Au). *International Journal of Hydrogen Energy* 51: 1002–1015. https://doi.org/10.1016/j.ijhydene.2023.07.135.

Liang, Z., Zhang, Y., Zhang, G. et al. (2023). Promotion effect of different lanthanide doping on Co/Al_2O_3 catalyst for dry reforming of methane. *International Journal of Hydrogen Energy* 48 (49): 18644–18656. https://doi.org/10.1016/j.ijhydene.2023.01.301.

Loktev, A.S., Mukhin, I.E., Bykov, M.A. et al. (2022). Novel high-performance catalysts for partial oxidation and dry reforming of methane to synthesis gas. *Petroleum Chemistry* 62 (5): 526–543. https://doi.org/10.1134/S0965544122020207.

Luisetto, I., Sarno, C., De Felicis, D. et al. (2017). Ni supported on γ-Al_2O_3 promoted by Ru for the dry reforming of methane in packed and monolithic reactors. *Fuel Processing Technology* 158: 130–140. https://doi.org/10.1016/j.fuproc.2016.12.015.

Lv, P., Yuan, Z., Wu, C. et al. (2007). Bio-syngas production from biomass catalytic gasification. *Energy Conversion and Management* 48 (4): 1132–1139. https://doi.org/10.1016/j.enconman.2006.10.014.

Maier, L., Schadel, B., Herrera Delgado, K. et al. (2011). Steam reforming of methane over nickel: development of a multi-step surface reaction mechanism. *Topics in Catalysis* 54 (13–15): 845–858. https://doi.org/10.1007/s11244-011-9702-1.

Mao, Y., Zhang, L., Zheng, X. et al. (2023). Coke-resistance over Rh–Ni bimetallic catalyst for low temperature dry reforming of methane. *International Journal of Hydrogen Energy* 48 (37): 13890–13901. https://doi.org/10.1016/j.ijhydene.2022.12.299.

Maschio, G., Lucchesi, A., and Stoppato, G. (1994). Production of syngas from biomass. *Bioresource Technology* 48 (2): 119–126. https://doi.org/10.1016/0960-8524(94)90198-8.

Matamba, T., Iglauer, S., and Keshavarz, A. (2022). A progress insight of the formation of hydrogen rich syngas from coal gasification. *Journal of the Energy Institute* 105 (August): 81–102. https://doi.org/10.1016/j.joei.2022.08.001.

Matsumura, Y. and Nakamori, T. (2004). Steam reforming of methane over nickel catalysts at low reaction temperature. *Applied Catalysis A: General* 258 (1): 107–114. https://doi.org/10.1016/j.apcata.2003.08.009.

Molino, A., Chianese, S., and Musmarra, D. (2016). Biomass gasification technology: the state of the art overview. *Journal of Energy Chemistry* 25 (1): 10–25. https://doi.org/10.1016/j.jechem.2015.11.005.

Moreno-Arciniegas, L.-S., Rodríguez-Corredor, F.-E., Afanador-Rey, L.-E. et al. (2009). Syngas obtainment from the gasification of asphaltenes of the San Fernando crude oil. *CT&F - Ciencia, Tecnología y Futuro* 3: 189–204.

Moschopedis, S.E., Fryer, J.F., and Speight, J.G. (1976). Investigation of asphaltene molecular weights. *Fuel* 55 (3): 227–232. https://doi.org/10.1016/0016-2361(76)90093-4.

Movasati, A., Alavi, S.M., and Mazloom, G. (2019). Dry reforming of methane over CeO_2-$ZnAl_2O_4$ supported Ni and Ni-Co nano-catalysts. *Fuel* 236 (July 2017): 1254–1262. https://doi.org/10.1016/j.fuel.2018.09.069.

Murmura, M.A., Diana, M., Spera, R. et al. (2016). Modeling of autothermal methane steam reforming: comparison of reactor configurations. *Chemical Engineering and Processing: Process Intensification* 109: 125–135. https://doi.org/10.1016/j.cep.2016.08.019.

Naidu, B.N., Kumar, K.D.P.L., Saini, H. et al. (2022). Coke deposition over Ni-based catalysts for dry reforming of methane: effects of MgO-Al_2O_3 support and ceria, lanthana promoters. *Journal of Environmental Chemical Engineering* 10 (1): 106980. https://doi.org/10.1016/j.jece.2021.106980.

Nezhad, M.Z., Rowshanzamir, S., and Eikani, M.H. (2009). Autothermal reforming of methane to synthesis gas: modeling and simulation. *International Journal of Hydrogen Energy* 34 (3): 1292–1300. https://doi.org/10.1016/j.ijhydene.2008.11.091.

Nisar, A., Khan, S., Hameed, M. et al. (2021). Bio-conversion of CO_2 into biofuels and other value-added chemicals via metabolic engineering. *Microbiological Research* 251 (June): 126813. https://doi.org/10.1016/j.micres.2021.126813.

Okoye-Chine, C.G., Otun, K., Shiba, N. et al. (2022). Conversion of carbon dioxide into fuels – a review. *Journal of CO_2 Utilization* 62 (May): 102099. https://doi.org/10.1016/j.jcou.2022.102099.

Olah, G.A., Prakash, G.K.S., and Goeppert, A. (2011). Anthropogenic chemical carbon cycle for a sustainable future. *Journal of the American Chemical Society* 133 (33): 12881–12898. https://doi.org/10.1021/ja202642y.

Oni, A.O., Anaya, K., Giwa, T. et al. (2022). Comparative assessment of blue hydrogen from steam methane reforming, autothermal reforming, and natural gas decomposition technologies for natural gas-producing regions. *Energy Conversion and Management* 254: 115245. https://doi.org/10.1016/j.enconman.2022.115245.

Osman, M., Zaabout, A., Cloete, S.W.P. et al. (2021). Pressurized chemical looping methane reforming to syngas for efficient methanol production: experimental and process simulation study. *Advances in Applied Energy* 4 (April): 100069. https://doi.org/10.1016/j.adapen.2021.100069.

Overa, S., Ko, B.H., Zhao, Y. et al. (2022). Electrochemical approaches for CO_2 conversion to chemicals: a journey toward practical applications. *Accounts of Chemical Research* 55 (5): 638–648. https://doi.org/10.1021/acs.accounts.1c00674.

Palmer, C., Upham, D.C., Smart, S. et al. (2020). Dry reforming of methane catalysed by molten metal alloys. *Nature Catalysis* 3 (1): 83–89. https://doi.org/10.1038/s41929-019-0416-2.

Panzone, C., Philippe, R., Chappaz, A. et al. (2020). Power-to-liquid catalytic CO_2 valorization into fuels and chemicals: focus on the Fischer-Tropsch route. *Journal of CO_2 Utilization* 38 (January): 314–347. https://doi.org/10.1016/j.jcou.2020.02.009.

Paris Agreement (2015). UNFCCC, COP report no. 21. https://unfccc.int/process-and-meetings/the-paris-agreement (accessed 12 December 2015).

Park, H.R., Pawar, A.U., Pal, U. et al. (2021). Enhanced solar photoreduction of CO_2 to liquid fuel over rGO grafted NiO-CeO_2 heterostructure nanocomposite. *Nano Energy* 79 (July 2020): 105483. https://doi.org/10.1016/j.nanoen.2020.105483.

Pei, P., Korom, S.F., Ling, K. et al. (2016). Cost comparison of syngas production from natural gas conversion and underground coal gasification. *Mitigation and Adaptation Strategies for Global Change* 21 (4): 629–643. https://doi.org/10.1007/s11027-014-9588-x.

Pham, T.T.P., Ro, K.S., Chen, L. et al. (2020). Microwave-assisted dry reforming of methane for syngas production: a review. *Environmental Chemistry Letters* 18: 1987–2019. https://doi.org/10.1007/s10311-020-01055-0.

Pomilla, F.R., Brunetti, A., Marcì, G. et al. (2018). CO_2 to liquid fuels: photocatalytic conversion in a continuous membrane reactor. *ACS Sustainable Chemistry and Engineering* 6 (7): 8743–8753. https://doi.org/10.1021/acssuschemeng.8b01073.

Prieto, G. (2017). Carbon dioxide hydrogenation into higher hydrocarbons and oxygenates: thermodynamic and kinetic bounds and progress with heterogeneous and homogeneous catalysis. *ChemSusChem* 10 (6): 1056–1070. https://doi.org/10.1002/cssc.201601591.

Rahbari, A., Ramdin, M., van den Broeke, L.J.P. et al. (2018). Combined steam reforming of methane and formic acid to produce syngas with an adjustable H_2:CO ratio. *Industrial and Engineering Chemistry Research* 57 (31): 10663–10674. https://doi.org/10.1021/acs.iecr.8b02443.

Raju, A.S.K., Park, C.S., and Norbeck, J.M. (2009). Synthesis gas production using steam hydrogasification and steam reforming. *Fuel Processing Technology* 90 (2): 330–336. https://doi.org/10.1016/j.fuproc.2008.09.011.

Ranjekar, A.M. and Yadav, G.D. (2021). Dry reforming of methane for syngas production: a review and assessment of catalyst development and efficacy. *Journal of the Indian Chemical Society* 98 (1): 100002. https://doi.org/10.1016/j.jics.2021.100002.

Rao, V.U.S., Stiegel, V.J., Cinquegrane, G.J. et al. (1992). Iron-based catalysts for slurry-phase Fischer-Tropsch process: technology review. *Fuel Processing Technology* 30 (1): 83–107. https://doi.org/10.1016/0378-3820(92)90077-4.

Reddy, K.P., Kim, D., Hong, S. et al. (2022). Tuning CO_2 hydrogenation selectivity through reaction-driven restructuring on Cu-Ni bimetal catalysts. *ACS Applied Materials and Interfaces* 15 (7): 9373–9381. https://doi.org/10.1021/acsami.2c20832.

Remiro-Buenamañana, S. and García, H. (2019). Photoassisted CO_2 conversion to fuels. *ChemCatChem* 11 (1): 342–356. https://doi.org/10.1002/cctc.201801409.

Requies, J., Cabrero, M.A., Barrio, V.L. et al. (2005). Partial oxidation of methane to syngas over Ni/MgO and Ni/La_2O_3 catalysts. *Applied Catalysis A: General* 289 (2): 214–223. https://doi.org/10.1016/j.apcata.2005.05.002.

Rui, N., Sun, K., Shen, C. et al. (2020). Density functional theoretical study of Au_4/In_2O_3 catalyst for CO_2 hydrogenation to methanol: the strong metal-support interaction and its effect. *Journal of CO_2 Utilization* 42: 101313. https://doi.org/10.1016/j.jcou.2020.101313.

Ruiz, J.A.C., Passos, F.B., Bueno, J.M.C. et al. (2008). Syngas production by autothermal reforming of methane on supported platinum catalysts. *Applied Catalysis A: General* 334 (1–2): 259–267. https://doi.org/10.1016/j.apcata.2007.10.011.

Sakakura, T., Choi, J., and Yasuda, H. (2007). Transformation of carbon dioxide. *Chemical Reviews* 107 (6): 2365–2387. https://doi.org/10.1021/cr068357u.

Salhi, N., Boulahouache, A., Petit, C. et al. (2011). Steam reforming of methane to syngas over $NiAl_2O_4$ spinel catalysts. *International Journal of Hydrogen Energy* 36 (17): 11433–11439. https://doi.org/10.1016/j.ijhydene.2010.11.071.

Satterfield, C.N., Hanlon, R.T., Tung, S.E. et al. (1986). Effect of water on the iron-catalyzed Fischer-Tropsch synthesis. *Industrial & Engineering Chemistry Product Research and Development* 25 (1): 407–414. https://doi.org/10.1021/i300023a007.

Schanke, D., Hilmen, A.M., Bergene, E. et al. (1995). Study of the deactivation mechanism of Al_2O_3-supported cobalt Fischer-Tropsch catalysts. *Catalysis Letters* 34 (3–4): 269–284. https://doi.org/10.1007/BF00806876.

Schulz, H. and Claeys, M. (1999). Reactions of α-olefins of different chain length added during Fischer–Tropsch synthesis on a cobalt catalyst in a slurry reactor. *Applied Catalysis A: General* 186 (1–2): 71–90. https://doi.org/10.1016/S0926-860X(99)00165-9.

Sheu, E.Y. (2002). Petroleum asphaltene – properties, characterization, and issues. *Energy & Fuels* 16 (1): 74–82. https://doi.org/10.1021/ef010160b.

Shi, T., Liu, S., and Zhang, Y.H.P.J. (2019). CO_2 fixation for malate synthesis energized by starch via in vitro metabolic engineering. *Metabolic Engineering* 55 (May): 152–160. https://doi.org/10.1016/j.ymben.2019.07.005.

Shi, Z., Yang, H., Gao, P. et al. (2018). Direct conversion of CO_2 to long-chain hydrocarbon fuels over K-promoted $CoCu/TiO_2$ catalysts. *Catalysis Today* 311 (August 2017): 65–73. https://doi.org/10.1016/j.cattod.2017.09.053.

Simsek, E., Avci, A.K., and Önsan, Z.I. (2011). Investigation of catalyst performance and microstructured reactor configuration for syngas production by methane steam reforming. *Catalysis Today* 178 (1): 157–163. https://doi.org/10.1016/j.cattod.2011.08.021.

Singh, S., Jain, S., Venkateswaran, P.S. et al. (2015). Hydrogen: a sustainable fuel for future of the transport sector. *Renewable and Sustainable Energy Reviews* 51: 623–633. https://doi.org/10.1016/j.rser.2015.06.040.

Song, Y., Ozdemir, E., Ramesh, S. et al. (2020). Dry reforming of methane by stable Ni–Mo nanocatalysts on single-crystalline MgO. *Science* 367 (6479): 777–781. https://doi.org/10.1126/science.aav2412.

Speight, J.G. (ed.) (2020). Hydrocarbons from synthesis gas. In: *Handbook of Industrial Hydrocarbon Processes*, 2e, 343–386. Elsevier. https://doi.org/10.1016/B978-0-12-809923-0.00008-4.

Steynberg, A. and Dry, M. (eds.) (2004). *Fischer-Tropsch Technology*. Amsterdam, Boston: Elsevier.

Steynberg, A.P. (2004). Chapter 1 - Introduction to Fischer-Tropsch technology. In: *Studies in Surface Science and Catalysis* (ed. A. Steynberg and M. Dry), 152: 1–63. Elsevier. https://doi.org/10.1016/S0167-2991(04)80458-0.

Tada, S. and Satokawa, S. (2018). Effect of Ag loading on CO_2-to-methanol hydrogenation over $Ag/CuO/ZrO_2$. *Catalysis Communications* 113 (February): 41–45. https://doi.org/10.1016/j.catcom.2018.05.009.

Tada, S., Watanabe, F., Kiyota, K. et al. (2017). Ag addition to CuO-ZrO_2 catalysts promotes methanol synthesis via CO_2 hydrogenation. *Journal of Catalysis* 351: 107–118. https://doi.org/10.1016/j.jcat.2017.04.021.

Tamošiūnas, A., Gimžauskaitė, D., Aikas, M. et al. (2023). Biomass gasification to syngas in thermal water vapor arc discharge plasma. *Biomass Conversion and Biorefinery* 13 (18): 16373–16384. https://doi.org/10.1007/s13399-023-03828-3.

Tang, S., Lin, J., and Tan, K.L. (1998). Partial oxidation of methane to syngas over Ni/MgO, Ni/CaO and Ni/CeO_2. *Catalysis Letters* 51: 169–175. https://doi.org/10.1023/A:1019034412036.

Teh, L.P., Setiabudi, H.D., Timmiati, S.N. et al. (2021). Recent progress in ceria-based catalysts for the dry reforming of methane: a review. *Chemical Engineering Science* 242: 116606. https://doi.org/10.1016/j.ces.2021.116606.

Usman, M., Humayun, M., Garba, M.D. et al. (2021). Electrochemical reduction of CO_2: a review of cobalt based catalysts for carbon dioxide conversion to fuels. *Nanomaterials* 11 (8): 2029. https://doi.org/10.3390/nano11082029.

Wallace, P., Jonson, K., and Thacker, P. (2001). Integration of solvent deasphalting and gasification. US, 6241874, 05/06/2001.

Wang, G., He, C.-T., Huang, R. et al. (2020). Photoinduction of Cu single atoms decorated on UiO-66-NH_2 for enhanced photocatalytic reduction of CO_2 to liquid fuels. *Journal of the American Chemical Society* 142 (45): 19339–19345. https://doi.org/10.1021/jacs.0c09599.

Wang, W., Jiang, X., Wang, X. et al. (2018). Fe–Cu bimetallic catalysts for selective CO_2 hydrogenation to olefin-rich C_2+ hydrocarbons. *Industrial & Engineering Chemistry Research* 57 (13): 4535–4542. https://doi.org/10.1021/acs.iecr.8b00016.

Wang, W., Wang, S., Ma, X. et al. (2011). Recent advances in catalytic hydrogenation of carbon dioxide. *Chemical Society Reviews* 40 (7): 3703. https://doi.org/10.1039/c1cs15008a.

Wang, X., Yang, G., Zhang, J. et al. (2016). Synthesis of isoalkanes over a core (Fe–Zn–Zr)–shell (zeolite) catalyst by CO_2 hydrogenation. *Chemical Communications* 52 (46): 7352–7355. https://doi.org/10.1039/C6CC01965J.

Wang, Y., Li, Y., Liu, J. et al. (2021). $BiPO_4$-derived 2D nanosheets for efficient electrocatalytic reduction of CO_2 to liquid fuel. *Angewandte Chemie* 133 (14): 7759–7763. https://doi.org/10.1002/ange.202014341.

Wei, J., Ge, Q., Yao, R. et al. (2017). Directly converting CO_2 into a gasoline fuel. *Nature Communications* 8 (1): 15174. https://doi.org/10.1038/ncomms15174.

Wei, J., Sun, J., Wen, Z. et al. (2016). New insights into the effect of sodium on Fe_3O_4-based nanocatalysts for CO_2 hydrogenation to light olefins. *Catalysis Science & Technology* 6 (13): 4786–4793. https://doi.org/10.1039/C6CY00160B.

Xu, X., Xie, K., Hu, J. et al. (2023). The metal-organic frameworks as unique platform for photocatalytic CO_2 conversion to liquid fuels. *Journal of Environmental Chemical Engineering* 11 (5): 110424. https://doi.org/10.1016/j.jece.2023.110424.

Yang, J., Ma, W., Chen, D. et al. (2014). Fischer-Tropsch synthesis: a review of the effect of CO conversion on methane selectivity. *Applied Catalysis A: General* 470: 250–260. https://doi.org/10.1016/j.apcata.2013.10.061.

Yao, B., Ma, W., Gonzalez-Cortes, S. et al. (2017). Thermodynamic study of hydrocarbon synthesis from carbon dioxide and hydrogen. *Greenhouse Gases: Science and Technology* 7(5): 942–957. https://doi.org/10.1002/ghg.1694.

Yao, B., Xiao, T., Makgae, O.A. et al. (2020). Transforming carbon dioxide into jet fuel using an organic combustion-synthesized Fe-Mn-K catalyst. *Nature Communications* 11 (1): 6395. https://doi.org/10.1038/s41467-020-20214-z.

Yentekakis, I.V., Panagiotopoulou, P., and Artemakis, G. (2021). A review of recent efforts to promote dry reforming of methane (DRM) to syngas production via bimetallic catalyst formulations. *Applied Catalysis B: Environmental* 296 (February): 120210. https://doi.org/10.1016/j.apcatb.2021.120210.

York, A.P.E., Xiao, T., and Green, M.L.H. (2003). Brief overview of the partial oxidation of methane to synthesis gas. *Topics in Catalysis* 22 (3–4): 345–358. https://doi.org/10.1023/A:1023552709642.

Yung, M.M., Jablonski, W.S., and Magrini-Bair, K.A. (2009). Review of catalytic conditioning of biomass-derived syngas. *Energy & Fuels* 23 (4): 1874–1887. https://doi.org/10.1021/ef800830n.

Zahid, U., Khalafalla, S.S., Alibrahim, H.A. et al. (2023). Techno-economic evaluation of simultaneous methanol and hydrogen production via autothermal reforming of natural gas. *Energy Conversion and Management* 296: 117681. https://doi.org/10.1016/j.enconman.2023.117681.

Zhang, H., Sun, Z., and Hu, Y.H. (2021). Steam reforming of methane: current states of catalyst design and process upgrading. *Renewable and Sustainable Energy Reviews* 149: 111330. https://doi.org/10.1016/j.rser.2021.111330.

Zhao, Y., Liu, X., Liu, Z. et al. (2021). Spontaneously Sn-doped Bi/BiOx core-shell nanowires toward high-performance CO_2 electroreduction to liquid fuel. *Nano Letters* 21 (16): 6907–6913. https://doi.org/10.1021/acs.nanolett.1c02053.

Zhou, W., Cheng, K., Kang, J. et al. (2019). New horizon in C1 chemistry: breaking the selectivity limitation in transformation of syngas and hydrogenation of CO_2 into hydrocarbon chemicals and fuels. *Chemical Society Reviews* 48 (12): 3193–3228. https://doi.org/10.1039/c8cs00502h.

Żogała, A. (2014). Equilibrium simulations of coal gasification – factors affecting syngas composition. *Journal of Sustainable Mining* 13 (2): 30–38. https://doi.org/10.7424/jsm140205.

13

Syngas Production Beyond Petroleum Feedstock Toward LCA

*Amol A. Inamdar, Francisco O. Aguirre, Olubayo O. Oginni, Heidi A. Taboada, and Lucy Mar Camacho**

* *Department of Environmental Engineering, Texas A&M University-Kingsville, Kingsville, TX, USA*

13.1 Introduction

Syngas, a synthesis gas, is a useful and important feedstock for producing chemicals, including ammonia for fertilizer, urea, and methanol as a solvent and chemical intermediary. Syngas is also an important feedstock for the production of synthetic fuels, including diesel by the Fisher–Tropsch process, and gasoline. Syngas is a mixture of carbon monoxide and hydrogen obtained from methane, a petrochemical product from the refinery of crude oil, and steam (Devold, 2013). The most common process for producing syngas is steam methane reforming (SMR). Other primary processes include carbon dioxide, methane reforming (dry reforming), and partial oxidation methane reforming (POM).

The SMR process produces 3 moles of hydrogen per mole of methane (3:1) compared to the POM and dry reforming, which produce a H_2/CO ratio of 2:1 and 1:1, respectively. This indicates that dry reforming, which can be carried out homogenously or catalytically, is the most energy efficient of the three processes (Hoek and Kersten, 2004). However, the amount of hydrogen production can increase in the SMR process using the water–gas shift (WGS) reaction. In this reaction, carbon monoxide reacts with steam and produces additional carbon dioxide and hydrogen.

The dry reforming process consumes two molecules of carbon monoxide, a greenhouse gas (GHG); however, it suffers from rapid catalyst deactivation due to carbon deposition (Hu and Ruckenstein, 2004). In addition to being an important

Cleaner Petroleum Production and Refining Technologies, First Edition, Edited by M. R. Riazi and Harvey W. Yarranton.

chemical feedstock, syngas is pivotal in advancing sustainable energy systems, including power-to-gas (PtG) systems, to enable the conversion of renewable electricity into chemical energy (Benarimo et al., 2023).

13.2 Other Syngas Production Pathways

New routes have been proposed to obtain syngas. They use non-refinery raw materials and may include thermochemical processes, such as gasification, pyrolysis, and reforming, and the biochemical processes of fermentation and anaerobic digestion (AD). Thermochemical technologies involve the use of thermal (heat) and a catalyst to convert carbonaceous materials into syngas. Gasification uses partial oxidation to convert solid or liquid carbon-rich feedstocks, such as wood chips, switchgrass, and corn stover, into syngas (Cormos, 2011; Hreiz et al., 2011). Key factors influencing the conversion efficiency in gasification include temperature, pressure, gas composition, and the heat source (Bridgwater, 2003), with gasifying agents such as air, steam, or CO_2 (Rey et al., 2024b). Pyrolysis is one of the emerging technologies used to generate biogas or syngas biofuel from biomass. The biofuel is obtained from the conversion of biomass by thermal decomposition at 400–800 °C in the absence of oxygen (Beneroso and Fernandez, 2016). High temperature, low heating rates, and extended gas residence times are often needed to generate a high quantity of syngas during pyrolysis. The pyrolysis process is often used for difficult biodegradable waste biomass such as digestate, wood biomass, and agricultural and municipal waste (Lee et al., 2019).

Biochemical techniques involve the use of biological processes to convert organic materials into syngas or intermediate gases that can be upgraded to syngas. The process of producing biogas from waste materials without oxygen is known as AD, or biomethanation (Sharma et al., 2020). Methane and carbon dioxide (CO_2) are the main constituents of the resultant biogas, with methane accounting for around 60% of the mixture. Syngas generated often consists of 50–70% CH_4, 30–50% CO_2, and other smaller impurities (H_2O, H_2S, N_2, volatile compounds (VOCs) or siloxanes. The synergistic integration of thermochemical and biochemical processes such as pyrolysis and AD has the potential to significantly enhance energy recovery from organic waste materials (Andreides et al., 2022). The combination offers a promising strategy for large-scale conversion of waste biomass into valuable products.

The typical composition and ratio of the main gases generated by these processes, including carbon monoxide (CO), hydrogen (H_2), methane (CH_4), carbon dioxide (CO_2), and steam (H_2O), can change based on the feedstock (Ruiz et al., 2013; Ongen et al., 2016; Plis and Wilk, 2011; Aydin et al., 2019). Other hydrocarbons that may be generated in addition to these main constituents include ethane

(C_2H_6) and ethylene (C_2H_4). These lighter hydrocarbons are frequently produced as byproducts of catalytic processes or during the thermal breakdown of organic materials, and they may be found in lower amounts (Khlifi et al., 2022; Luo et al., 2012). Additionally, propane (C_3H_8), a three-carbon alkane that can be produced during specific cracking processes or as a minor feedstock component, may be present in syngas (Ongen et al., 2016; Luo et al., 2012). Acetylene (C_2H_2), a very reactive two-carbon molecule, is another potential component. Acetylene can contribute to the variety of chemicals in the syngas mixture by forming under certain high-temperature settings or during incomplete combustion processes (Zheng et al., 2016; Kruesi et al., 2013).

13.3 Valuable Syngas Subproducts and Their Environmental Footprint

Syngas is a critical feedstock for industrial processes. Beyond fuel production, syngas generates valuable subproducts, including ammonia, crucial in producing nitrogen-based fertilizers, industrial chemicals, and methanol.

13.3.1 Ammonia Production

Ammonia is pivotal in producing nitrogen-based fertilizers and various industrial chemicals, impacting agricultural productivity and industrial processes. Faria (2021) underlined that 80% of global ammonia production supports fertilizer manufacturing. The Haber–Bosch process has substantially improved crop yields, contributing to food security. Ammonia synthesis via the Haber–Bosch process traditionally relies on hydrogen from syngas produced through fossil fuel methods such as SMR (Figure 13.1) (Ikpe et al., 2024). This energy-intensive process generates significant carbon emissions. Alternatives are being explored to make ammonia production more sustainable. Keller and Sharma (2023) introduced a process utilizing ammonia cracking and reverse water–gas shift (rWGS) reactions to produce syngas from ammonia and CO_2, presenting a more energy-efficient approach. Nemmour et al. (2023) explored renewable hydrogen via water electrolysis, potentially replacing syngas-derived hydrogen for ammonia production, significantly reducing its carbon footprint. Bianchi and Bosio (2024) investigate solid oxide cells to produce green ammonia using renewable energy, decarbonizing the Haber–Bosch process. Erfani et al. (2024) reviewed reactor design and gas separation improvements for ammonia synthesis to increase efficiency and lower emissions. Pinzón et al. (2024) highlighted ammonia's growing role as a hydrogen carrier in green ammonia production, positioning it as a key component in decarbonized energy systems.

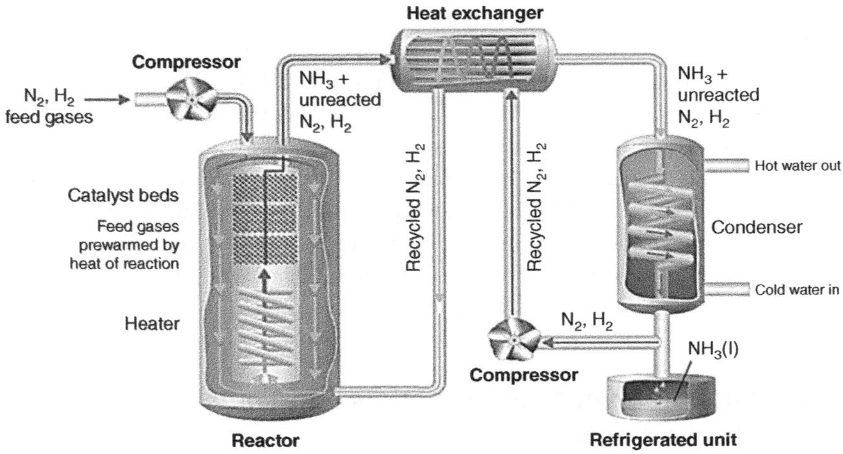

Figure 13.1 Typical flowsheet for ammonia synthesis via Haber–Bosch process Ikpe et al. (2024)/with permission of IKSAD Publications.

One of the concerns of producing ammonia from syngas is the environmental impact that it may cause for being an energy-intensive process and CO_2 generator, and different alternatives have been proposed to reduce this impact. Chen et al. (2022) reported that natural gas-based ammonia production results in high GHG emissions in China's fertilizer sector. Galusnyak et al. (2023) compared green and blue ammonia production, highlighting the reduced emissions of green ammonia despite its higher energy consumption. Pozo et al. (2023) explored the potential for net-negative emissions through coal-biomass co-gasification combined with carbon capture technologies. Bennett and Serrenho (2024) discussed how SMR generates substantial CO_2 emissions per ton of ammonia produced, encouraging the transition to renewable hydrogen to reduce the environmental impact. Khademi and Lotfi-Varnoosfaderani (2021) proposed using glycerol steam reforming as a renewable hydrogen source for ammonia synthesis, reducing GHG emissions.

Zheng et al. (2022) suggested bio-oil gasification as a renewable syngas source, providing a lower-carbon route to ammonia production. Zhou et al. (2022) presented a coal-based ammonia synthesis process that reduces carbon emissions by integrating efficient CO_2 capture. Atsbha et al. (2023) explored ammonia recovery from wastewater as a sustainable fertilizer source, emphasizing circular economy benefits. Hassan et al. (2024) focused on urea-coated fertilizers, demonstrating how ammonia-based formulations enhance nitrogen use efficiency and reduce GHG emissions. Irfan et al. (2024) compared conventional ammonia production methods with biomass gasification and electrolysis-based green ammonia production. Their findings highlight the importance of renewable hydrogen sources for decarbonizing ammonia synthesis. Joseph et al. (2024) addressed the technical and ecological challenges of scaling up green ammonia as a carbon-neutral energy carrier. Together, these studies underscore the need to reduce ammonia's energy intensity and carbon footprint by transitioning to renewable production methods.

13.3.2 Production of Methanol and Other Subproducts

Likewise, methanol synthesis has seen significant advancements in reducing its environmental footprint through testing and implementing other routes different from the traditional syngas-based production. Wolday and Ramteke (2024) emphasized optimizing the process through surrogate models to minimize energy use and CO_2 emissions, highlighting improved resource efficiency. Lozano and Lozano (2018) explored using biomass-derived syngas, showing that renewable feedstocks like agricultural residues can significantly lower CO_2 emissions. Kim et al. (2022) presented carbon-neutral pathways using hydrogen and CO_2 electrolysis, further reducing emissions. Finally, Liu et al. (2024) advocated incorporating biomass in coal-based methanol production, lowering environmental impact while improving energy efficiency.

Various other byproducts from syngas production, such as urea, play a significant role in industrial applications. Lin and Wu (2018) highlighted using biobutanol for hydrogen-rich syngas production, demonstrating that urea can enhance syngas yield while lowering costs. Wu and Lin (2018) explored the autothermal reforming of biodiesel mixed with aqueous urea, achieving high syngas efficiency with reduced sulfur content. Wang et al. (2021) investigated the gasification of urea-formaldehyde plastics in supercritical water, producing hydrogen-rich syngas and delivering a sustainable solution for plastic waste management.

13.4 Feedstock Alternatives for Syngas Production

Different feedstocks have been proposed as alternatives for syngas production. Some of them offer increased energy efficiency and reduced carbon footprint. Biomass feedstocks such as municipal solid waste (MSW), corn straw, wheat straw, forest residues, and fossil-based feedstocks like coal and natural gas are extensively utilized for their unique properties and environmental implications. Biomass options offer renewable and sustainable pathways, with MSW improving energy density through pretreatment and forest residues contributing to renewable natural gas (RNG) production. Similarly, agricultural residues like wheat and corn straw excel in bioenergy applications due to their lignocellulosic composition.

On the other hand, fossil-based feedstocks like coal and natural gas remain crucial for syngas production, with advancements in carbon capture and hybrid gasification technologies enhancing their environmental compatibility.

13.4.1 Biomass Feedstocks and Fossil-based Feedstocks

MSW, when pretreated via torrefaction, improves its calorific value and energy density by reducing moisture content and altering the H/C and O/C ratios, making it a feasible renewable energy source (Abdulyekeen et al., 2021). Gasification of MSW blended with biomass enhances sustainable waste management by integrating biochemical and thermochemical processes, improving energy recovery, and reducing ecological impacts (Hameed et al., 2021). MSW also significantly contributes to bioethanol and RNG production when combined with lignocellulosic feedstocks, minimizing landfill use and GHG emissions, aligning with circular economy goals (Shakelly et al., 2023; Akbari and Kumar, 2024). Corn biomass, including straw, cobs, and starch, offers immense potential for whole-crop biorefinery applications, such as producing valuable bioproducts like pullulan. Adaptive laboratory evolution has improved pullulan production and its molecular weight using corn straw hydrolysates, demonstrating its viability as a renewable industrial feedstock (He et al., 2023). Additionally, pyrolysis of corn residues yields bio-oil with high calorific value, making it suitable for energy and chemical industries.

Wheat straw, a globally abundant lignocellulosic feedstock, is promising for bioenergy production. Pretreatment with KOH at 2% concentration significantly improves biodegradability and enhances biogas yields in AD due to its high cellulose and hemicellulose content (Rani et al., 2022). Co-pyrolysis of wheat straw with plastics like PET enhances oil yield and heating value while reducing oxygen content in pyrolysis oil, making it suitable for high-quality biofuels (Anis Kumar et al., 2024). Forest residues provide a cost-effective pathway for RNG production through gasification. They are competitive in yield and production cost compared to other feedstocks like corn stover and contribute to decarbonization by leveraging carbon credit incentives (Akbari and Kumar, 2024).

Coal gasification for syngas production uses steam gasification in circulating fluidized bed reactors at 870 °C and 10 bar, yielding syngas rich in CO and H_2, suitable for methanation and synthesis applications. WGS reactions with Fe—Cr and Cu—Zn-based catalysts efficiently adjust the H_2/CO ratio, enhancing syngas stability under varying conditions (Akbari and Kumar, 2024). Co-gasification of coal with biomass in a Top-Lit Updraft (TLUD) gasifier improves carbon conversion efficiency and reduces environmental impact compared to coal alone (Coronel et al., 2022). Integrating carbon capture technologies, such as the Selexol process, further enhances sustainability by reducing CO_2 emissions while maintaining high syngas quality. Natural gas reforming to syngas involves advanced technologies like TREMP, which utilizes nickel-based catalysts to achieve methane contents of around 93%. Syngas derived from coal-biomass blends can supplement natural gas in industrial burners with up to 15% syngas content without requiring significant modifications (Coronel et al., 2022). Hybrid gasification technologies integrate thermochemical and biochemical approaches to reduce emissions and enhance syngas quality. Additionally, retrofitting existing gasification plants with modern catalysts and process improvements enhances efficiency and reduces production costs. Utilizing coal and natural gas alongside renewable blends, such as biomass or MSW, further reduces environmental impacts by leveraging circular economy principles (Hameed et al., 2021).

13.4.2 Influence of Feedstock on Environmental Performance

The use of crop residues like wheat straw and corn stover for biochar production significantly reduces GHG emissions due to carbon sequestration during slow pyrolysis. Studies show that biochar has emission factors ranging from −1.10 to 0.68 tons CO_2 equivalent per ton, depending on the feedstock and carbonization method (Kavindi et al., 2025). However, the environmental impact depends on raw material prices and carbon capture efficiency. Methanation processes using natural gas achieve high methane yields but rely on nickel-based catalysts and significant energy input, leading to moderate emissions reductions (Coronel et al., 2022; Robbins et al. 2024). Increased pyrolysis temperatures reduce biochar yield but increase water use due to higher energy consumption. Moisture-rich feedstocks require extensive drying, further escalating pretreatment water use (Alherbawi et al., 2024) – natural gas reforming processes like dry methane reforming exhibit a lower water footprint than coal-based technologies.

However, integrating carbon capture technologies in fossil feedstock-based processes adds to water use, primarily during amine scrubbing and other separation stages (Cao et al., 2022). Pretreatment methods like torrefaction improve calorific value and reduce energy loss during thermochemical conversion. Hybrid systems combining biochemical and thermochemical pathways enhance energy recovery (Kavindi et al., 2025). Electrification of reactors in natural gas reforming, such as rWGS, increases energy efficiency when renewable electricity is used. However, energy-intensive processes like SMR depend on high operating temperatures and fossil fuel-based heat sources, reducing overall efficiency (Cao et al., 2022; Robbins et al., 2024).

These findings underline that feedstock choice is a critical determinant of environmental performance. Biomass feedstocks generally offer significant sustainability benefits, especially when integrated with advanced pretreatment and carbon capture technologies. Fossil-based feedstocks can achieve moderate environmental improvements with carbon capture and reactor electrification innovations but remain constrained by their inherently high carbon footprint.

13.5 Process Optimizations to Improve Environmental Performance

Syngas production, whether from biomass, fossil fuels, or waste materials, involves energy-rigorous processes that generate emissions, waste, and byproducts. Optimizing syngas production processes is essential for minimizing environmental impact and improving energy efficiency and resource utilization (Hsieh et al., 2018). The success of these optimizations can be measured through several key performance indicators (KPIs), including energy efficiency, GHG emissions, water consumption, waste generation, and economic feasibility. For example, a process that achieves higher syngas yield with lower energy input, reduced carbon emissions, and efficient byproduct utilization is considered superior. Comparative metrics such as global warming potential (GWP), water footprint, and energy return on investment (EROI) are often used to evaluate and benchmark different production methods (Guang Li et al., 2021; Khlifi et al., 2024).

13.5.1 Technological Innovations

Technological progressions in syngas production are essential for improving energy efficiency and reducing the environmental burden associated with gasification and reforming processes. Advancements have been focused on biofeedstocks to compensate for the environmental limitations of producing syngas from fossil-based feedstocks. Modern gasification technologies, such as chemical looping gasification (CLG), use oxygen carriers to provide a cleaner and more energy-efficient conversion of biomass into syngas. Unlike traditional methods that rely on molecular oxygen, CLG reduces energy consumption and emissions while simultaneously enhancing tar removal (Guang Li et al., 2021). Advanced reactor designs, such as dual fluidized beds and plasma-assisted gasifiers, further optimize heat transfer and reaction rates, achieving higher syngas yields and fewer impurities (Rey et al., 2024a). Innovations in catalysts have also transformed syngas production. Bifunctional catalysts, including core-shell structures, enhance the selectivity and efficiency of Fischer–Tropsch synthesis and methanol production. For example, zeolite-supported catalysts can significantly improve the conversion rates of CO and H_2 into desired products while minimizing byproduct formation (Mahmoudi et al., 2024). Carbon capture and utilization (CCU) are increasingly integrated into syngas processes. By capturing CO_2 directly from gasification reactors and reformers, these systems not only mitigate GHG emissions but also produce high-purity CO_2 for industrial applications, such as improved oil recovery and chemical synthesis (Fu et al., 2024).

13.5.2 Waste Minimization and Byproduct Utilization

Reducing waste, such as tar and particulate, and maximizing byproduct utilization and valorization, including CO_2 capture, are critical for improving the sustainability of syngas production. Tar removal remains a challenge in biomass gasification. Advanced cleaning systems, such as dry gas cleanup technologies and microwave-based tar cracking, minimize tar formation and convert these impurities into valuable syngas components (Rey et al., 2024a). The adoption of biochar-based catalysts in tar management offers a sustainable alternative to traditional metal-based catalysts, reducing environmental impacts by up to 93% (Frazier et al., 2015). Capturing CO_2 from syngas systems enables its use in producing chemicals such as urea or fuels like methanol. CO_2 can also serve as a feedstock for algae cultivation, which in turn produces bio-oils and other high-value bioproducts (Ayodele et al., 2019). These strategies not only reduce emissions but also contribute to a circular economy. Syngas byproducts, including ammonia, produced via the Haber–Bosch process, as well as methanol, are essential precursors in the chemical industry. Ammonia serves as a critical input for fertilizers and hydrogen storage (Khalid et al., 2024). Biochar, another byproduct, can be used as a soil amendment to improve carbon sequestration and reduce reliance on synthetic fertilizers (Frazier et al., 2015).

13.5.3 Renewable Energy Integration

The integration of renewable energy sources into syngas production systems can drastically reduce their environmental footprint by lowering reliance on fossil fuels. One of the proposed alternatives is solar-assisted gasification, which uses concentrated solar power (CSP) to supply the high temperatures needed for gasification, reducing reliance on fossil fuels and significantly lowering CO_2 emissions. CSP systems, capable of reaching temperatures above 1,000 °C, enable efficient conversion of feedstocks like biomass and municipal waste into syngas. This solar-driven approach can cut carbon emissions by up to 60% compared to conventional methods while enhancing thermal efficiency (Fu and Wang, 2024). Hybrid systems that combine solar energy with renewable electricity or waste heat recovery further optimize performance, ensuring continuous operation during periods of low solar availability. These systems improve process efficiency by integrating technologies like solid oxide fuel cells (SOFCs) or thermal storage for stable syngas production. While high capital costs and solar intermittency pose challenges, advancements in reactor designs and thermal storage solutions are making solar gasification more practical and cost-effective. As solar power costs decline, this technology offers a sustainable pathway for producing low-carbon syngas, helping industries achieve net-zero emissions (Raya-Imbernón et al., 2024).

Another alternative is the use of wind and solar energy as renewable electricity to power water electrolysis units used to produce green hydrogen that can be blended with syngas and enhance its calorific value. This approach not only reduces emissions but also aligns with global decarbonization goals (Khlifi et al., 2024). Advancements in co-electrolysis, such as methane-assisted solid oxide electrolysis cells (SOECs), simultaneously reduce water and CO_2 to produce syngas directly, achieving up to 70% lower electricity consumption compared to traditional methods (Cui et al., 2021). Low-temperature anion exchange membrane (AEM) electrolyzers can also be used for flexible and decentralized syngas production by dynamically adjusting the H_2:CO ratio with operational changes (Raya-Imbernón et al., 2024). Finally, hybrid systems integrating syngas production with renewable power sources, such as solar and wind, can further optimize energy use. For example, combining gasification with SOFCs and organic Rankine cycles (ORC) significantly improves overall system efficiency and reduces carbon footprints (Fu and Wang, 2024).

13.6 Environmental Impact of Syngas Production

13.6.1 Global Warming Potential

Syngas, a mixture of carbon monoxide and hydrogen (Chinnici et al., 2019), used as an intermediate for ammonia and methanol production, is also the major source for pure hydrogen and carbon monoxide for chemical processes, and it is used for the production of synthetic fuel, either as diesel (by the Fisher–Tropsch process) or gasoline (Devold, 2013). The ideal ratio of hydrogen to carbon monoxide for methanol synthesis via the Fisher–Tropsch process is 2:1. The ratio of hydrogen to carbon monoxide in syngas can be adjusted by the WGS equilibrium reaction to exchange carbon monoxide and water with carbon dioxide and hydrogen (Arts et al., 2018; Devold, 2013). The carbon monoxide and hydrogen mixture often also contains carbon dioxide. Therefore, the potential for carbon oxide emissions exists. Carbon oxide is a primary GHG that significantly contributes to climate change when emitted in excess. Therefore, reducing carbon oxide emissions is critical for mitigating its climate change impact (Arts et al., 2018). It has been reported that carbon dioxide emissions contribute 72–77% to greenhouse effects compared to the contributions from methane (14–18%), nitrous oxides (8–9%), and other gases (less than 1%) (Devold, 2013).

An alternative option to further reduce the carbon content in syngas is by adding ammonia (NH_3) into the mixture. Ammonia is a carbon-free fuel, which has drawn interest due to its better storage and transport properties than hydrogen (H_2), and it has been proposed as a blending gas to syngas. However, the combustion may generate excessive NO_x emissions (Qian et al., 2024). To obtain clean and low-carbon combustion and less NO_x formation, a blended fuel of ammonia/syngas ratio can be used with moderate/intense low-oxygen dilution (MILD) combustion in co-flow (Liao et al., 2024). Studies conducted on the mechanisms of NO generation from syngas under MILD conditions in counter-flow flame suggested that a high CO/H_2 ratio and pressure greater than 10 atm improved the NO emission reduction (Shi et al., 2022). Liao et al. (2024) found that when the ratio of ammonia to syngas is 3:7 and the oxygen proportion increases to 15%, high-temperature combustion, instead of MILD, is reached, and it causes NH formation from the decomposition of ammonia, which participates more actively in the consumption process of NO and N_2O (Liao et al., 2024). Another study found that increasing the proportion of syngas or increasing the methane reduced the flow temperature gradient, which improves the MILD combustion stability. It has also been found that increasing the ammonia ratio decreased the concentration of OH radicals and NO_x generation (Mousavi et al., 2024).

Syngas production from fossil resources emits large amounts of GHG (Bachmann et al., 2023). These emissions can be reduced by using other resources such as biomass, biowaste, or carbon dioxide as the feedstock, even though a different environmental impact may arise (Bachmann et al., 2023). Syngas production by biomass gasification emits more direct GHG emissions than SMR, but reduces the global warming impact (GWI) compared to fossil-based production. However, heat supply based on biomass emits more direct GHG emissions than heat provided by natural gas or low-carbon electricity. This contrast indicates that overall, GHG emissions are reduced by using biomass for syngas production (Bachmann et al., 2023). Although biomass solely cultivated for syngas production reduces the GWI of syngas, it leads to trade-offs in freshwater, eutrophication, and land use, mainly due to biomass cultivation. An alternative is to use biowaste as feedstock, especially if the biowaste is processed by AD in combination with low-carbon electricity, which helps to reduce both the syngas GWI and GHG emissions. Syngas production using CO_2-based resources can also be used as long as low-carbon electricity is used (Bachmann et al., 2023). It is worth mentioning that if carbon-free electricity is used in the production of syngas, more GHG will be generated due to the GHG emissions from cultivation of the biomass. Electrochemical carbon dioxide reduction is the most promising syngas production process to reduce GWI, but it requires further advancement in technology readiness (Bachmann et al., 2023).

13.6.2 Water Consumption and Water Footprint

Syngas production from biomass as an alternative to fossil-based feedstocks has been the focus of recent efforts. However, the major impact of using biomass for syngas production results from the use of water, which is crucial for biofuel production (Li et al., 2021). It has been estimated that water consumption increases by up to 58% (Bachmann et al., 2023). A study conducted on the water consumption of syngas from corn straw and wheat straw through CLG (Hu et al., 2021) reported that the main water consumption is crop growth stage, accounting for 99% of the total water consumption. It also reported that the water consumption due to crop transportation and syngas production can be ignored (Li et al., 2021).

Some examples of water consumption include a typical coal-to-liquid/synthetic natural gas (CTL/SNG) plant, where 70–80% of the total water consumption and emission of the entire plant are due to the syngas production unit, or a typical coal-to-synthetic natural gas (CtSNG) system, where only about 4% of the water consumption is converted into products, and the left is lost in different forms during different processes (Wang et al., 2021). These examples show the importance of considering water saving (Xu et al., 2015). Independently of the raw syngas source and production process, water can be used through a water factory to provide water for the combined heat and power plant (CHP) and the cleanup unit. Water can also be provided through a recirculating cooling water station (RCWS), and, although RCWS is the main cause of water consumption, it could also provide the most potential for water saving. By adopting hybrid cooling systems (HCS), the water consumed in RCWS can be reduced by 60% (Wang et al., 2021).

13.6.3 Energy Consumption

Alternative and clean energy are major areas of research due to the need for sustainable energy development. This includes the production and purification of hydrogen and syngas with a focus on supplying the increasing demands for fuels and energy, increasing the efficiency of energy utilization for fuel and electricity production, eliminating pollutants, and reducing GHG emissions (Song, 2010). Syngas plays a major role in the generation of power, industrial heat, fuels, chemical feedstocks, and chemical products. Therefore, syngas production goes hand in hand with purification requirements and energy consumption.

Alternative ways have been explored to save energy while generating pure syngas for ammonia or methanol production. A study conducted by Luu et al. (2015) co-fed captured CO_2 from a carbon capture plant (CCP) with methane from natural gas (NG) into a dry methane reformer (DMR), and added hydrogen to produce syngas for methanol synthesis as an alternative to transitioning into the methanol economy. The method reduced the demand of methane by 22%. It also reduced the use of other resources, including water, hydrogen, and oxygen, while still maintaining the appropriate composition for the methanol reaction (Luu et al., 2015).

Syngas production from renewable energy and from other surplus resources or waste from industrial, domestic, urban, and agricultural systems has been explored as an alternative to conventional fossil fuels to enable a net reduction in energy footprint or in carbon dioxide emissions. Syngas produced from gasification of biomass sources (Ciferno and Marano, 2002), such as coppiced wood or wood from sustainable forestry operations, can become a carbon-neutral gas if the carbon dioxide emitted after burning the syngas is effectively re-adsorbed during the photosynthetic growth of the next crop. Syngas production from agricultural straws can also become carbon-neutral since the gasification process converts the lignin, which is 25–30% of the biomass. Syngas can also be produced from food waste by gasification. Food waste is biomass waste that usually accumulates in landfills and is converted into GHG methane. Syngas could also be obtained by using concentrated solar energy as a high-grade heat source for the production process (Murray and Fletcher, 1994). Syngas production without the constraints associated with biomass cultivation could also be possible in a co-electrolysis process run on renewable electricity (Graves et al., 2011). The electrochemical conversion of steam and carbon dioxide using renewable-generated electricity to syngas – called power-to-syngas – such as co-electrolysis, for the production of high-value chemicals can be an important enabling step for transitioning the energy system (Brynolf et al., 2018).

Syngas derived from hydrocarbons is the major source for pure hydrogen (Arts et al., 2018). However, hydrogen can be produced from other hydrocarbons, including natural gas or ethanol, and also by electrolysis. Hydrogen production via syngas has a 75–80% efficiency, compared with the hydrogen production from water by electrolysis, which has an efficiency of only 25% at normal conditions or 50% at high temperature and high pressure (Devold, 2013). Hydrogen produced by steam reforming of ethanol (SRE) occurs through the reaction of ethanol and water over a catalyst. The reaction occurs at high temperatures (300–800 °C), it is highly endothermic (standard enthalpy of 173.3 kJ/mol), and it produces carbon dioxide in addition to hydrogen. Stoichiometric steam-ethanol molar ratios higher than the required 3.0 are used in this process because bioethanol, or the crude oil (fermentation broth), contains only 12 vol% of ethanol and about 86 vol% of water (molar ratio of 18) in order to minimize the cost of ethanol distillation, even though it represents a higher energy cost of the system due to the extra steam that is generated. An advantage of this process is that the excess of water contributes to the WGS, or the reaction of carbon monoxide and water vapor, and it forms carbon dioxide and one additional mole of hydrogen (Song, 2010).

The source and the intended end use determine the level of purification of the syngas. Contaminants present in raw syngas produced by oxygen-blown gasification, including particulate, alkali metals, and tars, can cause corrosion, fouling, plugging, catalyst poisoning, and catalyst carbonization. Raw gas produced from in situ gasification of coal can additionally contain other contaminants, such as mercaptans, heavy metals, ammonia, alkali salts, activated carbon, BTX, trace volatile metals, non-water-soluble contaminants, and sulfur compounds (Erikson, 2011). Some of the contaminants present in the raw syngas can be highly reactive and impact the performance of the process and life of the equipment. However, the selection of the purification process can increase processing and energy cost. The extra cost can be minimized by understanding the required level of purity required by the process (Pozo et al., 2023).

13.6.4 Resource Depletion and Land Use

An alternative for syngas production, different than using traditional fossil-based generation, is by using bioconversion methods. However, a major concern is diverting agricultural land areas away from food production (Devold, 2013; Bachmann et al., 2023). The use of land has been estimated to increase by 66%, although the increase will occur only if the biomass is required for the gasification process. Additional land won't be required if syngas is produced from biowaste by AD using wind power. In this case, land use can be reduced by 20% compared to the land used by the traditional fossil fuel-based syngas production. Land use can be considerably increased (by up to 260%) if CCU-based syngas is produced since the high electricity demand causes it to be increased (Bachmann et al., 2023).

13.7 Life Cycle Assessment of Syngas Production

13.7.1 Life Cycle Assessment Framework

Life cycle assessment (LCA) is a crucial tool for assessing the environmental effects of syngas production across its entire lifecycle. By calculating resource use, energy consumption, emissions, and water footprints, LCA identifies trade-offs between production methods and optimizes environmental outcomes (Guang Li et al., 2021; Khlifi et al., 2024). Commonly, syngas is evaluated based on a functional unit of "1 cubic meter (m^3) of syngas" or "1 megajoule (MJ) of energy produced." The system boundaries include processes from feedstock cultivation or extraction to syngas production (cradle-to-gate), excluding transportation and downstream applications (Guang Li et al., 2021; Frazier et al., 2015), and the full lifecycle, from feedstock cultivation to the utilization and disposal of syngas and its derivatives (cradle-to-grave), such as methanol and ammonia (Fu and Wang, 2024; Rey et al., 2024a). Clearly defining the system boundaries ensures consistency when comparing different production pathways. For example, biomass gasification often demonstrates a lower carbon footprint than fossil-based methods but may have higher water and land use impacts (Ayodele et al., 2019; Guang Li et al., 2021).

Life cycle inventory (LCI) involves data collection on all inputs, such as raw materials (e.g., biomass, natural gas, MSW), water, catalysts, and energy for processing (electricity, heat) (Wardhani et al., 2024; Fu and Wang, 2024), and outputs or products (syngas, subproducts like ammonia and methanol) and emissions (e.g., CO_2, methane, tars, and particulates) (Mahmoudi et al., 2024; Khalid et al., 2024). Feedstock selection is pivotal in determining the LCA outcomes for bioenergy and syngas production (Kavindi, et al., 2025). The key LCI stages include feedstock preparation, syngas production, and waste management. Biomass cultivation and preprocessing can dominate water and land use impacts. For example, water consumption for syngas from wheat straw is about 2,170 l/m^3, compared to 1,140 l for corn straw (Guang Li et al., 2021). Processes like gasification and pyrolysis require significant energy inputs, with emissions varying based on operational parameters (Rey et al., 2024a; Khlifi et al., 2024). Effective handling of residues, such as spent catalysts and tar, is critical to minimizing environmental burdens (Frazier et al., 2015; Rey et al., 2024a).

Finally, life cycle impact assessments (LCIA) convert LCI data into potential environmental impacts such as GWP, water footprint, energy use, and air quality and toxicity. Biomass-derived syngas often exhibits lower GWP compared to fossil-based syngas, though emissions from fertilizer use can offset some advantages (Ayodele et al., 2019; Guang Li et al., 2021). Syngas production via CLG can be water-intensive. Crop growth is often the largest contributor to water use (Guang Li et al., 2021). Integrating renewable energy into syngas production processes can significantly reduce environmental burdens (Fu and Wang, 2024; Khalid et al., 2024). Syngas production can emit NO_x, SO_x, and other pollutants. Cleaner technologies are essential for reducing toxicity risks (Rey et al., 2024a; Frazier et al., 2015).

Interpreting the LCA through hotspot analysis, scenario analysis, or sensitivity analysis helps identify environmental hotspots and evaluate potential improvements. For example, cultivation and syngas production often dominate environmental impacts, particularly in water and energy use (Guang Li et al., 2021; Fu and Wang, 2024); exploring alternatives, such as using biochar as a catalyst or implementing carbon capture, can enhance sustainability (Frazier et al., 2015; Rey et al., 2024a); and varying key parameters, such as feedstock type or gasification conditions, can significantly affect results (Wardhani et al., 2024; Khlifi et al., 2024). Biomass-derived syngas generally has a lower carbon footprint compared to fossil-derived syngas, but its sustainability can be influenced by water and land use during feedstock cultivation (Guang Li et al., 2021). For example, ammonia production via the Haber–Bosch process, a key application of syngas, is energy-intensive and relies heavily on renewable hydrogen integration to reduce emissions (Ayodele et al., 2019). The interplay between feedstock characteristics, processing technologies, and environmental impacts underscores the importance of strategic feedstock selection in optimizing LCA results (Shakelly et al., 2023).

It is worth mentioning that LCA studies for the production of syngas as a final product are indirectly reported in the literature through the analysis of processes using the major constituent-based syngas as the feedstock (example, CO_2-based syngas). This is because syngas is not sold directly as a final product but as an intermediate in the chemical industry (Arts et al., 2018).

13.7.2 LCA of Syngas Versus Natural Gas and Biomass-derived Fuel

Syngas, derived from biomass, fossil fuels, or waste streams, presents significant potential as an alternative to conventional fuels like natural gas, biomass-derived fuels, and methane. However, to ensure that the selection is a viable option, it is crucial to evaluate the life cycle impact of syngas, focusing on emissions, energy efficiency, and resource use. GHG emissions are a critical factor in evaluating syngas and natural gas. Syngas production, particularly from gasification, tends to emit more CO_2 but results in lower CH_4 emissions compared to natural gas systems, where methane leakage can be significant (Patel et al., 2016; Cherubini and Ulgiati, 2010). Combustion of syngas may also produce higher NO_x emissions, depending on feedstock and production processes (Basu, 2010).

In the case of natural gas, extraction and combustion processes achieve higher conversion efficiencies, resulting in an increase of the lower heating value (LHV) (~50 MJ/kg) compared to syngas (~10–15 MJ/kg) (Spath and Mann, 2000). However, syngas can recover additional energy when integrated into CHP systems, where both thermal and electrical energy are utilized. CHP systems improve overall energy efficiency to levels exceeding 80% (Basu, 2010). Likewise, syngas production via gasification has high water demand, whereas natural gas extraction methods such as hydraulic fracturing can result in substantial land and water impacts (IPCC, 2019; Patel et al., 2016). Table 13.1 summarizes the differences between syngas, natural gas, and biomass-derived fuel.

The comparison between syngas and biomass-derived fuels like biodiesel, bioethanol, and biogas is important to identify feedstock advantages, as syngas can utilize non-food biomass or waste, reducing competition with agricultural resources. Additionally, such comparisons provide critical information to guide policy development, including biofuel mandates, renewable energy subsidies, and investment priorities. Biomass-derived fuels are considered carbon-neutral, as the

Table 13.1 LCIA of syngas as compared to natural gas and biomass-derived fuel.

Parameter	Syngas	Natural gas	Biodiesel/bioethanol/ biogas
GHG emissions	Moderate CO_2, lower CH_4	Lower CO_2, higher CH_4 (Patel et al., 2016)	Low to carbon-neutral (Cherubini and Ulgiati, 2010)
Energy efficiency	Lower LHV	Higher LHV (Spath and Mann, 2000)	Moderate (Sharma et al., 2013)
Resource use	Higher water demand	Land/water impacts from fracking (IPCC, 2019)	High land/water use (Luo et al., 2009)

CO_2 emitted during combustion is offset by carbon sequestration during feedstock growth (Cherubini and Ulgiati, 2010). However, syngas emissions depend on feedstock and production processes (Ahmed and Gupta, 2011). Combustion emissions, such as NO_x and PM, are comparable between syngas and biogas (Luo et al., 2009). While biodiesel and bioethanol have moderate energy yields, syngas production can achieve higher efficiencies when coupled with CHP systems. Additionally, syngas processes can use waste biomass as feedstock, reducing energy inputs compared to biofuel crop cultivation. Biodiesel and bioethanol production involve significant land use and water consumption for crop cultivation, whereas syngas production can utilize non-food biomass or waste materials. Table 13.1 summarizes the LCIA of syngas as compared with natural gas and biomass-derived fuel.

13.7.3 PtG Syngas Versus Methane and SNG

Syngas and methane in PtG systems utilize excess renewable energy to produce gaseous fuels for storage or use. Methane production in PtG systems generally results in lower CO_2 emissions due to its simpler structure and synthesis pathway (Brynolf et al., 2018; Fasihi et al., 2016). In contrast, syngas production, particularly through electrolysis and gasification, can result in higher emissions unless waste biomass or renewable inputs are used (Götz et al., 2016). Methane synthesis achieves higher energy conversion efficiency compared to syngas production, which requires additional processing steps. However, syngas provides greater flexibility for downstream applications, such as Fischer–Tropsch synthesis for liquid fuels (Bailera et al., 2017). Syngas production pathways in PtG systems tend to have higher resource inputs, including electricity and water, compared to methane synthesis (Bailera et al., 2017). Table 13.2 summarizes the LCIA of PtG syngas and PtG methane.

Table 13.2 LCIA of power-to-gas syngas as compared to power-to-gas methane.

Parameter	Power-to-gas syngas	Power-to-gas methane
Emissions	Moderate CO_2, flexible inputs	Lower CO_2, simpler synthesis (Brynolf et al., 2018)
Energy efficiency	Lower efficiency	Higher efficiency (Fasihi et al., 2016)
Resource use	Higher inputs required	Moderate inputs (Bailera et al., 2017)

A comparative LCA of power-to-synthetic natural gas (power-to-SNG) and power-to-syngas was conducted by Bachmann et al. (2023) to evaluate the environmental benefit of integrating renewable electricity and carbon (PtG) into the chemical industry. The Sabatier reaction was used to produce SNG and the dry reforming of methane (DRM) to produce syngas. The motivation for this study was to use electricity to produce hydrogen, which was further converted with CO_2 as the renewable carbon source. The resulting products were used as feedstock for the chemical industry to replace current fossil-based feedstock. The threshold values for electricity supply were used to compare the environmental benefit of the power-to-SNG and power-to-syngas. The study concluded that the syngas production by DMR has the largest environmental potential and benefit, which makes syngas the more desirable product than methane if renewable electricity is limited (Bachmann et al., 2023; Shakelly et al., 2023).

13.8 Final Considerations from Syngas LCA

Biomass-based syngas can reduce GHG emissions significantly but requires careful consideration of feedstock displacement effects. Utilizing CO_2 through electrochemical reduction offers a viable pathway, but only if powered by renewable energy sources (Bachmann et al., 2023). Technologies like high temperature (HT)-co-electrolysis demonstrate high energy efficiencies but require substantial renewable electricity. In energy-limited regions, this creates competition with other renewable applications such as electrification or heating (Schreiber et al., 2020). Implementing flexible operation with intermittent renewable energy sources can balance energy demands and minimize carbon footprints. Biomass gasification and fermentation processes are water-intensive, with significant requirements for cooling, steam generation, and bioreactor operation. This water demand could exacerbate regional water stress, particularly in arid areas (Mukhti et al., 2023). Integrating bio-waste and biomass gasification into a circular economy framework can reduce resource competition while maximizing syngas yield.

References

Abdulyekeen, K.A., Umar, A.A., Patah, M.F.A. et al. (2021). Torrefaction of biomass: production of enhanced solid biofuel from municipal solid waste and other types of biomass *Renewable and Sustainable Energy Reviews* 150: 111436.

Ahmed, A. and Gupta, A.K. (2011). Syngas and hydrogen production from biomass-derived biofuels. *International Journal of Hydrogen Energy* 36 (9): 5602–5611.

Akbari, M. and Kumar, A. (2024). The development of data-intensive techno-economic models for the comparison of renewable natural gas production from six different biomass feedstocks for the decarbonization of energy demand sectors. *Fuel* 358: 130107.

Alherbawi, M., Parthasarathy, P., Elkhalifa, S. et al. (2024). Techno-economic and environmental analyses of the pyrolysis of food waste to produce bioproducts. *Heliyon* 10 (6): e27713.

Andreides, D., Fliegerova, K.O., Pokorna, D. et al. (2022). Biological conversion of carbon monoxide and hydrogen by anaerobic culture: prospect of anaerobic digestion and thermochemical processes combination. *Biotechnology Advances* 58: 107886.

Anis Kumar, M., Swarnalatha, A.P., Shwetha, J. et al. (2024). Pyrolysis behaviour and synergistic effect in co-pyrolysis of wheat straw and polyethylene terephthalate: a study on product distribution and oil characterization. *Heliyon* 10 (17): e37255.

Arts, J., Muller, E., and Thenert, K. (2018). Sustainable conversion of carbon dioxide: an integrated review of catalysis and life cycle assessment. *Chemical Reviews* 118 (2): 434–504.

Atsbha, M.W., Farkash, L., Bartal, G. et al. (2023). Extracting liquid fertilizer from ammonia-laden pharmaceutical wastewater using hollow fiber membrane contactors. *Journal of Cleaner Production* 418: 138183.

Aydin, E.S., Yucel, O., and Sadikoglu, H. (2019). Experimental study on hydrogen-rich syngas production via gasification of pine cone particles and wood pellets in a fixed bed downdraft gasifier. *International Journal of Hydrogen Energy* 44: 17389–17396.

Ayodele, B.V. et al. (2019). Process analysis and life cycle assessment of hydrogen production from biomass gasification and power-to-gas. *Renewable and Sustainable Energy Reviews* 101: 280–290.

Bachmann, M., Völker, S., Kleinekorte, J. et al. (2023). Syngas from what? Comparative life cycle assessment for syngas production from biomass, CO_2, and steel mill off-gases. *Sustainable Chemistry and Engineering* 11 (14): 5356–5366.

Bailera, M., Lisbona, P., Romeo, L.M. et al. (2017). Power to gas projects review: lab, pilot and demo plants for storing renewable energy and CO_2. *Renewable and Sustainable Energy Reviews* 69: 292–312.

Basu, P. (2010). *Biomass Gasification and Pyrolysis: Practical Design and Theory*. Academic Press.

Benarimo, T., Manzolini, G., and Manca, F. (2023). Power-to-gas and its integration into renewable energy systems: advances and challenges. *Energy Conversion and Management* 278: 116568.

Beneroso, D. and Fernandez, B.F. (2016). Chapter 5: Microwave technology for syngas production from renewable sources. In: *Syngas* (ed. R. Myers), 117–148. Nova Science Publishers Inc.

Bennett, A. and Serrenho, A.C. (2024). A systematic comparison of the energy and emissions intensity of hydrogen production pathways in the United Kingdom. *International Journal of Hydrogen Energy* 89: 364–374.

Bianchi, F.R. and Bosio, B. (2024). Modelling of green ammonia production based on solid oxide cells as electrolyser and oxygen separator for Haber-Bosch loop decarbonization. *International Journal of Hydrogen Energy* 95: 1183–1193.

Bridgwater, A.V. (2003). Renewable fuels and chemicals by thermal processing of biomass. *Chemical Engineering Journal* 91 (2–3): 87–102.

Brynolf, S., Taljegard, M., Grahn, M. et al. (2018). Electrofuels for the transport sector: a review of production costs. *Renewable and Sustainable Energy Reviews* 81: 1887–1905.

Cao, G., Handler, R.M., Luyben, W.L. et al. (2022). CO_2 conversion to syngas via electrification of endothermal reactors: process design and environmental impact analysis. *Energy Conversion and Management* 265: 115763.

Chen, Y., Lyu, Y., Yang, X. et al. (2022). Performance comparison of urea production using one set of integrated indicators considering energy use, economic cost and emissions' impacts: a case from China. *Energy* 254: 124489.

Cherubini, F. and Ulgiati, S. (2010). GHG balances of bioenergy systems. *Environmental Science & Policy* 13 (3): 79–87.

Chinnici, A., Feng Tian, Z., Han Lim, J. et al. (2019). Thermal performance analysis of a syngas-fuelled hybrid solar receiver combustor operated in the MILD combustion regime. *Combustion Science and Technology* 191: 2–17.

Ciferno, J.P. and Marano, J.J. (2002). *Benchmarking biomass gasification technologies for fuels*. Chemicals and Hydrogen Production National Energy Technology Laboratory (NETL) Report.

Cormos, C.C. (2011). Evaluation of power generation schemes based on hydrogen fuelled combined cycle with carbon capture and storage (CCS). *International Journal of Hydrogen Energy* 36: 3726–3738.

Coronel, D.A.Q., Rodas, Y.A.L., Corredor, L. et al. (2022). Co-gasification of biomass and coal in a top-lit updraft fixed bed gasifier: syngas composition and its interchangeability with natural gas for combustion applications. *Fuel* 316: 123394.

Cui, H., Turn, S.Q., and Keffer, V.I. (2021). Methane-assisted solid oxide electrolysis cells for syngas production: a review. *Renewable and Sustainable Energy Reviews* 135: 110239.

Devold, H. (2013). An introduction to oil and gas production, transport, refining and petrochemical industry. In: *Oil and Gas Production Handbook*, 162. Etterstad, Oslo: ©2006 - 2013 ABB Oil and Gas.

Erfani, N., Baharudin, L., and Watson, M. (2024). Recent advances and intensifications in Haber-Bosch ammonia synthesis process. *Chemical Engineering and Processing: Process Intensification* 204: 109962.

Erikson, J. (2011). *UCG syngas composition, cleanup and conversion technologies.* Paper presented at the IEA Underground Coal Gasification Workshop.

Faria, J.A. (2021). Renaissance of ammonia synthesis for sustainable production of energy and fertilizers. *Current Opinion in Green and Sustainable Chemistry* 29: 100466.

Fasihi, M., Bogdanov, D., and Breyer, C. (2016). Techno-economic assessment of power-to-gas syngas. *Energy Conversion and Management* 115: 151–163.

Frazier, E.A., Wright, D.T., and Tan, E.C. (2015). Bioconversion of biomass to hydrogen and syngas: process developments and insights. *Biotechnology Advances* 33 (7): 1245–1260.

Fu, Q., Li, Z., and Wang, J. (2024). Environmental assessment of syngas production via chemical looping gasification. *Energy Reports* 10: 123–135.

Galusnyak, S.C., Petrescu, L., Sandu, V.-C. et al. (2023). Environmental impact assessment of green ammonia coupled with urea and ammonium nitrate production. *Journal of Environmental Management* 343: 118215.

Götz, M., Lefebvre, J., and Mörs, F. (2016). Renewable power-to-gas systems: efficiency and LCA insights. *Renewable Energy* 85: 1371–1380.

Graves, C., Ebbesen, S.D., and Mogensen, M. (2011). Sustainable hydrocarbon fuels by recycling CO_2 and H_2O with renewable or nuclear energy. *Renewable and Sustainable Energy Reviews* 15 (1): 1–23.

Guang Li, M. et al. (2021). A comprehensive assessment of biomass-derived syngas and its applications. *Journal of Cleaner Production* 305: 127201.

Hameed, Z., Aslam, M., Khan, Z. et al. (2021). Gasification of municipal solid waste blends with biomass for energy production and resources recovery: current status, hybrid technologies and innovative prospects. *Renewable and Sustainable Energy Reviews* 136: 110375.

Hassan, M.U., Guoqin, H., Arif, M.S. et al. (2024). Can urea-coated fertilizers be an effective means of reducing greenhouse gas emissions and improving crop productivity? *Journal of Environmental Management* 367: 121927.

He, C., Zhang, X., Zhang, Z. et al. (2023). Whole-crop biorefinery of corn biomass for pullulan production by *Aureobasidium pullulans. Bioresource Technology* 370: 128517.

Hoek, A. and Kersten, L.B.J.M. (2004). The shell middle distillate synthesis process: technology, products and perspective. *Studies in Surface Science and Catalysis* 147: 25–30.

Hreiz, R., Gentric, C., and Midoux, N. (2011). Numerical investigation of swirling flow in cylindrical cyclones. *Chemical Engineering Research and Design* 89: 2521–2539.

Hsieh, T.-L., Zhang, Y., Xu, D. et al. (2022). Chemical looping gasification for producing high-purity, H_2-rich syngas in a cocurrent moving bed reducer with coal and methane co-feeds. *Industrial & Engineering Chemistry Research* 57 (6): 2018–2029.

Hu, J., Zhang, T., and Zhang, Q. et al. (2021). Application of calcium oxide/ferric oxide composite oxygen carrier for corn straw chemical looping Gasification. *Bioresource Technology* 330: 125011.

Hu, Y.H. and Ruckenstein, E. (2004). Catalytic conversion of methane to synthesis gas by partial oxidation and CO_2 reforming. *Advances in Catalysis* 48: 297–345.

Ikpe, A.E., Wilson, E.O., and Usungurua, E.O. (2024). Dynamics of ammonia synthesis from industrial reactors: a gaze towards production diversity. *10th International Black Sea Coastline Countries Scientific Research Conference*, 61–84. Samsun, Turkiye.

IPCC (2019). *Climate Change and Land: Greenhouse Gas Emissions of Fossil Fuels*. IPCC.

Irfan, H.M., Iqbal, K., Taipabu, M.I. et al. (2024). Decarbonization frameworks to industrial-scale ammonia production: techno-economic and environmental implications. *International Journal of Hydrogen Energy* 78: 580–593.

Joseph S.S., Melvin, S.S., Godwin, G. et al. (2024). Production and utilization of green ammonia for decarbonizing the energy sector with a discrete focus on sustainable development goals and environmental impact and technical hurdles. *Fuel* 360: 130626.

Kavindi, G.A.G., Tang, L., and Sasaki, Y. (2025). Assessing GHG emission reduction in biomass-derived biochar production via slow pyrolysis: a cradle-to-gate LCA approach. *Resources, Conservation & Recycling* 212: 107900.

Keller, M. and Sharma, A. (2023). Syngas production from NH_3 and CO_2 through chemical looping ammonia cracking – reverse water gas shift. *Journal of CO_2 Utilization* 76: 102588.

Khademi, M.H. and Varnoosfaderani, M.L. (2021). Sustainable ammonia production from steam reforming of biomass-derived glycerol in a heat-integrated intensified process: modeling and feasibility study. *Journal of Cleaner Production* 324: 129241.

Khalid, S., Ahmed, M.A., and Baloch, H. (2024). Comparative pyrolysis methods for syngas production: a review of technologies and efficiencies. *Energy Reports* 10: 345–358.

Khlifi, H., Bouazizi, N., and Aoun, M. (2024). Thermochemical processes for biomass gasification: environmental and economic perspectives. *Renewable Energy* 201: 1234–1250.

Khlifi, S., Lajili, M., Perré, P. et al. (2022). Numerical study of turbulent combustion of a lignocellulosic gas mixture in an updraft fixed bed reactor. *Sustainability* 14: 16587.

Kim, H., Byun, M., Lee, B. et al. (2022). Carbon-neutral methanol synthesis as carbon dioxide utilization at different scales: economic and environmental perspectives. *Energy Conversion and Management* 252: 115119.

Kruesi, M., Jovanovic, Z.R., dos Santos, E.C. et al. (2013). Solar-driven steam-based gasification of sugarcane bagasse in a combined drop-tube and fixed-bed reactor-thermodynamic, kinetic, and experimental analyses. *Biomass Bioenergy* 52: 173–183.

Lee, S.Y., Sankaran, R., Chew, K.W. et al. (2019). Waste to bioenergy: a review on the recent conversion technologies. *BMC Energy* 1 (1): 4.

Li, G., Ma, S., Liu, F. et al. (2021). Life cycle water footprint assessment of syngas production from biomass chemical looping gasification. *Bioresource Technology* 342: 125940.

Liao, H., Ding, C., Hu, F. et al. (2024). Numerical study on the effects of adding NH_3 into syngas/NH_3 mixtures on combustion characteristics and NO_x emissions in diluted hot co-flows. *International Journal of Hydrogen Energy* 84: 915–928.

Lin, K.W. and Wu, H.W. (2018). Hydrogen-rich syngas production of urea blended with biobutanol by a thermodynamic analysis. *International Journal of Hydrogen Energy* 43: 17562–17573.

Liu, L., Li, S., Zhuang, Y. et al. (2024). Exploring biomass addition in methanol production: a life cycle analysis of efficiency and environmental impacts. *Journal of Cleaner Production* 468: 143072.

Lozano, F.J. and Lozano, R. (2018). Assessing the potential sustainability benefits of agricultural residues: biomass conversion to syngas for energy generation or to chemicals production. *Journal of Cleaner Production* 172: 4162–4169.

Luo, L., van der Voet, E., and Huppes, G. (2009). Life cycle assessment and sustainability of biodiesel production. *Journal of Cleaner Production* 17 (2): 150–158.

Luo, S., Zhou, Y., and Yi, C. (2012). Syngas production by catalytic steam gasification of municipal solid waste in fixed-bed reactor. *Energy* 44: 391–395.

Luu, M.T., Milani, D., Bahadori, A. et al. (2015). Comparative study of CO_2 utilization in methanol synthesis with various syngas production technologies. *Journal of CO_2 Utilization* 12: 62–76.

Mahmoudi, H., Hosseini, S.E., and Wahid, M.A. (2024). Life cycle assessment of syngas production from biomass: a case study. *Journal of Environmental Management* 280: 111827.

Mousavi, S.M., Lee, B.J., Kim, J. et al. (2024). On the effects of adding syngas to an ammonia-MILD combustion regime – a computational study of the reaction zone structure. *International Journal of Hydrogen Energy* 52 (2): 26–40.

Mukhti, R., Setiadi, T., and Kresnowati, M.T.A.P. (2023). Challenges in syngas fermentation for bioethanol production: syngas composition. *Engineering Chemistry* 1: 9–19.

Murray, J.P. and Fletcher, E.A. (1994). Reaction of steam with cellulose in a fluidized-bed using concentrated sunlight. *Energy* 19 (10): 1083–1098.

Nemmour, A., Inayat, A., Janajreh, I. et al. (2023). Green hydrogen-based E-fuels (E-methane, E-methanol, E-ammonia) to support clean energy transition: a literature review. *International Journal of Hydrogen Energy* 48: 29011–29033.

Ongen, A., Ozcan, H.K., and Ozbas, E.E. (2016). Gasification of biomass and treatment sludge in a fixed bed gasifier. *International Journal of Hydrogen Energy* 41: 8146–8153.

Patel, M., Zhang, X., and Kumar, A. (2016). Techno-economic and life cycle assessment of natural gas and syngas pathways. *Applied Energy* 168: 146–157.

Pinzón, M., García-Carpintero, R., de la Osa, A.R. et al. (2024). Ammonia as a hydrogen carrier: an energy approach. *Energy Conversion and Management* 321: 118998.

Plis, P. and Wilk, R.K. (2011). Theoretical and experimental investigation of biomass gasification process in a fixed bed gasifier. *Energy* 36: 3838–3845.

Pozo, C.A.D., Cloete, S., and Alvaro, A.J. (2023). Ammonia from solid fuels: a cost-effective route to energy security with negative CO_2 emissions. *Energy* 278: 127880.

Qian, Y., Pei, X., Zheng, L. et al. (2024). Improving thermal efficiency of an ammonia-diesel dual-fuel compression ignition engine with the addition of premixed low-proportion hydrogen. *International Journal of Hydrogen Energy* 58: 707–716.

Rani, P., Bansal, M., and Pathak, V.V. (2022). Experimental and kinetic studies for improvement of biogas production from KOH pretreated wheat straw. *Current Research in Green and Sustainable Chemistry* 5: 100283.

Raya-Imbernón, V., García-López, C., and Sánchez-Sánchez, M. (2024). Advances in solar-assisted gasification for syngas production: a review. *Solar Energy* 230: 123–145.

Rey, G., Martínez, I., and García, L. (2024a). Environmental impact assessment of syngas production from municipal solid waste. *Waste Management* 134: 10–20.

Rey, J.R.C., Longo, A., Rijo, B. et al. (2024b). A review of cleaning technologies for biomass-derived syngas. *Fuel* 377: 132776.

Robbins B., Gaona A., Tavasoli A. et al. (2024). Life cycle environmental performance of methanol production through photocatalytic dry methane reforming. *Journal of CO_2 Utilization* 79: 102638.

Ruiz, J.A., Juárez, M.C., Morales, M.P. et al. (2013). Biomass gasification for electricity generation: review of current technology barriers. *Renewable and Sustainable Energy Reviews* 18: 174–183.

Schreiber, A., Peschel, A., Hentschel, B. et al. (2020). Life cycle assessment of power-to-syngas: comparing high temperature co-electrolysis and steam methane reforming. *Frontiers in Energy Research* 8: 533850.

Shakelly, N., Pérez-Cardona, J.R., Deng, S. et al. (2023). Comparative life cycle assessment of bioethanol production from different generations of biomass and waste feedstocks. *Procedia, CIRP Life Cycle Engineering Conference* 116: 630–635.

Sharma, S., Basu, S., Shetti, N.P. et al. (2020). Waste-to-energy nexus: a sustainable development. *Environmental Pollution* 267: 115501.

Sharma, S.D., McLennan, K., Dolan, M. et al. (2013). Design and performance evaluation of dry cleaning process for syngas. *Fuel* 108: 42–53.

Shi, G., Li, P., Hu, F. et al. (2022). NO mechanisms of syngas MILD combustion diluted with N_2, CO_2, and H_2O. *International Journal of Hydrogen Energy* 47 (166): 49–64.

Song, C. (2010). Introduction to hydrogen and syngas production and purification technologies. In: *Hydrogen and Syngas Production and Purification Technologies* (ed. K. Liu, C. Song, and V. Subramani), 1–13. Hoboken, NJ: John Wiley & Sons, American Institute of Chemical Engineers.

Spath, P.L. and Mann, M.K. (2000). *Life Cycle Assessment of Natural Gas Combined-Cycle Power Generation*. NREL Technical Report.

Wang, W., Bai, B., Wei, W. et al. (2021). Hydrogen-rich syngas production by gasification of urea-formaldehyde plastics in supercritical water. *International Journal of Hydrogen Energy* 46: 35121–35129.

Wardhani, GK., Prawisudha, P., and Yoshikawa, K. (2024). Sensitivity analysis of biomass gasification for syngas production: environmental and economic aspects. *Bioresource Technology Reports* 15: 100719.

Wolday, A.K. and Ramteke, M. (2024). Surrogate model-based optimization of methanol synthesis process for multiple objectives: a pathway towards achieving sustainable development goals. *Chemical Engineering Research and Design* 204: 172–182.

Wu, H.W. and Lin, K.W. (2018). Thermodynamic analysis of hydrogen-rich syngas production with a mixture of aqueous urea and biodiesel. *International Journal of Hydrogen Energy* 43: 6804–6814.

Xu, J., Yang, Y., and Li, Y.W. (2015). Recent development in converting coal to clean fuels in China. *Fuel* 152: 122–130.

Zheng, J.-L., Zhu, Y.-H., Sun, G.-T. et al. (2022). Bio-oil gasification for production of the raw gas as ammonia syngas. *Fuel* 327: 125029.

Zheng, X., Chen, C., Ying, Z. et al. (2016). Experimental study on gasification performance of bamboo and PE from municipal solid waste in a bench-scale fixed bed reactor. *Energy Conversion and Management* 117: 393–399.

Zhou, H., Ma, Y., Yang, Q. et al. (2022). A new scheme for ammonia and fertilizer generation by coal direct chemical looping hydrogen process: concept design, parameter optimization, and performance analysis. *Journal of Cleaner Production* 362: 132445.

14

Investment Required for Upgrading Petroleum Refineries – Simplified Approach

Thomas Murphy[1], George Hoekstra[2], and Marcio Wagner da Silva[3]

[1] *The Whitman School of Management and the College of Law, Syracuse University, Syracuse, NY, USA*
[2] *Hoekstra Trading LLC, Bolingbrook, IL, USA*
[3] *Process Engineering, Petrobras, Sao Paulo, Brazil*

Purpose

This chapter compares methods for evaluating proposed upgrades to petroleum refineries to meet the changing demands for transportation fuels and petrochemicals. Investments in new processes and equipment will provide flexibility, synergy, and growth options necessary for environmental regulation and improvement in safety and sustainability, together with frequent disruptions in crude oil supply and quality, and seasonal demands for refined products and petrochemical derivatives, the need for capital investment and upgrading of refineries is clear.

Traditional measures of attractiveness for a capital investment project include Discounted Cash Flow (DCF) (internal rate of return, net present value [NPV]); Capital Asset Pricing Model (CAPM); Heuristics (simple rules of thumb, like payback period or market multiples); Factor Models (arbitrage pricing theory), or based on asset characteristics (like storage, pipelines, or geological reservoir characteristics); and Behavioral Finance (wait to see what others decide to do).

The causes of today's global warming are many and include (1) the release of greenhouse gas emissions from fossil fuels and sources inside the earth itself, and (2) the loss of natural carbon dioxide sequestration from photosynthesis on the land and in the oceans. The regulatory environment is helping to reduce emissions, like the Tier-3 gasoline specifications. The supply and demand dynamics with crude oil supply have made it more difficult to plan and process heavy sour crude.

Cleaner Petroleum Production and Refining Technologies, First Edition, Edited by M. R. Riazi and Harvey W. Yarranton.

Upgrading petroleum refineries to meet the demand for transportation fuels and petrochemicals will require new investment and a change in consumer habits, together with rebuilding and replanting our photosynthesis carbon dioxide environment.

Recent assessments examining the future of fossil fuels were conducted by Assessment of the future of fossil fuels is being conducted to reduce carbon dioxide emissions to net zero by 2050 meeting the terms of the 2015 Paris Accord. This will allow less developed countries to focus on building their economies.

This is the message that the authors encouraged chemical engineering students at the 2024 AIChE Networking FEST: it is an exciting time to become a chemical engineer.

14.1 Introduction

This chapter is written to provide a strategic response to the risk and uncertainty facing the petrochemical refining industry today. It builds upon a tripod of industry values – "Safety, Environment, Profit" – and outlines the leading edge of quantitative methodologies to help management decide the way forward. The authors have decades of experience in the petrochemical refining industry and as leaders in quantitative modeling of valuation, risk, and strategy. Each of us has studied finance and quantitative modeling in business school after becoming a practicing chemical engineer. Our experience includes environmental remediation, wastewater treatment at a large petrochemical plant, and seeing the effects of hydrofluoric acid (HF) contamination and per- and polyfluoroalkyl (PFAS) forever chemicals.

Financial models are often used without questioning whether they fit the asset being valued, or the company, or even the industry in general. This was one of the causes of the 2008 stock market crash involving collateralized mortgage obligations (CMOs), which were bundled into portfolios and valued using a waterfall model that assumed diversification across geographic regions. The turmoil in the financial markets in 2008 was quickly followed by accusations of greed and mismanagement. Risky trading schemes and complex valuation models involving CMOs, and other complex structures were blamed. A massive government bailout of the banking and insurance industries was necessary, including at insurance giant AIG (American International Group, Inc.), to prevent a total financial meltdown. These systemic institutions were deemed too big to fail (Murphy, 2010).

Valuation, risk, and strategy of alternative petroleum and petrochemical refining alternatives is a multidisciplinary process. It combines chemical engineering, quantitative valuation and risk models, and advanced technology assessments using real options and systems engineering principles that have been evaluated and vetted in critical applications over the past 30 years.

Figure 14.1 is an outline of this chapter with the process steps to be considered in each case. Caution must be taken at each step because the solution to a multivariate optimization problem is not found by simply running a huge Monte Carlo simulation of all the inputs. These inputs, themselves, interact with each other in a system of equations resulting in partial derivatives. This is captured with an N-by-N matrix, which is the multivariate calculus problem with constraints.

Showing how this is done is the objective. Each petroleum refinery is different.

Figure 14.2 shows how flexibility in contracts and flexibility in processing equipment in a refinery interact with each other. Options are not always value-additive.

- Discounted cash flow methods (NPV and IRR)
- Heuristics and rules of thumb (payback period, market multiples)
- Behavioral finance (follow what others decide)
- Better methods – real option valuation and system dynamics
- Sequential investments
- Decision nodes
- Risk, uncertainty, and extreme events
- Analytical equations
- Lattice models – starting with binomial models
- Monte Carlo simulations and machine learning advancements
- Estimating model inputs
- Strategic investments in petrochemical refineries
- Example from coal-to-liquids technology

Figure 14.1 Limitation of traditional valuation models and better methods to determine projects fitting your strategy.

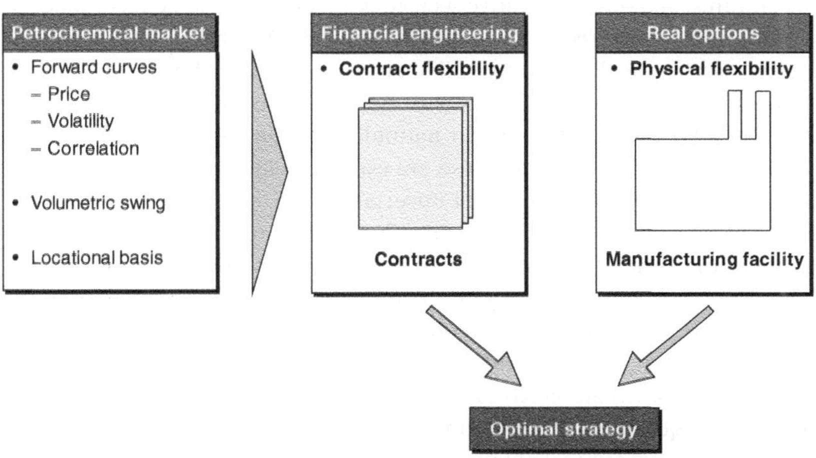

Figure 14.2 Flexibility exists in both contracts and physical assets.

- Examples: petroleum and petrochemical refinery
  - Has internal flexibility
  - Follows locational and seasonal demand
  - Crude oil supply variability
  - Flexibility increases value of the business
  - Crack-spread options
- Examples: chlorine / caustic – electrochemical unit
  - Electrolysis of brine solution in an ECU
  - Splitting a sodium chloride molecule
  - Zero internal flexibility
  - However, does have significant external flexibility
  - Ability to ramp-up and ramp-down during off-peak and on-peak hours
  - Supply and demand controls market prices
  - Lack of internal flexibility may reduce the value of the business
  - Energy hedging and contracting strategy

Figure 14.3 Real option assessments.

A flexible real option in the manufacturing process may replace the need for financial hedging.

Petroleum and petrochemical refineries have internal flexibility built into their operation to compensate for differences in crude oil supply and seasonal demands for products (gasoline in the summer months and heating oil in the winter) (Figure 14.3).

14.2 Derivatives

Options, futures, and other derivative securities are essential elements of modern finance and can reduce risk when properly applied. Derivatives are embedded in financial assets, tangible and intangible property, structured products, trading and hedging strategies, and contracts with flexible terms and conditions (Foote and Murphy, 2001). Real options occur naturally and can also be built-in as flexibility and growth opportunities. These are especially important in an environment characterized by a high degree of uncertainty.

Insurance can be modeled as a put option having an exercise price equal to the face value of the policy. Any loss in value of the asset is covered by an increase in value of the put option. Reinsurance contracts, commonly known as excess-of-loss contracts, are combinations of (both long and short) call option positions on layers of potential losses. Manufacturing companies often rely on the futures market to lock in raw material costs, and real options exist in the flexible operation of physical equipment. An oil refinery having flexibility to switch inputs and outputs has an embedded real option to switch, for example, between heating oil in the winter and gasoline in the driving seasons. Flexibility increases the value of every asset under conditions of uncertainty. Long-term contracts contain option-like features, including volumetric swing components when demand is uncertain. Contract prices can be tied to an index or a basket of commodities. Technology and processing equipment are often priced as intellectual property licenses. The value of these derivatives is determined by using complex quantitative models (Murphy, 2023b, Hagelin, 2009).

Chlorine and caustic are produced in an electrochemical unit that has zero internal flexibility, splitting salt solution (brine) into two units of chlorine and two units of sodium hydroxide. There is zero internal flexibility; however, there is a tremendous ability to ramp up the operation during daily off-peak power prices and vice versa (Figure 14.4).

Chemical production is enhanced with stand-alone processes like air separation units (ASU) that add oxygen and hydrogen and produce both electricity and steam for the crude oil refinery process. Internal flexibility increases the value of the refining process (Figure 14.5).

This example is from a consulting project valuing the divestiture of a chlor-alkali division of a large corporation, including earn-out payments over a five-year period.

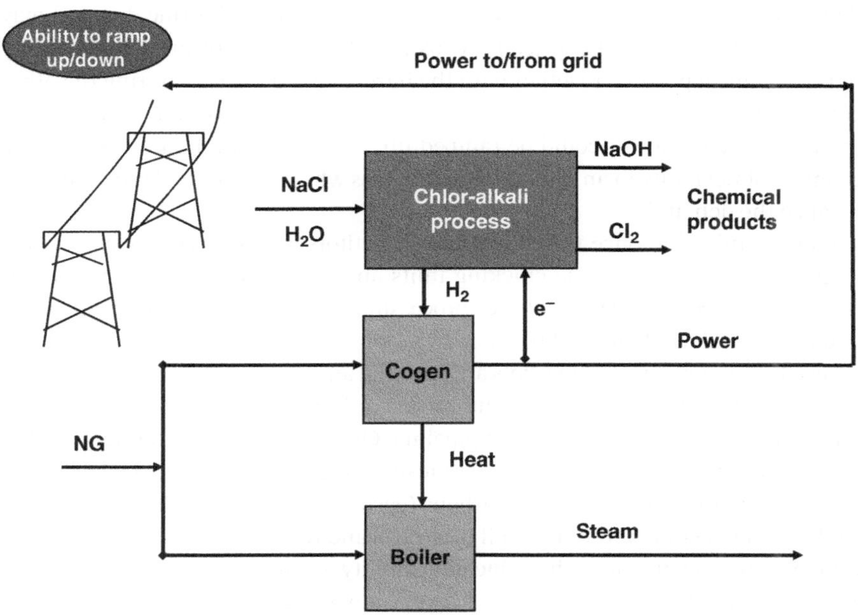

Figure 14.4 Chemical process with no internal flexibility.

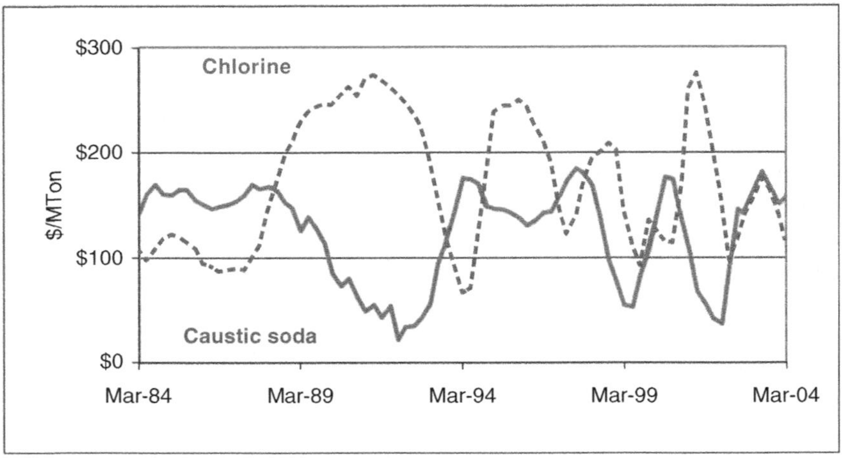

Figure 14.5 Twenty-year historical prices for chlorine and caustic soda – outputs are in countercyclical markets. *Source*: CMAI.

A basic chemical industry, the chlor-alkali process, uses electricity to break sodium chloride (NaCl) as a brine solution into chlorine and caustic soda. If demand for these products is not balanced, an increase in chlorine demand may cause a drop in the market price for caustic soda. The lack of internal flexibility in the chemical process, together with the burden and danger of storing the other components, depresses the net market prices. The process has some external flexibility, as electric power could be ramped up during off-peak hours. See also the example of flexibility in an industrial gas process where storage and pipelines help balance the demand.

Process units that once were thought competitors are now seen as synergistic. These include FCC catalytic cracking units and hydrolysis units for bottom-of-the-barrel processing. This includes processing steps to reduce sulfur emissions and nitrous oxides (Figure 14.6).

Gasoline and heating oil markets are seasonal and countercyclical (Figure 14.7).

This example is from a consulting project valuing a topping refinery along a remote portion of a major crude oil pipeline. One of the simplest refinery configurations consists of tankage and an atmospheric distillation unit, side strippers, and recovery units for light hydrocarbons (gasoline and diesel). The refinery pays a toll for the fraction of the crude oil it extracts and returns all other portions back into the crude oil pipeline. The value of flexibility in a modern petroleum and petrochemical refinery is much greater because it operates as a systems architecture and is adaptable to wide ranges of inputs and outputs.

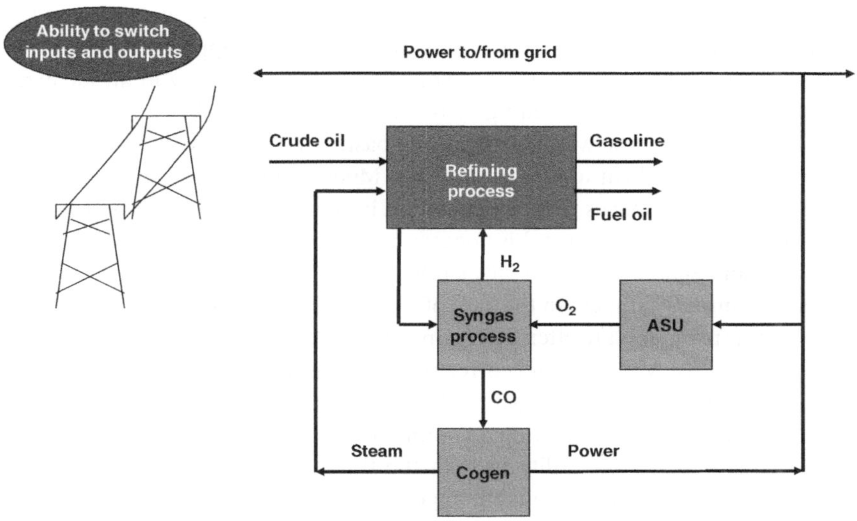

Figure 14.6 Chemical process with internal flexibility.

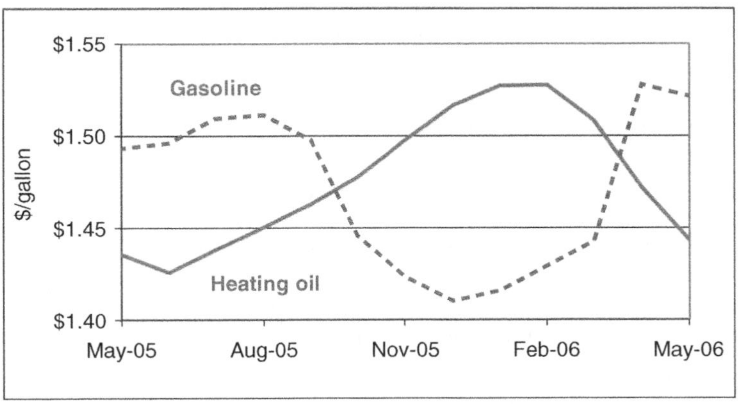

Figure 14.7 Market prices for gasoline and heating oil. *Source*: NYMEX.

14.3 Model Risk

All models are imperfect because they are based on assumptions that may not always hold up. Quantitative models are used to measure value and risk to help managers make better decisions. In every case, it is critical that management and quantitative modeling experts work together to understand the asset being modeled and the assumptions behind the model. Assessing model risk is a key element in preventing upsets by providing financial audit support.

Engineering students learn the laws of physics and chemistry in an idealized setting (in a vacuum) before adding real-world imperfections (like an atmosphere or friction). This is how we move from the basics of Newton's laws to chemical reactor kinetics, or heat and mass transfer. Model imperfections are especially true in the social sciences, which depend on human behavior, not scientific law. Unfortunately, we often forget this once we begin applying an analytical equation, like Black and Scholes (1973) or Black (1976), to value an option, or a more complicated quantitative model in the case of a credit default swap (CDS) or CMO.

The inputs to models are often determined by another model, which compounds the potential error. Volatility and correlation are forecasts based on historical and/or market-implied values. Forecast methodologies are prone to errors themselves. Poon and Granger (2003) provide a comprehensive review of different methods being used to forecast volatility in both the financial and commodity markets. The results can vary widely, causing huge differences in valuation and risk. One quantitative method does not fit every situation. This is the analysis and reasoning that we learned in law school: that the ruling will depend on the facts of the case.

To avoid being blindsided by errors, a practice must benchmark multiple valuation models, and "twist and turn" model inputs over a range to determine the extent of variation in the results. This fundamental idea is the basis of "stress testing," which should have been the norm prior to the financial meltdown. Questioning whether the models and inputs are appropriate will reduce financial risk.

Some of the errors that we have observed in industry include:

- Not incorporating all relevant market data when forecasting model inputs. (Leaving out certain time periods, as if the events that occurred then would never happen again, like the energy and dot com crashes.)
- Using over-the-counter (OTC) options rather than exchange-traded options (guaranteed by the exchange) as the basis for implied volatility forecasts without correcting for counterparty credit risk. (This overvalued a natural gas option portfolio owned by a major public utility in the United Kingdom.)
- Correlation Is Not Causation. (ENRON's suggestion to Air Liquide America in Houston to hedge power prices inside Texas (ERCOT: The Electric Reliability Council of Texas) using Florida orange juice futures because they determined a 0.90 R-squared using an Ordinary Least Squares Linear Regression. Dr. Murphy was Air Liquide's Energy Manager, and he told the Vice President of Energy to reject ENRON's proposal. At this time, just before ENRON's downfall, they were on the front cover of Time Magazine as "the smartest people in the room" (Murphy, 2010).

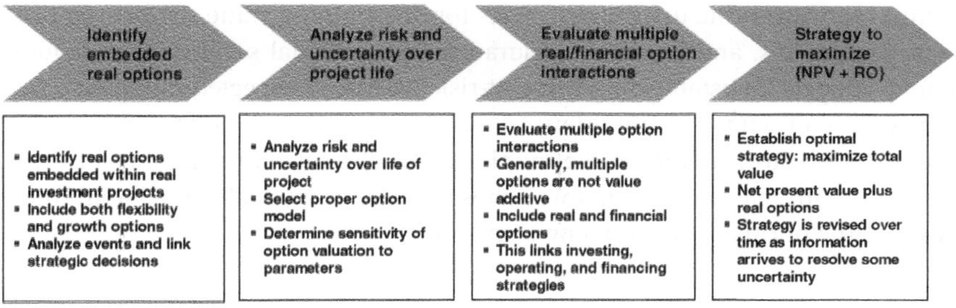

Figure 14.8 Real option assessments.

- Misapplication and abuse of models can occur when management does not incorporate proper risk management practices. Some financial models are quickly adopted before their limitations are fully understood.

Real option assessments are performed to identify and value flexibility. Real options are not necessarily value-additive (Figure 14.8).

Figure 14.10 is a basic flowchart showing how a real option assessment of a petroleum refinery can be integrated with corporate strategy, including financial options. Assessment of risk, uncertainty, and extreme events at each location will depend on market demand and assets in place. Financial options and contracts will interact with real options.

14.4 Quantitative Models

All models are imperfect and fundamentally flawed in some way. Forgetting this fact has led to many financial meltdowns. The long history of model-related valuation losses includes ENRON, Long Term Capital, Bear Stearns, and Lehman Brothers. Managers use models to guide decision-making in an uncertain environment. Good management recognizes the importance of and limitations of modeling. However, the need for a reality check has often been forgotten in the case of financial models.

The mathematician Edward E.P. Box is often quoted "All models are imperfect, but some are useful."

A model is composed of inputs, processing algorithms and heuristics, and outputs. For example, a forward curve model represents the price of commodities for later delivery. The inputs to a forward curve model include historical prices, volumes, and other drivers of forward value, exchange-traded futures, broker quotes, and fundamental supply and demand projections. Algorithms and heuristics may include technical filter algorithms, splines, expert analysis of the distribution of potential forward value drivers, and rules for determining value set forth in contracts, regulations, and practices. Insurance and financial service firms require complex models for establishing value-at-risk (VaR) and expected-tail-loss (ETL).

The choice of data is as important as the choice of model. Volatility measures include both historical and market implied. Blends using combinations of these may be set with reasons why that was chosen, which will establish an audit trail. These are the norms in an audit environment.

14.5 Model Audits and Critique

Models should be reviewed by independent qualified quantitative financial experts to avoid conflicts of interest and opinions that may become embedded in the organization. A fresh, independent perspective is required, starting with transparency and incorporating all information and data that are available. This avoids the situation described as "letting the fox guard the henhouse." Energy trading and valuation models are extraordinarily complex. A model risk audit and critique should include answering the following questions:

- Review the appropriateness of the methodology. Does the model, and its assumptions, fit the situation?
- Is the model transparent? Are the market data objective, verifiable, and with liquidity?
- Examine the operation of the model through an initial diagnostic. Start by reviewing the derivation of the model. For example, the Black–Scholes option pricing model contains seven restrictive assumptions and was derived using a delta hedging argument. These assumptions may not hold.
- Examine the modules and connections. This includes examining the model code, whether that's Excel/Python, C/C++, MATLAB, SAS.
- Examine the complexity of the algorithms and procedures.
- Perform a model "twist and turn" to determine when the model is likely to fail.
- Stress test model inputs to assess sensitivity.
- Vetting model forecasts with out-of-sample data.
- What happens if the model assumptions do not hold?
- What are the sources of the data and how were they collected? This includes input prices, interest rates, volatilities, and correlations.
- Is it a typical model used by a top tier organization? Is it consistent with the prevailing practice? Is it leading practice? Which methodology is driving the market?

Benchmark the model using alternative methods. For example, run a Monte Carlo simulation to benchmark the results from an analytic formula.

14.6 Systems Thinking

Leading-edge methods in systems thinking are being applied to incorporate correlative measures that are not confirmed causal measures. Both real options and systems dynamics were started around the same time at MIT and Harvard. Both have a large group of followers. The pioneering work on real options was performed by Lenos Trigeorgis at Harvard University. His doctoral dissertation (Trigeorgis, 1986) included advisors at the Massachusetts Institute of Technology. This multidisciplinary approach made it very appealing to chemical engineering and quantitative modeling applications.

All three co-authors of this chapter studied finance and quantitative modeling in business school and were already practicing chemical engineers. The real option methodology is very intuitive because chemical processes, unit operations, and equipment have embedded flexibility and often growth option potential.

Dr. Murphy subsequently worked for Lenos Trigeorgis in the Real Options Group with other experts from around the world. He went back to MIT in 2023 for a six-month postgraduate program in Technology and Innovation with System Architecture and is currently enrolled in another MIT program. Both methods have direct applications to petrochemical refining and are better when they are run together. The speed of computing is rapidly increasing, and artificial intelligence (AI) and machine learning (ML) methods include both correlation and causation measures.

Research and development projects are often valued by looking at when the investment occurs over time and when the information arrives to predict if the R&D will be successful. Clinical trials often indicate the success of pharmaceutical research at the end of the project (Figure 14.9).

Low flexibility:

- Where most of the information (Ω) arrives at the end of the project, or most of the investment (I) occurs up-front, so little can be saved from abandoning it.

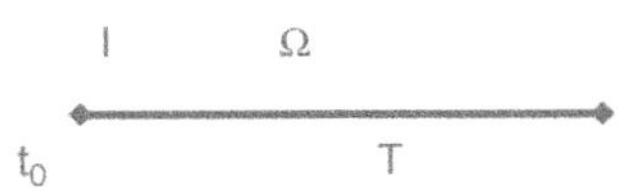

Examples:

- Double-blind clinical trial of a new drug, where results are revealed at the conclusion.
- Building a prototype airplane where the investment is committed up front, and information arrives at the end (does it fly?).
- Early basic research where the infrastructure has a high initial cost, and information arrives at the end of the project (superconductor–supercollider project).

Figure 14.9 Basic R&D project example – low flexibility.

Research & Development projects are well suited to real option models because of flexibility options that can be exercised as information arrives during the life of the project. Included are expected and unexpected discoveries. Applications of these models are applied in the valuation of patents and joint R&D ventures (Murphy, 1995).

A highly flexible R&D project has information and discoveries that arrive before all the investment has been expended. This is one of the possibilities in our example of coal-to-liquid (CTL) investments as a participation option.

If the investment occurs over the life of the R&D, an early determination that the results do not appear as attractive as initially thought. This has happened on projects like coal-to-liquids technology and is currently happening on artificial intelligence and machine learning (Figure 14.10).

The Design Structure Matrix (DSM) methodology utilized in the systems dynamics section of the MIT program is remarkably like quantitative modeling concepts most of us have applied in our own careers. Universities are Teach and Learn Systems, Macroeconomics is a System for Growth of the Economy, Hospitals are Health Care Systems, and Companies are Systems. MIT Professor Edward Crawley observed that "Systems Exhibit Emergence," which has a value that is more than the sum of its parts.

Systems thinking begins by identifying the entities that form a petroleum refinery and their functions, which create higher functionality, performance, reliability, safety, and environmental protection. The refinery system is the set of interrelated entities, which perform a function that is greater than the sum of its parts. This increase in value is called the emergence of the system. This is analogous to structuring flexibility and growth-related real options (Murphy, 2023a).

High flexibility:

- Where most of the information (Ω) arrives at the beginning of the project, or randomly through out the life of the project, and the investment (I) occurs over time, so that savings are possible by abandoning the project.

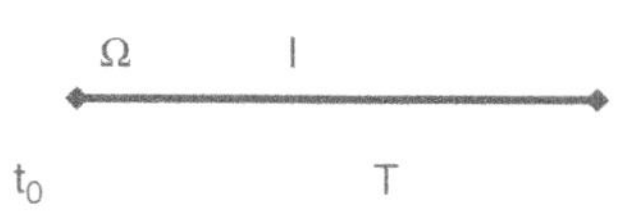

Examples:

- Open clinical trial of a new drug where information arrives and is reviewed continuously.
- Incremental new product development where modifications are made over time and investment is spread out.
- Early stage research at the bench level where there is a low initial cost infrastructure with information arriving randomly over time.

Figure 14.10 Basic R&D project example – high flexibility.

The systems approach focuses on finding the optimal addition, or subtraction, of processing steps at a petroleum refinery. There is no shortcut to this process; you cannot just throw it all into a Monte Carlo simulation or a waterfall model to get the optimal strategy.

A Design Structure Matrix (DSM) framework provides a concise visualization and comprehensive framework for the purpose of analysis and optimization to find the best solution to a complex problem.

Our experiences with matrix methodologies are twofold:

Chemical and biomolecular engineering systems are extraordinarily complex, with many competing reactions, elements, and factors that will affect the desired outcomes. This is true for improving petroleum refineries to make cleaner fuels for transportation vectors and to transition the industry into the chemical industry for making the products that we need to protect the environment. Each petroleum refinery is dependent on the characteristics of its crude oil supply and environmental regulations (Wagner et al., February 2023) (Hoekstra et al., 2023). Petroleum refining is evaluated as a system, and solutions are evaluated for each situation to maximize benefits and reduce capital investment.

Novel biomolecular materials, like green carbon nanostructures that are made from sustainable resources like agriculture and renewable forestry, have net zero carbon emissions. They are also being developed to clean up toxic materials, including PFAS forever chemicals. The chemical kinetics and reactor designs are so complex that not all parameters are fully understood. A DSM framework (we call this a canonical correlation matrix, which has symmetry along the diagonal) is utilized to measure the impact of parameters over a range of possibilities. In pyrolysis reactions this extends the visualization of processing parameters beyond the Arrhenius equation, where parameters of temperature, pressure, and concentrations are not enough.

Quantitative models are used to analyze value, risk, and strategy of both tangible and intangible assets in a firm. Complex machinery, like a jet engine for an aircraft, can be evaluated for growth and flexibility (real options) and can be better understood in a concise, clear visualization framework, which relates systems emergence benefits through Black and Scholes option and Black spread-option models. Spark-spread options for generating electricity with jet engines, or crack-spread options for determining the value of chemical derivatives and refinery modifications. These are clear examples of the benefits of a systems approach.

Matrix models can be simplified through rearranging the independent and dependent variables of an N-by-N matrix. Which factors that are most important can be assessed iteratively. Solutions to complex option models are often done by simulations. Setting up the data parameters is as critical as selection of the specific model. Simulations and clustering through Analysis of Variance Models (ANOVA) will provide visuals of value and risk for each strategy.

14.7 Assessment of Model Risk Fits into a DSM Framework

The systems dynamics approach uses correlation, not causation, like a real options approach. It models the correlation and interaction of variables being affected. Determining the key variables in a multivariate setting. Many of the problems with analytical methods in modeling come from not respecting the multivariate calculus of systems dynamics. It is critical for determining valuation, risk, and strategy.

Cluster analysis is looking at how data forms into clusters.

Deep learning and critical thinking are important habits that are acquired when working on innovative technologies.

Canonical form of a covariance matrix is the framework for an assessment of how a System is stacked up. In refining, the PINE assessment may question taking paraffinic molecules off for blending with ethanol for increasing octane rating. Then sending naphtha to a Fluid Catalytic Cracking (FCC) unit for catalytic treatment. This may also be beneficial in less developed countries where the expense and benefit may be in question.

Design System Matrix is a module cluster approach in an N-by-N matrix having a diagonal of correlations and covariances showing how the modules are dependent on each other. These models are not ordinary differential equations that can be separated into two sides and solved with an analytical solution; they are partial differential equations in a Hessian matrix equation form. This must be done properly to ensure that the optimization result is the desired global value.

Multi-attribute Utility (MAU) model is a logical step from traditional real option valuation to include the diverse views of refinery stakeholders that include process safety, maintenance, and employee relations.

Growth options are extremely important during acquisitions and divestitures. During M&A activity, new opportunities to become involved in technological advancements are critical to a company's strategy. It is also especially important for both sides of a transaction to see the synergies with their existing businesses.

A recent article by Dr. da Silva shows that the synergy between FCC units and hydrocracking units, which were thought to be competitors, is now considered to be synergistic partners for meeting the new mix of petroleum and petrochemical demand safely (Wagner Da Silva et al., 2024).

14.8 Real Option Strategy Assessment Example – Evaluating CTL Technology

This example shows how real option methodologies analyze potential strategies for investment for upgrading petroleum refineries to meet changing demands in transportation vectors, improve the environment, and make the petrochemicals necessary for modern materials needed in today's economy. The process is what the CEO of a portfolio of refineries will need to go through in deciding to strategically invest and gain a first-mover advantage, which historically has paid for the investment in the first year or two.

All these examples and assessments are from actual projects since 2001 for strategic investments, including Merger and Acquisitions and other projects, and shall remain confidential. They are general enough in scope that they represent comparable projects for today's landscape.

The valuation of Intellectual Property from Research & Development became a passion, as evidenced by my doctoral advisor Theodore Hagelin at Syracuse University – College of Law being granted a U.S. Patent on a Method for Valuing a Patent (Hagelin, 2009). This was an extension of Chapter 6 in my Ph.D. dissertation to include the valuation of technology licenses (Murphy, 1995).

Strategic options include deciding which strategy works best for your firm. A joint venture offers advantages over a solo R&D venture. Wait and see what other companies are doing and find a way to help (Figure 14.11).

Figure 14.12 outlines five alternative strategic responses to a major change in the petroleum industry. It was taken from our experience of modeling large-scale research and development projects with a real options approach (Murphy, 1995). The value of these alternatives is like the structure of a SWORD (Stock Warrant Off Balance Sheet Financing of R&D) that considers a combination of DCF and the option value from flexibility and growth opportunities. The strategy limits the investment in processing equipment until some risk and uncertainty is resolved.

Early carbon tax legislation and house emissions as headed in slightly different directions to reduce global warming. These included Bingman – Spector and Lieberman – Warner. Similar to today's legislation on renewable fuel standards and greenhouse emissions. The electricity load of data warehouses for AI assumes that predicting the future based on historical data (correlation, not causation) is a major assumption that quantitative modeling experts say is a huge assumption (Figure 14.12).

Discounted cash flow results over time help show the net present value (NPV) and internal rate of return (IRR) of a project that is no longer in a growth mode (Figure 14.13).

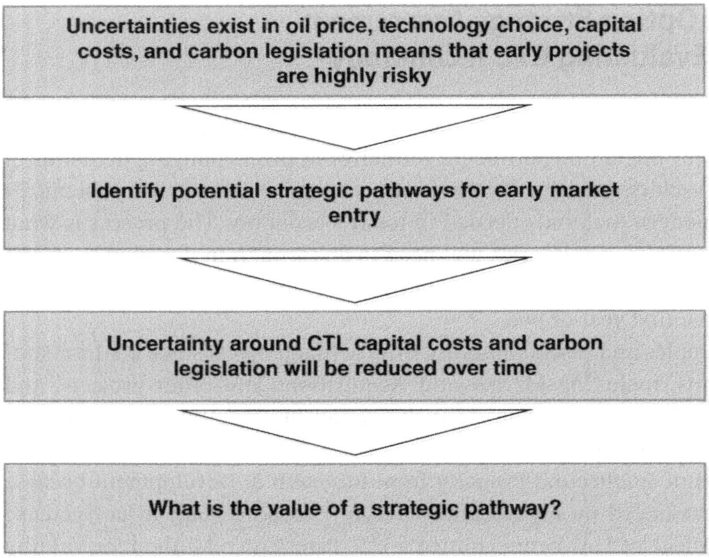

Figure 14.11 Quantitative model assessment of alternatives – the example from coal to liquids participation.

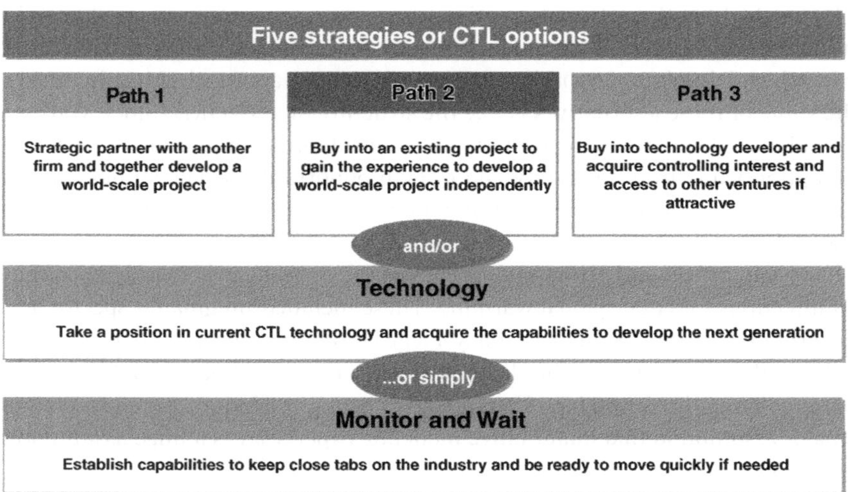

Figure 14.12 Strategic options for CTL project participation.

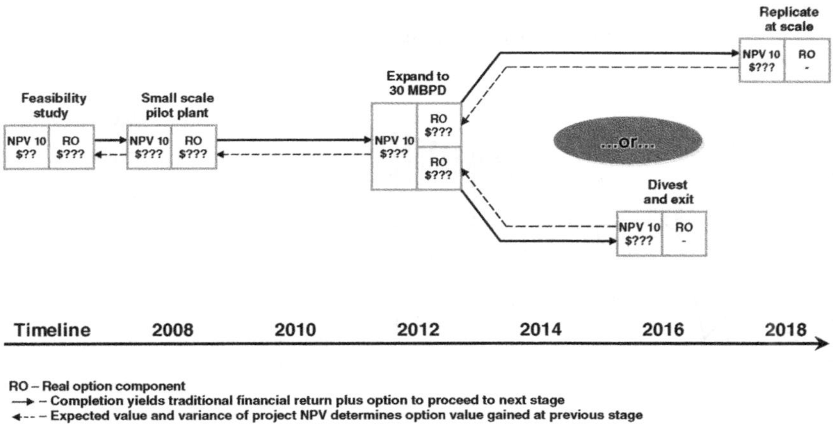

Figure 14.13 Path 2 option diagram: initial investment buys a valuable option going forward.

Model assumptions and methodology

- Model starts with current CTL technology and applies learning curve
  - CTL performance conservative average of public and private studies
  - Economies of scale and learning curvesimilar to LNG and oil sands experience
- All scenarios run assuming carbon tax legislation
  - Low case is representative of Bingman-Spector
  - High case is representative of Lieberman-Warner
  - Extreme carbon prices stimulate development of CCS
- Coal and electricity prices calculated with MRN-NEEMS
  - Forward curves used to predict generation capacity additions
  - Resulting commodity prices consistent with scenarios

Figure 14.14 The NPV at each stage was found with a DCF model incorporating Monte Carlo simulation of key variables.

Exercise discretion over critical assumptions and the choice of financial modeling methodologies. Every model is flawed in some way. Using various methods and models will reduce over-dependence. As the mathematician George E. P. Box said, "All models are flawed, but some are useful." (Figure 14.14).

Include capital costs and the learning curve over time. These will significantly affect the value of early projects (Figure 14.15).

Include capital inflation and economies of scale. Early projects will face headwinds from later projects (Figure 14.16).

The key uncertainties illustrated in this example are like what we would expect to be important in a refining or petrochemical project to meet changes in the

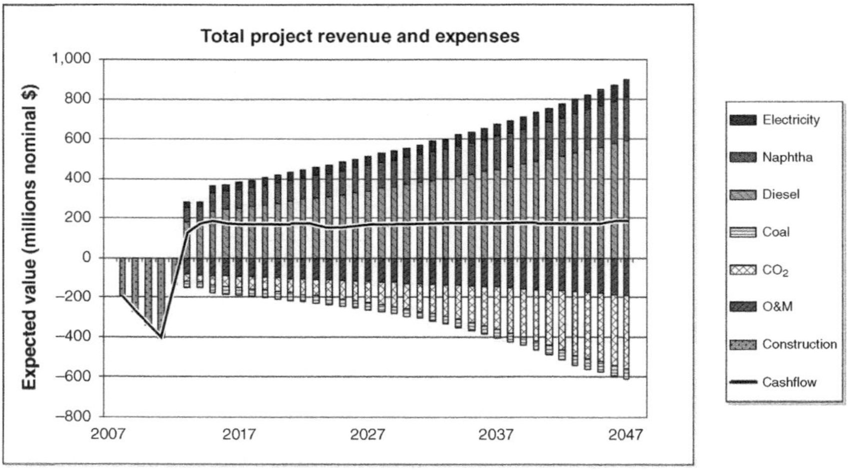

Figure 14.15 Example of DCF expected value results.

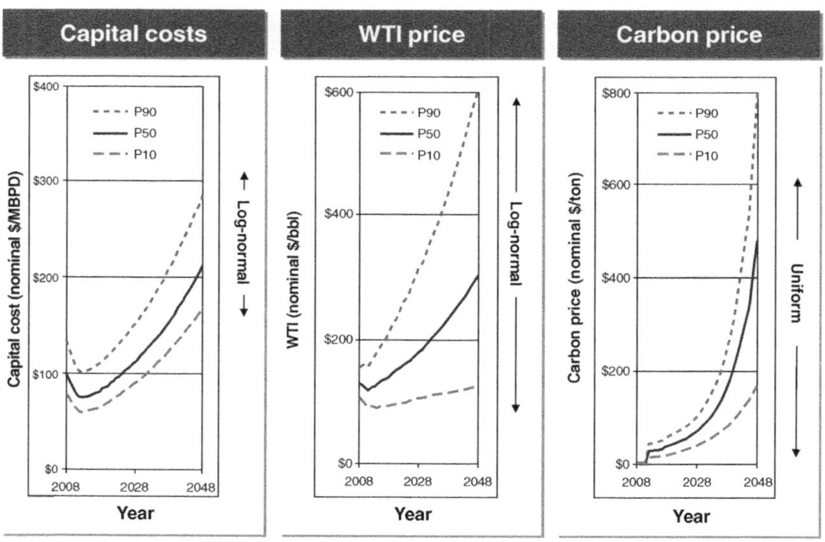

Figure 14.16 Distributions of critical uncertainties.

transportation vectors and chemicals required for advanced materials. The P90, P50, and P10 values are from the company's forward curves and included the work to establish consistency with best practices in quantitative modeling and audit practice.

Invest in an optimal growth strategy for your company. Benchmark which models and model inputs fit your competitive advantage (Figure 14.17).

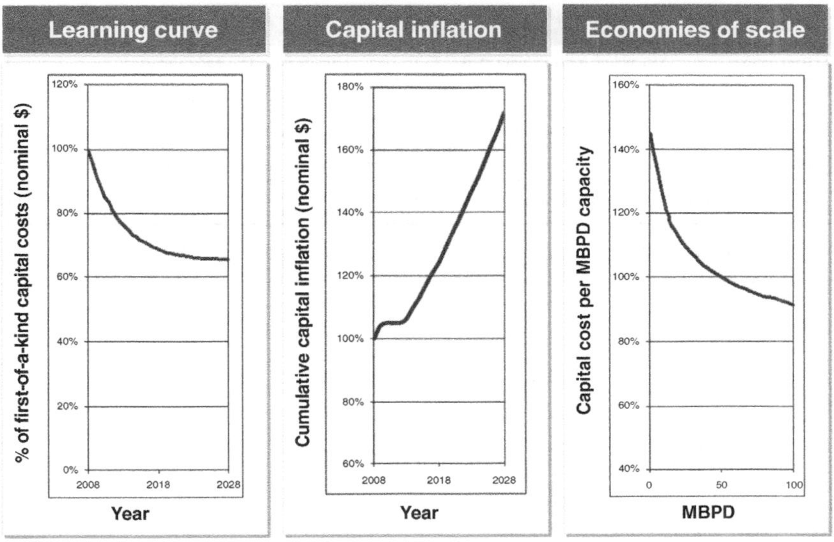

Figure 14.17 Capital costs are a function of three variables.

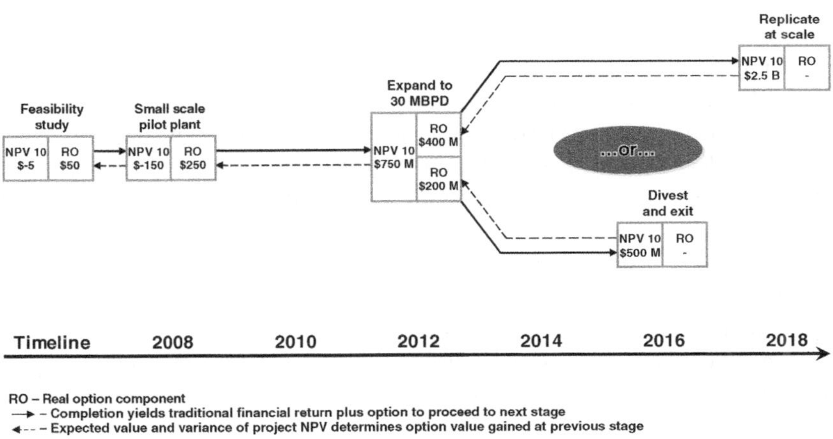

Figure 14.18 Path 2 option diagram: initial investment buys a valuable option going forward.

Learn from the top chemical engineers and financial modeling experts. Invest in your company's future by supporting and retaining your employees (Figure 14.18).

A binomial model, like Cox et al. (1979), is outlined in Figure 14.18. It shows both the real option and DCF values of a stagewise investment over time under

	Criteria	Pros	Cons
Attractiveness	Real option	$50 million for feasibility study	Highly uncertain early stages
	Initial commitment		Almost $1 billion commitment for stage one
	Timing	Project well defined; FID scheduled for 7/2009	This may be a "bleeding edge" project
	Flexibility	Ability to learn from project then go it alone	No flexibility options or research opportunities
	Expansion potential	Initial expansion from 10 to 30 MBPD	Business model favors series of small projects with project financing and IPO capacity

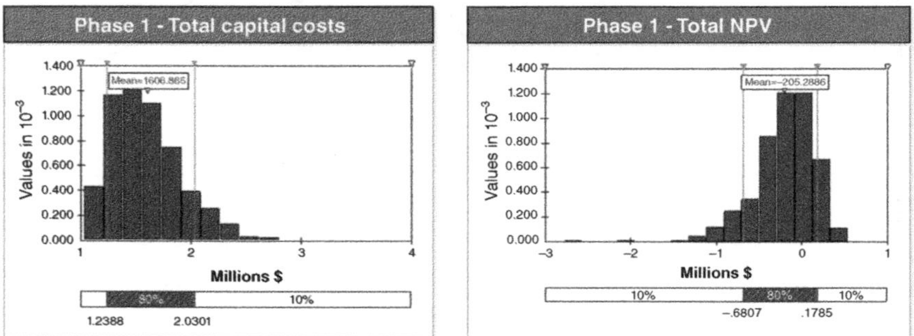

Figure 14.19 Path 2 sample attractiveness.

conditions of uncertainty. The combined value (NPV + ROV) shows that being active with technology early on can be a great strategy. Today we have extended these models with modified lattice approaches that incorporate the benefits from a system approach developed at MIT. These values of sequential capitalization structures are developed in Chapter 5 of the author's Ph.D. dissertation (Murphy, 1995). Each petroleum refinery location will be able to assess the value of its strategy going forward with the help of independent and objective quantitative modeling experts.

The probability distributions of these simulations show significant skew on both total capital costs and total NPV to help select a more appropriate strategy for each petroleum and petrochemical refining company at each location. This approach allows the CEO to examine direction without the expense of deciding on technology and equipment offerings, as would happen with a start-from-scratch refinery (Figure 14.19).

14.9 Environmental Compliance: Environmental Credits and Offsets

In the category of environmental compliance, Hoekstra Trading LLC applied options pricing theory to calculate the theoretical price of a Renewable Identification Number (RIN), which is an environmental credit that provides flexible compliance options to obligated parties for meeting renewable fuel volume mandates in the United States. The Hoekstra theoretical value (denoted D4T) has been used by refiners and other obligated parties in their renewable fuel compliance and credit strategies since 2021. The D4T value is calculated weekly and has recently been licensed to Bloomberg Finance LLC for distribution to Bloomberg users through the Bloomberg network (Hoekstra, 2023). To our knowledge, this is the first time the theoretical value of an environmental credit, calculated using options pricing theory, has been licensed for distribution through a major media outlet.

14.10 Conclusion

Derivatives can be found embedded in every asset class, liability, and business. They provide valuable flexibility and growth under conditions of uncertainty over time. Quantitative models are used to evaluate these derivatives and to better understand and manage their risks. This is true for insurance companies, energy providers, and financial services firms. Models must be transparent and based on assumptions that fit the application and assumptions. The inputs used in these models must include relevant information, both historical and market implied. Proper risk management practice is better when it includes a multidisciplinary perspective.

It takes a combination of perspectives and experiences to do this work, combining chemical engineering, quantitative modeling, and industry experience to be highly effective. Like recent financial market crashes, overreliance on the latest models and simulation packages that tell you what to do, remember to dig deep and see if they understand the multivariate calculus, correlation versus causation, the Durbin–Watson statistic, the central limit theorem, and what to do when the economics of carbon markets and environmental regulation change.

This is not something you want from someone just starting out. The authors here have over 80 years of combined experience and are at the leading edge of developments, including artificial intelligence, regulatory credits, machine learning, and petroleum and petrochemical processes.

References

Black, Fischer (1976). The pricing of commodity contracts, *Journal of Financial Economics*, 3, 167–179.

Black, F. and Scholes, M. (1973). The pricing of options and corporate liabilities, *Journal of Political Economy*, (May-June), 637–659.

Cox, J., Ross, S., and Rubinstein, M. (1979). Option pricing, a simplified approach. *Journal of Financial Economics* 7 (3): 229.

Epstein, A. (2009). *Fossil Future: Why Global Human Flourishing Requires More Oil. Coal. And Natural Gas – Not Less.*

Foote, W. and Murphy, T, (2001). A flexible strategy, a real option model of an energy-intensive company shows the optimal strategy for managing energy price risk. *Energy and Power Risk Management, Risk International*, August 2001.

Hagelin, T. (2009). Method for valuing intellectual property. US Patent 7,493,262 B2, February 17, 2009.

Hoekstra, G. (2023). Bloomberg licensing agreement. Press release (26 December). https://hoekstratrading.com/hoekstra-trading-announces-hoekstra-bloomberg-licensing-agreement

Hoekstra, G., Wagner de Silva, M., and Murphy, T. (2023), Tier 3 gasoline production – the value of flexibility in the volatile fuel markets. *Hydrocarbon Processing*, July 2023.

Murphy, T. (1995). Financing research and development projects: real-options, spillover, and SWORD securities. Unpublished Ph.D. dissertation, Syracuse University. Available at UMI dissertation abstracts number 9625859 or from the university. Copyright 1995 by Murphy, Thomas.

Murphy, T. (2010). *Derivatives and model risk*. Unpublished paper presented at the Association of American Law Schools, Annual Meeting in New Orleans, January 2010.

Murphy, T. (August 2023a) MIT Technology and Innovation Acceleration Program, graded assignments in petroleum and petrochemical refining with Marcio Wagner da Silva and George Hoekstra. 5.0/5.0 GPA.

Murphy, T. (2023b). *Quantitative modeling risks, financial markets, technology, ML & AI Mumbai Stock Exchange* – Bombay Broker's Forum, October 2023.

Poon, S. and Granger, C. (2003) Forecasting financial market volatility – a review, *Journal of Economic Literature*, June 2003.

Trigeorgis, L. (1986). Valuing investment opportunities: an options approach to strategic capital budgeting. Unpublished doctoral dissertation. Harvard University.

Wagner Da Silva, M. (2024). Process optimization - meet gasoline quality and process safety requirements through new alkylation technologies. *Hydrocarbon Processing*, (August 2024).

Wagner da Silva, M., Hoekstra, G., Murphy, T., Tier 3 gasoline production: challenges and opportunities for refineries. *Hydrocarbon Processing*, February 2023.

Index

Pages in *italics* refer to figures and pages in **bold** refer to tables.

Cleaner Petroleum Production and Refining Technologies, First Edition, Edited by M. R. Riazi and Harvey W. Yarranton.

h

i

j

k

l

t

Printed and bound by CPI Group (UK) Ltd, Croydon, CR0 4YY
17/09/2026
14964311-0007